The Vertebrate Skeleton

by Sidney H. Reynolds

Copyright © 8/14/2015
Jefferson Publication

ISBN-13: 978-1516906314

Printed in the United States of America

Contents

CHAPTER I.
INTRODUCTORY ACCOUNT OF THE SKELETON IN GENERAL.

By the term **skeleton** is meant the hard structures whose function is to support or to protect the softer tissues of the animal body.

The skeleton is divisible into

A. The Exoskeleton, which is external;

B. The Endoskeleton, which is as a rule internal; though in some cases, e.g. the antlers of deer, endoskeletal structures become, as development proceeds, external.

In Invertebrates the hard, supporting structures of the body are mainly **exoskeletal**, in Vertebrates they are mainly **endoskeletal**; but the endoskeleton includes, especially in the skull, a number of elements, the **dermal** or **membrane** bones, which are shown by development to have been originally of external origin. These membrane bones are so intimately related to the true endoskeleton that they will be described with it. The simplest and lowest types of both vertebrate and invertebrate animals have unsegmented skeletons; with the need for flexibility however segmentation arose both in the case of the invertebrate exoskeleton and the vertebrate endoskeleton. The exoskeleton in vertebrates is phylogenetically older than the endoskeleton, as is indicated by both palaeontology and embryology. Palaeontological evidence is afforded by the fact that all the lower groups of vertebrates—Fish, Amphibia, and Reptiles—had in former geological periods a greater proportion of species protected by well-developed dermal armour than is the case at present. Embryological evidence tends the same way, inasmuch as dermal ossifications appear much earlier in the developing animal than do the ossifications in the endoskeleton.

Skeletal structures may be derived from each of the three germinal layers. Thus **hairs** and **feathers** are **epiblastic** in origin, **bones** are **mesoblastic**, and the **notochord** is **hypoblastic**.

3

The different types of skeletal structures may now be considered and classified more fully.

A. Exoskeletal structures.

I. Epiblastic (epidermal).

Exoskeletal structures of epiblastic origin may be developed on both the inner and outer surfaces of the Malpighian layer of the epidermis . Those developed on the outer surface include **hairs**, **feathers**, **scales**, **nails**, **beaks** and **tortoiseshell**; and are specially found in vertebrates higher than fishes. Those developed on the inner surface of the Malpighian layer include only the **enamel** of teeth and some kinds of scales. With the exception of feathers, which are partly formed from the horny layer, all these parts are mainly derived from the Malpighian layer of the epidermis.

Hairs are slender, elongated structures which arise by the proliferation of cells from the Malpighian layer of the epidermis. These cells in the case of each hair form a short papilla, which sinks inwards and becomes imbedded at the bottom of a follicle in the dermis. Each hair is normally composed of an inner cellular pithy portion containing much air, and an outer denser cortical portion of a horny nature. Sometimes, as in Deer, the hair is mainly formed of the pithy portion, and is then easily broken. Sometimes the horny part predominates, as in the bristles of Pigs. A highly vascular dermal papilla projects into the base of the hair.

Feathers, like hairs, arise from epidermal papillae which become imbedded in pits in the dermis. But the feather germ differs from the hair germ, in the fact that it first grows out like a cone on the surface of the epidermis, and that the horny as well as the Malpighian layer takes part in its formation.

Nails, **claws**, **hoofs**, and the **horns of Oxen** are also epidermal, as are such structures as the **scales** of reptiles, of birds' feet, and of *Manis* among mammals, the **rattle** of the rattlesnake, the **nasal horns** of *Rhinoceros*, and the **baleen** of whales. All these structures will be described later.

Nails arise in the interior of the epidermis by the thickening and cornification of the stratum lucidum. The outer border of the nail soon becomes free, and growth takes place by additions to the inner surface and attached end.

When a nail tapers to a sharp point it is called a **claw**. In many cases the nails more or less surround the ends of the digits by which they are borne.

Horny **beaks** of epidermal origin occur casing the jaw-bones in several widely distinct groups of animals. Thus among reptiles they are found in Chelonia (tortoises and turtles) as well as in some extinct forms; they occur in all living birds, in *Ornithorhynchus* among mammals, and in the larvae of many Amphibia.

In a few animals, such as Lampreys and *Ornithorhynchus*, the jaws bear horny tooth-like structures of epidermal origin.

The **enamel** of teeth and of placoid scales is also epiblastic in origin , and it may be well at this point to give some account of the structure of teeth, though they are partly mesoblastic in origin. The simplest teeth are those met with in sharks and dogfish, where they are merely the slightly modified scales developed in the integument of the mouth. They pass by quite insensible gradations into normal placoid scales, such as cover the general surface of the body. A **placoid scale** is developed on a papilla of the dermis which projects outwards and backwards, and is covered by the columnar Malpighian layer of the epidermis. The outer layer of the dermal papilla then gradually becomes converted into dentine and bone, while enamel is developed on the inner side of the Malpighian layer, forming a cap to the scale. The Malpighian and horny layers of the epidermis get rubbed off the enamel cap, so that it comes to project freely on the surface of the body.

As regards their attachment teeth may be (1) attached to the fibrous integument of the mouth, or (2) fixed to the jaws or other bones of the mouth, or (3) planted in grooves, or (4) in definite sockets in the jaw-bones (see p. 107).

Teeth in general consist of three tissues, **enamel**, **dentine** and **cement**, enclosing a central pulp-cavity containing blood-vessels and nerves. Enamel is, however, often absent, as in all living Edentates.

Enamel generally forms the outermost layer of the crown or visible part of the tooth; it is the hardest tissue occurring in the animal body and consists of prismatic fibres arranged at right angles to the surface of the tooth. It is characterised by its bluish-white translucent appearance.

II. Mesoblastic (mesodermal).

Dentine or **ivory** generally forms the main mass of a tooth. It is a hard, white substance allied to bone. When examined microscopically dentine is seen to be traversed by great numbers of nearly parallel branching tubules which radiate outwards from the pulp-cavity. In fishes as a rule, and sometimes in other animals, a variety of dentine containing blood-vessels occurs, this is called **vasodentine**.

Cement or **crusta petrosa** forms the outermost layer of the root of the tooth. In composition and structure it is practically identical with bone. In the more complicated mammalian teeth, besides enveloping the root, it fills up the spaces between the folds of the enamel.

The hard parts of a tooth commonly enclose a central pulp-cavity into which projects the pulp, a papilla of the dermis including blood-vessels and nerves. As long as growth continues the outer layers of this pulp become successively calcified and added to the substance of the dentine. In young growing teeth the pulp-cavity remains widely open, but in mammals the

4

general rule is that as a tooth gets older and the crown becomes fully formed, the remainder of the pulp becomes converted into one or more tapering roots which are imbedded in the alveolar cavities of the jaws. The opening of the pulp-cavity is then reduced to a minute perforation at the base of each root. A tooth of this kind is called a **rooted** tooth.

But it is not only in young teeth that the pulp-cavity sometimes remains widely open; for some teeth, such as the tusks of Elephants and the incisor teeth of Rodents, form no roots and continue to grow throughout the animal's life. Such teeth are said to be rootless or to have persistent pulps.

An intermediate condition is seen in some teeth, such as the grinding teeth of Horses. These teeth grow for a very long time, their crowns wearing away as fast as their bases are produced; finally however definite roots are formed and growth ceases.

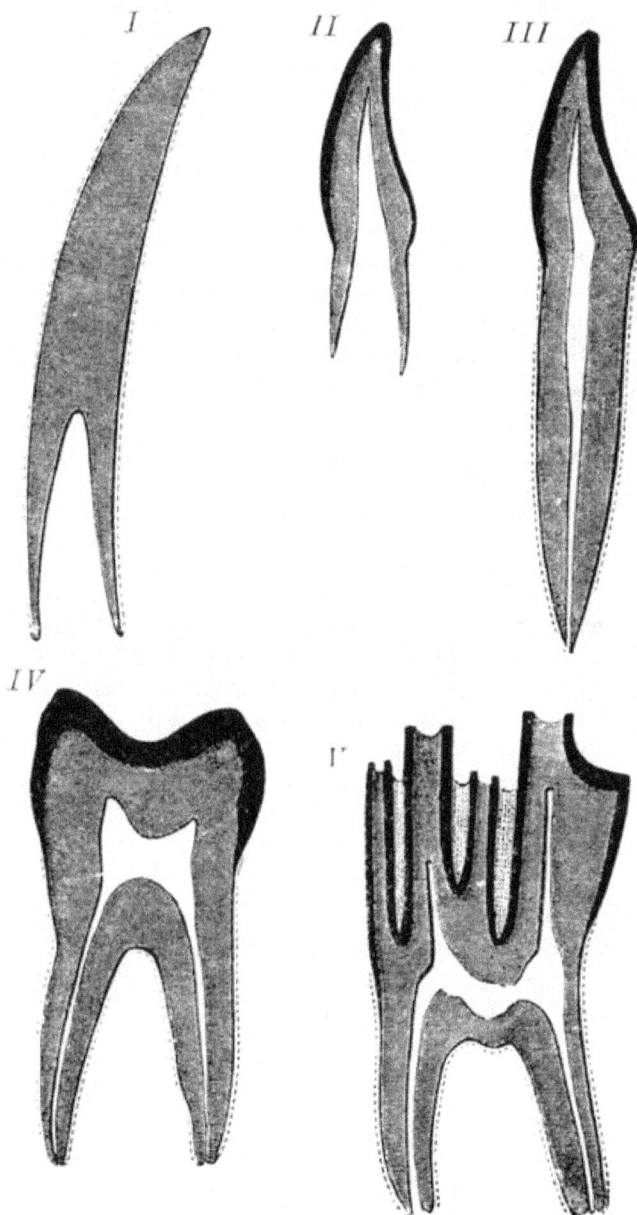

Fig. 1. Diagrammatic sections of various forms of teeth
(from Flower).

I. Incisor or tusk of elephant, with pulp-cavity persistently open at base. II. Human incisor during development with root imperfectly formed, and pulp-cavity widely open at base. III. Completely formed human incisor, with pulp-cavity contracted to a small aperture at the end of the root. IV. Human molar with broad crown and two roots. V. Molar of Ox, with the enamel covering the crown, deeply folded and the depressions filled with cement. The surface is worn by use, otherwise the enamel coating would be continuous at the top of the ridges. In all the figures the enamel is black, the pulp white, the dentine represented by horizontal lines, and the cement by dots.

The teeth of any animal may be **homodont**, that is, all having the same general character, or **heterodont**, that is, having different forms adapted to different functions. The dentition is heterodont in a few reptiles and the majority of mammals.

Succession of teeth. In most fishes, and many amphibians and reptiles the teeth can be renewed indefinitely. In sharks, for example, numerous rows of reserve teeth are to be seen folded back behind those in use (see fig. 15). The majority of mammals have only two sets of teeth, and are said to be **diphyodont**; some have only a single series (**monophyodont**).

Development of teeth. A brief sketch of the method in which development of teeth takes place in the higher vertebrates may here be given. Along the surface of the jaws a thickening of the epiblastic epithelium takes place, giving rise to a ridge, which sinks inwards into the tissue of the jaw, and it is known as the primary enamel organ. At the points where teeth are to be developed special ingrowths of this primary enamel organ take place, and into each there projects a vascular dental papilla from the surrounding mesoblast of the jaw. Each ingrowth of the enamel organ forms an **enamel cap**, which gradually embraces the dental papilla, and at the same time appears to be pushed on one side, owing to the growth not being uniform. The external layer of the dental papilla is composed of long nucleated cells, the **odontoblasts**, and it is by these that the dentine is formed. Similarly the internal layer of the enamel organ is formed of columnar enamel cells, which give rise to the enamel. The mesoblastic cells surrounding the base of the tooth give rise to the cement.

Bone is in many cases exoskeletal, but it will be most conveniently described with the endoskeleton.

The **scales of fish** are wholly or in part mesoblastic in origin, being totally different from those of reptiles. The **cycloid** and **ctenoid** scales of Teleosteans (see p. 105) are thin plates coated with epidermis. They are sometimes bony, but as a rule are simply calcified. **Ganoid** scales are flat plates of bone coated with an enamel-like substance, and articulating together with a peg and socket arrangement; they are probably identical with enlarged and flattened placoid scales.

The **armour plates** of fossil Ganoids, Labyrinthodonts, and Dinosaurs, and of living Crocodiles, some Lizards and Armadillos, are composed of bone. They are always covered by a layer of epidermis.

The **antlers of deer** are also composed of bone; they will be more fully described in the chapter on mammals. It may perhaps be well to mention them here, though they really belong to the endoskeleton, being outgrowths from the frontal bones.

B. Endoskeletal structures.

I. Hypoblastic.

(*a*) The **notochord** is an elastic rod formed of large vacuolated cells, and is surrounded by a membranous sheath of mesoblastic origin. It is the primitive endoskeleton in the Chordata, all of which possess it at some period of their existence; while in many of the lower forms it persists throughout life. Even in the highest Chordata it is the sole representative of the axial skeleton for a considerable part of the early embryonic life. A simple unsegmented notochord persists throughout life in the Cephalochordata, Cyclostomata, and some Pisces, such as Sturgeons and Chimaeroids.

(*b*) The enamel of the pharyngeal teeth of the Salmon and many other Teleosteans is hypoblastic in origin. The epiblast of the stomodaeum, in which the other teeth are developed, passes into the hypoblast of the mesenteron in which these pharyngeal teeth are formed.

II. Mesoblastic.

The most primitive type of a mesoblastic endoskeleton consists of a membranous sheath surrounding the notochord, as in *Myxine* and its allies. The first stage of complication is by the development of cartilage in the notochordal sheath, as in *Petromyzon*. Often the cartilage becomes calcified in places, as in the vertebral centra of *Scyllium* and other Elasmobranchs. Lastly, the formation of bone takes place; it generally constitutes the most important of the endoskeletal structures.

Bone may be formed in two ways:—

(1) by the direct ossification of pre-existing cartilage, when it is known as **cartilage bone** or **endochondral bone**;

(2) by independent ossification in connective tissue; it is then known as **membrane** or **dermal** or **periosteal bone**.

With the exception of the *clavicle* all the bones of the trunk and limbs, together with a large proportion of those of the skull, are preformed in the embryo in cartilage, and are grouped as cartilage bones; while the clavicle and most of the roofing and jaw-bones of the skull are not preformed in cartilage, being developed simply in connection with a membrane. Hence it is customary to draw a very strong line of distinction between these two kinds of bone; in reality however this distinction is often exaggerated, and the two kinds pass into one another, and as will be shown immediately, the permanent osseous tissue of many of those which are generally regarded as typical cartilage bones, is really to a great extent of periosteal origin. The palatine bone, for instance, of the higher vertebrates in general is preceded by a cartilaginous bar, but is itself almost entirely a membrane bone.

Before describing the development of bone it will be well to briefly describe the structure of adult bone and cartilage.

The commonest kind of **cartilage**, and that which preforms so many of the bones of the embryo, is **hyaline** cartilage. It consists of oval nucleated cells occupying cavities (**lacunae**) in a clear intercellular semitransparent matrix, which is probably secreted by the cells. Sometimes one cell is seen in each lacuna, sometimes shortly after cell-division a lacuna may contain two or more cells. The free surface of the cartilage is invested by a fibrous membrane, the **perichondrium**.

Bone consists of a series of lamellae of ossified substance between which are oval spaces, the **lacunae**, giving rise to numerous fine channels, the **canaliculi**, which radiate off in all directions. The lacunae are occupied by the **bone cells** which correspond to cartilage cells, from which if the bone is young, processes pass off into the canaliculi. It is obvious that the ossified substance of bone is intercellular in character, and corresponds to the matrix of cartilage.

Bone may be compact, or loose and spongy in character, when it is known as **cancellous bone**. In compact bone many of the lamellae are arranged concentrically round cavities, the **Haversian canals**, which in life are occupied by blood-vessels. Each Haversian canal with its lamellae forms a **Haversian system**. In spongy bone instead of Haversian canals there occur large irregular spaces filled with marrow, which consists chiefly of blood-vessels and fatty tissue. The centre of a long bone is generally occupied by one large continuous marrow cavity. The whole bone is surrounded by a fibrous connective tissue membrane, **the periosteum**.

The development of bone.

Periosteal ossification. An example of a bone entirely formed in this way is afforded by the parietal. The first trace of ossification is shown by the appearance, below the membrane which occupies the place of the bone in the early embryo, of calcareous spicules of bony matter, which are laid down round themselves by certain large cells, the **osteoblasts**. These osteoblasts gradually get surrounded by the matter which they secrete and become converted into bone cells, and in this way a mass of spongy bone is gradually produced. Meanwhile a definite periosteum has been formed round the developing bone, and on its inner side fresh osteoblasts are produced, and these with the others gradually render the bone larger and more and more compact. Finally, the middle layer of the bone becomes again hollowed out and rendered spongy by the absorption of part of the bony matter.

Endochondral ossification. This is best studied in the case of a long bone like the femur or humerus. Such a long bone consists of a shaft, which forms the main part, and two terminal portions, which form the **epiphyses**, or portions ossifying from centres distinct from that forming the shaft or main part of the bone.

In the earliest stage the future bone consists of hyaline cartilage surrounded by a vascular sheath, the perichondrium.

Then, starting from the centre, the cartilage becomes permeated by a number of channels into which pass vessels from the perichondrium and osteoblasts. In this way the centre of the developing shaft becomes converted into a mass of cavities separated by bands or trabeculae of cartilage. This cartilage next becomes calcified, but as yet is not converted into true bone. The osteoblasts in connection with the cavities now begin to deposit true endochondral spongy bone, and then after a time this becomes absorbed by certain large cells, the osteoclasts, and resolved into marrow or vascular tissue loaded with fat. So that the centre of the shaft passes from the condition of hyaline cartilage to that of calcified cartilage, thence to the condition of spongy bone, and finally to that of marrow. At the same time beneath the perichondrium osteoblasts are developed which also begin to give rise to spongy bone. The perichondrium thus becomes the periosteum, and the bone produced by it, is periosteal or membrane bone. So that while a continuous marrow cavity is gradually being formed in the centre of the shaft, the layer of periosteal bone round the margin is gradually thickening, and becoming more and more compact by the narrowing down of its cavities to the size of Haversian canals. The absorption of endochondral and formation of periosteal bone goes on, till in time it comes about that the whole of the shaft, except its terminations, is of periosteal origin. At the extremities of the shaft, however, and at the epiphyses, each of which is for a long time separated from the shaft by a pad of cartilage, the ossification is mainly endochondral, the periosteal bone being represented only by a thin layer.

Until the adult condition is reached and growth ceases, the pad of cartilage between the epiphysis and the shaft continues to grow, its outer (epiphysial) half growing by the formation of fresh cartilage as fast as its inner half is encroached on by the growth of bone from the shaft. The terminal or articular surfaces of the bone remain throughout life covered by layers of articular cartilage.

Even after the adult condition is reached the bone is subject to continual change, processes of absorption and fresh formation going on for a time and tending to render the bone more compact.

Methods in which bones are united to one another.

The various bones composing the endoskeleton are united to one another either by **sutures** or by movable **joints**.

When two bones are suturally united, their edges fit closely together and often interlock, being also bound together by the periosteum.

In many cases this sutural union passes into fusion or **ankylosis**, ossification extending completely from one bone to the other with the obliteration of the intervening suture. This feature is especially well marked in the cranium of most birds.

The various kinds of joints or articulations may be subdivided into imperfect joints and perfect joints.

In **imperfect joints**, such as the intervertebral joints of mammals, the two contiguous surfaces are united by a mass of fibrous tissue which allows only a limited amount of motion.

In **perfect joints** the contiguous articular surfaces are covered with cartilage, and between them lies a synovial membrane which secretes a viscid lubricating fluid.

The amount of motion possible varies according to the nature of the articular surfaces; these include—

a. **ball and socket joints**, like the hip and shoulder, in which the end of one bone works in a cup provided by another, and movements can take place in a variety of planes.

b. **hinge joints**, like the elbow and knee, in which as in ball-and-socket joints one bone works in a cup provided by another, but movements can take place in one plane only.

THE ENDOSKELETON.

The endoskeleton is divisible into **axial** and **appendicular** parts; and the **axial** skeleton into—

1. the spinal column,

2. the skull {*a.* the cranium,
{*b.* the jaws and visceral skeleton,

3. the ribs and sternum.

I. The Axial Skeleton.

1. The Spinal column.

The spinal column in the simplest cases consists of an unsegmented rod, the notochord, surrounded by the **skeletogenous layer**, a sheath of mesoblastic origin, which also envelops the nerve cord. Several intermediate stages connect this simple spinal column with the vertebral column characteristic of higher vertebrates. A typical vertebral column may be said to consist of (1) a series of cartilaginous or bony blocks, the vertebral **centra**, which arise in the sheath surrounding the notochord. They cause the notochord to become constricted and to atrophy to a varying extent, though a remnant of it persists, either permanently or for a long period, within each centrum or between successive centra. (2) From the dorsal surface of each centrum arise a pair of processes which grow round the spinal cord and unite above it, forming a **dorsal** or **neural arch**. (3) A similar pair of processes arising from the ventral surface of the centrum form the **ventral** or **haemal arch**. To the ventral arch the ribs strictly belong, and it tends to surround the ventral blood-vessels and the body cavity with the alimentary canal and other viscera.

A **neural spine** or spinous process commonly projects upwards from the dorsal surface of the neural arch, and a pair of **transverse processes** project outwards from its sides. When, as is commonly the case, the two halves of the haemal arch do not meet, the ventral surface of the centrum often bears a downwardly-projecting **hypapophysis**.

The character of the surfaces by which vertebral centra articulate with one another varies much. Sometimes both surfaces are concave, and the vertebra is then said to be **amphicoelous**; sometimes a centrum is convex in front and concave behind, the vertebra is then **opisthocoelous**, sometimes concave in front and convex behind, when the vertebra is **procoelous**. Again, in many vertebrae both faces of the centra are flat, while in others they are saddle-shaped, as in the neck vertebrae of living birds, or biconvex, as in the case of the first caudal vertebra of crocodiles.

In the higher vertebrates pads of fibrocartilage—the **intervertebral discs**—are commonly interposed between successive centra, these or parts of them often ossify, especially in the trunk and tail, and are then known as **inter centra**.

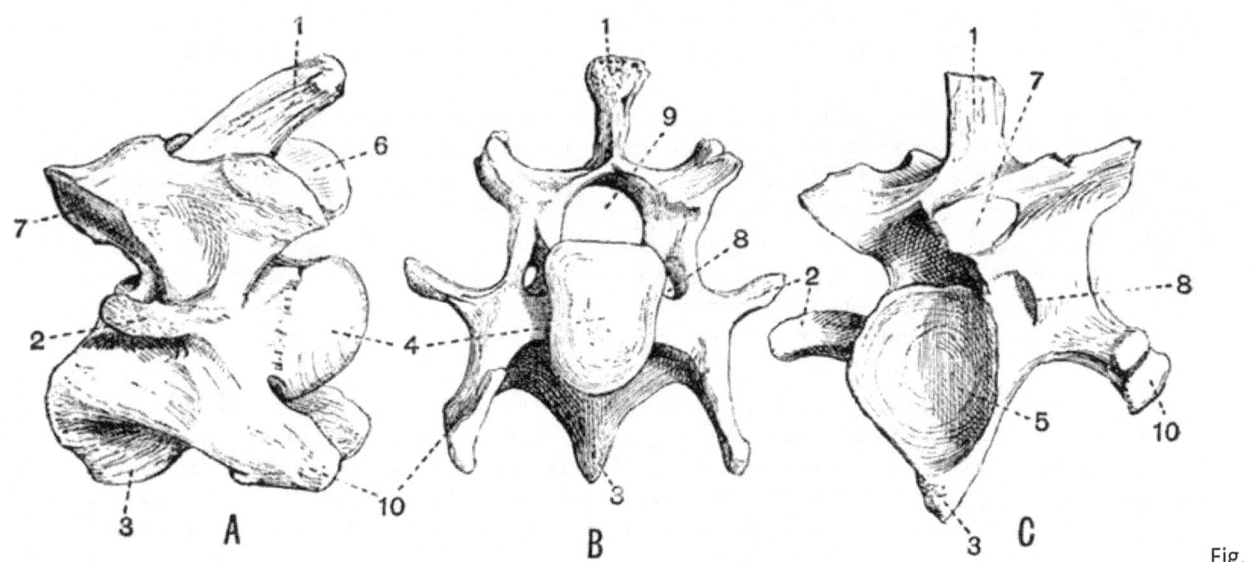

Fig. 2.

Cervical vertebrae of an Ox (*Bos taurus*).
A, is the fifth; B, the fourth; C, the third. X ¼ (Camb. Mus.)

1. neural spine.
2. transverse process.
3. hypapophysis.
4. convex anterior face of the centrum.
5. concave posterior face of the centrum.

6. prezygapophysis.
7. postzygapophysis.
8. vertebrarterial canal.
9. neural canal.
10. inferior lamella of transverse process.

The vertebrae of the higher forms can generally be arranged in the following five groups, each marked by certain special characteristics:

1. The **cervical** or **neck vertebrae**. These connect the skull with the thorax, and are characterised by relatively great freedom of movement. They often bear small ribs, but are distinguished from the succeeding thoracic vertebrae by the fact that their ribs do not reach the sternum. The first cervical vertebra which articulates with the skull is called the **atlas**, but a study of the nerve exits shows that the first vertebra is not serially homologous throughout the Ichthyopsida, so that it is best to reserve the term atlas for the first vertebra in Sauropsida and Mammalia.

2. The **thoracic vertebrae** (often called dorsal) bear movably articulated ribs which unite ventrally with the sternum.

3. The **lumbar vertebrae** are generally large, and are often more movable on one another than are the thoracic vertebrae. They bear no ribs.

4. The **sacral vertebrae** are characterised by the fact that they are firmly fused together, and are united with the pelvic girdle by means of their transverse processes and rudimentary ribs.

5. The **caudal** or **tail vertebrae** succeed the sacral. The anterior ones are often fused with one another and with the sacrals, but they differ from true sacral vertebrae in that there are no rudimentary ribs between their transverse processes and the pelvic girdle. They often bear V-shaped **chevron bones**.

In fish and snakes the vertebral column is divisible into only two regions, an anterior trunk region, whose vertebrae bear ribs, and a posterior tail region, whose vertebrae are ribless.

2. The Skull.

Before giving a general account of the adult skull it will be well to briefly describe its development.

General development of the Cranium.

Shortly after its appearance, the central nervous system becomes surrounded by a membranous mesodermal investment which in the region of the spinal cord is called the **skeletogenous layer** or **perichordal sheath**, while in the region of the brain it is called the **membranous cranium**. Ventral to the central nervous system is the notochord, which extends far into the region of the future cranium, and like the nervous system, is enclosed by the skeletogenous layer. The primitive cartilaginous cranium is formed by histological differentiation within the substance of the membranous cranium and always consists of the following parts:

(*a*) the **parachordals**. These are a pair of flat curved plates of cartilage, each of which has its inner edge grooved where it comes in contact with the notochord. The parachordals, together with the notochord, form a continuous plate, which is known

9

as the **basilar plate**. The basilar plate is the primitive floor below the hind- and mid-brain. In front the parachordals abut upon another pair of cartilaginous bars, the trabeculae, the two pairs of structures being sometimes continuous with one another from the first;

(*b*) the **trabeculae** which meet behind and embrace the front end of the notochord. Further forwards they at first diverge from one another, and then converge again, enclosing a space, the **pituitary space**. After a time they generally fuse with one another in the middle line, and, with the parachordals behind, form an almost continuous basal plate. The trabeculae generally appear before the parachordals. They form the primitive floor below the fore-brain;

(*c*) the cartilaginous **capsules** of the three pairs of **sense organs**. At a very early stage of development involutions of the surface epiblast give rise to the three pairs of special sense organs—the olfactory or nasal organs in front, the optic in the middle, and the auditory behind. The olfactory and auditory organs always become enclosed in definite cartilaginous capsules, the eyes often as in the Salmon, become enclosed in cartilaginous sclerotic capsules, while sometimes, as in mammals, their protecting capsules are fibrous.

Each pair of sense capsules comes into relation with part of the primitive cranium, and greatly modifies it. Thus the auditory or periotic capsules press on the parachordals till they come to be more or less imbedded in them. Perhaps owing to the pressure of the nasal capsules the trabeculae fuse in front, and then grow out into an anterior pair of processes, the **cornua trabeculae**, and a posterior pair, the **antorbital processes**, which together almost completely surround the nasal capsules. The sclerotic capsules of the eyes greatly modify the cranium, although they never become completely united with it.

The cartilaginous cranium formed of the basal plate, together with the sense capsules, does not long remain merely as a floor. Its sides grow vertically upwards, forming the **exoccipital** region of the cranium behind, and the **alisphenoidal** and **orbitosphenoidal** regions further forwards. In many forms, such as Elasmobranchs, all these upgrowths meet round the brain, roofing it in and forming an almost complete cartilaginous cranium. But in most vertebrata, while in the occipital region, the cartilaginous cranium is completed dorsally, in the alisphenoidal and orbitosphenoidal regions the cartilage merely forms the lateral walls of the cranium, the greater part of the brain having dorsal to it a wide space, closed by merely membranous tissue in connection with which the large frontal and parietal bones are subsequently formed.

The Skull includes

a. the cranium,

b. the jaws and visceral skeleton.

The **cranium** can be further subdivided into

(1) an axial portion, the **cranium proper** or **brain case**;

(2) **the sense capsules.** The capsules of the auditory and olfactory sense organs are always present, and as has been already mentioned, in many animals the eye likewise is included in a cartilaginous capsule.

(1) The cranium proper or brain case.

The cranium varies much in form and structure. In lower vertebrates, such as Sharks and Lampreys, it remains entirely cartilaginous and membranous, retaining throughout life much of the character of the embryonic rudiment of the cranium of higher forms. The dogfish's cranium, described on pp. 73 to 76, is a good instance of a cranium of this type. But in the majority of vertebrates the cartilage becomes more or less replaced by cartilage bone, while membrane bones are also largely developed and supplant the cartilage.

The cranium of most vertebrates includes a very large number of bones whose arrangement varies much, but one can distinguish a definite **basicranial axis** formed of the basi-occipital, basisphenoid, and presphenoid bones, which is a continuation forwards of the axis of the vertebral column. From the basicranial axis a wide arch arises, composed of a number of bones, which form the sides and roof of the brain-case These bones are arranged in such a manner that if both cartilage and membrane bones are included they can be divided into three rings or segments. The hinder one of these segments is the occipital, the middle the parietal, and the anterior one the frontal.

The occipital segment is formed of four cartilage bones, the **basi-occipital** below, two **exoccipitals** at the sides, and the **supra-occipital** above. The parietal segment is formed of the **basisphenoid** below, two **alisphenoids** at the sides and two membrane bones, the *parietals* above, and the frontal segment in like manner consists of the **presphenoid** below, the two **orbitosphenoids** at the sides, and two membrane bones, the *frontals*, above. The parietals and frontals, being membrane bones, are not comparable to the supra-occipital, in the way that the presphenoid and basisphenoid are to the basi-occipital.

The cartilage bones of the occipital segments are derived from the parachordals of the embryonic skull, those of the parietal and frontal segments from the trabeculae.

In front of the presphenoid the basicranial axis is continued by the **mesethmoid**.

(2) The sense capsules.

These enclose and protect the special sense organs.

(*a*) **Auditory capsule.**

The basisphenoid is always continuous with the basi-occipital, but the alisphenoid is not continuous with the exoccipital as the **periotic** or **auditory capsule** is interposed between them. Each periotic capsule has three principal ossifications; an anterior bone, the **pro-otic**, a posterior bone, the **opisthotic**, and a superior bone, the **epi-otic**.

These bones may severally unite, or instead of uniting with one another they may unite with the neighbouring bones. Thus the epi-otic often unites with the supra-occipital, and the opisthotic with the exoccipital.

Two other bones developed in the walls of the auditory capsule are sometimes added, as in Teleosteans; these are the **pterotic** and **sphenotic**.

(b) Optic capsule.

The eye is frequently enclosed in a cartilaginous sclerotic capsule, and in this a number of scale-like bones are often developed.

Several membrane bones are commonly formed around the orbit or cavity for the eye. The most constant of these is the *lachrymal* which lies in the anterior corner; frequently too, as in Teleosteans, there is a *supra-orbital* lying in the upper part of the orbit, or as in many Reptiles, a *postorbital* lying in the posterior part of the orbit.

(c) Nasal capsule.

In relation to the nasal capsules various bones occur.

The basicranial axis in front of the presphenoid is ossified, as the **mesethmoid**, dorsal to which there sometimes, as in Teleosteans, occur a *median ethmoid* and a pair of **lateral ethmoids**. Two pairs of membrane bones very commonly occur in this region, viz. the *nasals* which lie dorsal to the mesethmoid, and the *vomers* (sometimes there is only one) which lie ventral to it.

The part of the skull lying immediately in front of the cranial cavity and in relation to the nasal capsules constitutes the **ethmoidal region**.

There remain certain other membrane bones which are often found connected with the cranium. Of these, one of the largest is the *parasphenoid* which, in Ichthyopsids, is found underlying the basicranial axis. *Prefrontals* often, as in most reptiles, occur lying partly at the sides and partly in front of the frontal, and *postfrontals* similarly occur behind the orbit lying partly behind the frontals and partly at their sides. Lastly a *squamosal bone* is, as in Mammals, very commonly developed, and lies external and partly dorsal to the auditory capsules.

The Jaws and Visceral Skeleton.

In the most primitive fish these consist of a series of cartilaginous rings or arches placed one behind another and encircling the anterior end of the alimentary canal. Originally they are mainly concerned with branchial respiration.

The first or **maxillo-mandibular** arch forms the upper jaw and the lower jaw or mandible.

The second or **hyoid** arch bears gills and often assists in attaching the jaws to the cranium. The remaining arches may bear gills, though the last is commonly without them.

The above condition is only found in fishes, in higher animals the visceral skeleton is greatly reduced and modified.

The first or maxillo-mandibular arch is divisible into a dorsal portion, the **palato-pterygo-quadrate bar**, which forms the primitive upper jaw and enters into very close relations with the cranium, and a ventral portion, **Meckel's cartilage**, which forms the primitive lower jaw. The cartilaginous rudiments of both these portions disappear to a greater or less extent and become partly ossified, partly replaced by or enveloped in membrane bone.

The posterior part of the palato-pterygo-quadrate bar becomes ossified to form the **quadrate**, the anterior part to form the palatine and pterygoid, or the two latter may be formed partially or entirely of periosteal bone, developed round the cartilaginous bar. Two pairs of important membrane bones, the *premaxillae* and *maxillae* form the anterior part of the upper jaw, and behind the maxilla lies another membrane bone, the *jugal* or *malar*, which is connected with the quadrate by a *quadratojugal*. The premaxillae have a large share in bounding the external nasal openings or anterior nares.

In lower vertebrates the nasal passage leads directly into the front part of the mouth cavity and opens by the posterior nares. In some higher vertebrates, such as mammals and crocodiles, processes arise from the premaxillae and palatines, and sometimes from the pterygoids, which meet their fellows in the middle line and form the palate, shutting off the nasal passage from the mouth cavity and causing the posterior nares to open far back.

The cartilage of the lower jaw is in all animals with ossified skeletons, except the Mammalia, partly replaced by cartilage bone forming the **articular**, partly overlain by a series of membrane bones the *dentary, splenial, angular, supra-angular* and *coronoid.* In many sharks large paired accessory cartilages occur at the sides of the jaws; and in a few reptiles and some Amphibia, such as the Frog, the ossified representative of the anterior of these structures occurs forming the **mento-meckelian** bone. In mammals the lower jaw includes but a single bone.

The quadrate in all animals with ossified skeletons, except the Mammalia, forms the suspensorium of the mandible or the skeletal link between the jaw and the cranium; in the Mammalia, however, the mandible articulates with the squamosal, while the quadrate is greatly reduced, and is now generally considered to be represented by the tympanic ring of the ear.

11

The second visceral or hyoid arch in fishes consists of two pieces of cartilage, a proximal piece the **hyomandibular**, and a distal piece the **cerato-hyal**. The cerato-hyals of the two sides are commonly united by a median ventral plate, the **basi-hyal**. The hyoid arch bears gills on its posterior border, but its most important function in most fishes is to act as the suspensorium. In higher vertebrates the representative of the hyomandibular is much reduced in size, and comes into relation with the ear forming the **auditory ossicles**; the cerato-hyal looses its attachment to the hyomandibular and becomes directly attached to the cranium, forming a large part of the hyoid apparatus of most higher vertebrates.

Behind the hyoid arch come the branchial arches. They are best developed in fishes, in which they are commonly five in number and bear gills. Their ventral ends are united in pairs by median pieces, the **copulae**.

In higher vertebrates they become greatly reduced, and all except the first and second completely disappear. In the highest vertebrates, the mammals, the second has disappeared, but in birds and many reptiles it is comparatively well developed.

3. The Ribs and Sternum.

The **ribs** are a series of segmentally arranged cartilaginous or bony rods, attached to the vertebrae; they tend to surround the body cavity, and to protect the organs contained within it. Ribs are very frequently found attached to the transverse processes of the vertebrae, but a study of their origin in fish shows that they are really the cut off terminations of the ventral arch, not of the transverse processes which are outgrowths from the dorsal arch. In the tail their function is to surround and protect structures like the ventral blood-vessels which do not vary much in size, consequently they meet one another, and form a series of complete ventral or haemal arches. But the trunk contains organs like the lungs and stomach which are liable to vary much in size at different times, consequently the halves of the haemal arch do not meet ventrally, and then the ribs become detached from the rest of the haemal arch. Having once become detached, they are able to shift about and unite themselves to various points of the vertebra. They frequently, as has been already mentioned, become entirely attached to the transverse process, or they may be attached to the transverse process by a dorsal or **tubercular** portion and to the centrum or to the ventral arch by a ventral or **capitular** portion.

In all animals above fishes the distal ends of the thoracic ribs unite with a median breast bone or sternum which generally has the form of a segmented rod. The **sternum** is really formed by the fusion of the distal ends of a series of ribs. In many animals elements of the shoulder girdle enter into close relation with the rib elements of the sternum.

II. The Appendicular Skeleton.

This consists of the skeleton of the anterior or **pectoral**, and the posterior or **pelvic** limbs, and their girdles. In every case (except in Chelonia) the parts of the appendicular skeleton lie external to the ribs.

1. The Limb girdles.

The Pectoral girdle. In the simplest case the pectoral or shoulder girdle consists of a hoop of cartilage incomplete dorsally. It is attached by muscle to the vertebral column, and is divided on either side into dorsal and ventral portions by a cavity, the **glenoid cavity**, at the point where the anterior limb articulates. In higher fishes this hoop is distinctly divided into right and left halves; it becomes more or less ossified, and a pair of important bones, the clavicles, are developed in connection with its ventral portion.

In higher vertebrates ossification sets up in the cartilage and gives rise on each side to a dorsal bone, the **scapula**, and frequently to an anterior ventral bone, the **precoracoid**, and a posterior ventral bone, the **coracoid**. The precoracoid is often not ossified, and upon it is developed the clavicle which more or less replaces it. In some forms a **T** shaped *interclavicle* occurs, in others **epicoracoids** are found in front of the coracoids. In all vertebrata above fish, except the great majority of mammals, the coracoids are large and articulate with the sternum. But in mammals the coracoids are nearly always quite vestigial, and the pectoral girdle is attached to the axial skeleton by the clavicle or sometimes by muscles and ligaments only.

The **Pelvic girdle** like the pectoral consists primitively of a simple rod or hoop of cartilage, which in vertebrata above fishes is divided into dorsal and ventral portions, by a cavity, the **acetabulum**, with which the posterior limb articulates. In the pelvic girdle as in the pectoral one dorsal, and (commonly) two ventral ossifications take place. The dorsal bone is the **ilium** and corresponds to the scapula. The posterior ventral bone is the **ischium** corresponding to the coracoid. The anterior ventral bone is the **pubis** and is generally compared to the precoracoid, but in some cases a fourth pelvic element, the **acetabular** or **cotyloid** bone is found, and this may correspond to the precoracoid.

The pelvic girdle differs from the pectoral in the fact that the dorsal bones—the ilia—are nearly always firmly united to transverse processes of the sacral vertebrae, by means of rudimentary ribs. The pubes and ischia generally meet in ventral symphyses.

2. The Limbs.

It will be most convenient to defer a discussion of the limbs of fishes to chap. VIII.

All vertebrates above fishes have the limbs divisible into three main segments:—

Anterior or Fore limb.   **Posterior or Hind limb.**

Proximal segment.	upper arm or *brachium*.	thigh.
Middle segment.	fore-arm or *antibrachium*.	shin or *crus*.
Distal segment.	*manus*.	*pes*.

The proximal segments each contain one bone, the **humerus** in the case of the upper arm, and the **femur** in the case of the thigh. The middle segments each contain two bones, the **radius** and **ulna** in the case of the fore-arm, and the **tibia** and **fibula** in the case of the shin.

The manus and pes are further subdivided into

(*a*) two or three proximal rows of bones forming the wrist or **carpus** in the case of the manus, and the ankle or **tarsus** in the case of the pes.

(*b*) a middle row called respectively the **metacarpus** and **metatarsus**.

(*c*) a number of distal bones called the **phalanges** which form the skeleton of the fingers and toes, or **digits**.

Typically the manus and pes both have five digits (pentedactylate). The first digit of the manus is commonly called the **pollex**, and the first digit of the pes the **hallux**.

In a very simple **carpus** such as that of *Chelydra*, there are nine bones. They are arranged in a proximal row of three, the radiale, intermedium, and ulnare,—the first being on the radial side of the limb, and a distal row of five called respectively carpale 1, 2, 3, 4, 5, beginning on the radial side. Between these two rows is a single bone the centrale, or there may be two.

Similarly there are nine bones in a simple **tarsus** such as that of *Salamandra*. They form a proximal row of three, the tibiale, intermedium and fibulare, and a distal row of five, called respectively tarsale 1, 2, 3, 4, 5, beginning on the tibial side. Between the two rows there is a centrale as in the carpus, or there may be two.

The following names derived from human anatomy are commonly applied to the various carpal and tarsal bones:

Carpus.	**Tarsus.**
radiale = scaphoid	tibiale}
intermedium = lunar	intermedium } astragalus
ulnare = cuneiform	fibulare = calcaneum
centrale = central	centrale = navicular
carpale 1 = trapezium	tarsale 1 = internal cuneiform
"2 = trapezoid	"2 = middle "
"3 = magnum	"3 = external "
"4 }	"4 }
"5 } = unciform	"5 }= cuboid

Note. The above is the view commonly accepted concerning the homology of the carpal and tarsal bones. But with regard to the proximal row of tarsal bones there is difference of opinion. All anatomists are agreed that the calcaneum is the fibulare and that the intermedium is contained in the astragalus, but while the majority regard the astragalus as the fused tibiale and intermedium, Baur considers that a small bone found on the tibial side of the tarsus in *Procavia*, many Rodents, Insectivores, and the male *Ornithorhynchus*, is the vestigial tibiale, and regards the astragalus as the intermedium alone. He also considers that the mammalian scaphoid represents a centrale.

Modifications in the positions of the limbs.

In their primitive position the limbs are straight and are extended parallel to one another at right angles to the axis of the trunk. Each limb then has a dorsal surface, a ventral surface, an anterior or **pre-axial** edge, and a posterior or **postaxial** edge.

In the anterior limb the radius and the pollex are pre-axial, the ulna and the fifth finger are postaxial. In the posterior limb the tibia and the hallux are pre-axial, the fibula and the fifth toe are postaxial. The Cetacea and various extinct reptiles, such as *Ichthyosaurus* and *Plesiosaurus*, have their limbs in practically this primitive position.

The first modification from it is produced by the bending ventrally of the middle segments of both limbs upon the proximal segments, while the distal segment is bent in the opposite direction on the middle segment. Then the ventral surfaces of the antibrachium and crus come to look inwards, and their dorsal surfaces to look outwards. The brachium and manus, thigh and pes still have their dorsal surfaces facing upwards and their ventral surfaces facing downwards as before, and the relations of their pre- and postaxial borders remain as they were. Many Amphibians and Reptiles, such as tortoises, carry their limbs in this position.

In all higher vertebrates, however, a further change takes place, each limb is rotated as a whole from its proximal end, the rotation taking place in opposite directions in the fore and hind limbs respectively. The anterior limb is rotated backwards from the shoulder, so that the brachium lies nearly parallel to the body, and the elbow points backwards, the antibrachium downwards, and the manus backwards; the pre-axial surface of the whole limb with the radius and pollex now faces outwards, and the postaxial surface with the ulna and fifth finger now faces inwards. In the Walrus and, to a certain extent, in the Sea lions the anterior limb remains throughout life in this position. The posterior limb is also rotated, but the rotation in this case takes place forwards, so that the thigh lies nearly parallel to the body, the knee-joint pointing forwards; the crus downwards and the pes forwards. The pre-axial surface of the whole limb with the tibia and hallux looks towards the middle of the body, the postaxial surface with the fibula and fifth toe looks outwards. This is the position in which the hind limb is carried in nearly all mammals.

In nearly all mammals a further change takes place in the position of the anterior limb. The radius and ulna have hitherto been parallel to one another, but now the lower end of the radius, carrying with it the manus, comes to be rotated forwards round the ulna, so that the manus, as well as the pes, comes to be forwardly-directed, and its pre-axial surface faces inwards.

In the majority of mammals the radius and ulna are permanently fixed in this, which is known as the **prone** position, but in man and some other mammals the manus can be pronated or turned into this position at will. When the radius and ulna are parallel throughout their whole length the manus is said to be in the **supine** position.

The **extensor** side of a limb is that to which the muscles which straighten it are attached, the **flexor** side is that to which the muscles which bend it are attached.

CHAPTER II.
CLASSIFICATION.

The following classification includes *only the forms mentioned in the succeeding pages.* The relative value of some of the terms employed in classification is not identical throughout the book. This remark applies specially to the term *group*, which is a convenient one, owing to its not having such a hard and fast zoological meaning as has the term *family*, for instance. The term *group* is applied in this book to divisions of the animal kingdom of very different classificatory importance.

PHYLUM CHORDATA.

SUBPHYLUM A. HEMICHORDATA.

- Balanoglossus.
- Cephalodiscus.
- Rhabdopleura.
- ? Phoronis.
- (? Actinotrocha—larval Phoronis).

SUBPHYLUM B. UROCHORDATA (TUNICATA).

- Group Larvacea and others.

SUBPHYLUM C. CEPHALOCHORDATA.

- Amphioxus—lancelet.

Note. In this chapter all the generic names printed in italics are those of extinct animals.

SUBPHYLUM D. VERTEBRATA.

DIVISION (I). CYCLOSTOMATA.

- Order 1. Marsipobranchii.
- Family **Myxinoidei**. Myxine—hag-fish.
- Bdellostoma.

- Family **Petromyzontidae**. Petromyzon—lamprey.
- (Ammocoetes—larval lamprey.)
- Family **Palaeospondylidae**. *Palaeospondylus.*
- Order 2. Ostracodermi.
- Suborder 1. Heterostraci.
- Family **Pteraspidae**. *Pteraspis.*
- Suborder 2. Osteostraci.
- Family **Cephalaspidae**. *Cephalaspis.*
- Suborder 3. Antiarcha.
- Family **Asterolepidae**. *Pterichthys.*
- *Asterolepis.*

DIVISION (II). GNATHOSTOMATA.

A. ICHTHYOPSIDA.

CLASS I. PISCES.

- Order 1. Elasmobranchii.
- Suborder (1). Ichthyotomi.
- Family **Pleuracanthidae**. *Xenacanthus.*
- Suborder (2). Pleuropterygii.
- *Cladoselache.*
- Suborder (3). Selachii.
- Group Squalidae.
- Family **Notidanidae**. Heptanchus.
- Hexanchus.
- Chlamydoselache—frill-gilled shark.
- Family **Cochliodontidae**. *Cochliodus.*

- Family **Cestraciontidae**. Cestracion—Port Jackson shark.
- *Acrodus.*
- Family **Scylliidae**. Scyllium—spotted dogfish.
- Family **Lamnidae**. Odontaspis.
- Family **Carcharidae**. Galeus—tope.
- Family **Spinacidae**. Acanthias—spiny dogfish.
- Scymnus.
- Family **Squatinidae**. Squatina (Rhina)—angel fish.
- Group Batoidei.
- Family **Pristidae**. Pristis—saw-fish.
- Family **Raiidae**. Raia—skate.
- Family **Myliobatidae**. Myliobatis—eagle ray.
- Family **Trygonidae**. Trygon—sting ray.
- Family **Torpedinidae**. Torpedo—electric ray.
- Suborder (4). Acanthodii.
- Family **Acanthodidae**. *Acanthodes.*
- Family **Diplacanthidae**. *Diplacanthus.*
- Order 2. Holocephali.
- Family **Chimaeridae**. Chimaera—rabbit fish.

- Harriotta.
- Callorhynchus.
- *Ischyodus.*
- Order 3. Ganoidei.
- Suborder (1). Chondrostei.
- Family **Palaeoniscidae**. *Palaeoniscus.*
- *Trissolepis.*
- Family **Acipenseridae**. Acipenser—sturgeon.
- Scaphirhynchus.
- Family **Polyodontidae**. Polyodon (Spatularia)—spoon-beaked sturgeon.
- Psephurus—slender-beaked sturgeon.

- Suborder (2). Crossopterygii.
- Family **Holoptychiidae**. *Holoptychius.*
- Family **Rhizodontidae**. *Rhizodus.*
- Family **Osteolepidae**. *Osteolepis.*
- Family **Polypteridae**. Polypterus—bichir.
- Calamoichthys—reed-fish.
- Suborder (3). Holostei.
- Family **Lepidosteidae**. Lepidosteus—gar pike.
- Family **Semionotidae**. *Lepidotus.*
- Family **Amiidae**. Amia—bow-fin.
- Order 4. Teleostei.
- Suborder (1). Plectognathi.
- Family **Balistidae**. Balistes—file-fish.
- Family **Gymnodontidae**. Diodon—globe-fish.
- Family **Ostracionidae**. Ostracion—coffer-fish.
- Suborder (2). Physostomi.
- Family **Siluridae**.—cat-fishes.
- Family **Cyprinidae**. Cyprinus—carp.
- Family **Esocidae**. Esox—pike.
- Family **Salmonidae**. Salmo—salmon.
- Family **Clupeidae**. Clupeus—herring.
- Exocaetus—'flying fish'.
- Family **Muraenidae**. Anguilla—eel.
- Suborder (3). Anacanthini.
- Family **Gadidae**. Gadus—cod, haddock, whiting.
- Family **Pleuronectidae**. Solea—sole.
- Suborder (4). Pharyngognathi.
- Family **Labridae**. Labrus—wrasse.
- Scarus—parrot fish.

- Suborder (5). Acanthopterygii.
- Family **Cataphracti**. Dactylopterus—flying gurnard.
- Family **Percidae**. Perca—perch.

- Order 5. Dipnoi.
- Suborder (1). Sirenoidei.
- Family **Dipteridae**. *Dipterus.*
- Family **Monopneumona**. Ceratodus—barramunda.
- Family **Dipneumona**. Protopterus—African mud-fish.
- Lepidosiren.
- Suborder (2). Arthrodira.
- Family **Coccosteidae**. *Coccosteus.*
- *Dinichthys.*

Note. Palaeontological research has disclosed the existence of a great number of forms which seem to connect with one another almost all the orders of fishes as usually recognised. Forms connecting the living Ganoids with the Teleosteans have been especially numerous, so that these terms Ganoid and Teleostean can hardly be any longer used in a precise and scientific sense. This has rendered the subject of the classification of fishes a very difficult one. Though unsuitable for adoption in a work like the present, by far the most natural classification hitherto proposed seems to be that of Smith Woodward. He considers that the course of development of fishes has followed two distinct lines, the autostylic and hyostylic (see p. 119), and groups the various forms as follows:

Hyostylic.	Autostylic.
Subclass 1. Elasmobranchii.	Subclass 3. Holocephali.
1. Ichthyotomi.	1. (unknown).
2. Selachii.	2. Chimaeroidei.
3. Acanthodii.	3. (unknown).
Subclass 2. Teleostomi.	Subclass 4. Dipnoi.
1. Crossopterygii (Palaeozoic and Mesozoic).	1. Sirenoidei.
2. Crossopterygii (Cainozoic).	2. (unknown).
3. Actinopterygii.	3. Arthrodira.

The primitive forms in each of these four subclasses have the fins archipterygia (see p. 127).

CLASS II. AMPHIBIA.
- Order 1. Urodela.
- Suborder (1). Ichthyoidea.
- Group A. ,Perennibranchiata.
- Family **Menobranchidae**. Menobranchus.
- Family **Proteidae**. Proteus—olm.
- Family **Sirenidae**. Siren.
- Group B. Derotremata.
- Family **Amphiumidae**. Megalobatrachus.
- Cryptobranchus (Menopoma).
- Amphiuma.
- Suborder (2). Salamandrina.
- Family **Salamandridae**. Salamandra—salamander.
- Molge—newt.
- Onychodactylus.

- Amblystoma.
- (Siredon—axolotl, larval Amblystoma).
- Batrachoseps.
- Spelerpes (Gyrinophilus).
- Order 2. Labyrinthodontia.
- Group **Lepospondyli**. *Branchiosaurus.*
- Group **Temnospondyli**. *Archegosaurus.*
- *Nyrania.*
- *Euchirosaurus.*
- Group **Stereospondyli**. *Capitosaurus.*
- *Mastodonsaurus.*
- Order 3.Gymnophiona.
- Family **Caeciliidae**. Siphonops.
- Epicrium.

- Order 4. Anura.
- Suborder (1). Aglossa.
- Family **Xenopidae**. Xenopus.
- Family **Pipidae**. Pipa—Surinam toad.
- Suborder (2). Phaneroglossa.
- Group Arcifera.
- Family **Discoglossidae**. Discoglossus—painted frog.
- Bombinator—fire-bellied frog.
- Alytes—midwife frog.
- Family **Pelobatidae**. Pelobates—toad frog.
- Family **Hylidae**. Hyla—green tree-frog.
- Family **Bufonidae**. Bufo—toad.
- Docidophryne.
- Family **Cystignathidae**. Ceratophrys—horned frog.
- Group Firmisternia.
- Family **Ranidae**. Rana—common and edible frogs.
- Family **Engystomatidae**. Brachycephalus.

B. SAUROPSIDA.

CLASS I. REPTILIA.

- Order 1. Theromorpha.
- Group **Anomodontia**. *Dicynodon.*
- *Udenodon.*
- Group **Placodontia**. *Placodus.*
- Group **Pariasauria**. *Pariasaurus.*
- *Elginia.*
- Group **Theriodontia**. *Dimetrodon.*
- *Galesaurus.*
- *Cynognathus.*

- Order 2. Sauropterygia.

- Family **Mesosauridae**. *Mesosaurus.*
- Family **Nothosauridae**. *Nothosaurus.*
- Family **Plesiosauridae**. *Plesiosaurus.*
- *Pliosaurus.*
- Order 3. Chelonia.
- Suborder (1). Trionychia.
- Family **Trionychidae**. Trionyx—snapping turtle.
- Suborder (2). Cryptodira.
- Family **Dermochelydidae**. Dermochelys (Sphargis)—leathery
- turtle.
- Family **Chelonidae**. Chelone—green turtle.
- Family **Chelydridae**. Chelydra—terrapin.
- Family **Chersidae**. Testudo—tortoise.
- Suborder (3). Pleurodira.
- Family **Chelydae**. Chelys.
- Order 4. Ichthyosauria.
- Family **Ichthyosauridae**. *Ichthyosaurus.*
- Order 5. Rhynchocephalia.
- Suborder (1). Rhynchocephalia vera.
- Family **Sphenodontidae**. Sphenodon (Hatteria).
- Family **Rhynchosauridae**. *Hyperodapedon.*
- Suborder (2). Proganosauria.
- Family **Proterosauridae**. *Proterosaurus.*
- Order 6. Squamata.
- Suborder (1). Lacertilia.
- Group **Lacertilia vera**.
- Family **Geckonidae**. Gecko.
- Family **Pygopodidae**. Lialis—scale-foot.

- Family **Agamidae**. Draco—flying lizard.
- Agama.
- Family **Iguanidae**. Iguana.
- Family **Anguidae**. Ophisaurus (Bipes, Pseudopus).
- Anguis—blindworm.
- Family **Varanidae**. Varanus—monitor.
- Family **Amphisbaenidae**. Chirotes.
- Amphisbaena.
- Family **Scincidae**. Tiliqua (Cyclodus).
- Scincus—skink.
- Chalcides (Seps).
- Group **Rhiptoglossa**.
- Family **Chamaeleonidae**. Chamaeleon.
- Suborder (2). Ophidia.
- Family **Typhlopidae**. Typhlops—blind snake.
- Family **Boidae**. Python.

- Family **Colubridae**. Tropidonotus—ringed snake.
- Family **Hydrophidae**—sea snakes.
- Family **Crotalidae**. Crotalus—rattlesnake.
- Suborder (3). Pythonomorpha.
- Family **Mosasauridae**. *Mosasaurus.*
- Order 7. Dinosauria.
- Suborder (1). Sauropoda.
- Family **Atlantosauridae**. *Brontosaurus.*
- Family **Cetiosauridae**. *Morosaurus.*
- Suborder (2). Theropoda.
- Family **Megalosauridae**. *Megalosaurus (Ceratosaurus).*
- Family **Compsognathidae**. *Compsognathus.*

- Suborder (3). Orthopoda.
- Section (*a*). Stegosauria.
- Family **Scelidosauridae**. *Polacanthus.*
- Family **Stegosauridae**. *Stegosaurus.*
- Section (*b*). Ceratopsia.
- Family **Ceratopsidae**. *Polyonax (Ceratops).*
- Section (*c*). Ornithopoda.
- Family **Camptosauridae**. *Hypsilophodon.*
- Family **Iguanodontidae**. *Iguanodon.*
- Family **Hadrosauridae**. *Hadrosaurus.*
- Order 8. Crocodilia.
- Suborder (1). Parasuchia.
- Family **Phytosauridae**. *Phytosaurus (Belodon).*
- Suborder (2). Eusuchia.
- Family **Teleosauridae**. *Teleosaurus.*
- *Metriorhynchus.*
- Family **Goniopholidae**. *Goniopholis.*
- Family **Alligatoridae**. Alligator.
- Caiman.
- Jacare.
- Family **Crocodilidae**. Crocodilus.
- Family **Garialidae**. Garialis (Gavialis).
- Order 9. Pterosauria.
- Family **Pterodactylidae**. *Pterodactylus.*
- Family **Rhamphorhynchidae**. *Rhamphorhynchus.*
- Family **Pteranodontidae**. *Pteranodon.*

CLASS II. AVES.
- Subclass (I). Archaeornithes.
- *Archaeopteryx.*
- Subclass (II). Neornithes.
- Order 1. Ratitae.

- Group **Æpyornithes**. *Æpyornis.*
- Group **Apteryges**. Apteryx—kiwi.
- Group **Dinornithes**. Moas.
- Group **Megistanes**. Casuarius—cassowary.
- Dromaeus—emeu.
- Group **Rheornithes**. Rhea—American ostrich.
- Group **Struthiornithes**. Struthio—ostrich.
- Order 2. Odontolcae.
- *Hesperornis.*
- Order 3. Carinatae.
- Group **Ichthyornithiformes**.
- *Ichthyornis.*
- *Apatornis.*
- *Odontopteryx.*
- Group **Colymbiformes**.
- Subgroup Colymbi—divers.
- Group **Sphenisciformes**.
- Subgroup Sphenisci—penguins.

- Group **Ciconiiformes**.
- Subgroup Steganopodes. Sula—gannet.
- Pelicanus—pelican.
- Phaëthon—frigate bird.
- Phalacrocorax—cormorant.
- Subgroup Ardeae. Ardea—heron
- Subgroup Ciconiae. Leptoptilus—adjutant.
- Ciconia—white stork.
- Group **Anseriformes**.
- Subgroup Palamedeae. Palamedea }
- } screamers.
- Chauna   }
- Subgroup Anseres. Anas—wild duck.
- Anser—goose.
- Plectropterus—spur-winged goose.
- Cygnus—swan.
- Mergus—merganser.
- Group **Falconiformes**.
- Subgroup Cathartae. Cathartes—American vulture.
- Subgroup Accipitres. Falco—falcon.
- Vultur—vulture.
- Harpagus.
- Gypogeranus—secretary bird.
- Group **Tinamiformes**.
- Subgroup Tinami. Tinamus.
- Group **Galliformes**.

- Subgroup Galli. Gallus—fowl.
- Pavo—peacock.
- Subgroup Opisthocomi. Opisthocomus—hoatzin.
- Group **Gruiformes**.
- Gruidae—cranes.
- Group **Stereornithes**. *Phororhacos.*

- Group **Charadriiformes**.
- Subgroup Limicolae. Charadriidae—plovers.
- Parra—jacana.
- Subgroup Lari. Laridae—gulls.
- Alcidae—auks.
- Subgroup Pteroclidae. Pterocles—sandgrouse.
- Subgroup Columbidae. Columbae—pigeons.
- *Didus*—dodo.
- *Pezophaps*—solitaire.
- Group **Cuculiformes**.
- Subgroup Cuculi. Scythrops.
- Subgroup Psittaci. Stringops—owl-parrot.
- Group **Coraciiformes**.
- Subgroup Coraciae. Coracias—roller.
- Buceros—hornbill.
- Upupa—hoopoe.
- Subgroup Striges. Owls.
- Subgroup Cypseli. Cypselidae—swifts.
- Trochilidae—humming-birds.
- Subgroup Trogonidae. Trogons.
- Subgroup Pici. Rhamphastos—toucan.
- Picus—woodpecker.
- Group **Passeriformes**. Crows, finches, larks, warblers,
- and many others.

C. MAMMALIA.

Class MAMMALIA.

- Subclass (I). Ornithodelphia or Prototheria.
- Order. Monotremata.

- Family **Ornithorhynchidae**. Ornithorhynchus—duck-bill.
- Family **Echidnidae**. Echidna—spiny ant-eater.
- Group **Multituberculata**. *Tritylodon.*
- Subclass (II). Didelphia or Metatheria.
- Order. Marsupialia.
- Suborder (1). Polyprotodontia.
- Family **Amphitheriidae**. *Phascolotherium.*
- Family **Didelphyidae**. Didelphys—opossum.
- Family **Dasyuridae**. Thylacinus—Tasmanian wolf.

- Sarcophilus—Tasmanian devil.
- Dasyurus.
- Family **Peramelidae**. Perameles—bandicoot.
- Choeropus.
- Family **Notoryctidae**. Notoryctes—marsupial mole.
- Suborder (2). Diprotodontia.
- Family **Phascolomyidae**. Phascolomys—wombat.
- Family Phalangeridae. Tarsipes.
- Phalanger—cuscus.
- Phascolarctus—koala.
- *Thylacoleo.*
- Family **Diprotodontidae**. *Diprotodon.*
- Family **Nototheriidae**. *Nototherium.*
- Family **Macropodidae**. Macropus—kangaroo.
- Family **Epanorthidae**. Coenolestes.
- Subclass (III). Monodelphia or Eutheria.
- Order 1. Edentata.
- Family **Bradypodidae**. Bradypus }
- }—sloths.
- Choloepus }

- Family **Megatheriidae**. *Megatherium*—ground sloth.
- Family **Myrmecophagidae**. Myrmecophaga—great ant-eater.
- Cycloturus—two-toed ant-eater.
- Family **Dasypodidae**. Chlamydophorus }
- Dasypus }—armadillos.
- Priodon }
- Tatusia }
- Family **Glyptodontidae**. *Glyptodon.*
- Family **Manidae**. Manis—pangolin.
- Family **Orycteropodidae**. Orycteropus—aard vark.
- Order 2. Sirenia.
- Family **Manatidae**. Manatus—manatee.
- Family **Rhytinidae**. *Rhytina*—Steller's sea-cow.
- Family **Halicoridae**. Halicore—dugong.
- Family **Halitheriidae**. *Halitherium.*
- Order 3. Cetacea.
- Suborder (1). Archaeoceti.
- Family **Zeuglodontidae**. *Zeuglodon.*
- Suborder (2). Mystacoceti or Balaenoidea.
- Family **Balaenidae**. Balaena—right whale.
- Megaptera—humpbacked whale.
- Balaenoptera—rorqual.
- Suborder (3). Odontoceti.
- Family **Physeteridae**. Physeter—sperm whale.

- Hyperoödon—bottlenose.
- Ziphius.
- Mesoplodon.
- Family **Physodontidae**. *Physodon.*

- Family **Squalodontidae**. *Squalodon.*
- Family **Platanistidae**. Platanista—Gangetic dolphin.
- Inia.
- Pontoporia.
- Family **Delphinidae**. Monodon—narwhal.
- Phocaena—porpoise.
- Orca—killer.
- Globicephalus—Ca'ing whale.
- Grampus.
- Lagenorhynchus.
- Delphinus—dolphin.
- Tursiops.
- Prodelphinus.
- Order 4. Ungulata.
- Division A. Ungulata vera.
- Suborder (1). Artiodactyla.
- Section (*a*). Suina.
- Family **Hippopotamidae**. Hippopotamus.
- Family **Suidae**. Sus—pig.
- Babirussa.
- Phacochaerus—wart hog.
- *Hyotherium.*
- Family **Cotylopidae**. *Cotylops (Oreodon).*
- *Cyclopidius.*
- Family **Agriochoeridae**. *Agriochoerus.*
- Family **Anoplotheriidae**. *Anoplotherium.*
- Section (*b*). Tylopoda.
- Family **Camelidae**. Camelus—camel.
- Auchenia—llama.
- Section (*c*). Tragulina.
- Family **Tragulidae**. Dorcatherium (Hyomoschus)—chevrotain.

Section (*d*). Ruminantia or Pecora.
- Family **Cervidae**. Moschus—musk deer.
- Cervus—deer.
- Cervulus—muntjac.
- Hydropotes—Chinese water deer.
- Family **Giraffidae**. Giraffa—giraffe.
- *Sivatherium.*
- Family **Antilocapridae**. Antilocapra—prongbuck.

- Family **Bovidae**. Tetraceros—four-horned antelope.
- Gazella—gazelle.
- Bos—ox.
- Bison.
- Bubalus—buffalo.
- Suborder (2). Perissodactyla.
- Family **Tapiridae**. Tapirus—tapir.
- Family **Lophiodontidae**. *Lophiodon.*
- *Hyracotherium.*
- Family **Palaeotheriidae**. *Palaeotherium.*
- Family **Equidae**. *Hipparion.*
- Equus—horse.
- Family **Rhinocerotidae**. Rhinoceros.
- *Elasmotherium.*
- Family **Titanotheriidae**. *Titanotherium (Brontops).*
- *Palaeosyops.*
- Family **Chalicotheriidae**. *Chalicotherium.*
- Family **Macraucheniidae**. *Macrauchenia.*
- Division B. Subungulata.
- Suborder (1). Toxodontia.
- Family **Astrapotheriidae**. *Astrapotherium.*
- Family **Nesodontidae**. *Nesodon.*
- Family **Toxodontidae**. *Toxodon.*
- Family **Typotheriidae**. *Typotherium.*

- Suborder (2). Condylarthra.
- Family **Phenacodontidae**. *Phenacodus.*
- Suborder (3). Hyracoidea.
- Family **Hyracidae**. Procavia (Hyrax).
- Suborder (4). Amblypoda.
- Family **Coryphodontidae**. *Coryphodon.*
- Family **Uintatheriidae**. *Uintatherium (Dinoceras).*
- Suborder (5). Proboscidea.
- Family **Dinotheriidae**. *Dinotherium.*
- Family **Elephantidae**. *Mastodon.*
- Elephas—elephant.
- Group Tillodontia.
- Order 5. Rodentia.
- Suborder (1). Simplicidentata.
- Section Sciuromorpha.
- Family **Castoridae**. Castor—beaver.
- Section Myomorpha.
- Family **Lophiomyidae**. Lophiomys.
- Family **Muridae**. Hydromys.
- Acanthomys—spiny mouse.

- Mus—mouse.
- Family **Spalacidae**. Bathyergus.
- Family **Dipodidae**. Dipus—jerboa.
- Pedetes—Cape jumping-hare.
- Section Hystricomorpha.
- Family **Hystricidae**. Hystrix—porcupine.
- Family **Chinchillidae**. Chinchilla.
- Lagostomus—viscacha.

- Family **Dasyproctidae**. Coelogenys—paca.
- Dasyprocta—agouti.
- Family **Caviidae**. Cavia—guinea-pig.
- Hydrochaerus—capybara.
- Suborder (2). Duplicidentata.
- Family **Leporidae**. Lepus—hare and rabbit.
- Order 6. Carnivora.
- Suborder (1). Creodonta.
- Family **Hyaenodontidae**. *Hyaenodon.*
- Suborder (2). Carnivora vera or Fissipedia.
- Section Æluroidea.
- Family **Felidae**. Felis—cat, lion, tiger.
- *Machaerodus*—sabre-toothed lion.
- Family **Viverridae**. Viverra—civet.
- Paradoxurus—palm civet.
- Family **Protelidae**. Proteles—aard wolf.
- Family **Hyaenidae**. Hyaena.
- Section Cynoidea.
- Family **Canidae**. Canis—dog, wolf, fox.
- Section Arctoidea.
- Family **Ursidae**. Ursus—bear.
- Family **Mustelidae**. Latax—sea otter.
- Suborder (3). Pinnipedia.
- Family **Otariidae**. Otaria—sea lion.
- Family **Trichechidae**. Trichechus—walrus.
- Family **Phocidae**. Ogmorhinus—sea leopard.
- Order 7. Insectivora.
- Suborder (1). Dermoptera.
- Family **Galeopithecidae**. Galeopithecus—'flying lemur'.

- Suborder (2). Insectivora vera.
- Family **Macroscelidae**. Macroscelides—jumping shrew.
- Family **Erinaceidae**. Erinaceus—hedgehog.
- Gymnura.
- Family **Soricidae**. Sorex—shrew.
- Family **Talpidae**. Talpa—mole.

- Family **Potamogalidae**. Potamogale.
- Family **Solenodontidae**. Solenodon.
- Family **Centetidae**. Microgale.
- Centetes—tenrec.
- Family **Chrysochloridae**. Chrysochloris—golden mole.
- Order 8. Chiroptera.
- Suborder (1). Megachiroptera.
- Family **Pteropidae**. Pteropus—flying fox.
- Suborder (2). Microchiroptera.
- Family **Rhinolophidae**. Horseshoe bats.
- Family **Phyllostomatidae**. Desmodus—vampire.
- Order 9. Primates.
- Suborder (1). Lemuroidea.
- Family **Tarsiidae**. Tarsius—tarsier.
- Family **Chiromyidae**. Chiromys—aye aye.
- Suborder (2). Anthropoidea.
- Family **Hapalidae**. Hapale—marmoset.
- Family **Cebidae**. Mycetes—howling monkey.
- Ateles—spider monkey.
- Family **Cercopithecidae**. Cynocephalus—baboon.
- Macacus.
- Colobus.
- Family **Simiidae**. Hylobates—gibbon.
- Simia—orang.
- Gorilla.
- Anthropopithecus—chimpanzee.
- Family **Hominidae**. Homo—man.

CHAPTER III.
SKELETON OF HEMICHORDATA, UROCHORDATA, AND CEPHALOCHORDATA.

SUBPHYLUM A. HEMICHORDATA.

The subphylum includes three genera, *Balanoglossus*, *Cephalodiscus* and *Rhabdopleura*; and perhaps a fourth, *Phoronis*.

The skeletal structures found in *Balanoglossus* are all endoskeletal. They include:

(1) The **notochord**. This arises as a diverticulum from the alimentary canal which grows forwards into the proboscis and extends beyond the front end of the central nervous system. It is hypoblastic in origin and arises in the same way as does the notochord of *Amphioxus*. Its cells become highly vacuolated and take on the typical notochordal structure. The cavity of the primitive diverticulum becomes obliterated in front, but behind it opens throughout life into the alimentary canal.

(2) The **axial skeletal rods**. These are a pair of chitinous rods which lie ventral to the notochord and in the collar region unite to form a single mass.

(3) The **branchial skeleton**. The gill bars separating the gill slits from one another are strengthened by chitinous rods in a way closely similar to that in *Amphioxus*. But between one primary forked rod and the next there are two secondary unforked rods—not one, as in *Amphioxus*.

27

(4) The **chondroid tissue**. This is of mesoblastic origin and may be regarded as an imperfect sheath for the notochord.

In *Cephalodiscus* and *Rhabdopleura* as in *Balanoglossus* the notochord forms a small diverticulum growing forwards from the alimentary canal into the proboscis stalk.

Recent researches on *Phoronis* show the existence in the collar region of the larva (*Actinotrocha*) of a paired organ, which is regarded by its discoverer as representing a double notochord.

SUBPHYLUM B. UROCHORDATA (Tunicata).

Skeletal structures of epiblastic and hypoblastic origin occur in the Urochordata. Most Tunicates are invested by a thick gelatinous test which often contains calcareous spicules, and serves as a supporting organ for the soft body. The cells of this test are mesodermal in origin.

In larval Tunicata and in adults of the group Larvacea the tail is supported by a typical notochord, which is confined to the tail. In all Tunicata except Larvacea all trace of the notochord is lost in the adult.

SUBPHYLUM C. CEPHALOCHORDATA.

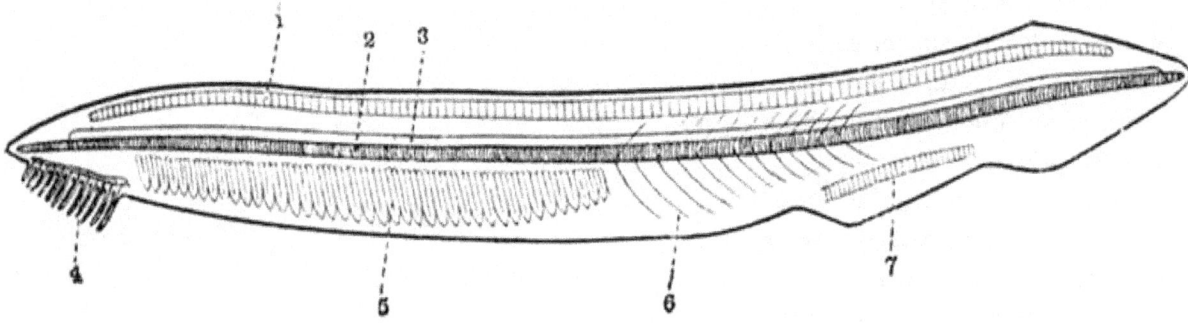

Fig. 3. Diagram of the skeleton of *Amphioxus lanceolatus* × 3 (after a drawing in the Index collection at the Brit. Mus.).

1. skeleton of dorsal fin.

2. notochord.

3. neural tube.

4. buccal skeleton.

5. branchial skeleton.

6. septa separating the myotomes.

7. skeleton of ventral fin.

This subphylum includes the well-known genus *Amphioxus*. In *Amphioxus* the skeleton is very simple. It contains no trace of cartilage or bone and remains throughout life in a condition corresponding to a very early stage in Vertebrata. The skeleton of *Amphioxus* is partly hypoblastic, partly mesoblastic in origin.

(*a*) **Hypoblastic skeleton.**

The **notochord** (fig. 3, 2) is an elastic rod extending along the whole length of the body past the anterior end of the nerve cord. It lies ventral to the nerve cord, and shows no trace of segmentation. It is chiefly made up of greatly vacuolated cells containing lymph, but near the dorsal and ventral surfaces the cells are less vacuolated. The notochord is immediately surrounded by a structureless cuticular layer, the *chordal sheath*, and outside this comes the mesoblastic *skeletogenous layer*, which also surrounds the nerve cord.

The **branchial skeleton**. This consists of a series of chitinous elastic rods which strengthen the gill bars and are alternately forked and unforked ventrally. The forked rods are primary, and are U-shaped in section, the unforked rods are secondary, and are circular in section. All these rods are united at intervals by transverse rods.

(*b*) **Mesoblastic skeleton.**

The **buccal skeleton**. On each side of the mouth there is a curved bar resembling the notochord in structure. The bars are segmented, and each segment bears a smaller rod which supports a tentacle, the whole forming the buccal skeleton (fig. 3, 4).

The notochord is enclosed in a thick **sheath** of connective tissue continuous with a thinner sheath round the nerve cord. The sheaths of the notochord and nerve cord together form the skeletogenous layer, and prolongations of it form the myomeres or septa between the myotomes or segments of the great lateral muscles of the body.

The **skeleton of each median fin** consists of small cubical masses of a gelatinous substance arranged in rows (fig. 3, 1 and 7), and serving to strengthen the fins.

CHAPTER IV.
SUBPHYLUM D. VERTEBRATA.

The animals included in this great group all possess an internal axial skeleton forming the vertebral column or back-bone; and a dorsal spinal cord. The vertebral column is developed from the skeletogenous layer, which surrounds the spinal cord together with the notochord and its sheath; and in the great majority of cases the notochord becomes more or less modified and reduced in the adult. In some cases the notochord remains unmodified and the skeletogenous layer surrounding it is not segmented to form vertebrae, but in every case the neural arches which protect the spinal cord are segmented. The notochord never extends further forwards than the mid-brain.

All true vertebrates possess a cranium or skeletal box enclosing the brain.

(I.) Cyclostomata.

The mouth in living forms is suctorial and is not supported by jaws. In some fossil forms the character of the mouth is unknown.

Order I. Marsipobranchii.

In these animals limbs and limb girdles are always completely absent. They have no exoskeleton except horny teeth.

The endoskeleton, excluding the notochord, is entirely cartilaginous or membranous. The axial skeleton consists of a cartilaginous cranium without jaws, succeeded by a thick persistent notochord enveloped in a sheath. The notochord in living forms is unsegmented, but segmented cartilaginous neural arches are present in some cases. A complicated series of cartilaginous elements occurs in relation to the mouth, gills, and sense organs. The median fins are supported by cartilaginous pieces, the radiale. The order includes the Lampreys and Hags.

Order II. Ostracodermi.

The forms included in this group have long been extinct, being known only from beds of Upper Silurian and Lower Devonian age. They differ much from all other known animals. The exoskeleton is always greatly developed and includes (1) large bony plates covering the anterior region; (2) scales covering the posterior region. The plates are deeply marked by canals belonging to dermal sense organs. Jaws are unknown, and arches for the support of the appendicular skeleton are rudimentary or absent. The tail is heterocercal (see p. 60).

Suborder (1). Heterostraci.

The exoskeleton consists principally of calcifications forming dorsal and ventral shields which cover the head and abdominal region; the dorsal shield is formed of a few plates firmly united, the ventral shield of a single plate. The shields are composed of three layers, the middle layer being traversed by canals belonging to the dermal sense organs which open to the exterior by a series of pores. The tail is sometimes covered by scales. The orbits are widely separated and laterally placed. Paired appendages are absent. These curious forms are found in beds of Upper Silurian and Lower Devonian age. One of the best known genera is *Pteraspis*.

Suborder (2). Osteostraci.

The exoskeleton as in the Heterostraci consists of shields and scales, the shields being divisible into three layers. The anterior part of the body is covered dorsally by a single large shield which differs from those of the Heterostraci in having the inner layer ossified. The middle layer contains canals for the passage of blood vessels, but the exoskeleton shows no impressions of dermal sense organs. The posterior part of the body is covered by large quadrangular scales. Paired appendages are absent, but median dorsal and caudal fins occur supported by scales, not fin-rays. *Cephalaspis*, the best known of these animals, occurs in beds of Lower Devonian age.

Suborder (3). Antiarcha.

The exoskeleton is formed of bony plates, the dorsal and ventral shields each consisting of several symmetrically arranged pieces. The tail may be covered with small scales or may be naked. The head is articulated with the trunk, and its angles are drawn out into a pair of segmented paddle-like appendages, covered with dermal plates. The orbits are close together. A dorsal fin and traces of mouth parts occur in *Pterichthys*, but the endoskeleton is unknown. The best known forms *Pterichthys* and *Asterolepis* occur in beds of Lower Devonian age.

General account of the skeleton of

Marsipobranchii.

The Marsipobranchii are worm-like animals. The living forms include two families, the Myxinoidei (Hags)—genera *Myxine* and *Bdellostoma*—and the Petromyzontidae (Lampreys).

Three species of *Petromyzon* are known, *P. fluviatilis*, *P. marinus* and *P. planeri*. The larval forms were for a long time thought to belong to a separate genus and were called *Ammocoetes*.

The Myxinoids, although very highly specialised in their own way, are at distinctly a lower stage of development than the adult Lamprey, and come nearer to the larval Lamprey or Ammocoete.

Spinal column.

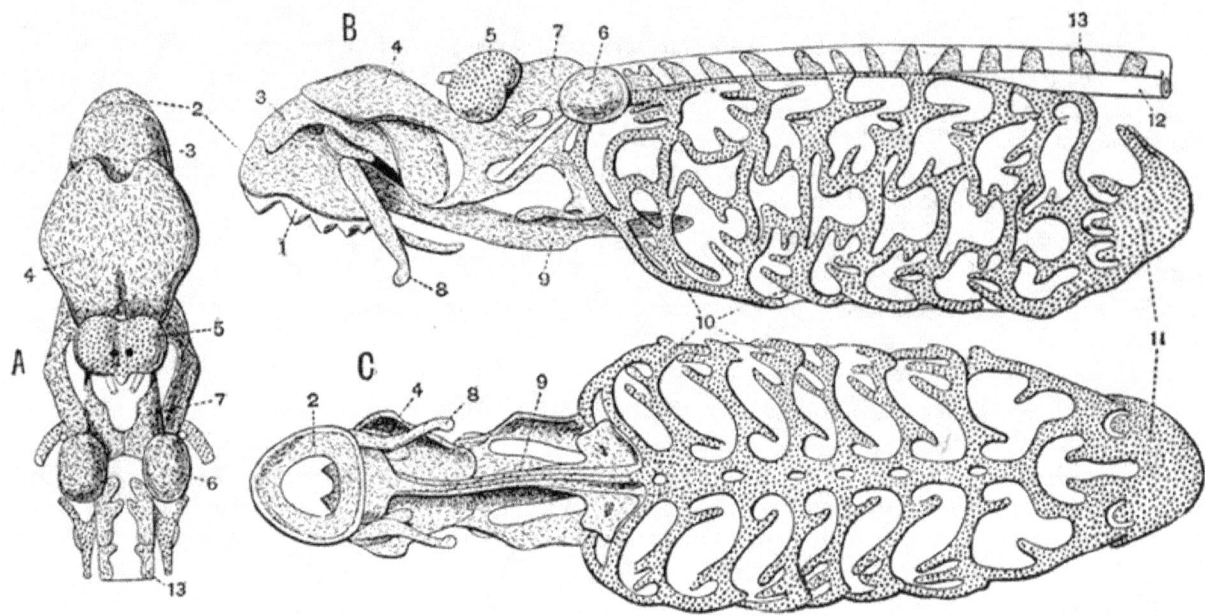

Fig. 4. A, dorsal; B, lateral and C, ventral view of the skull of *Petromyzon marinus* × 1 (after Parker).

1. horny teeth.
2. labial cartilage.
3. anterior dorsal cartilage.
4. posterior dorsal cartilage.
5. nasal capsule.
6. auditory capsule.
7. dorsal portion of trabeculae.

8. lateral distal mandibular.
9. lingual cartilage.
10. branchial basket.
11. cartilaginous cup supporting pericardium.
12. sheath of notochord.
13. neural plate.

In Myxinoids and larval lampreys, the notochord is enclosed in a thick chordal sheath, in connection with which in the tail region there occur cartilaginous pieces forming neural arch elements. In the trunk region, however, no cartilage occurs in connection with the spinal column, the only cartilage present being that forming the radiale of the dorsal fin. On the other hand in most species of lamprey (*Petromyzon*) cartilaginous pieces forming imperfect neural arches (fig. 4, B, 13) are found lying in the tough skeletogenous layer dorsal to the notochord, and extending throughout the whole length of the trunk and tail. Two of these pieces, which are probably homologous with the neural plates (see p. 72) of Elasmobranchs, occur to each *neuromere*, or segment as determined by the spinal nerves. The dorsal and caudal fins are supported by paired cartilaginous radiale which are connected proximally with the skeletogenous layer.

The Skull.

In Myxinoids the cranium is a mere cartilaginous floor without side walls or roof, and the trabeculae end without growing forwards into cornua. In Lampreys the trabeculae grow forwards and send up plates of cartilage which meet above (fig. 4, 7) and form side walls and a roof for part of the brain case. In Lampreys a labial suctorial apparatus is well developed, including a large ring-like piece of cartilage (fig. 4, 2) which supports the oral funnel and bears a large armament of horny teeth. In Myxinoids on the other hand the labial skeleton is small and consists merely of barbels round the mouth.

The olfactory organ of Myxinoids has a very curious skeleton. It is covered with a kind of grating of cartilage which is prolonged in front into a tube composed of a series of imperfect cartilaginous rings. In Lampreys the olfactory organ opens merely by a short membranous passage. In correlation with the small development of the labial suctorial apparatus in Myxinoids the lingual apparatus is very greatly developed. The tongue in *Myxine* has been said to 'dominate the whole body' (Parker). It is supported by a great median cartilaginous bar which when followed forwards first becomes bifid and still further forwards becomes four-cleft.

The horny teeth in Myxinoids are chiefly borne on the very large supralingual apparatus. They form a double series arranged in the form of an arch. In *Myxine* there are seven large teeth and nine small ones on each side. In *Bdellostoma* the teeth of the two rows are more equal in size. In *Bdellostoma* and *Myxine* it has been shown that imperfect calcified teeth occur below the horny teeth.

In Lampreys the lingual apparatus (fig. 4, C, 9) is well developed, but not excessively so. It consists of a long median cartilaginous bar which ends in front with a semicircular piece of cartilage supporting the median part of the tongue.

In both Myxinoids and Lampreys there is a complicated branchial basket apparatus, but while in Myxinoids the basket apparatus is interbranchial, formed deep within the head near the hypoblastic lining of the throat, in Lampreys it is extra-branchial and formed outside the head cavities (fig. 4, 10). The two sides of the basket apparatus in *Myxine* are not symmetrical. In the interbranchial basket apparatus of Myxinoids the hyoid and first and second branchial arches can be recognised. Traces of the interbranchial skeleton of Myxinoids can be detected in Lampreys, and similarly in Myxinoids, there are indications of the extra-branchial skeleton of *Petromyzon*. The branchial basket in Lampreys forms at its posterior end a kind of cup which supports the pericardium (fig. 4, 11).

A remarkable Cyclostome named *Palaeospondylus* has recently been described from the Scottish Old Red Sandstone. It differs however from all living Cyclostomes, in having a spinal column formed of distinct vertebrae with well-developed neural arches. The caudal fin is well developed and the dorsal radiale are forked as in lampreys. The skull is well calcified and the auditory capsules are specially large. The mouth is very similar to that of lampreys, being circular and without jaws; it is provided with barbels or cirri. There is no trace of limbs and the average length is only about 1-1½ inches.

CHAPTER V.
(II.) GNATHOSTOMATA.

The mouth is supported by definite jaws.

ICHTHYOPSIDA.

The epiblastic exoskeleton is generally unimportant, the mesoblastic exoskeleton is usually well developed.

The notochord with its membranous sheath (1) may remain unmodified, or (2) may be replaced by bone or cartilage derived from the skeletogenous layer, or (3) may be calcified to a varying extent.

The first vertebra is not homologous throughout the whole series and so is not strictly comparable to the atlas of Sauropsids and Mammals.

The centra of the vertebrae have no epiphyses. The skull may be (*a*) incomplete and membranous, or (*b*) more or less cartilaginous, or (*c*) bony. Membrane bones are not included in the cranial walls, and there are large unossified tracts in the skull. When membrane bones are developed in connection with the skull, a large parasphenoid occurs. The basisphenoid is always small or absent. The skull may be immovably fixed to the vertebral column, or may articulate with it by a single or double occipital condyle. When the occipital condyle is double, it is formed by the exoccipitals, and the basi-occipital is small or unossified. The mandible may be (*a*) cartilaginous, (*b*) partially ossified, or (*c*) membrane bones may be developed in connection with it,—if so, there is usually more than one membrane bone developed in connection with each half.

There are at least four pairs of branchial arches present during development. The sternum, if present, is not costal in origin.

Class I. Pisces.

The exoskeleton is in the form of scales, which may be entirely mesoblastic or dermal in origin (e.g. *cycloid* and *ctenoid* scales), or may be formed of both mesoblast and epiblast (e.g. *placoid* and *ganoid* scales). Large bony plates may be derived from both these types of scale. In general fish with a greatly developed dermal armour have the endoskeleton poorly developed; and the converse also holds good.

The integument of the dorsal and ventral surfaces is commonly prolonged into longitudinal unpaired fins, supported by an internal skeleton. These fins are distinguished according to their position as dorsal, caudal and anal fins. The dorsal and anal fins are used chiefly as directing organs, the caudal fin is however a most important organ of propulsion.

Three types of tail are found in fishes, viz.:—

1. The **diphycercal**, in which the axis is straight and the tail is one-bladed and symmetrical, an equal proportion of radiale being attached to the upper and lower surfaces of the axis.

2. The **heterocercal**, in which the tail is asymmetrical and the axis is bent upwards, the proportion of radiale or of fin-rays attached to its upper surface being much smaller than that attached to its lower surface.

3. The **homocercal**, in which the tail though externally symmetrical, so far resembling the diphycercal type, is internally really heterocercal, the great majority of the radiale or of the fin-rays being attached to the lower surface of the axis.

The cranium in the simplest cases (e.g. Selachii) forms a cartilaginous box enclosing the brain and sense organs; in bony fishes it is greatly complicated. When palatine or pterygoid bones are present they are formed by the ossification of cartilage; in Sauropsida and Mammalia they are laid down as membrane bones. There is no tympanic cavity or auditory ossicle in relation to the ear.

There are two principal types of suspensorium by means of which the jaws are attached to the cranium:—

(1) The **Autostylic**. This is the primitive condition in which the mandibular arch articulates with the base of the cranium in front of the hyoid and in a similar manner.

(2) The **Hyostylic**. In this case the mandibular arch becomes connected with the hyomandibular and supported by the hyoid arch. These terms are more fully discussed in Chapter VIII.

There is always an internal framework supporting the gills; it usually consists of the hyoid arch and five, rarely six or seven, pairs of branchial arches. The limbs are represented by two pairs of fins, the pectoral and the pelvic; they are not divided into proximal, middle and distal portions. The ribs do not unite with a median ventral sternum, or meet in the mid-ventral line in any other way in the trunk region.

Order I. Elasmobranchii.

The exoskeleton is in the form of placoid scales which are sometimes so numerous as to give the whole skin a rough surface forming shagreen. In some cases the placoid scales are enlarged to form plates or spines capped or coated with enamel. These spines may be imbedded in the flesh in front of the paired or unpaired fins, or may be attached to the tail. They are specially characteristic of the suborder Acanthodii. The endoskeleton is cartilaginous and true bone is never found. Much of the skeleton, especially of the vertebral column, is however often calcified, this being especially well seen in the anterior part of the vertebral column of Rays (Raiidae). In living forms cartilaginous biconcave vertebrae are always well developed, but in some extinct forms the notochord persists unconstricted. Neural and haemal arches are however always developed; they sometimes remain separate, sometimes fuse with the centra. Ribs are often wanting and when present are often not separated off from the vertebrae. The cranium is a simple cartilaginous box whose most prominent parts are the capsules which enclose the sense organs. The skull is sometimes immovably fixed to the vertebral column, sometimes articulates with it by means of two condyles. There is no operculum and no representative of the maxilla or premaxillae. The teeth are very variable. Large pectoral and pelvic fins always occur.

The Elasmobranchii may be divided into four suborders:—

(1) Ichthyotomi.

(2) Pleuropterygii.

(3) Selachii.

(4) Acanthodii.

Suborder (1). Ichthyotomi.

The members of this suborder range from the Devonian to the Permian and so have long been extinct.

The endoskeletal cartilage has granular calcifications evenly distributed throughout it. The notochord is unconstricted, but the neural and haemal arches are well-developed, and the neural spines are long and slender. There is a continuous dorsal fin with separate basalia and radiale. The tail is diphycercal, and the pectoral fins are typical archipterygia. The pelvic fins of the male are prolonged to form claspers.

The best known of these primitive Elasmobranchs are the Pleuracanthidae.

Suborder (2). Pleuropterygii.

This suborder was formed for the reception of *Cladoselache*, an Elasmobranch found in the Lower Carboniferous of Ohio.

The exoskeleton is in the form of small, thickly-studded dermal denticles. The vertebral centra are unossified, and the tail is strongly heterocercal. There were certainly five, perhaps seven gill slits, and the suspensorium is apparently hyostylic. The paired fins are, according to the view which derives them by concentration from continuous lateral folds, the most primitive known (see p. 129) and claspers are absent.

Suborder (3). Selachii.

Cartilaginous or partially calcified biconcave vertebrae are always well developed; they constrict the notochord intervertebrally. The neural and haemal arches and spines are stout and intercalary cartilages (interdorsalia) are present. The tail is heterocercal, but in some cases (*Squatina*) approaches the diphycercal condition. In most cases the suspensorium is hyostylic, the jaws being attached to the cranium by means of the hyomandibular, and the palato-pterygo-quadrate bar not being fused to the cranium. There are generally five pairs of branchial arches, and gill rays are borne on the posterior surface of the hyoid arch, and on both the anterior and posterior surfaces of the first four branchial arches. The Notidanidae differ from most Selachians in two respects, first as regards the suspensorium,—Meckel's cartilage articulating directly with the palato-pterygo-quadrate bar, and not being connected with the hyoid arch; and secondly as regards the number of branchial arches,—six pairs occurring in *Hexanchus* and seven in *Heptanchus*.

The pectoral fins are without the segmented axis of the archipterygium. In most cases they are sharply marked off from the body and lie almost at right angles to it; but in the Rays they have the form of lateral expansions in the same plane as the body, from which they are not sharply marked off. The pelvic fins in the male bear long grooved cartilaginous rods which are accessory copulatory organs or claspers.

There are two principal groups of Selachii, the Squalidae or Sharks and Dogfish, and the Batoidei or Skates and Rays. The Squalidae have the shape of ordinary fish, the pectoral fins are vertically placed and the body ends in a powerful heterocercal tail. The Batoidei have flattened bodies owing to the great size and horizontal position of the pectoral fins. The tail is long and thin and is often armed with spines. The teeth in Selachii differ much in character in the different forms, and are always arranged in numerous rows. They are generally pointed and triangular or conical in the Squalidae, while in the Batoidei they are often broad and flattened.

Suborder (4). Acanthodii.

The fishes included in this group are all extinct and in some respects are intermediate between Elasmobranchii and Ganoidei. The body is elongated and closely covered with small scales consisting of dentine enamelled at the surface. The notochord is persistent and the calcification of the endoskeletal cartilage is only superficial. The tail is heterocercal. The jaws bear small conical teeth, or in some cases are toothless. The skeleton of all the fins differs from that of modern Elasmobranchs in having the cartilaginous radiale much reduced, and the fins are nearly always each provided with an anterior spine, which except in the case of the pectoral fins is merely inserted between the muscles. These spines are really enormous dermal fin-rays; the pectoral fin-spine is articulated to the pectoral girdle.

The suborder includes many well-known extinct forms like *Acanthodes* and *Diplacanthus*; it ranges from the Devonian to the Permian.

PISCES, HOLOCEPHALI.

Order II. Holocephali.

This order includes a single suborder only.

Suborder. Chimaeroidei.

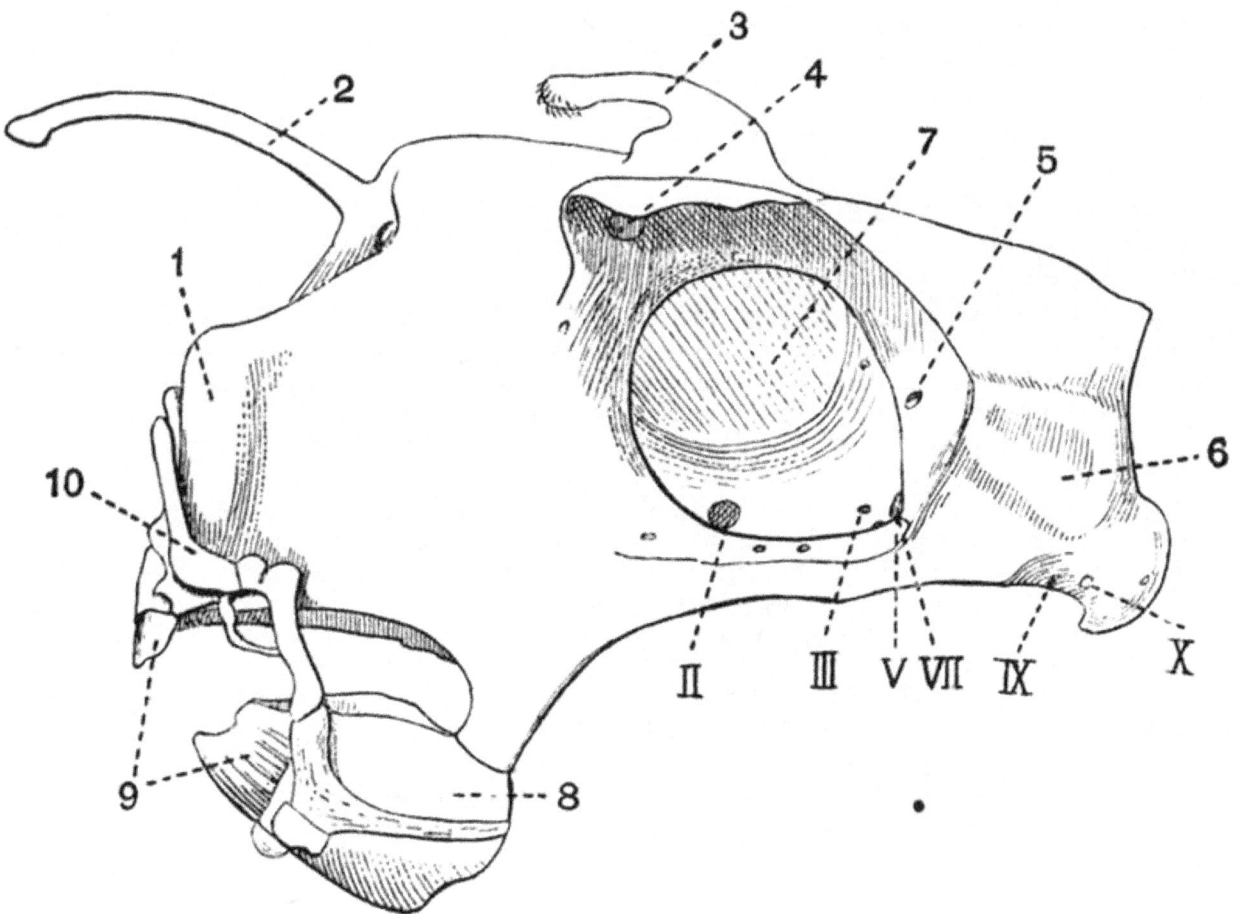

33

Fig. 5. Skull of a male *Chimaera monstrosa* (after Hubrecht).

1. nasal capsule.

2. cartilaginous appendage to the fronto-nasal region.

3. erectile appendage.

4. foramen by which the ophthalmic nerves leave the orbit.

5. foramen by which the ophthalmic branch of the Vth nerve enters the orbit.

6. auditory capsule.

7. interorbital septum.

8. mandible articulating with an outgrowth from the posterior part of the palato-pterygo-quadrate.

9. teeth.

10. labial cartilage.

II. III. V. VII. IX. X. foramina for the passage of cranial nerves.

These singular fish have the skin smooth and in living forms almost or quite scaleless. The palato-pterygo-quadrate bar and hyomandibular are fused to the cranium, and Meckel's cartilage articulates directly with the part corresponding to the quadrate. The skull is distinctly articulated with the spinal column, the notochord is persistent and unconstricted, and the skeletogenous layer shows no trace of metameric segmentation, though in the neural arches this segmentation is readily traceable. The neural arches of the first few vertebrae are fused together and completely surround the notochord, while they do not in other parts of the body. The tail is diphycercal. Of the living genera, in *Callorhynchus* there is no trace of calcification in the skeletogenous layer, while in *Chimaera* rings of calcification are found, there being three to five for each vertebra as indicated by the foramina for the exit of the spinal nerves. The pelvic fins are produced into claspers. Besides the living genera *Chimaera*, *Harriotta* and *Callorhynchus* a fair number of fossil forms are known, e.g. *Ischyodus*.

Order III. Ganoidei.

The fishes included under the term Ganoidei form a very heterogeneous group, some of which closely approach the Dipnoi, others the Elasmobranchii, others the Teleostei. The great majority of them are extinct, only eight living genera being known; these are all inhabitants of the northern hemisphere, and with the exception of *Acipenser*, which is both fluviatile and marine, are entirely confined to fresh water.

The following is a list of the living genera of Ganoids with their respective habitats:—

Acipenser. Rivers and seas of the northern hemisphere.

Scaphirhynchus. Mississippi and rivers of Central Asia.

Polyodon (*Spatularia*). Mississippi.

Psephurus. Yan-tse-kiang, and Hoangho.

Polypterus. Rivers of tropical Africa.

Calamoichthys. Some rivers of West Africa.

Lepidosteus. Freshwaters of Central and North America and Cuba.

Amia. Rivers of Carolina.

The exoskeleton is very variable, thus the body may be:—

(*a*) Naked or with minute stellate ossifications as in the Polyodontidae. (*b*) Partially covered with large detached bony plates as in *Scaphirhynchus* and *Acipenser*. (*c*) Entirely covered with rhomboidal ganoid scales as in *Lepidosteus*, *Polypterus*, *Palaeoniscus* and many extinct forms. (*d*) Covered with rounded scales shaped like the cycloid scales of Teleosteans as in *Amia*. (*e*) Having the trunk and part of the tail covered with rhomboidal scales, and the remainder of the tail with rounded scales as in *Trissolepis*.

The teeth also are very variable. The endoskeleton shows every stage of transition from an almost entirely cartilaginous state as in *Acipenser* to a purely bony state as in *Lepidosteus*. Sometimes, as in *Acipenser*, the notochord persists, and its sheath is unsegmented; sometimes, as in *Lepidosteus*, there are fully formed vertebrae. The tail may be heterocercal, as in *Acipenser*, or diphycercal as in *Polypterus*. The cartilaginous cranium is always covered with external membrane bone to a greater or less extent, and the suspensorium is markedly hyostylic. The pectoral girdle is formed of two parts, one endoskeletal and cartilaginous, corresponding with the pectoral girdle of Elasmobranchs, and one exoskeletal and formed of membrane bones, corresponding with the clavicular bones of Teleosteans. The pelvic fins are always abdominal. The fins often, as in *Polypterus*, have spines (fulcra) attached to their anterior borders.

The order Ganoidei may be divided into three suborders:

(1) Chondrostei. Living genera *Acipenser*, *Scaphirhynchus*, *Polyodon* and *Psephurus*.

(2) Crossopterygii. Living genera *Polypterus* and *Calamoichthys*.

(3) Holostei. Living genera *Lepidosteus* and *Amia*.

Suborder (1). Chondrostei.

The skull is immovably fixed to the vertebral column. By far the greater part of the skeleton is cartilaginous. The notochord is persistent and unconstricted, its sheath is membranous, but cartilaginous neural and haemal arches are developed. Intercalary pieces (interdorsalia) occur between the neural arches, and similar pieces (interventralia) between the haemal arches. The cranium is covered with membrane bone, and teeth are but slightly developed. The tail is heterocercal. Gill rays occur on the hyoid arch, and the gills are protected by a bony operculum attached to the hyomandibular. The skin (1) may be almost or quite naked, (2) may carry bony plates arranged in rows, or may be covered (3) with rhomboidal scales, or (4) partly with rhomboidal, partly with cycloidal scales.

Suborder (2). Crossopterygii.

The exoskeleton has the form of cycloidal or rhomboidal scales. The condition of the vertebral column differs in the different genera. Sometimes, as in *Polypterus*, there are well-developed ossified vertebrae; sometimes, as in many extinct forms, the notochord persists and is unconstricted. The tail may be diphycercal or heterocercal. The pectoral and sometimes the pelvic fins consist of an endoskeletal axis bearing a fringe of dermal rays.

Suborder (3). Holostei.

The exoskeleton has the form of cycloidal or rhomboidal scales. The notochord is constricted and its sheath is segmented and ossified, forming distinct vertebrae, which are generally biconcave, sometimes opisthocoelous (*Lepidosteus*). The cartilaginous cranium is largely replaced by bone, and in connection with it we find not only membrane bone, but cartilage bone, as the basi-occipital, exoccipitals, and pro-otic are ossified. The tail is heterocercal. The suspensorium resembles that of Teleosteans, consisting of a proximal ossification, the hyomandibular, which is movably articulated to the skull and a distal ossification, the symplectic. The two are separated by some unossified cartilage. The cartilaginous upper and lower jaws are to a great extent surrounded and replaced by a series of membrane bones.

Order IV. Teleostei.

The exoskeleton is sometimes absent but generally consists of overlapping cycloid or ctenoid scales. Bony plates are sometimes present, as in the Siluridae, or the body may be encased in a complete armour of calcified plates, as in *Ostracion*. Enamel is however never present, and the plates are entirely mesodermal. The skeleton is bony, but in the skull much cartilage generally remains. The vertebral centra are usually deeply biconcave, and the tail is of the masked heterocercal type distinguished as *homocercal*. In the skull the occipital region is always completely ossified, while the sphenoidal region is generally less ossified. The skull has usually a very large number of membrane bones developed in connection with it. The teeth vary much in character in the different members of the order, but are as a rule numerous and pointed, and are ankylosed to the bone. The suspensorium is hyostylic and the jaws have much the same arrangement as in the Holostei. There are five pairs of branchial arches, of which all except the last bear gill rays. A series of dermal opercular bones is developed in connection with these arches. The pectoral girdle consists almost entirely of dermal clavicular bones. The pelvic girdle has disappeared, its place being taken by the enlarged and ossified dermal fin-rays of the pelvic fins.

The group includes the vast majority of living fish (see p. 33).

Order V. Dipnoi.

The exoskeleton is of two types; dermal bones are largely developed in the head region, while the tail and posterior part of the body may be naked or may be covered with overlapping scales. The cranium remains chiefly cartilaginous, the palato-pterygo-quadrate bar is fused with the cranium, and the suspensorium is autostylic. The gill clefts are feebly developed and open into a cavity covered by an operculum. The notochord is persistent and unconstricted, and the limbs are archipterygia. The pelvic fins are without claspers.

Suborder (1). Sirenoidei.

The head has well developed membrane bones. The trunk is covered with overlapping scales and bears no bony plates. Three pairs of teeth are present, two in the upper and one in the lower jaw, the two principal pairs of teeth are borne on the palato-pterygoids and splenials, while the third pair are found in the vomerine region. The tail is diphycercal in living forms. In the extinct Dipteridae it is heterocercal. The pectoral girdle includes both membrane and cartilage bones. The pelvic girdle consists of a single bilaterally symmetrical piece of cartilage.

This suborder is represented by the living genera *Ceratodus*, *Protopterus* and *Lepidosiren*, and among extinct forms by the Dipteridae and others.

Suborder (2). Arthrodira.

Bony plates are developed not only on the head but also on the anterior part of the trunk, where they consist of a dorsal, a ventral, and a pair of lateral plates which articulate with the cranial shield. The posterior part of the trunk is naked. The tail is diphycercal. The jaws are shear-like, and their margins are usually provided with pointed teeth whose bases fuse with the tissue of the jaw and constitute dental plates. There seem to have been three pairs of these plates, arranged as in the Sirenoidei, the principal ones in the upper jaw being borne on the palato-pterygoids. Small pelvic fins are present, but pectoral fins are unknown.

The Arthrodira occur chiefly in beds of Devonian and Carboniferous age. Two of the best known genera are *Coccosteus* from the European Devonian and *Dinichthys*, a large predatory form from the lower Carboniferous of Ohio.

CHAPTER VI.
THE SKELETON OF THE DOGFISH.

Scyllium canicula.

I. EXOSKELETON.

The exoskeleton of the dogfish is mainly composed of placoid scales, each of which consists of a little bony base imbedded in the skin, bearing a small backwardly-directed spine formed of dentine capped with enamel. The scales are larger on the dorsal than on the ventral surface, and on the jaws they are specially large and regularly arranged in rows, there forming the teeth. The margins of the jaws or lips are without scales.

A second exoskeletal structure is found in the fins, all of which, both paired and unpaired, have, in addition to their cartilaginous endoskeleton, large numbers of long slender horny fibres, the fin-rays, which are of exoskeletal origin.

II. ENDOSKELETON.

The endoskeleton of the dogfish consists almost entirely of cartilage, which however may become calcified in places, e.g. the centrum of each vertebra is lined by a layer of calcified tissue.

The endoskeleton is divisible into an **axial** portion consisting of the vertebral column, skull, and skeleton of the median fins, and an **appendicular** portion consisting of the skeleton of the paired fins and their girdles.

1. The Axial Skeleton.

A. The Vertebral Column and Ribs.

The vertebral column consists of a series of some hundred and thirty vertebrae, each of which is united with its predecessor and successor in such a way as to allow a large amount of flexibility.

These vertebrae are developed round an unsegmented rod, the **notochord**, which forms the axial support of the embryo. The notochord remains continuous throughout the whole vertebral column, but is greatly constricted opposite the middle of each vertebra, and thus rendered moniliform. The vertebrae are divided into two groups, an anterior group of trunk vertebrae, and a posterior group of caudal or tail vertebrae.

A typical vertebra consists of a middle portion, the **centrum**, a dorsal portion, the **dorsal** or **neural arch**, which surrounds the spinal cord, and a ventral portion, the **ventral** or **haemal arch**, which similarly encloses a space.

The tail vertebrae of the dogfish have this typical arrangement, the trunk vertebrae have the haemal arches modified.

Each **centrum** is a short cylinder of cartilage surrounding an hourglass-shaped cavity occupied by the notochord. The **neural arches** are composed of three separate elements, the **vertebral neural plates** (basidorsalia), **intervertebral neural plates** (interdorsalia), and **neural spines** (supradorsalia).

The **vertebral neural plates** are in the adult fused with their respective centra, and are notched behind for the exit of the ventral (motor) roots of the spinal nerves. The **intervertebral neural plates** are polygonal pieces alternating with the vertebral neural plates; they are notched behind, but at a more dorsal level than are the vertebral neural plates, for the exit of the dorsal or sensory roots of the spinal nerves.

The **neural spines** are small patches of cartilage filling up the gaps between the dorsal ends of the neural plates.

The **haemal arches** (basiventralia) differ much in the trunk and tail portions of the vertebral column. In the trunk portion the centra are flattened below, and the two halves of the haemal arch diverge from one another as blunt **ventri-lateral processes** to which short cartilaginous rods, the **ribs**, are attached. Further back at about vertebra 37, the two halves of the haemal arch project downwards and meet forming a complete arch. Further back still, towards the hind end of the tail, the haemal arches bear median **haemal spines** (ventrispinalia).

B. The Skull.

The skull of the dogfish remains cartilaginous throughout the life of the animal, and has consequently a far more simple structure than have the skulls of higher animals, in which complication has been produced by the development of bone.

The skull consists of the following parts:—

(1) a dorsal portion, the **cranium**, which lodges the brain, and to the sides of which the capsules of the auditory and olfactory sense organs are united. The cranium may be compared to an unsegmented continuation of the vertebral column;

(2) a number of ventral structures, disconnected or only loosely connected with the cranium. These together constitute the **visceral skeleton** forming the jaws and supporting the gills.

(1) The Cranium.

The **Cranium** is an oblong box, with a flattened floor and a more irregular roof. Its sides are expanded in front owing to the olfactory capsules, and behind owing to the auditory capsules, while in the middle they are deeply hollowed to form the orbits.

(*a*) On the dorsal surface of the cranium the following points should be noticed. First at the anterior end, the large thin-walled **nasal** or **olfactory capsules** (fig. 6, 1), each of which is drawn out into a narrow cartilaginous process.

The olfactory capsules have no ventral walls, and are separated from one another by the **internasal septum**, which is drawn out into a third slender process. These three processes together constitute the **rostrum** (fig. 6, 2).

Behind the olfactory capsules comes a large, nearly circular, hole, the **anterior fontanelle**, slightly behind which are the two **ophthalmic foramina**. The dorsal and ventral boundaries of the orbits are respectively formed by the prominent **supra-orbital** and **suborbital ridges**. Behind are the **auditory capsules** (fig. 6, 8), each of which is marked by a pair of prominent ridges, converging towards the middle line to a pair of apertures. These apertures communicate with two canals, the **aqueductus vestibuli**, which lead into the internal ear. The two ridges lodge respectively the **anterior and posterior vertical semicircular canals** of the ear.

(*b*) The principal structures to be noted in a side view of the cranium are contained in the **orbit** or eye-cavity. Near the base of the orbit at its anterior end is seen the small **orbitonasal foramen** (fig. 6, 7), for the passage of blood-vessels, not nerves. Above it is the large **ophthalmic foramen** (fig. 6, 5) so prominent in a dorsal view of the skull; through it the ophthalmic branches of the fifth and seventh nerves pass. Slightly further back near the ventral surface is the large **optic foramen** (fig. 6, II.) for the passage of the second nerve. Vertically above the optic foramen, near the dorsal surface, is the very small **foramen for the fourth nerve** (fig. 6, IV.). Behind and a little above the optic foramen is another small aperture, the **foramen for the third nerve**. Behind and slightly below this is the large **foramen for the sixth and main branches of the fifth and seventh nerves** (fig. 6, V.). In front of and slightly below this foramen are seen two other small apertures; the more anterior and ventral of these (fig. 6, 4) is for the passage of a vessel connecting the efferent artery of the hyoid gill with the internal carotid artery inside the skull, the more posterior and dorsal is for the **interorbital canal** (fig. 6, 3) which unites the two orbital sinuses. Above and very slightly in front of the large foramen for the sixth and main parts of the fifth and seventh nerves, are two small foramina (fig. 6, Va., and VIIa.), through which the **ophthalmic branches of the fifth and seventh nerves** enter the orbit. Behind and slightly below the large foramen just mentioned is a small hole through which the external carotid enters the orbit (fig. 6, 9).

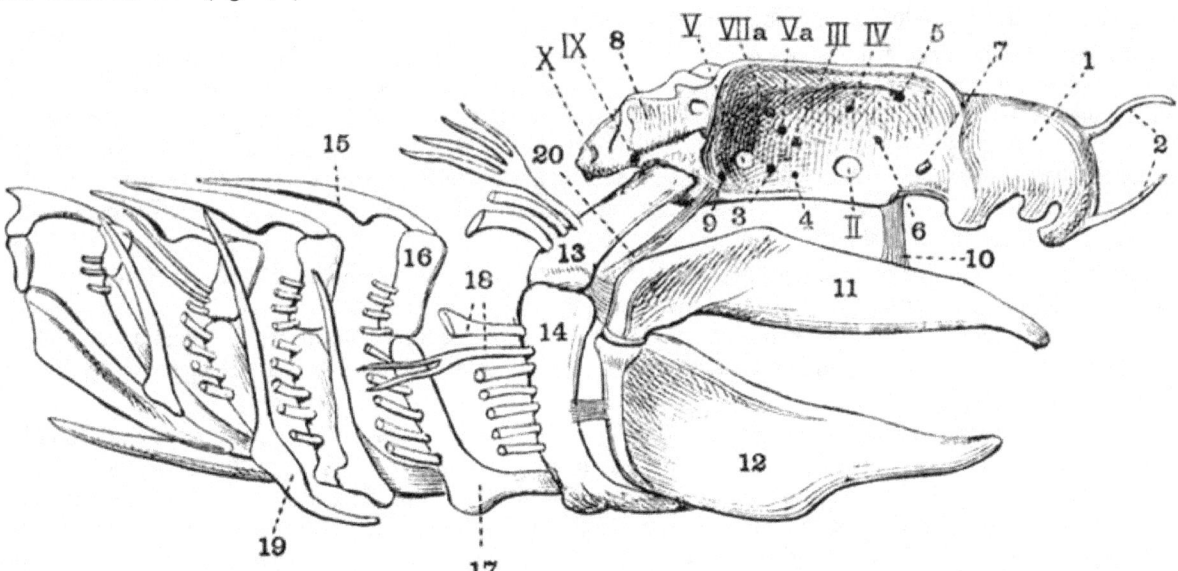

Fig. 6. Lateral view of the skull of a Dogfish (*Scyllium canicula*) × 2/3.

1. nasal capsule.
2. rostrum.
3. interorbital canal.
4. foramen for hyoidean artery.
10. ethmo-palatine ligament.
11. palato-pterygo-quadrate bar.
12. Meckel's cartilage.
13. hyomandibular.

37

5. foramen for the exit of the ophthalmic branches of Vth and VIIth nerves.

6. foramen through which the external carotid leaves the orbit.

7. orbitonasal foramen.

8. auditory capsule.

9. foramen through which the external carotid enters the orbit.

14. cerato-hyal.

15. pharyngo-branchial.

16. epi-branchial.

17. cerato-branchial.

18. gill filaments, nearly all have been cut off short for the sake of clearness.

19. extra-branchial

20. pre-spiracular ligament.

II. III. IV. V. Va. VIIa. foramina for passage of cranial nerves.

Behind the orbit is the **auditory capsule**. This is marked below by a prominent **surface for the articulation of the hyomandibular**, above which is the deep **postorbital groove** for the passage of a blood-vessel, connecting the orbital and anterior cardinal sinuses.

(*c*) Passing to the posterior end of the cranium: in the centre is seen the large **foramen magnum** through which the brain and spinal cord communicate. The **notochord** enters the skull just below this foramen, and on each side of the notochord is a projection, the **occipital condyle**, by which the first vertebra articulates with the skull.

External to the condyles are the prominent **pneumogastric foramina** for the passage of the tenth nerves, and further to the sides, just beyond the posterior vertical semicircular canals, are a pair of deep pits in which lie the **foramina for the ninth nerves** (fig. 6, IX).

(*d*) The broad and flat ventral surface of the cranium is continued in front as the **internasal septum** and terminated laterally by the **suborbital ridges**. At a little behind the middle it is traversed by two shallow grooves along which the internal carotid arteries run. At the divergent ends of these grooves are seen two small apertures through which the external carotids enter the orbit (fig. 6, 9), and at the point where they meet is a single small aperture through which the internal carotid enters the cranium.

(2) The Visceral Skeleton.

The **Visceral skeleton** forms a series of seven cartilaginous arches or hoops, surrounding the anterior part of the alimentary canal, and enclosing a wide but rather shallow space.

(*a*) The first or **mandibular arch** is the largest of the series, and forms the upper and lower jaws. Each half of the upper jaw or **palato-pterygo-quadrate** bar is formed by a thick cartilaginous rod which meets its fellow in the middle line in front, the two being united by ligament. Each half is connected to the cranium just in front of the orbit by the **ethmo-palatine ligament** (fig. 6, 10), and at its hind end articulates with one of the halves of the lower jaw. Each half of the lower jaw or **Meckel's cartilage** (fig. 6, 12) is a cartilaginous bar, wide behind but narrow in front, where it is united to its fellow by a median ligament. Imbedded in the tissue external to the upper jaw are a pair of **labial cartilages**, and a similar but smaller pair are imbedded in the tissue external to the lower jaw.

The jaws are developed from a structure whose dorsal and ventral portions subsequently become of very different importance. The ventral portion forms both upper and lower jaws, the former being developed as an outgrowth from the latter. The dorsal portion forms only the **pre-spiracular ligament** (fig. 6, 20), a strong fibrous band containing a nodule of cartilage, and running from the anterior part of the auditory capsule to the point where the jaws are connected with the hyomandibular.

(*b*) The **hyoid arch** consists of a pair of cartilaginous rods which are attached at their dorsal ends to the cranium, and are united ventrally by a broad median plate of cartilage, the **basi-hyal**. Each rod is divided into a dorsal portion, the **hyomandibular** and a ventral portion, the **cerato-hyal**. The **hyomandibular** (fig. 6, 13) is a short stout rod of cartilage projecting outwards, and somewhat backwards and downwards from the cranium, with which it articulates behind the orbit and below the postorbital groove. Its distal end articulates with a rather long slender bar, the **cerato-hyal** (fig. 6, 14), which is in its turn attached to the side of the **basi-hyal**. The **basi-hyal** is a broad plate, rounded in front and drawn out behind into two processes to which the two halves of the first branchial arch are attached. The posterior surfaces of both hyomandibular and cerato-hyal bear slender cartilaginous processes, the **gill rays**. The hyoid arch forms the main **suspensorium** or means by which the jaws are attached to the cranium. This attachment is chiefly brought about by a series of short ligaments which connect the posterior ends of both upper and lower jaws with the hyomandibular, but there is also a ligament connecting the lower jaw with the cerato-hyal. The attachment of the jaws to the cranium is also partially effected by the pre-spiracular and ethmo-palatine ligaments.

(*c*) Each of the five **branchial arches** is a hoop, incomplete above and formed of four or more pieces of cartilage. The most dorsal elements, the **pharyngo-branchials**, are flattened, pointed plates whose free inner ends run obliquely backwards, and terminate below the vertebral column. They are connected at their outer ends with the short broad **epi-branchials** (fig. 6, 16) which lie at the sides of the pharynx. From the epi-branchials arise the long **cerato-branchials** (fig. 6, 17) which run forwards and inwards along the ventral wall of the pharynx. The first four cerato-branchials are connected with small rods, the **hypo-branchials**, which run backwards to meet one another in the middle line. The last two pairs of hypo-branchials and the fifth cerato-branchials are connected with a broad median plate, the **basibranchial**. Along the outer sides of the second, third and fourth cerato-branchials are found elongated curved rods, the **extra-branchials** (fig. 6, 19). The epi-branchials and cerato-branchials bear gill rays along their posterior borders.

C. The Skeleton of the Median Fins.

The **dorsal fins** have a skeleton consisting of a series of short cartilaginous rods, the **basals** or basalia, which slope obliquely backwards. Their bases are imbedded in the muscles of the back, while their free ends bear a number of small polygonal cartilaginous plates, the **radials** or radiale. Associated with this cartilaginous skeleton are a number of long slender horny fibres, the fin-rays, which have been already referred to in connection with the exoskeleton. The skeleton of the other median fins mainly consists of these fibres, the cartilaginous portion being reduced or absent.

2. The Appendicular Skeleton.

This includes the skeleton of the two pairs of limbs and of their respective girdles.

The Pectoral girdle forms a crescent-shaped hoop of cartilage, incomplete above and lying just behind the visceral skeleton. The mid-ventral part of the hoop is the thinnest portion, and is drawn out in front into a short rounded process which is cupped dorsally and supports part of the floor of the pericardium (fig. 7, 1). On each side of this flattened mid-ventral portion the arch becomes very thick and bears on its outer border a surface with which the three basal cartilages of the fin articulate. The dorsal ends or scapular portions of the girdle form a pair of gradually tapering horns.

The Pectoral fin articulates with the pectoral girdle by means of three basalia or basal cartilages, the **propterygium, meso-pterygium** and **meta-pterygium**. The most anterior and the smallest of these is the **propterygium** (fig. 7, 5), while the most posterior one, the **meta-pterygium** (fig. 7, 3), is much the largest. Along the outer borders of the three basalia are arranged a series of close set cartilaginous pieces, the **radiale**. The propterygium supports only a single radial, which is however much larger than any of the others. The meso-pterygium also supports only a single radial which divides distally.

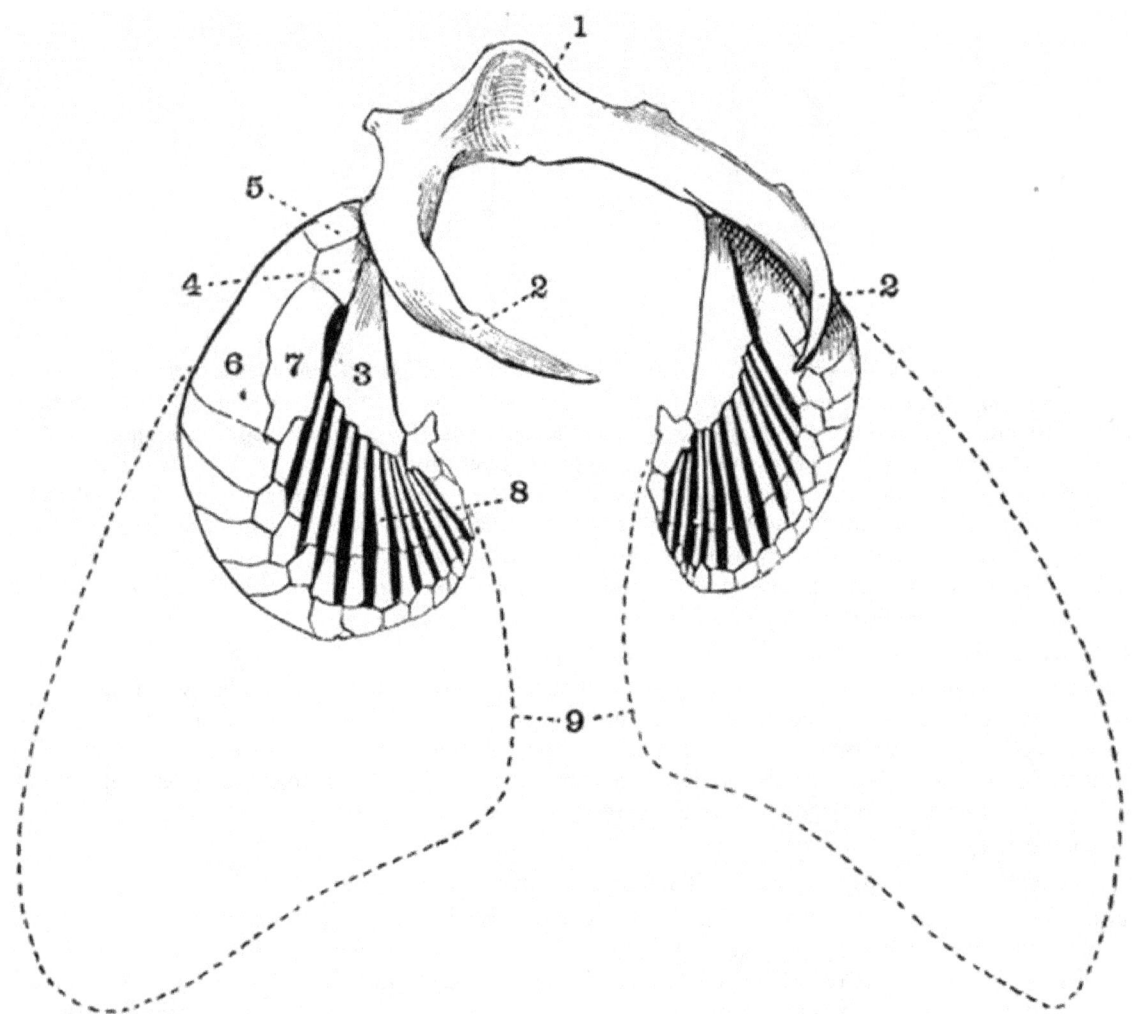

Fig. 7. Semidorsal view of the pectoral girdle and fins
of a Dogfish (*Scyllium canicula*) × 2/3.
The gaps between the radiale are blackened.

1. hollow in the mid-ventral part
of the pectoral girdle which
supports the pericardium.

2. dorsal (scapular portion) of
pectoral girdle.

3. meta-pterygium.

4. meso-pterygium.

5. propterygium.

6. propterygial radial.

7. meso-pterygial radial.

8. meta-pterygial radial.

9. outline of the distal part of
the fin which is supported
by horny fin-rays.

The meta-pterygium bears about twelve long narrow radials, the first nine of which are traversed by a transverse joint at about two-thirds of the way from their origin. Succeeding the radials are a series of small polygonal pieces of cartilage arranged in one or more rows and attached to the ends of the radials, and finally the fin is completed by the dermal fin-rays.

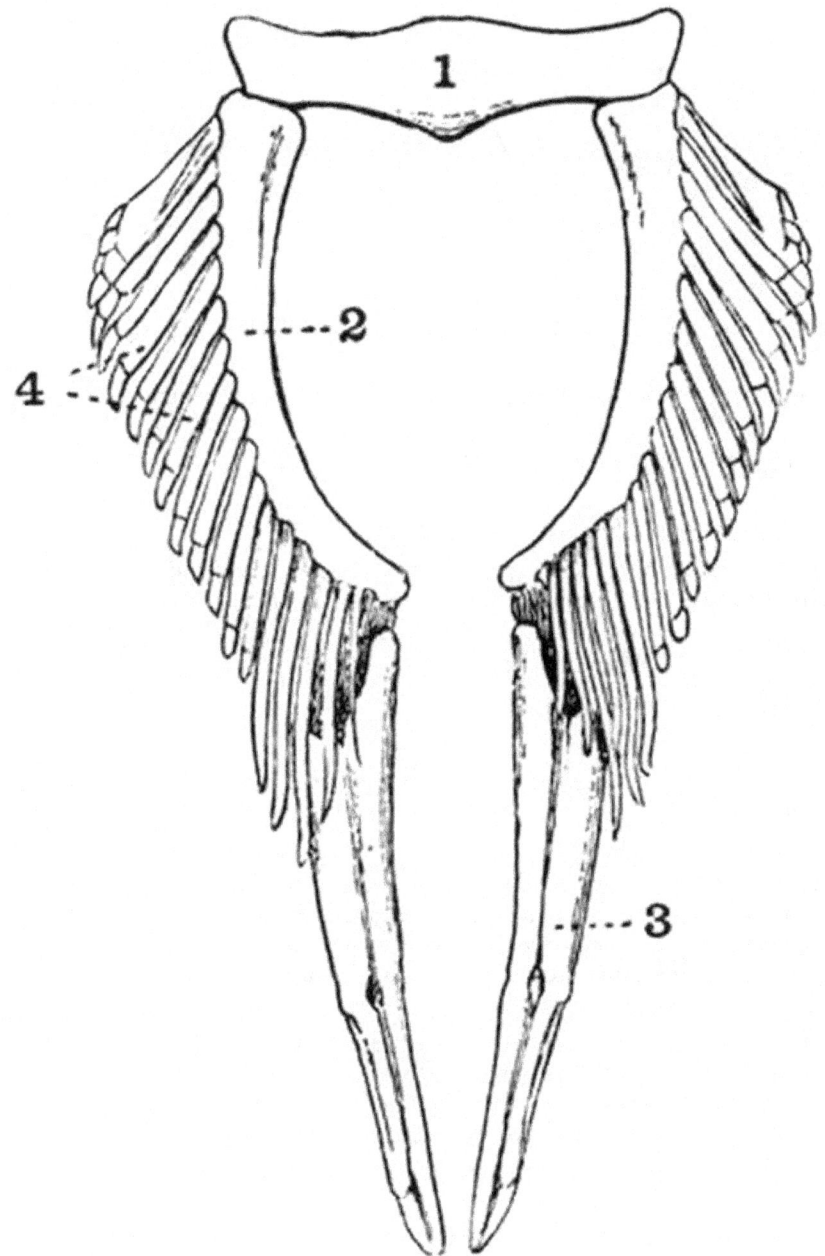

Fig. 8. Dorsal view of the pelvic girdle and fins of a male Dogfish (*Scyllium canicula*).

1. pelvic girdle.

2. basi-pterygium.

3. clasper.

4. radiale.

The Pelvic Girdle is much smaller than the pectoral. It is formed of a stout nearly straight bar of cartilage placed transversely across the ventral region of the body. The bar has no dorsal or lateral extensions, and is terminated by short blunt processes. It bears on its posterior surface a pair of facets with which the pelvic fins articulate.

The Pelvic Fin is smaller and more simply constructed than is the pectoral. It consists of a long, somewhat curved rod, the **basi-pterygium** (fig. 8, 2), running directly backwards on the inner side of the fin, and articulating in front with the pelvic girdle. From its outer side arise a series of about fourteen parallel cartilaginous radials which bear smaller polygonal pieces. The anterior one or two of these radials may articulate independently with the pelvic girdle. In the adult male dogfish the distal end of the basi-pterygium bears a stout rod nearly as long as itself, and grooved on the dorsal surface. This is the skeleton of the **clasper** (fig. 8, 3).

CHAPTER VII.
THE SKELETON OF THE CODFISH. (Gadus morrhua.)

I. EXOSKELETON.

The exoskeleton includes

(1) **Scales.** These are of the type known as **cycloid** and consist of flat rounded plates composed of concentrically arranged laminae of calcified matter, with the posterior margin entire. The anterior end of each scale is imbedded in the skin and is overlapped by the preceding scales.

(2) The **teeth.** These are small, pointed, calcified structures arranged in large groups on the premaxillae, mandible, vomer, and superior and inferior pharyngeal bones.

(3) The **fin-rays.** These are delicate, nearly straight bony rods which support the fins.

II. ENDOSKELETON.

The endoskeleton of the Codfish, though partially cartilaginous, is mainly ossified.

It is divisible into an **axial portion**, including the skull, vertebral column, ribs, and skeleton of the median fins, and an **appendicular portion**, including the skeleton of the paired fins and their girdles.

1. The Axial Skeleton.

A. The Vertebral Column.

This consists of a series of some fifty-two vertebrae, all completely ossified.

It is divisible into two regions only, viz. the **trunk** region, the vertebrae of which bear movable ribs, and the **caudal** or **tail** region, the vertebrae of which do not bear movable ribs.

Trunk vertebrae.

These are seventeen in number; the ninth may be described as typical of them all. It consists of a short deeply biconcave **centrum** whose two cavities communicate by a narrow central canal. From the dorsal surface of the anterior half of the centrum arise two strong plates, the dorsal or **neural processes**, which are directed obliquely backwards and meet forming the dorsal or **neural arch**. This is produced into a long backwardly-directed dorsal or **neural spine**.

From the lower part of the anterior edge of each neural arch arise a pair of blunt triangular projections which overhang the posterior half of the preceding centrum, and bear a pair of flattened surfaces which correspond to the anterior or **prezygapophyses** of most vertebrae, they differ however from ordinary prezygapophyses in the fact that they look downwards and outwards. From the posterior end of the centrum arise a pair of short blunt processes each of which bears an upwardly- and inwardly-directed articulating surface corresponding to a **postzygapophysis**.

The two halves of the ventral arch form a pair of large **ventri-lateral processes** which arise from the anterior half of the centrum and pass outwards and slightly backwards and downwards.

Behind these there arises on each vertebra a second outgrowth which is small and flattened, and like the ventri-lateral process serves to protect the air-bladder. The surface of the centrum is marked by more or less wedge-shaped depressions, one in the mid-dorsal line, and two on the ventral surface immediately mesiad to the bases of the ventri-lateral process. There are also a number of smaller depressions.

The space between one centrum and the next is in the fresh skeleton filled up by the gelatinous remains of the **notochord**.

The first few vertebrae differ from the others in having very short centra and no ventri-lateral processes.

The first vertebra comes into very close relation to the posterior part of the skull, articulating with the exoccipitals. In the next few vertebrae the centra gradually lengthen, and at the fourth or fifth vertebra the ventri-lateral processes appear and gradually increase in size as followed back. They likewise gradually come to arise at a lower level on the centrum, and also become more and more downwardly directed, till at the last trunk vertebra they nearly meet.

The **neural spines** of the anterior trunk vertebrae are much longer than those of the posterior ones, that of the first vertebra being the largest and longest of all, and articulating with the skull. The spinal nerves pass out through wide notches or spaces between the successive neural arches.

Caudal vertebrae.

The caudal vertebrae are about thirty-five in number, each consists of a centrum with a slender backwardly-directed dorsal or neural arch, similar to those of the posterior trunk vertebrae. The two halves of the ventral or haemal arch however do not form outwardly-directed ventri-lateral processes, but arise on the ventral surface of the centrum, and passing downwards meet and enclose a space; they thus form a complete canal, and are prolonged into a backwardly-directed ventral or **haemal spine**. The anterior haemal arches are much larger than the corresponding neural arches, but when followed back they gradually decrease in size, till at about the twenty-fourth caudal vertebra they are nearly as small as the neural arches. The last caudal vertebra is succeeded by a much flattened **hypural** bone or **urostyle**, which together with the posterior neural and haemal spines supports the tail-fin.

B. The Ribs.

The **ribs** are slender, more or less cylindrical bones attached to the poster-dorsal faces of the ventri-lateral processes of all the trunk vertebrae except the first and second. The earlier ones are thicker and more curved; the later ones thinner and more nearly straight. The ribs are homologous with the distal parts of the haemal arches of the caudal vertebrae.

Associated with the ribs are a second series of rib-like bones, the **intermuscular bones**. These are slender, curved bones which arise from the ribs or from the ventri-lateral processes at a distance of about an inch from the centra, and curve upwards, outwards and backwards. In the anterior region where the ventri-lateral processes are short they arise from the ribs, further back they arise from the ventri-lateral processes.

C. The Unpaired or Median Fins.

These are six in number, three being **dorsal**, one **caudal** and two **anal**.

The **dorsal** and **anal** fins each consist of two sets of structures, the **fin-rays** and the **interspinous bones**. Each fin-ray forms a delicate, nearly straight, bony rod which becomes thickened and bifurcated at its proximal or vertebral end, while distally it is transversely jointed and flexible, frequently also becoming more or less flattened.

The first dorsal fin has thirteen rays, the second, sixteen to nineteen, the third, seventeen to nineteen. The first anal fin has about twenty-two, the second anal fourteen. In each fin the posterior rays rapidly decrease in size when followed back.

The **interspinous bones** of the dorsal and anal fins alternate with the neural and haemal spines respectively, and form short, forwardly-projecting bones, each attached proximally to the base of the corresponding fin-ray.

The **caudal fin** consists of a series of about forty-three rays which radiate from the posterior end of the vertebral column, being connected with the urostyle or hypural bone, and with the posterior neural and haemal spines without the intervention of interspinous bones. Like the other fin-rays those forming the caudal fin are transversely jointed, and are widened and frayed out distally.

The tail-fin in the Cod is **homocercal**, i.e. it appears to be symmetrically developed round the posterior end of the vertebral column, though in reality a much greater proportion is attached below the end of the vertebral column than above it. It is a masked heterocercal tail.

The Skull.

Owing to the fact that very little cartilage remains in the skull of the adult Codfish, its relation to the completely cartilaginous skull of the Dogfish is not easily seen. Before describing it therefore, the skull of the Salmon will be described, as it forms an intermediate type.

THE SKULL OF THE SALMON.

The Salmon's skull consists of (1) the **chondrocranium**, which remains partly cartilaginous and is partly converted into cartilage bone, especially in the occipital region, (2) a large series of plate-like membrane bones.

The Chondrocranium.

This is an elongated structure, wide behind owing to the fusion of the large auditory capsules with the cranium, and elongated and tapering considerably in front; in the middle it is much contracted by the large orbital cavities.

Dorsal surface of the Cranium.

In the centre of the posterior end of the dorsal surface is the **supra-occipital** (fig. 9, A, 1) with a prominent posterior ridge. It is separated by two tracts of unossified cartilage from the large series of bones connected with the **auditory organ**. The first of these is the **epi-otic** (fig. 9, 2), which is separated by only a narrow tract of cartilage from the supra-occipital, and is continuous laterally with the large **pterotic** (fig. 9, A, 3) which overlaps in front a smaller bone, the **sphenotic** (fig. 9, 4). Both epi-otic and pterotic are drawn out into rather prominent backwardly-projecting processes.

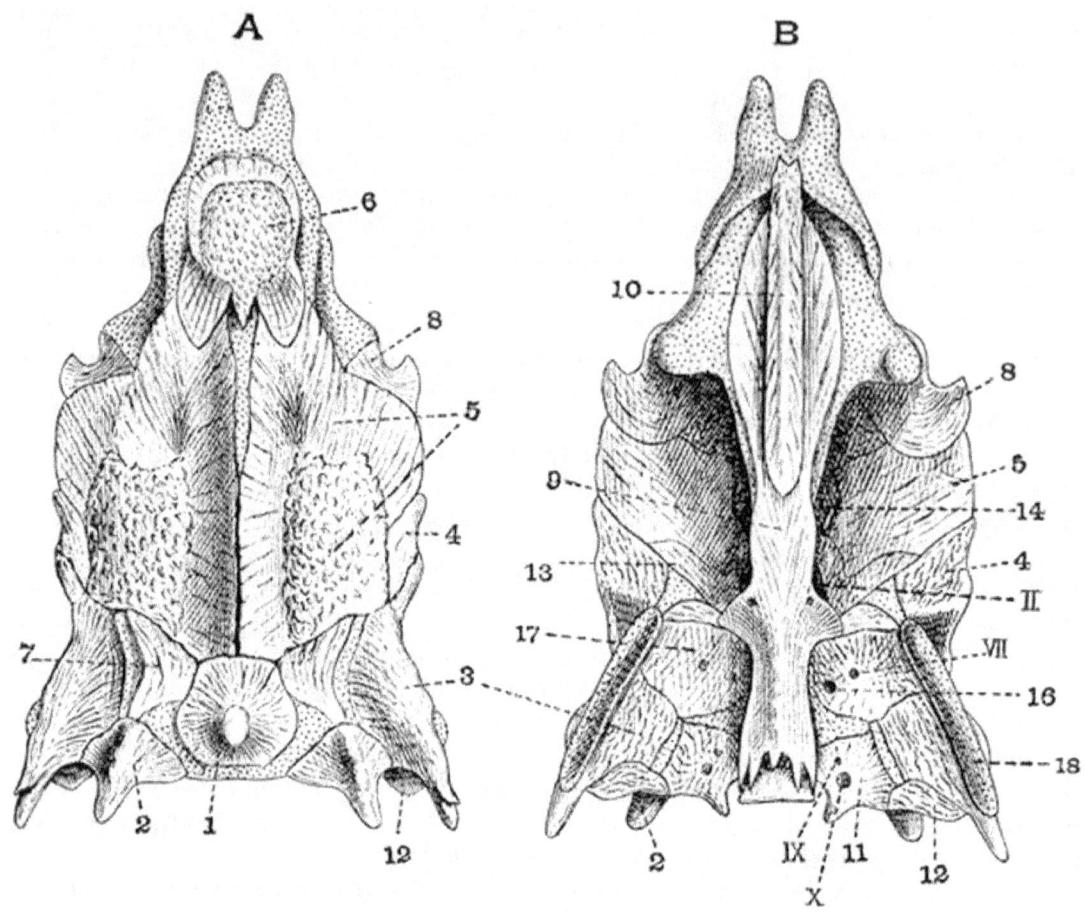

Fig. 9. A. dorsal and B. ventral view of the cranium of a Salmon (*Salmo salar*) from which most of the membrane bones have been removed (after Parker). Cartilage is dotted.

1. supra-occipital.
2. epi-otic.
3. pterotic.
4. sphenotic.
5. frontal.
6. median ethmoid.
7. parietal.
8. lateral ethmoid.
9. parasphenoid.
10. vomer.
11. exoccipital.

12. opisthotic.
13. alisphenoid.
14. orbitosphenoid.
16. foramen for passage of an artery.
17. pro-otic.
18. articular surface for hyomandibular.
II. VII. IX. X. foramina for the passage of cranial nerves.

The greater part of the remainder of the dorsal surface is formed of unossified cartilage which is pierced by three large vacuities or **fontanelles**. The anterior fontanelle is unpaired, and lies far forward near the anterior end of the long cartilaginous snout, the two larger posterior ones lie just in front of the supra-occipital and lead into the cranial cavity. In front of the orbit the skull widens again, and is marked by two considerable **lateral ethmoid** (fig. 9, 8) ossifications. In front of these are a pair of deep pits, the **nasal fossae**, at the base of which are a pair of foramina through which the olfactory nerves pass out; they communicate with a space, the **middle narial cavity**, seen in a longitudinal section of the skull.

The long cartilaginous snout is more or less bifid in front, especially in the male (fig. 9).

Posterior end of the Cranium.

The **foramen magnum** forms a large round hole leading into the cranial cavity, and is bounded laterally by the two **exoccipitals** and below by them, and to a very slight extent by the **basi-occipital**, the three bones together forming a concave **occipital condyle** by which the vertebral column articulates with the skull.

The exoccipitals are connected laterally with a fourth pair of auditory bones, the **opisthotics**, and just meet the epi-otics dorsolaterally, while dorsally they are separated by a wide tract of unossified cartilage from the supra-occipital.

The opisthotics are connected laterally with the pterotics.

Side of the Cranium.

At the posterior end is seen the **basi-occipital** in contact above with the **exoccipital**, which is pierced by a prominent foramen for the exit of the tenth nerve. In front of this lies a small foramen, sometimes double, for the ninth nerve.

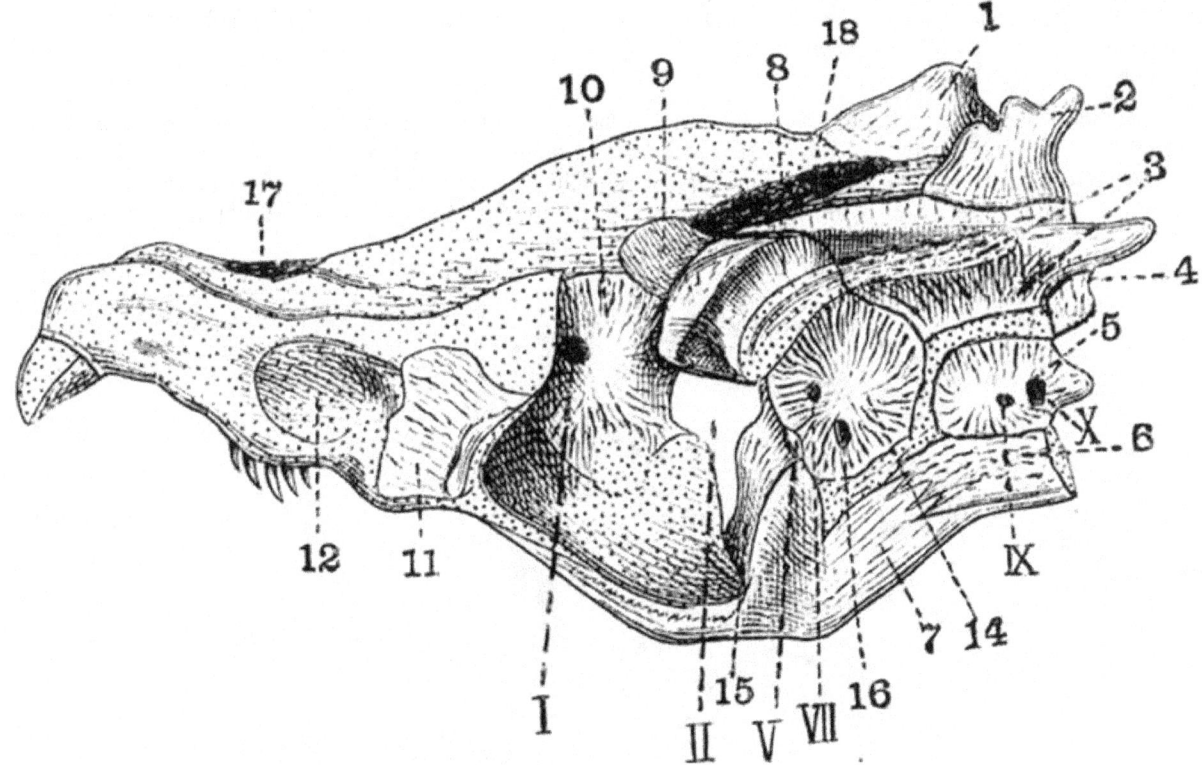

Fig. 10. Lateral view of the chondrocranium of a Salmon (*Salmo salar*) (after Parker). A few membrane bones are also shown. Cartilage is dotted.

1. supra-occipital.
2. epi-otic.
3. pterotic.
4. opisthotic.
5. exoccipital.
6. basi-occipital.
7. parasphenoid.
8. sphenotic.
9. alisphenoid.
10. orbitosphenoid.
11. lateral- or ectethmoid.
12. olfactory pit; the vomerine teeth are seen just below.
14. pro-otic.
15. basisphenoid.
16. foramen for the passage of an artery.
17. anterior fontanelle.
18. posterior fontanelle.
I. II. V. VII. IX. X. foramina for the passage of cranial nerves.

In front of the exoccipital is the large **pro-otic** pierced by two prominent foramina. Through the more dorsal of these (fig. 10, VII.) the facial nerve passes out, while the more ventral (fig. 10, 16) is for the passage of an artery. Dorsal to the exoccipital are the **opisthotic** and **pterotic**, and dorsal to the pro-otic is the **sphenotic**. The **pterotic** is marked by a prominent

45

groove often lined by cartilage, which is continued forwards along a tract of cartilage between the pro-otic and sphenotic. With this groove the hyomandibular articulates.

There are considerable ossifications in the sphenoidal region of the side of the cranium. The anterior boundary of the posterior fontanelle is formed by the large **alisphenoid**, which is continuous behind with the pro-otic and sphenotic, and below with a slender **basisphenoid**. Both in front of and behind the basisphenoid there are considerable vacuities in the walls of the cranium; through the posterior of these openings (fig. 10, V.) the main part of the trigeminal nerve passes out, and through the anterior one, the optic (fig. 10, II.). The alisphenoid is continuous in front with the **orbitosphenoid** (fig. 10, 10), which is pierced by the foramen for the exit of the first nerve (fig. 10, I.), and in front of the orbitosphenoid there is a large vacuity. The **lateral ethmoid** is seen in the side view as well as in the dorsal view. Further forwards are seen the olfactory pits, and the long cartilaginous snout.

A **ventral view** of the cartilaginous cranium shows much the same points as the side view. The basisphenoid appears on the surface immediately in front of the basi-occipital.

The Skull with membrane bones.

The **dorsal surface**. The greater part of the dorsal surface in front of the supra-occipital is overlaid by a pair of large rough *frontals* (figs. 9, A, 5, and 10, 5). They cover the posterior fontanelles and stretch over from the sphenotic to the lateral ethmoid, forming a roof for the orbit. They meet in the middle line behind, but in front are separated by a narrow tract of unossified cartilage, and are overlapped by the *median ethmoid* (figs. 9, A, 6, and 11, 6). At the sides of the supra-occipital behind the frontals are a pair of small *parietals* (figs. 9, A, 7, and 11, 7).

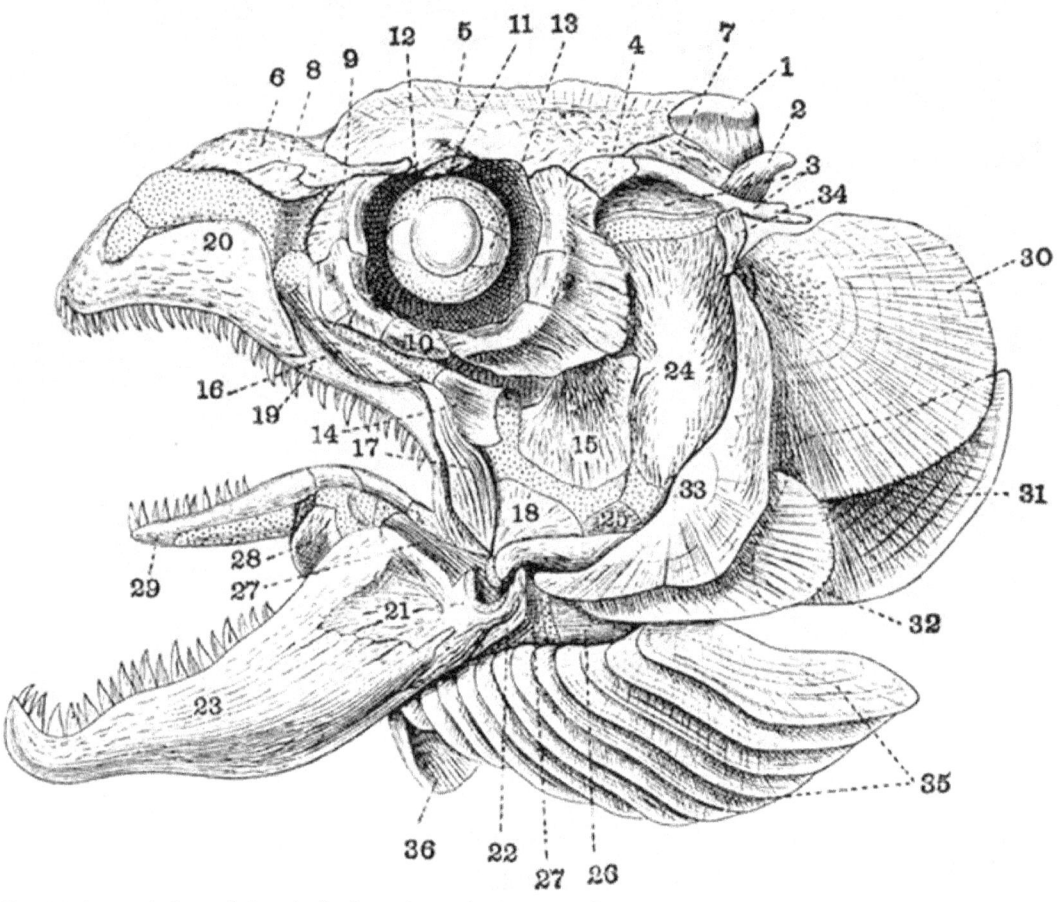

Fig. 11. Lateral view of the skull of a Salmon (*Salmo salar*) (after Parker). Cartilage is dotted.

1. supra-occipital.	19. maxillae
2. epi-otic.	20. premaxillae.
3. pterotic.	21. articular.
4. sphenotic.	22. angular.

5. frontal.

6. median ethmoid.

7. parietal.

8. nasal.

9. lachrymal.

10. suborbital.

11. supra-orbital.

12. cartilaginous sclerotic.

13. ossification in sclerotic.

14. meso-pterygoid.

15. meta-pterygoid.

16. palatine.

17. jugal.

18. quadrate.

23. dentary.

24. hyomandibular.

25. symplectic.

26. epi-hyal.

27. cerato-hyal.

28. hypo-hyal.

29. glosso-hyal.

30. opercular.

31. sub-opercular.

32. infra-opercular.

33. pre-opercular.

34. supratemporal

35. branchiostegal rays.

36. basi-branchiostegal.

In a **ventral view** the cranium is seen to be chiefly covered by two large membrane bones, the *parasphenoid* (fig. 9, B, 9) behind, the *vomer* in front. A view of the **posterior end** differs from that of the cartilaginous cranium only in the fact that the end of the *parasphenoid* appears lying ventral to the basi-occipital.

The **lateral view** differs very markedly from that of the cartilaginous cranium, there being a great development of membrane bone in connection with the jaws and branchial apparatus. Lying dorsally are seen the *median ethmoid, frontal, parietal*, and **supra-occipital** as before. Lying external to the middle of the *median ethmoid* is seen the small *nasal* (fig. 11, 8), and below the hinder part is the *lachrymal*. The *lachrymal* (fig. 11, 9) forms the first of a series of seven small bones which surround the orbit forming the **orbital ring**. Of these the one lying immediately in the mid-ventral line of the orbit is the *suborbital*, while the one lying in the mid-dorsal line and attached to the frontal is the *supra-orbital* (fig. 11, 11). The orbit has a cartilaginous *sclerotic* in which are two small ossifications (fig. 11, 13) laterally placed.

Bones of the upper jaw.

The **palato-pterygo-quadrate bar** is in a very different condition from that of the dogfish, it is partially cartilaginous, partially converted into cartilage bone, partially overlapped by membrane bone. It is narrow in front but becomes much broader and deeper when followed back. Its anterior end forms the **palatine** which bears teeth, and in front is completely ossified, while behind the cartilage is only sheathed by bone.

Just behind the palatine the outer part of the cartilage is ossified, forming two small bones, the **pterygoid** and **meso-pterygoid**, while behind them is a larger, somewhat square bone, the **meta-pterygoid** (fig. 11, 15).

Below the meta-pterygoid is a tract of unossified cartilage, and then comes the **quadrate** (fig. 11, 18).

The lower angle of the quadrate bears a cartilaginous **condyle** with which the mandible articulates. In front of the palatine the cartilaginous snout is overlapped by three membrane bones, the *jugal, maxilla* and *premaxillae*.

The *premaxillae* (fig. 11, 20), the largest of these, overlaps the maxilla behind; both bones bear teeth. The *jugal* (fig. 11, 17) lies above the maxilla and overlaps it in front.

The lower jaw.

The **lower jaw** is a strong bar and is like the upper jaw, partly cartilaginous, forming **Meckel's cartilage**, partly ossified, and sheathed to a considerable extent in membrane bone.

The outer side and posterior end is ossified, forming the large **articular** (fig. 11, 21), but the condyle is cartilaginous and the anterior part of the articular forms merely a splint on the outer side of Meckel's cartilage, which extends beyond it for a considerable distance. The angle of the jaw just below the condyle is formed by a small *angular* (fig. 11, 22), and the anterior two-thirds of the jaw is sheathed in the large tooth-bearing *dentary* (fig. 11, 23).

The Hyoid arch.

The **hyoid arch** has a number of ossifications in it and is closely connected with the mandibular arch.

The **hyomandibular** (fig. 11, 24) is a large bone which articulates with a shallow groove lined by cartilage and formed partly in the pterotic, partly in front of it. The hyomandibular is overlapped in front by the meta-pterygoid, while below it tapers and is succeeded by a small area of unossified cartilage followed by the forwardly-directed **symplectic** which fits into a groove in the quadrate.

47

The unossified tract between the hyomandibular and symplectic is continuous in front with a strong bar, which remains partly cartilaginous and is partly converted into cartilage bone. The proximal part is ossified, forming the **epi-hyal**, the middle part forms the **cerato-hyal** (fig. 11, 27), in front of which is the small **hypo-hyal**. The hyoid arches of the two sides are united by the large tooth-bearing **glosso-hyal** (fig. 11, 29). Attached to the lower surface of the hyoid arch are a series of twelve flat *branchiostegal rays* (fig. 11, 35). Each overlaps the one in front of it, the posterior one being the largest. The branchiostegal rays of the two sides are united in front by an unpaired membrane bone, the *basi-branchiostegal* (fig. 11, 36).

Opercular bones. Behind the hyomandibular there is a large bony plate, the **operculum**, formed of four large membrane bones. The anterior of these, the *pre-opercular* (fig. 11, 33), is crescentic in shape, and with its upper end a small *supratemporal* (fig. 11, 34) is connected.

Behind the upper part of the pre-opercular is the largest of the opercular bones, the *opercular proper*. Its lower edge overlaps the sub-opercular, and both opercular and sub-opercular are overlapped by the *infra-opercular* (fig. 11, 32) in front. The infra-opercular is in its turn overlapped by the *pre-opercular*.

Branchial arches.

There are five branchial arches, the first four of which bear gill rays. Each of the first three consists of a shorter upper portion directed obliquely backwards and outwards, and a longer lower portion forming a right angle with the upper and directed obliquely forwards and inwards. The greater part of each arch is ossified.

The upper part of either of the first two consists of a short tapering **pharyngo-branchial** directed inwards, and of a long **epi-branchial** tipped with cartilage at both ends. The junction of the upper and lower parts is formed by a cartilaginous hinge-joint between the epi-branchial and cerato-branchial. The **cerato-branchial** is a long bony rod separated by a short area of cartilage from the **hypo-branchial**, which is succeeded by the **basibranchial** meeting its fellow in the middle line. The **fourth arch** has a short epi-branchial and no ossified pharyngo-branchial, while the fifth is reduced to little more than the cerato-branchial, which bears a few teeth on its inner edge. All the branchial arches have projecting from their surfaces a number of little processes which act as strainers. The first and fourth arches have one series of these, the second and third have two.

THE SKULL OF THE CODFISH.

A full description having been already given of the Salmon's skull, that of the Codfish will be described in a briefer manner. The skull is very fully ossified, and the great number of plate-like bones render it a very complicated structure.

The Cranium.

At the posterior end of the dorsal surface is the large **supra-occipital**, which is drawn out behind into the large blade-like **occipital spine**. On each side of the supra-occipital are the small irregular *parietals*, while in front of it the roof of the skull is mainly formed by the very large unpaired *frontal*.

A complicated series of bones are developed in connection with the **auditory capsule**, which forms a large projecting mass united with the side of the cranium and drawn out behind into a pair of strong processes, the **epi-otic** and **parotic** processes. Both these processes are connected behind with a large V-shaped bone, the *post-temporal* (fig. 13, 1), which will be described when dealing with the pectoral girdle. The **epi-otic process** is formed by the **epi-otic**, which is continuous in front with the parietal. The **parotic process** is formed by two larger bones, a more dorsal one, the **pterotic**, and a more ventral and internal one, the **opisthotic**, which is continuous in front with the large **pro-otic**. Intervening between the pterotic and frontal is another rather large bone, the **sphenotic**, this articulates below with the pro-otic. The pterotic and sphenotic together give rise to a large concave surface by which the hyomandibular articulates with the cranium. Several of the cranial nerves pass out through the bones of the auditory capsule. The ninth leaves by a foramen near the posterior border of the opisthotic, the fifth and seventh by a notch in the anterior border of the pro-otic.

A number of bones are likewise developed in connection with the orbit forming the **orbital ring**. Of these the most anterior, the *lachrymal*, is much the largest, the others are five to seven in number, the most ventral being the *suborbital*. The sclerotic coat of the eye is cartilaginous.

Two pairs of bones and one unpaired bone are developed in connection with the **olfactory capsules**, of these, the *nasals* are narrow bones lying next the lachrymals, but nearer the middle line; they overlap the second pair of bones, the irregular **lateral ethmoids**. These meet one another in the middle line, and are overlapped behind by the frontal. They articulate laterally with the lachrymal and palatine, and ventrally with the parasphenoid.

In a **posterior view** the foramen magnum and the four bones which surround it and together form the occipital segment are well seen. On the ventral side is the **basi-occipital**, terminated posteriorly by a slightly concave surface which articulates with the centrum of the first vertebra. The sides of the foramen magnum are formed by the **exoccipitals**, a pair of very irregular bones, pierced by a pair of prominent foramina for the exit of the tenth nerves. The exoccipitals also bear a pair of surfaces for articulation with corresponding ones on the neural arch of the first vertebra. The most dorsal of the four bones is the supra-occipital.

On the ventral surface of the cranium in front of the basi-occipital is seen the *parasphenoid*, a very long narrow bone which underlies the greater part of the cranium. Behind, it articulates dorsally with the basi-occipital and dorsolaterally with the pro-otics and opisthotics, in front it articulates dorsally with the lateral ethmoid and ventrally with the vomer. At the sides of the parasphenoid are the small **alisphenoids** articulating above with the postfrontals, in front with the frontals, and behind with the pro-otics.

The *vomer* is an unpaired bone lying immediately in front of the parasphenoid. In front it terminates with a thickened curved margin bearing several rows of small teeth; behind it tapers out into a long process which underlies the anterior part of the parasphenoid. Immediately dorsal to the vomer is another median bone, the *median ethmoid*; this is truncated in front and tapers out behind into a process which fits into a groove on the ventral side of the frontal.

Bones in connection with the upper jaw.

These bear a close resemblance to those of the Salmon. The most anterior bone is the *premaxillae*, a thick curved bone meeting its fellow in the middle line. The point of junction of the two is drawn out into a short process, and the oral surface is thickly covered with small teeth. The dorsal ends of the premaxillae are seen in the fresh skull to meet a large patch of cartilage. Behind the premaxillae is the *maxilla*, a long rod-like toothless bone, somewhat expanded at the upper end where it articulates with the premaxillae and vomer.

Articulating in front with the anterior end of the maxilla and with the **lateral ethmoid** is a very irregular bone, the **palatine** (fig. 12, 1); it articulates behind with two flat bones, the **pterygoid** and **meso-pterygoid**. The pterygoid is united behind with two more bones, the **quadrate** (fig. 12, 4) and **meta-pterygoid**. The **quadrate** is a rather stout irregular bone, bearing on its lower surface a prominent saddle-shaped articulating surface for the mandible. The palatine, pterygoid and quadrate bones are the ossified representatives of the palato-pterygo-quadrate bar of the Dogfish.

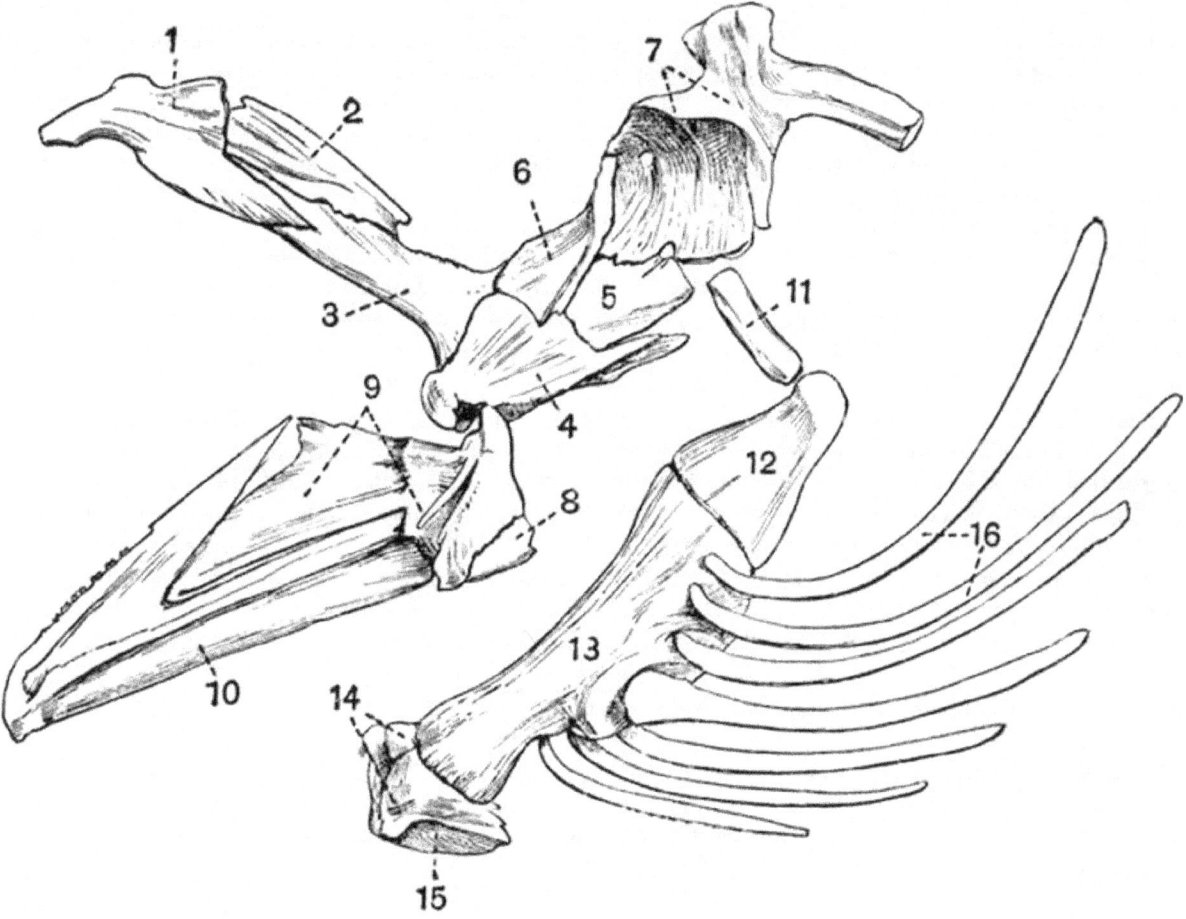

Fig. 12. Mandibular and hyoid arches of a Cod (*Gadus morrhua*) × ½ (Brit. Mus.).

1. palatine.	9. articular.
2. meso-pterygoid.	10. dentary.
3. pterygoid.	11. inter-hyal.

49

4. quadrate.	12. epi-hyal.
5. symplectic.	13. cerato-hyal.
6. meta-pterygoid.	14. hypo-hyal.
7. hyomandibular.	15. uro-hyal.
8. angular.	16. branchiostegal rays.

The quadrate is united behind with the **symplectic** (fig. 12, 5), and the meta-pterygoid with the symplectic and **hyomandibular**, both of which bones will be described immediately in connection with the hyoid arch.

The Lower jaw.

The **lower jaw or mandible** like that of the Salmon is partly cartilaginous, forming **Meckel's cartilage**, partly formed of cartilage bone, partly of membrane bone. Meckel's cartilage is of course not seen in the dried skull.

The lower jaw includes one cartilage bone, the **articular** (fig. 12, 9), this is a large bone connected by a saddle-shaped surface with the quadrate. Meckel's cartilage lies in a groove on its under surface, and projects beyond it in front. The *angular* is a small thick bone united to the lower surface of the articular at its posterior end. The *dentary* (fig. 12, 10) is a large tooth-bearing bone meeting its fellow in the middle line in front, while the articular fits into a deep notch at its posterior end.

The hyoid arch.

The **hyomandibular** (fig. 12, 7) is a large irregular bone, articulating by a prominent rounded head with the sphenotic and pterotic. It is united in front with the meta-pterygoid and symplectic, and sends off behind a strong process which articulates with the opercular. The **symplectic** is a long somewhat triangular bone drawn out in front into a process which fits into a groove on the inner surface of the quadrate. The distal portion of the hyoid arch is strongly developed and consists of first the **inter-hyal** (fig. 12, 11), a short bony rod, which articulates dorsally with a patch of cartilage intervening between the posterior part of the hyomandibular and the symplectic. Below it is united with the apex of the triangular **epi-hyal**, a bone suturally connected with the large **cerato-hyal** (fig. 12, 13) which unites distally with two small **hypo-hyals**. To the cerato-hyal are attached a series of seven strong curved cylindrical rods, the *branchiostegal rays*. The first of these is the smallest and they increase in size up to the last. The four dorsal ones are attached to the outer surface of the cerato-hyal, the three ventral ones to its inner surface. Interposed between the hypo-hyals of the two sides is an unpaired somewhat triangular plate, the uro-hyal or *basi-branchiostegal* (fig. 12, 15).

The branchial arches.

The **branchial arches** are five in number and consist of the following parts on each side. The dorsal end is formed of the **supra-pharyngeal** bone, a large irregular bone covered ventrally with teeth of a fair size, and representing the fused **pharyngo-branchials** of the four anterior arches. Its external surface is continuous with four small **epi-branchials** which pass horizontally backwards and outwards. Their distal ends meet four long **cerato-branchials** which are directed forwards and inwards and form the principal part of the arches.

Each of the first three cerato-branchials articulates ventrally with a **hypo-branchial**, and the hypo-branchials of the two sides are united in the middle line by an unpaired **basibranchial**. The third hypo-branchial is much flattened. The fourth cerato-branchial is united by cartilage with the posterior surface of the third hypo-branchial, which it meets near the middle line.

The fifth arch consists only of the cerato-branchial, a wide structure covered with teeth and generally called the **inferior pharyngeal bone**.

The skeleton of the **operculum** consists of the same four bones as in the Salmon, namely the *opercular*, the *infra-opercular*, the *pre-opercular* and the *sub-opercular*. Of these the anterior bone, the *pre-opercular*, is the largest, while the *infra-opercular* is the smallest. The *opercular* has a facet for articulation with the hyomandibular.

2. The Appendicular Skeleton.

The Pectoral girdle.

This is of a highly specialised type. Membrane bones are greatly developed, and the cartilage bones, the **scapula** and **coracoid**, are much reduced in size and importance.

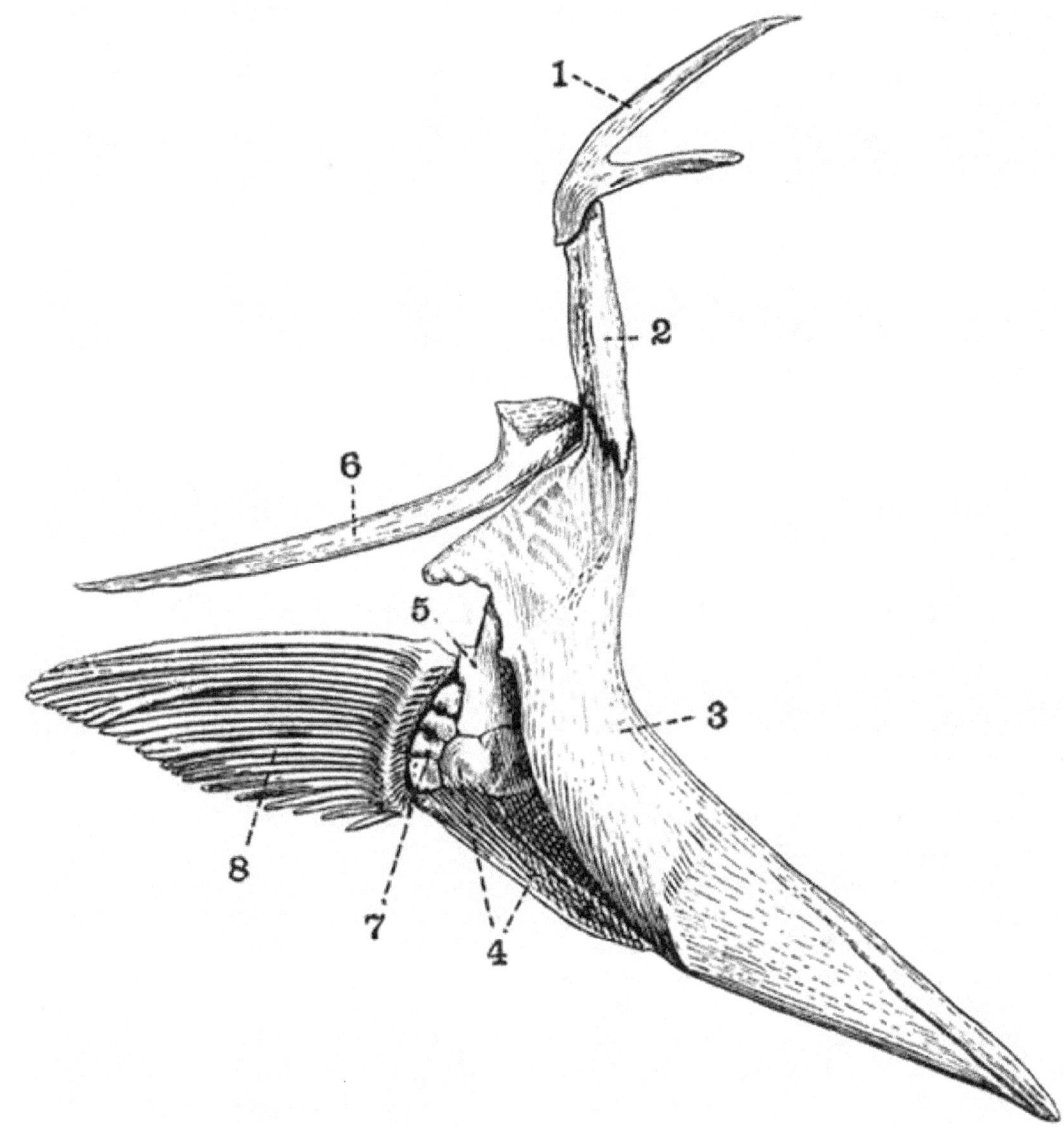

Fig. 13. The right half of the pectoral girdle and right pectoral fin of a Cod (*Gadus morrhua*) × ½ (Brit. Mus.).

1. post-temporal.	5. scapula.
2. supra-clavicle.	6. post-clavicle.
3. clavicle.	7. brachial ossicles.
4. coracoid.	8. dermal fin-rays.

The largest bone in the shoulder girdle is the *clavicle* (fig. 13, 3), which is irregularly crescent shaped, thick in front and tapering off behind. To the outer side of its upper part is attached a thick cylindrical bone, the *supra-clavicle*, which passes upwards and is connected with a strong **V** shaped bone, the *post-temporal*. The apex of the **V** meets the supra-clavicle, the inner limb articulates with the epi-otic process, the outer with the parotic process. Projecting downwards from the upper part of the clavicle is a long bony rod, flattened proximally and cylindrical and pointed distally, this is the *post-clavicle* (fig. 13, 6).

The **scapula** (fig. 13, 5) is a small irregular plate of bone attached to the inner side of the middle of the *clavicle*. The **coracoid** is a larger plate of similar character, irregularly triangular in shape, attached to the inner side of the clavicle immediately below the scapula. The scapula and coracoid bear the pectoral fin.

The Pectoral fins.

Each of these consists of four small irregular bones, the **brachial ossicles** (fig. 13, 7), bearing a series of about nineteen dermal *fin-rays*. The brachial ossicles represent the reduced and modified radiale and basalia of cartilaginous fish such as the dogfish. The fin-rays (fig. 13, 8) which form the whole external portion of the fin are long slender rods having essentially the same character as those of the unpaired fins.

The Pelvic girdle.

The **pelvic girdle** in the Cod as in other Teleosteans is entirely absent, its place being taken by the enlarged basi-pterygia of the fins.

The Pelvic fins.

These have a very anomalous position in the Cod, being attached to the throat in front of the pectoral girdle. Each consists of a basal portion, the **basi-pterygium**, and of a number of dermal rays. The basi-pterygium consists of an expanded ventral portion which meets its fellow below in the middle line, and to which the rays are attached, and of an inwardly-directed dorsal portion which also meets its fellow and is imbedded in the flesh. The rays are six in number and are long slender structures similar to those of the other fins.

CHAPTER VIII.
GENERAL ACCOUNT OF THE SKELETON IN FISHES.

EXOSKELETON.

The most primitive type of exoskeleton is that found in Elasmobranchs and formed of **placoid** scales; these are tooth-like structures consisting of dentine and bone capped with enamel, and have been already described (p. 4). In most Elasmobranchs they are small and their distribution is fairly uniform, but in the Thornback skate, *Raia clavata*, they have the form of larger, more scattered spines. In adult Holocephali and in *Polyodon* and *Torpedo* there is no exoskeleton, in young Holocephali, however, there are a few small dorsal ossifications.

The plates or scales of many Ganoids may have been formed by the gradual fusion of elements similar to these placoid scales, and often bear a number of little tooth-like processes. In *Lepidosteus*, *Polypterus*, and many extinct species, these *ganoid* scales, which are rhomboidal in form and united to one another by a peg and socket articulation, enclose the body in a complete armour. In *Trissolepis* part of the tail is covered by rhomboidal scales, while rounded scales cover the trunk and remainder of the tail. *Acipenser* and *Scaphirhynchus* have large dermal bony plates which are not rhomboidal in shape and do not cover the whole body. In *Acipenser* a single row extends along the middle of the back and two along each side.

The majority of Teleosteans have thin flattened scales which differ from those of Ganoids in being entirely mesodermal in origin, containing no enamel. There are two principal types of Teleostean scales, the cycloid and ctenoid. A **cycloid** scale is a flat thin scale with concentric markings and an entire posterior margin. A **ctenoid** scale differs in having its posterior margin pectinate. The Dipnoi have overlapping cycloid scales. The rounded scales of *Amia* and of many fossil ganoids such as *Holoptychius* are shaped like cycloid scales, but differ from them in being more or less coated with enamel. In Eels and some other Teleosteans the scales are completely degenerate and have almost disappeared. Some Teleosteans, like *Diodon hystrix*, have scales with triradiate roots from which arise long sharp spines directed backwards. These scales, which resemble teeth, contain no enamel; they become erect when the fish inflates its body into a globular form. Many Siluroids have dermal armour in the form of large bony plates which are confined to the anterior part of the body. In *Ostracion* the whole body is covered by hexagonal plates, closely united together.

The **fin-rays** are structures of dermal origin which entirely or partially support the unpaired fins, and assist the bony or cartilaginous endoskeleton in the support of the paired fins.

In Elasmobranchs, Dipnoi, and Chondrosteous ganoids the skeletons of the fins are, as a rule, about half of exoskeletal, half of endoskeletal origin, the proximal and inner portion being cartilaginous and endoskeletal, the distal and outer portion being exoskeletal, and consisting of horny or of more or less calcified fin-rays. In bony Ganoids and Teleosteans the endoskeletal parts are greatly reduced and the fins come to consist mainly of the fin-rays, which are ossified and frequently become flattened at their distal ends.

The fin-rays of the ventral part of the caudal fin are carried by the haemal arches; those of the dorsal and anal fins and of the dorsal part of the caudal fin generally by interspinous bones, which in adult Teleosteans alternate with the neural and haemal spines. In Dipnoi these interspinous bones articulate with the neural and haemal spines. In many Siluroids the anterior rays of the dorsal and pectoral fins are developed into large spines which often articulate with the endoskeleton, or are sometimes fused with the dermal armour plates. Similar spines may occur in Ganoids in front of both the dorsal and anal fins.

Polypterus has a small spine or *fulcrum* in front of each segment of the dorsal fin. Such spines are often found fossilised, and are known as *ichthyodorulites*.

Similar spines are found in many Elasmobranchs, but they are simply inserted in the flesh, not articulated to the endoskeleton. They also differ from the spines of Teleosteans and Ganoids in the fact that they are covered with enamel, and often have their edges serrated like teeth. In the extinct Acanthodii they generally occur in front of all the fins, paired and unpaired.

In *Trygon*, the Sting-ray, the tail bears a serrated spine which is used for purposes of offence and defence. Many ichthyodorulites may have been spines of this nature fixed to the tail, rather than spines situated in front of the fins. The spines, which are always found in front of the dorsal fin in Holocephali, agree with those of Elasmobranchs in containing enamel, and with those of Teleosteans in being articulated to the endoskeleton.

Teeth.

The teeth of fish are subject to a very large amount of variation, perhaps to more variation than are those of any other class of animals. Sometimes, as in adult Sturgeons, they are entirely absent, sometimes they are found on all the bones of the mouth, and also on the hyoid and branchial arches. The teeth are all originally developed in the mucous membrane of the mouth, but they afterwards generally become attached to firmer structures, especially to the jaws. In Elasmobranchs, however, they are generally simply imbedded in the tough fibrous integument of the mouth. Their attachment to the jaws may take place in three different ways.

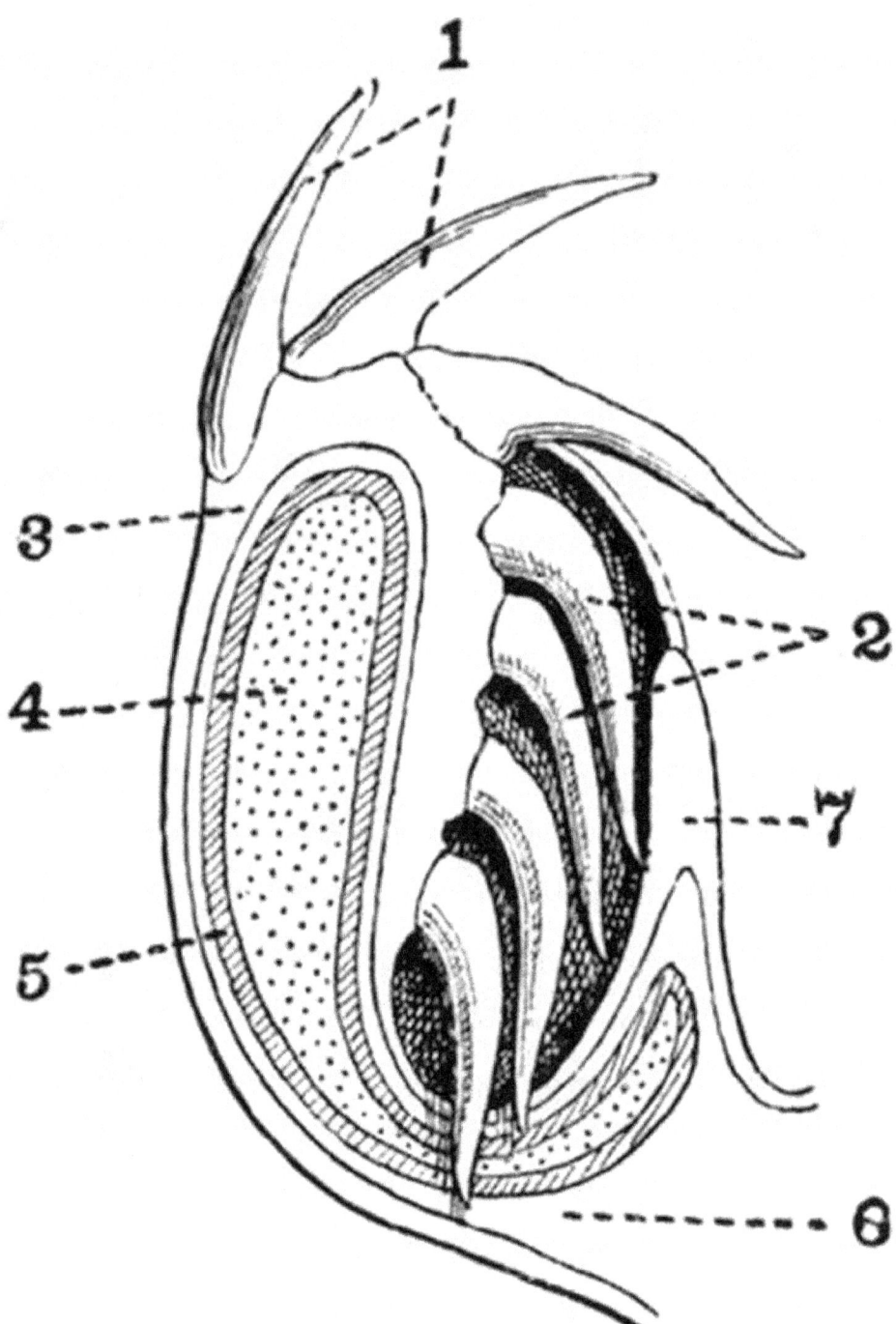

Fig. 14. Diagram of a section through the jaw of a Shark (*Odontaspis americanus*) showing the succession of teeth (Brit. Mus. from specimen and diagram).

1. teeth in use.
2. teeth in reserve.
3. skin.
4. cartilage of the jaw.
5. encrusting calcification of cartilage.
6. connective tissue.
7. mucous membrane of the mouth.

(1) By an elastic hinge-joint, as in the Angler (*Lophius*), and the Pike (*Esox lucius*). In the Angler the tooth is held by a fibrous band attaching its posterior end to the subjacent bone, in the Pike by uncalcified elastic rods in the pulp cavity.

(2) By ankylosis, i.e. by the complete union of the calcified tooth substance with the subjacent bone. This is the commonest method among fish.

(3) By implantation in sockets. This method is not very common among fish. The teeth are sometimes, as in *Lepidosteus*, ankylosed to the base of the socket. In this genus there is along each ramus of the mandible a median row of large teeth placed in perfect sockets, and two irregular lateral rows of small teeth ankylosed to the jaw.

Dentine, enamel and cement are all represented in the teeth of fishes, but the enamel is generally very thin, and cement is but rarely developed. Dentine forms the main bulk of the teeth; it is sometimes of the normal type, but generally differs from that in higher vertebrates in being vascular, and is known as *vasodentine*. A third type occurs, known as *osteodentine*; it is traversed by canals occupied by marrow, and is closely allied to bone.

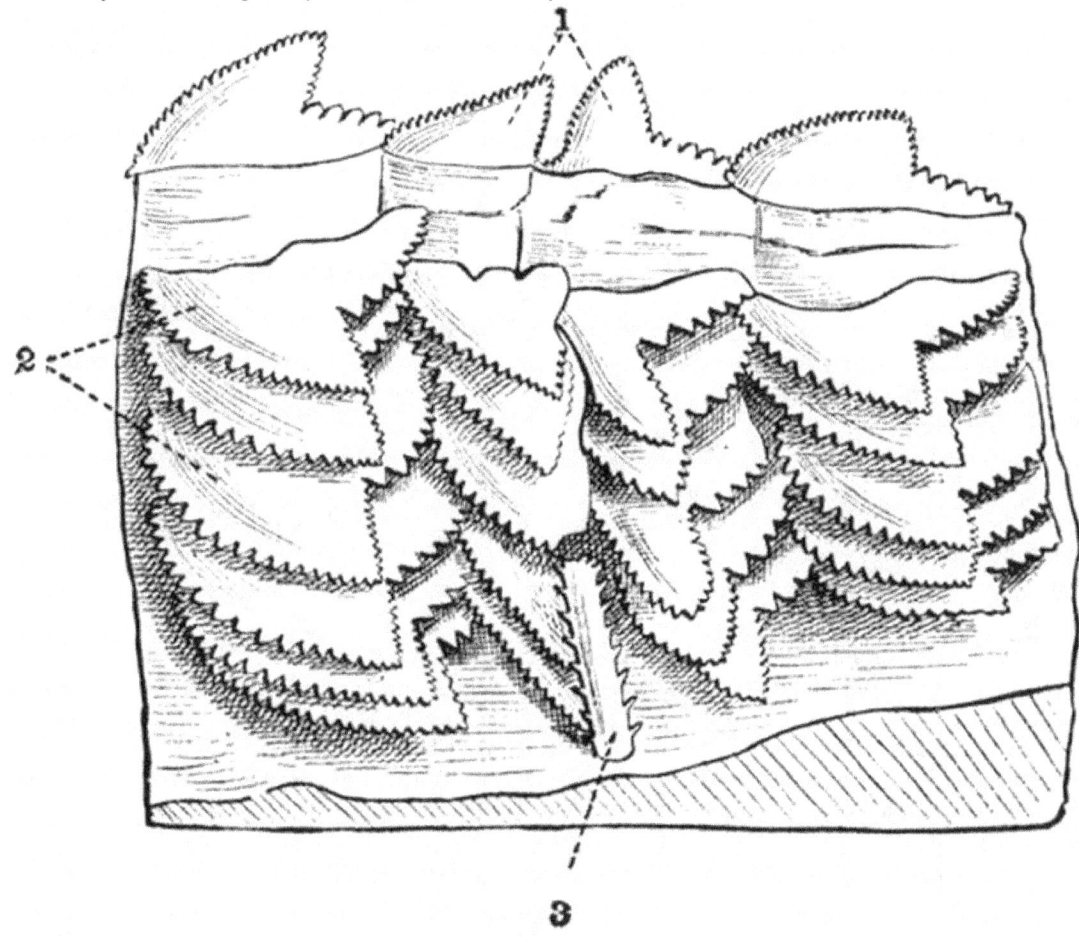

Fig. 15. Part of the lower jaw of a Shark (*Galeus*)
(from Owen after André).

1. teeth in use.

2. reserve teeth folded back.

3. part of the caudal spine of

a Sting-ray (*Trygon*) which has
pierced the jaw and affected the
growth of the teeth.

The teeth are generally continually renewed throughout life, but sometimes one set persists.

The teeth of Selachii are fundamentally identical with placoid scales. They are developed from a layer of dental germs which occurs all over the surface of the skin, except in the region of the lips. At this point the layer of tooth-producing germs extends back into the mouth, being projected by a fold of the mucous membrane (fig. 14, 7). Here new teeth are successively formed, and as they grow each is gradually brought into a position to take the place of its predecessor by the shifting outwards of the gum over the jaw. Owing to this arrangement sharks have practically an unlimited supply of teeth (figs. 14 and 15).

Two principal types of teeth are found in Elasmobranchs. In Sharks and Dogfish, on the one hand, the teeth are very numerous, simple, and sharp-pointed, and are with or without serrations and lateral cusps. Many Rays and fossil Elasmobranchs, on the other hand, have broad flattened teeth adapted for crushing shells. Intermediate conditions occur between these two extremes. Thus in *Cestracion* and many extinct sharks, such as *Acrodus*, while the median teeth are sharp, the lateral teeth are more or less flattened and adapted for crushing. In various species belonging to the genus *Raia* the teeth of the male are sharp, while those of the female are blunt. A very specialised dentition is met with in the Eagle-rays (Myliobatidae), in which the jaws are armed with flattened angular tooth-plates, arranged in seven rows, forming a compact pavement; the plates of the middle row are very wide and rectangular, those of the other rows are much smaller and hexagonal. Lastly, in *Cochliodus* the individual crushing teeth are fused, forming two pairs of spirally-coiled dental plates on each side of each jaw. *Pristis*, the Saw-fish, has a long flat cartilaginous snout, bearing a double row of persistently-growing teeth planted in sockets along its sides. Each tooth consists of a number of parallel dentinal columns, united at the base, but elsewhere distinct.

In the Holocephali—*Chimaera, Hariotta* and *Callorhynchus*—only three pairs of teeth or dental plates occur, two pairs in the upper jaw, one in the lower. These structures persist throughout life and grow continuously. The upper tooth structures are attached respectively to the ethmoid or vomerine region of the skull, and to the palato-pterygoids. The vomerine teeth are small, while those attached to the mandible and the palato-pterygoid region are large and bear several roughened ridges adapted for grinding food. The teeth of the two opposite sides of the jaw meet in a median symphysis. The teeth of *Chimaera* are more adapted for cutting, those of *Callorhynchus* for crushing. Many extinct forms are known, some of whose teeth are intermediate in structure between those of *Chimaera* and *Callorhynchus*.

The teeth of Ganoids are also extremely variable. Among living forms, the Holostei are more richly provided with teeth than are any other fishes, as they may occur on the premaxillae, maxillae, palatines, pterygoids, parasphenoid, vomers, dentaries, and splenials. Among the Chondrostei, on the other hand, the adult Acipenseridae are toothless; small teeth however occur in the larval sturgeon, and in *Polyodon* many small teeth are found attached merely to the mucous membrane of the jaws. Many fossil Ganoids have numerous flattened or knob-like teeth, borne on the maxillae, palatines, vomers and dentaries. Others have a distinctly heterodont dentition. Thus in *Lepidotus* the premaxillae bear chisel-like teeth, while knob-like teeth occur on the maxillae, palatines and vomers. In *Rhizodus* all the teeth are pointed, but while the majority are small a few very large ones are interspersed.

In Teleosteans, too, the teeth are eminently variable both in form and mode of arrangement. They may be simple and isolated, or compound, and may be borne on almost any of the bones bounding the mouth cavity, and also as in the Pike, on the hyoid and branchial arches. The splenial however never bears teeth and the pterygoid and parasphenoid only rarely, thus differing from the arrangement in the Holostei.

The isolated teeth are generally conical in form and are ankylosed to the bone that bears them. Such teeth are, with a few exceptions such as *Balistes*, not imbedded in sockets nor replaced vertically.

In some fish beak-like structures occur, formed partly of teeth, partly of the underlying jaw bones. These beaks are of two kinds: (1) In *Scarus*, the parrot fish, the premaxillae and dentaries bear numerous small, separately developed teeth, which are closely packed together and attached by their proximal ends to the bone, while their distal ends form a mosaic. Not only the teeth but the jaws which bear them are gradually worn away at the margins, while both grow continuously along their attached edge. (2) In Gymnodonts, e.g. *Diodon*, the beaks are formed by the coalescence of broad calcified horizontal plates, which when young are free and separated from one another by a considerable interval.

In some Teleosteans the differentiation of the teeth into biting teeth and crushing teeth is as complete as in *Lepidosteus*. Thus in the Wrasse (*Labrus*), the jaws bear conical slightly recurved teeth arranged in one or two rows, with some of the anterior ones much larger than the rest. The bones of the palate are toothless, while both upper and lower pharyngeal bones are paved with knob-like crushing teeth; such pharyngeal teeth occur also in the Carp but are attached only to the lower pharyngeal bone, the jaw bones proper being toothless.

In Dipnoi the arrangement of the teeth is very similar to that in Holocephali. The mandible bears a single pair of grinding teeth attached to the splenials, and a corresponding pair occur on the palato-pterygoids. In front of these there are a pair of small conical vomerine teeth loosely attached to the ethmoid cartilage. The palato-pterygoid teeth of *Ceratodus* are roughly semicircular in shape with a smooth convex inner border, and an outer border bearing a number of strongly marked ridges. The teeth of the extinct Dipteridae resemble those of *Ceratodus* but are more complicated.

ENDOSKELETON.

Spinal column.

The spinal column of fishes is divisible into only two regions, a caudal region in which the haemal arches or ribs meet one another ventrally, and a precaudal region in which they do not meet.

The various modifications of the spinal column in fishes can be best understood by comparing them with the arrangement in the simplest type known, namely *Amphioxus*. In *Amphioxus* the notochord is immediately surrounded by a structureless cuticular layer, the *chordal sheath*. Outside this is the *skeletogenous layer*, which in addition to surrounding the notochord and chordal sheath embraces the nerve cord dorsally, and laterally sends out septa forming the *myomeres*.

The Cartilaginous ganoids *Acipenser, Polyodon* and *Scaphirhynchus* are the simplest fishes as regards their spinal column. The notochord remains permanently unconstricted and is enclosed in a chordal sheath, external to which is the skeletogenous layer. In this layer the development of cartilaginous elements has taken place. In connection with each *neuromere*, or segment as determined by the points of exit of the spinal nerves, there are developed two pairs of ventral cartilages, the ventral arches (basiventralia) and intercalary pieces (interventralia); and at least two pairs of dorsal pieces, the neural arches (basidorsalia) and intercalary pieces (interdorsalia). The lateral parts of the skeletogenous layer do not become converted into cartilage, so there are no traces of vertebral centra. The ventral or haemal arches meet one another ventrally and send out processes to protect the ventral vessels. The neural arches do not meet, but are united by a longitudinal elastic band.

In Cartilaginous ganoids the only indications of metameric segmentation are found in the neural and haemal arches. The case is somewhat similar with the Holocephali and Dipnoi.

In the Holocephali the notochord grows persistently throughout life, and is of uniform diameter throughout the whole body except in the cervical region and in the gradually tapering tail. The chordal sheath is very thick and includes a well-marked zone of calcification which separates an outer zone of hyaline cartilage from an inner zone. There are also a number of cartilaginous pieces derived from the skeletogenous layer which are arranged in two series, a dorsal series forming the neural arches and a ventral series forming the haemal arches. These do not, except in the cervical region, meet one another laterally round the notochord and form centra. To each neuromere there occur a pair of basidorsals, a pair of interdorsals, and one or two supradorsals. In the tail the arrangement is irregular.

In the Dipnoi as in the Holocephali the notochord grows persistently and uniformly, and the chordal sheath is thick and cartilaginous though there are no metamerically arranged centra. The neural and haemal arches and spines are cartilaginous and interbasalia (intercalary pieces) are present. The basidorsalia and basiventralia do not in *Ceratodus* meet round the notochord and enclose it except in the anterior part of the cervical and posterior part of the caudal region.

In Elasmobranchii the chordal sheath is weak and the skeletogenous layer strong. Biconcave cartilaginous vertebrae are developed, and as is the case in most fishes, constrict the notochord *vertebrally*.

Two distinct types of vertebral column can be distinguished in Elasmobranchs:

1. In many extinct forms and in the living Notidanidae, *Cestracion*, and *Squatina*, the dorsal and ventral arches do not meet one another laterally round the centrum, and consequently readily come away from it.

2. In most living Elasmobranchs the arches meet laterally round the centrum.

The vertebrae are never ossified but endochondral calcification nearly always takes place, though it very rarely reaches the outer surface of the vertebrae. Elasmobranchs are sometimes subdivided into three groups according to the method in which this calcification takes place:

1. **Cyclospondyli** (*Scymnus, Acanthias*), in which the calcified matter is deposited as one ring in each vertebra.

2. **Tectospondyli** (*Squatina, Raia, Trygon*), in which there are several concentric rings of calcification.

3. **Asterospondyli** (Notidanidae, *Scyllium, Cestracion*), in which the calcified material instead of forming one simple ring, extends out in a more or less star-shaped manner.

In *Heptanchus* the length of the vertebral centra in the middle of the trunk is double that in the anterior and posterior portions, and as the length of the arches does not vary, the long centra carry more of them than do the short centra.

In many Rays the skull articulates with the vertebral column by distinct occipital condyles.

In Bony Ganoids the skeletogenous layer becomes calcified ectochondrally in such a way that the notochord is pinched in at intervals, and distinct vertebrae are produced. Ossification of the calcified cartilage rapidly follows. In *Amia* the vertebrae are biconcave, in *Lepidosteus* they are opisthocoelous, cup and ball joints being developed between the vertebrae in a manner unique among fishes. The notochord entirely disappears in the adult *Lepidosteus*, but at one stage in larval life it is expanded vertebrally and constricted intervertebrally in the manner usual in the higher vertebrata, but unknown elsewhere among fishes.

The tail of *Amia* is remarkable from the fact that as a rule to each neuromere, as determined by the exit of the spinal nerves, there are two centra, a posterior one which bears nothing, and an anterior one which bears the neural and haemal arches, these being throughout the vertebral column connected with the centra by cartilaginous discs.

In most Teleosteans but not in the Plectognathi the neural arches are continuous with the centra, which are nearly always deeply biconcave.

In some cases many of the anterior vertebrae are ankylosed together and to the skull. The vertebrae often articulate with one another by means of obliquely placed flattened surfaces, the zygapophyses. The centrum in early stages of development is partially cartilaginous, but the neural arches and spines in the trunk at any rate, pass directly from the membranous to the osseous condition.

Fins.

The most primitive fins are undoubtedly the unpaired ones, which probably originally arose as ridges or folds of skin along the mid-dorsal line of the body, and passed thence round the posterior end on to the ventral surface, partially corresponding in position and function to the keel of a ship.

In long 'fish' which pass through the water with an undulating motion such simple continuous fins may be the only ones found, as in *Myxine*. To support these median fins skeletal structures came to be developed; these show two very distinct forms, viz. cartilaginous endoskeletal pieces, the *radiale*, and horny exoskeletal fibres, the *fin-rays*. Mechanical reasons caused the fin to become concentrated at certain points and reduced at intervening regions. Thus a terminal caudal fin arose and became the chief organ of propulsion, and the dorsal and ventral fins became specialised to act as balancing organs.

In some of the earlier Elasmobranchs, the Pleuracanthidae, the endoskeletal cartilaginous radiale are directly continuous with outgrowths from the dorsal and ventral arches of the vertebrae, and form the main part of the fin. In later types of Elasmobranchs the horny exoskeletal fin-rays have comparatively greater prominence. In bony fish, as has been already stated, the horny fibres are replaced by bony rays of dermal origin, and at the same time complete reduction and disappearance of the cartilaginous radiale takes place.

The Caudal fin.

The caudal region of the spinal column in fishes is of special importance. It is distinctly marked off from the rest of the spinal column by the fact that the ventral or haemal arches meet one another and are commonly prolonged into spines, while in the trunk region they do not meet but commonly diverge from one another.

In some fish the terminal part of the caudal region of the spinal column retains the same direction as the rest of the spinal column. The blade of the caudal fin is then divided into two nearly equal portions, and is said to be **diphycercal**. This condition is generally regarded as the most primitive one; it occurs in the Ichthyotomi, Holocephali, all living Dipnoi, *Polypterus* and some extinct Crossopterygii, and a few Selachii and Teleostei. It occurs also in deep-sea fish belonging to almost every group, and under these conditions obviously cannot be regarded as primitive, but must be looked on as a feature induced by the peculiar conditions of life.

In the great majority of fish the terminal part of the caudal region of the spinal column is bent dorsalwards, and the part of the blade of the caudal fin which arises on the dorsal surface is much smaller than is that arising on the ventral surface. Such a fin is said to be **heterocercal**.

Strictly speaking all fish whose tails are not diphycercal have heterocercal tails, but the term is commonly applied to two-bladed tails in which the spinal column forms a definite axis running through the dorsal blade, while the ventral blade is enlarged and generally forms the functional part of the tail. Such heterocercal tails are found in nearly all Elasmobranchii, together with the living cartilaginous Ganoidei, and many extinct forms belonging to the same order; *Lepidosteus*, *Amia*, and the Dipteridae among Dipnoi, have tails which, though obviously heterocercal, are not two-bladed.

The vast majority of the Teleostei and some extinct Ganoidei have heterocercal tails of the modified type to which the term **homocercal** is applied. The hypural bones which support the lower half of the tail fin become much enlarged, and frequently unite to form a wedge-shaped bone which becomes ankylosed to the last ossified vertebral centrum. The fin-rays then become arranged in such a way as to produce a secondary appearance of symmetry. Some homocercal fish such as the Perch have the end of the notochord protected by a calcified or completely ossified sheath, the **urostyle**, to which several neural and haemal arches may be attached, and which becomes united with the centrum of the last vertebra; in others such as the Salmon the end of the notochord is protected only by laterally placed bony plates.

The Skull.

It is often impossible to draw a hard and fast line between the cranium and the vertebral column. This is the case for instance in *Acipenser* (fig. 18, 16) among Chondrostei, in *Amia* among Holostei, and in *Ceratodus* and *Protopterus* among Dipnoi. The occipital region of the skull in *Amia* is clearly formed of three cervical vertebrae whose centra have become absorbed into the cranium, while the neural arches and spines are still distinguishable.

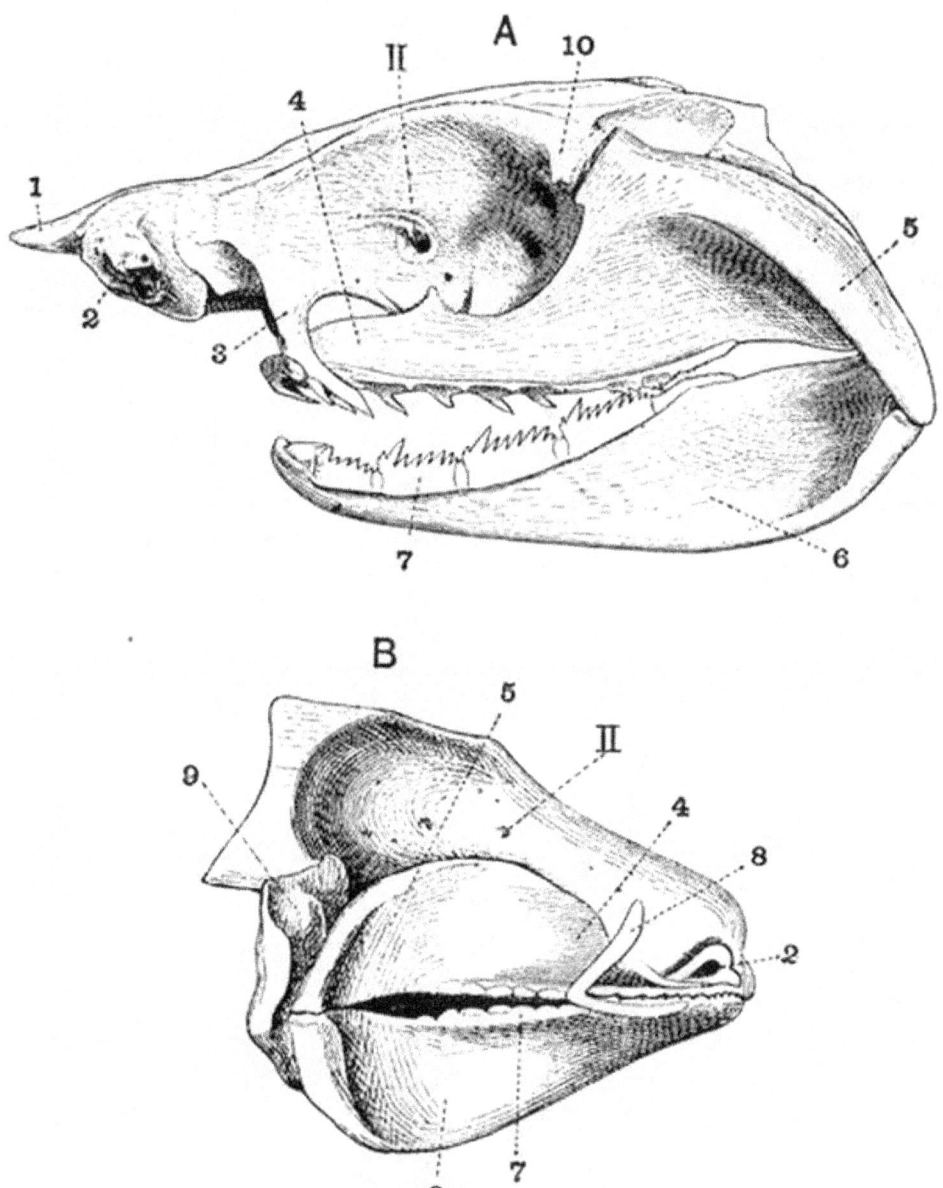

Fig. 16. A. Skull of *Notidanus* × ½ (Brit. Mus.). B. Skull of *Cestracion* × 1/3 (after Gegenbaur). In neither case are the branchial arches shown.

1. rostrum.
2. olfactory capsule.
3. ethmo-palatine process.
4. palatine portion of
palato-pterygo-quadrate bar.
5. quadrate portion of bar.

6. Meckel's cartilage.
7. teeth.
8. labial cartilage.
9. hyomandibular.
10. postorbital process.
II. optic foramen.

The simplest type of cranium is that found in Elasmobranchs: it consists of a simple cartilaginous box, which is generally immovably fixed to the vertebral column, though in some forms, like *Scymnus* and *Galeus*, a joint is indicated, and in others, such as the Rays, one is fairly well developed. The cranium in Elasmobranchs is never bony, though the cartilage is sometimes calcified. It is drawn out laterally into an antorbital process in front of the eye, and a postorbital process behind it. The nasal capsules are always cartilaginous, and the eye, as a general rule, has a cartilaginous sclerotic investment. The cranium is often prolonged in front into a rostrum which is enormously developed in *Pristis* and some Rays. The cartilaginous roof of the cranium is rendered incomplete by the presence of a large hole, the anterior fontanelle.

Two pairs of labial cartilages (fig. 16, B, 8) are often present. They lie imbedded in the cheeks outside the anterior region of the jaws, and are specially large in *Squatina*.

As regards the visceral arches the simplest and most primitive condition of the jaws is that of the Notidanidae, in which the mandibular and hyoid arches are entirely separate. In these primitive fishes the palato-pterygo-quadrate bar articulates with the postorbital process (fig. 16, 10), while further forwards it is united to the cranium by the ethmo-palatine ligament. The hyoid arch is small and is broadly overlapped by the mandibular arch. The term **autostylic** is used to describe this condition of the suspensorium. From this condition we pass in the one direction to that of *Cestracion* (fig. 16, B), in which the whole of the palato-pterygo-quadrate bar has become bound to the cranium, and in the other to that of *Scyllium*. In *Scyllium* (fig. 6), while the ethmo-palatine ligament is retained, the postorbital articulation of the palato-pterygo-quadrate has been given up, so that the palato-pterygo-quadrate comes to abut on the hyomandibular and is attached to it by ligaments. The pre-spiracular ligament (fig. 16, 20) running from the auditory capsule also assists in supporting the jaws.

Lastly we come to the purely **hyostylic** condition met with in Rays, in which the mandibular arch is entirely supported by the hyomandibular. In some Rays the hyoid is attached to the posterior face of the hyomandibular near its proximal end, and may even come to articulate with the cranium.

The **visceral arches of Elasmobranchs** may be summarised as follows:—

1. The **mandibular arch**, consisting of a much reduced dorsal portion, the pre-spiracular ligament, and a greatly developed ventral portion from which both upper and lower jaws are derived. The mandible (Meckel's cartilage) is the original lower member of the mandibular arch, and from it arises an outgrowth which forms the upper jaw or palato-pterygo-quadrate bar. In *Scymnus* this bears a few branchiostegal rays.

2. The **hyoid arch**, which consists of the hyomandibular and the hyoid, and bears branchiostegal rays on its posterior face.

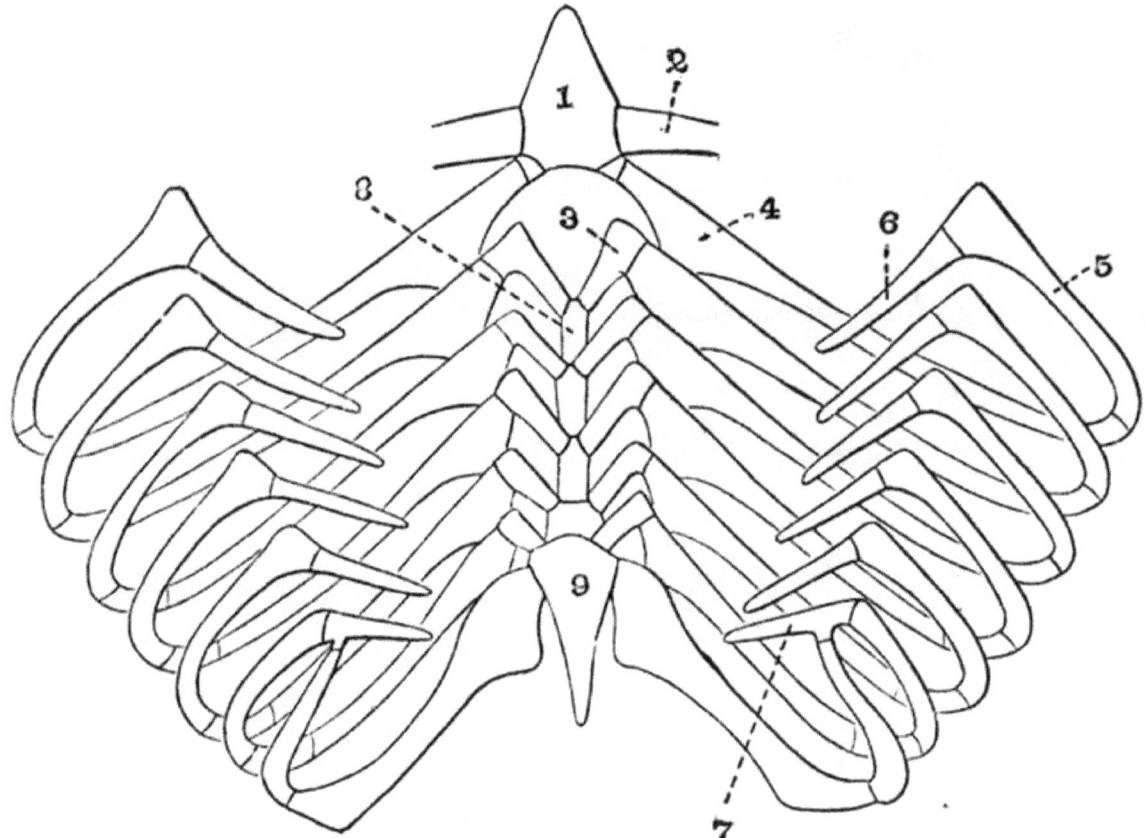

Fig. 17. Dorsal view of the Branchial arches of *Heptanchus*. (From Gegenbaur).

1. basi-hyal.
2. cerato-hyal.
3. second hypo-branchial.
4. first cerato-branchial.

7. pharyngo-branchial, common to the sixth and seventh arches.
8. basibranchial of second arch.

60

5. first epi-branchial.

6. first pharyngo-branchial.

9. basibranchial, common to the
sixth and seventh arches.

3. The **branchial arches**, generally five in number, all of which except the last bear branchiostegal rays. In the Notidanidae the number of branchial arches is increased beyond the normal series, thus in *Hexanchus* there are six, and in *Heptanchus* seven. There are six also in *Chlamydoselache* and *Protopterus*.

4. The so-called external branchial arches which are cartilaginous rods attached to all the visceral arches. They are especially large in *Cestracion*.

The skull in Holocephali is entirely cartilaginous. The palato-pterygo-quadrate bar is fixed to the cranium, and to it the mandible articulates. There is a well-marked joint between the skull and the spinal column.

In living Cartilaginous Ganoids the primitive cartilaginous cranium is very massive, and is greatly prolonged anteriorly, while posteriorly it merges into the spinal column. Although it is mainly cartilaginous a number of ossifications take place in the skull, and membrane bones are now found definitely developed, especially in connection with the roof of the cranium. In *Acipenser* (fig. 18) the ossifications in the cartilage include the pro-otic, which is pierced by the foramen for the fifth nerve, the alisphenoid, orbitosphenoid, ectethmoid, palatine, pterygoid, meso-pterygoid, hyomandibular (fig. 18, 11), cerato-hyal, all the cerato-branchials, and the first two epi-branchials. Most of these structures are, however, partly cartilaginous, though they include an ossified area. The membrane bones too of *Acipenser* are very well developed, they include a bone occupying the position of the supra-occipital, and form a complete dorsal cephalic shield. Resting on the ventral surface are a vomer and a very large parasphenoid (fig. 18, 3). There is a bony operculum attached to the hyomandibular, and membrane bones representing respectively the maxilla and dentary are attached to the jaws. The suspensorium is most markedly hyostylic. The palato-pterygo-quadrate bar has a very curious shape and is quite separate from the cranium. It is connected to the hyomandibular by a thick symplectic ligament containing a small bone homologous with the symplectic of Teleosteans.

Polyodon differs much from *Acipenser*, the membrane bones not being so well developed though they cover the great cartilaginous snout.

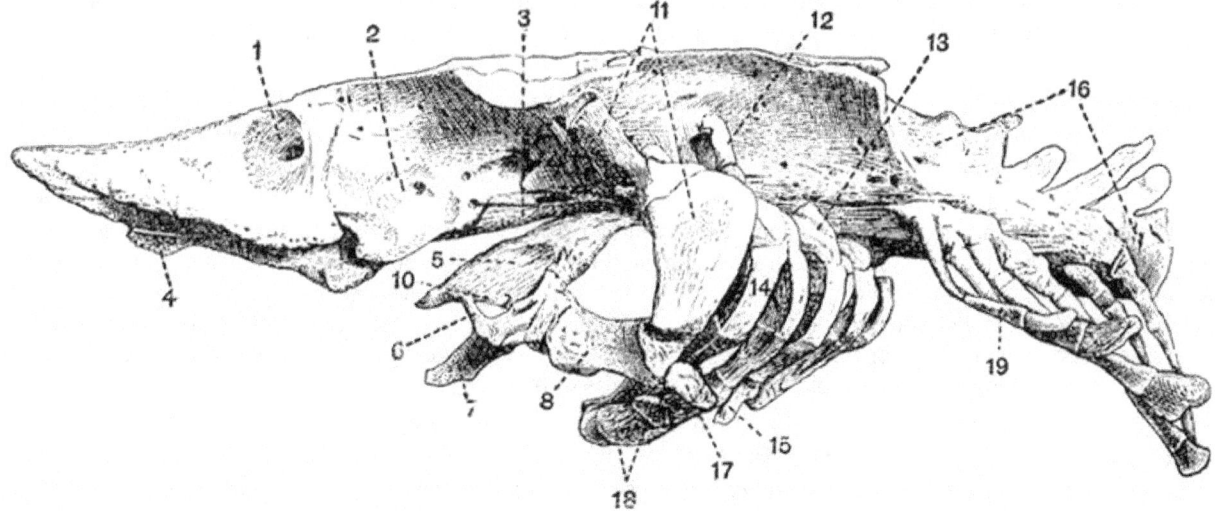

Fig. 18. Lateral view of the skull of a Sturgeon (*Acipenser sturio*). Nearly all the membrane bones have been removed (Brit. Mus.).

1. nasal cavity.

2. orbit.

3. parasphenoid.

4. vomer.

5. pterygoid.

6. maxilla. (The dotted line
running from 6 passes
into the mouth cavity.)

7. dentary.

8. symplectic.

10. palatine.

11. hyomandibular.

12. pharyngo-branchial.

13. epi-branchial.

14. cerato-branchial.

15. hypo-branchial.

16. coalesced anterior vertebrae.

17. inter-hyal.

18. cerato-hyal.

19. rib.

The skull in *Polypterus* (Crossopterygii) shows a great advance towards the condition met with in Teleostei. The cranium remains to a great extent unossified, and large dorsal and ventral fontanelles pierce its walls. It is covered by a great development of membrane bones, paired nasals, frontals, parietals, supra- and post-temporals, and dermo-supra-occipitals among others being present. The palato-pterygo-quadrate bar is fused to the cranium, and in connection with it the following paired membrane bones appear, palatine, ecto-, meso- and meta-pterygoid, and further forwards jugal, vomer, maxilla and premaxillae. The membrane bones developed in connection with each ramus of the mandible are the dentary, angular, and splenial, in addition to the cartilage bone the articular. Several large opercular bones occur. There are also a pair of large jugular or gular plates, and several large opercular bones.

In Bony Ganoids both cartilage bone and membrane bone is well developed. The pro-otics and exoccipitals are well ossified, but the supra-occipital and pterotics are not. Lateral ethmoids are developed, and there are ossifications in the sphenoidal region which vary in different forms. The place of the cartilaginous palato-pterygo-quadrate is taken by a series of bones, the quadrate behind and the palatine, ecto-, meso- and meta-pterygoids in front. In *Lepidosteus*, however, the palatine and pterygoid are membrane bones, as they are in *Polypterus* and the Frog. Paired maxillae, premaxillae, vomers and a parasphenoid occur forming the upper jaw and roof of the mouth, and a series of membrane bones are found investing the mandible and forming the operculum.

In *Amia* membrane bones are as freely developed as they are in Teleosteans; they include on each side a squamosal, four opercular bones, a lachrymal, a pre-orbital, one or two suborbitals, two large postorbitals and a supratemporal; while investing the mandible, besides the dentary, splenial, angular, and supra-angular, there is an unpaired jugular. The articular too is double and a mento-meckelian occurs. In *Amia* teeth are borne on the premaxillae, maxillae, vomers, palatines and pterygoids.

Bony Ganoids are the lowest animals in which squamosal bones are found, and they do not occur in Teleosteans.

The suspensorium in bony Ganoids, as in the Chondrostei, is hyostylic, and there are two ossifications in the hyomandibular cartilage, viz. the hyomandibular, and the symplectic.

The skull of Teleostei is very similar to those of *Lepidosteus* and *Amia*. Although the bony skull is greatly developed and very complicated, much of the original cartilaginous cranium often persists. Membrane bones are specially developed on the roof of the skull where they include the parietal, frontal, and nasal bones. The same bones are developed in connection with the upper jaw and roof of the mouth as in bony Ganoids, but only two membrane bones occur in the lower jaw, viz. the angular and dentary. A number of large ossifications take place in the cartilage of the auditory capsules. In some forms parts of the last pair of branchial arches are broadened out and form the pharyngeal bones which bear teeth. The opercular bones and those of the upper and lower jaws are quite comparable to those of bony Ganoids.

A full account of the Teleostean skull has been given in the case of the Salmon (pp. 87-96) and the Cod (pp. 96-101).

In Dipnoi the skull is mainly cartilaginous, but both cartilage- and membrane-bone occur also. Cartilage-bone is found in the ossified exoccipitals, while of membrane-bones *Protopterus* has among unpaired bones a fronto-parietal, a median ethmoid, and a parasphenoid, and among paired bones nasals and large supra-orbitals. The skull of *Ceratodus* (fig. 19) has an almost complete roof of membrane bones, including some whose homology is doubtful. The ethmo-vomerine region is always cartilaginous, but bears small teeth. The palato-pterygo-quadrate bar is ossified and firmly united to the cranium, and the mandible articulates directly with it (autostylic). Membrane bones are freely developed in connection with the mandible, dentary, splenial, and angular bones being all present. There are two opercular bones.

In the extinct Dipteridae the cranium is very completely covered with plates of dermal bone, and the skeleton in general is more ossified than is the case in recent Dipnoi.

Six pairs of branchial arches occur in *Protopterus*; *Ceratodus* and *Lepidosiren* have five, like most other fish. The branchial arches bear gill rakers.

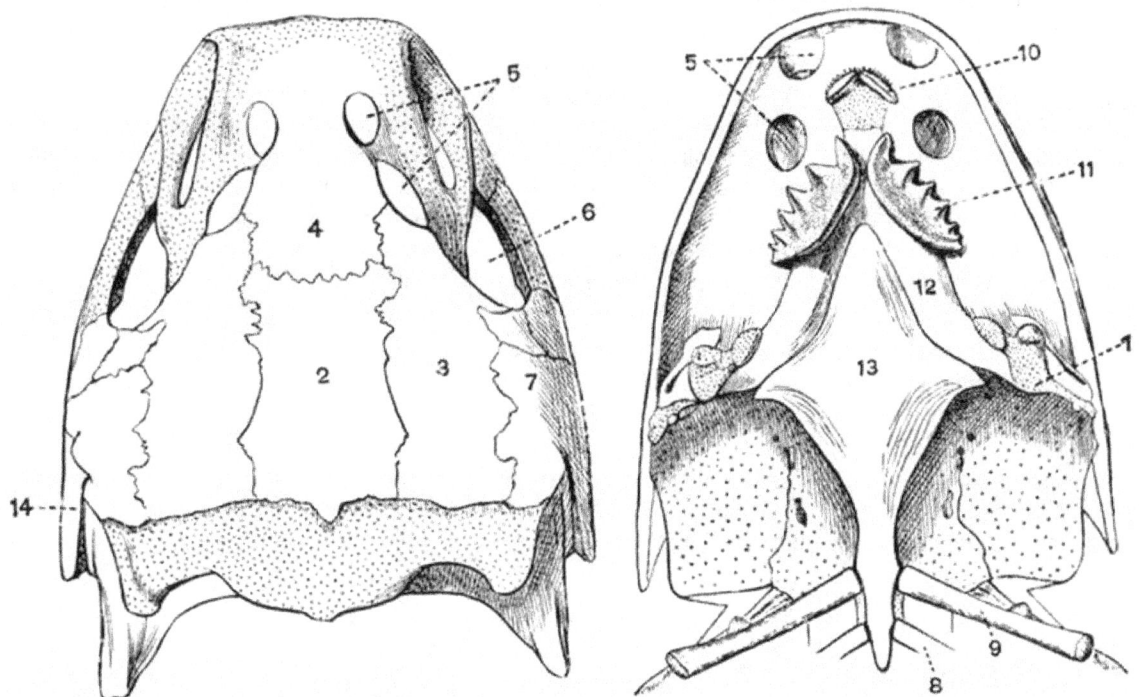

Fig. 19. Dorsal (to the left) and ventral (to the right) views of the cranium of *Ceratodus miolepis* (after Günther).

1. cartilaginous part of the quadrate with which the mandible articulates.
2. scleroparietal.
3. frontal.
4. ethmoid.
5. nares.
6. orbit.
7. pre-opercular (squamosal).
8. second rib.
9. first rib.
10. vomerine tooth.
11. palato-pterygoid tooth.
12. palato-pterygoid.
13. parasphenoid.
14. interopercular.

Ribs.

As has been already mentioned (p. 24), although ribs commonly appear to be the cut-off ends of the transverse processes, they are really elements derived from the ventral or haemal arch.

In Elasmobranchii and other cartilaginous fish they have the form of small cartilaginous structures imperfectly separated from the diverging halves of the ventral arch, and are often absent.

In Teleostei and bony Ganoids they often have different attachments in different parts of the body. In the tail region they are not differentiated from the two halves of the ventral arch, which meet in the middle line, and are prolonged into a haemal spine. In the posterior trunk region they sometimes form distinct processes diverging from the two halves of the ventral arch; while further forward they may shift their attachment so as to arise from the dorsal side of the two halves of the ventral arch and at some distance from their ends, which now diverge as ventri-lateral processes.

Appendicular Skeleton.

Pectoral girdle.

The simplest type of pectoral girdle is found in Elasmobranchs. It is entirely cartilaginous and consists of a curved ventrally-placed rod, ending dorsally in two horn-like scapular processes which are sometimes attached to the cranium or vertebral column. In Rays the shoulder girdle is very large, and has a distinct suprascapular portion forming a broad plate attached to the neural spines of the vertebrae. There is often a cup-like glenoid cavity for the articulation of the limb; this cavity is specially large in Rays and is much pierced by holes. In Dipnoi the cartilaginous girdle still occurs, but on it there is a deposit of membrane bone forming the clavicle, infraclavicle, and supra-clavicle. These bones, which with the exception of the clavicle, are unknown in higher vertebrates, are better developed in Ganoids, and best of all in Teleosteans. They are connected by the supratemporal with the epi-otic and opisthotic regions of the cranium. Owing to this development of dermal

63

bone, the original cartilaginous arch becomes much reduced, but ossifications representing the scapula and coracoid occur in bony Ganoids and Teleosteans.

Pelvic girdle.

In Elasmobranchs the pelvic girdle consists of a short ventral rod of cartilage representing the ischium and pubis, which does not send up dorsal iliac processes. In *Chimaera* the pelvic girdle has a flattened pointed iliac portion, and ventrally an unpaired movable cartilaginous plate which bears hooks and is supposed to be copulatory in function. Claspers of the usual type are present as well. The Dipnoi have a primitive kind of pelvis in the form of a cartilaginous plate lying in the mid-ventral line and drawn out into three horns anteriorly. In Ganoids the pelvis has almost entirely disappeared, though small cartilaginous vestiges of it remain in *Polypterus*. In Teleosteans even these vestiges are gone, and in these fish and Ganoids the place of the pelvis is taken by the enlarged basi-pterygia (meta-pterygia) of the fins.

Paired fins.

As regards the origin of the limbs or paired fins of fishes there are two principal views. One view, that of Gegenbaur, considers that limbs and their girdles are derived from visceral arches which have migrated backwards. The other view, which probably now has the greater number of supporters, considers that the paired fins of fishes are of essentially the same nature as the median fins.

According to Gegenbaur's view the **archipterygium** of *Ceratodus* (fig. 20) represents the lowest type of fin; it consists of a central cartilaginous axis bearing a large number of radiale. The dorsal or pre-axial radiale are more numerous than the ventral or postaxial, and at the margin of the fin the cartilaginous endoskeletal radiale are replaced by horny exoskeletal fin-rays.

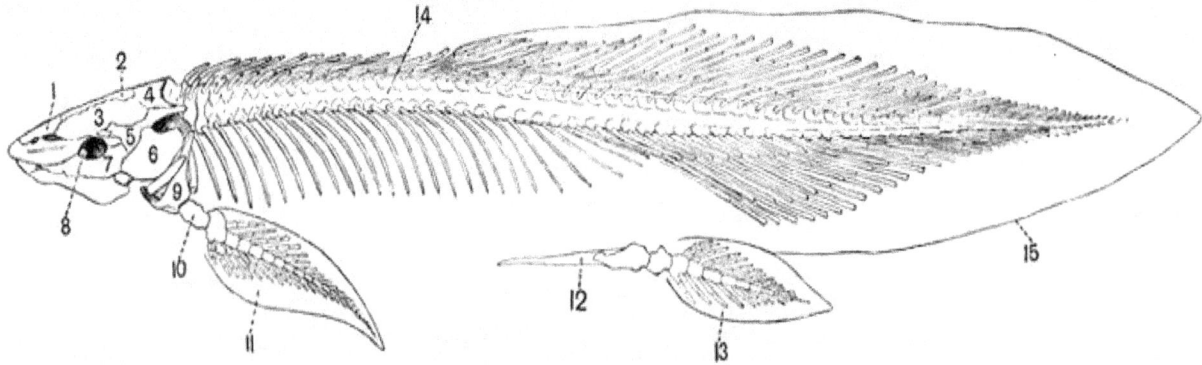

Fig. 20. Lateral view of the skeleton of *Ceratodus miolepis* (after Günther).

1. ethmoid.
2. scleroparietal.
3. frontal.
4. cartilaginous posterior part of cranium.
5. pre-opercular (squamosal).
6. opercular.
7. suborbital.
8. orbit.

9. pectoral girdle.
10. proximal cartilage of pectoral fin.
11. pectoral fin.
12. pelvic girdle.
13. pelvic fin.
14. spinal column.
15. caudal fin (diphycercal).

It is impossible here to give a full discussion of the rival views, but some of the points which support Gegenbaur's view may be mentioned. The fact that migration of visceral arches has to be assumed is no difficulty, as it is obvious that migration in the opposite direction has taken place in many Teleosteans such as the Cod, whose pelvic fins are attached to the throat in front of the pectorals. If migration did take place, the pelvic fins being older than the pectoral should be the more modified, and this is the case. Again, if the pectoral girdle is a modified branchial arch, it must at some period have carried a gill, and in *Protopterus* it does bear a vestigial gill.

According to the view more prevalent at the present time, the paired fins have been derived from two continuous folds of skin and their skeletal supports running forward from the anal region along the sides of the body, their character being similar to the fold that gave rise to the median fins. In support of this view it may be argued that the paired and unpaired fins are often identical in structure, and that some Elasmobranch embryos do show a ridge running between the pectoral and pelvic fins. Then from this continuous fold two pairs of smaller folds may have been specialised off, and in each a number of cartilaginous radiale may have been developed. The fin of *Cladoselache* from the Carboniferous of Ohio apparently illustrates this condition. It consists of certain basal pieces which do not project beyond the body wall and bear a number of unsegmented cartilaginous radiale, which show crowding together and are sometimes bifurcated distally; they extend throughout the whole fin from the body wall to the margin. From this fin the archipterygium might be easily derived by the enlargement of one of the middle radiale and the segmentation and partial fusion of them all.

Whether the archipterygium be a primitive or secondary type of fin, when it is once reached it is easy to derive all the other types from it. The fins of the other living Dipnoi,—*Protopterus* and *Lepidosiren*—are simply archipterygia from which the radiale have almost or completely disappeared, leaving only the segmented axes. Archipterygia too are found in the pectoral fins of the Ichthyotomi, but the postaxial radiale are much reduced.

The **ichthyopterygium**, or type of fin, characteristic of many modern Elasmobranchs such as *Scyllium*, may have been derived from the archipterygium by the gradual reduction of the rays on the postaxial side of the axis and their condensation on the pre-axial side. The Ichthyotomi such as *Xenacanthus* show one stage in the reduction of the postaxial rays, and a further stage is seen in the Notidanidae and some other sharks like *Scymnus* and *Acanthias*, in which a few postaxial rays still remain. The condensation of the pre-axial rays when further continued leads to one of the rays getting an attachment to the girdle. Thus the fin comes to articulate with the girdle by two basalia or basal pieces; a third attachment is formed in the same way and the three basalia are called respectively pro-, meso-, and meta-pterygia. By some authors the meta-pterygium and by others the meso-pterygium is regarded as homologous with the axis of the archipterygium.

The pectoral fins of Elasmobranchs vary very much in their mode of attachment. In some of the sharks, including the Notidanidae and *Scyllium*, all three basalia articulate with the pectoral girdle, while in others such as *Cestracion* the meta-pterygium is excluded. In Rays the propterygium and the meta-pterygium are long and narrow and diverge much from one another; other basalia work their way in between the meso-pterygium and meta-pterygium, and come to articulate with the pectoral girdle. Sometimes they fuse and form a second meso-pterygium. The radiale are greatly elongated and are segmented.

In *Chimaera* all three basalia are present, but the meso-pterygium is shifted and does not articulate with the pectoral girdle.

In *Acipenser* and *Polyodon* the pectoral fin is built on the same type as in Elasmobranchs, but becomes modified from the fact that the propterygium is replaced by dermal bone which forms a large **marginal ray**. Extra meso-pterygia are formed in the same way as in Rays.

In *Polypterus* the pro- and meta-pterygia have ossified while the meso-pterygium remains chiefly cartilaginous; the fin-rays are also chiefly ossified.

In *Amia*, *Lepidosteus*, and certain Teleosteans like *Salmo*, not only the propterygium but the meso-pterygium is almost suppressed by the marginal ray.

In the great majority of Teleosteans a still further stage is reached, the endoskeletal elements, the basalia and radiale are almost entirely suppressed and the fin comes to consist entirely of ossified fin-rays of dermal origin.

In some Teleosteans—*Exocaetus*, a herring, and *Dactylopterus*, a gurnard—the pectoral fins are so enormously developed that by means of them the fish is able to fly through the air for considerable distances. The skeleton of these great fins is almost entirely composed of dermal bone.

Pelvic fin.

The pelvic fin is almost always further removed from the archipterygial condition, and is in general more modified than is the pectoral. Thus in the Ichthyotomi, while the pectoral fins are archipterygia similar to those of *Ceratodus*, the pelvic fins consist of an axis bearing rays on the postaxial side only, and prolonged distally into a clasper. In Dipnoi however the pelvic fins are very similar to the pectoral. In Elasmobranchs the meso-pterygium is missing, the propterygium is small or absent, and the fin is mainly composed of the meta-pterygium (generally called basi-pterygium) and its radiale. The males in Elasmobranchii and Holocephali have the distal end of the meta-pterygium prolonged into a clasper.

In Ganoids and in Teleosteans the loss of the pelvic girdle causes the pelvic fin to be still further removed from the primitive state. There is always a large basi-pterygium which lies imbedded in the muscles and meets its fellow at its proximal end. In Cartilaginous Ganoids it has a secondary segmentation. Its relation to its fellow is subject to much variation in Teleosteans, sometimes as in the Perch the two are in contact throughout, sometimes as in the Salmon they meet distally as well as proximally, but are elsewhere separated by a space, sometimes as in the Pike and Bony Ganoids they diverge widely. The radiale are articulated to the basi-pterygium. In Cartilaginous Ganoids and *Polypterus* they are well developed, in other Ganoids and in Teleosteans they are in the main replaced by dermal fin-rays.

In some Teleosteans such as the Cod the pelvic fins have migrated from their usual position and come to be attached to the throat in front of the pectoral fins. Fish with this arrangement are grouped together as **jugulares**.

65

CHAPTER IX.
CLASS II. AMPHIBIA.

Amphibia differ markedly from Pisces in the fact that in the more abundant and familiar forms the skin is naked, and that when the integument is prolonged into median fins they are devoid of fin-rays. The notochord may persist, but bony vertebral centra are always developed. These are sometimes biconcave, sometimes procoelous, sometimes opisthocoelous. There is only one sacral vertebra, except in rare cases. The cartilaginous cranium persists to a considerable extent but is more or less replaced by cartilage bone, and overlain by membrane bone. The basi-occipital is not completely ossified, and the skull articulates with the vertebral column by means of two occipital condyles formed by the exoccipitals.

There is a large parasphenoid, but there are no ossifications in the basisphenoidal, presphenoidal, and alisphenoidal regions. In most cases the epi-otics and opisthotics are ossified continuously with the exoccipitals.

The palato-pterygo-quadrate bar is firmly united to the cranium, so the skull is autostylic. The palatines and pterygoids are membrane bones. Teeth are nearly always borne on the vomers and commonly on the maxillae and premaxillae. There are no sternal ribs, and the sternum is very intimately related to the pectoral girdle. There are no obturator foramina. The limbs are as in the higher vertebrata, divisible into upper arm, fore-arm, and manus (wrist and hand), and into thigh, shin, and pes (ankle and foot) respectively. The posterior limb is, as a rule, pentedactylate, but in nearly every case the pollex is vestigial or absent.

Order 1. Urodela.

The Urodela are elongated animals with a naked skin, a persistent tail, and generally four short limbs.

The vertebral centra are opisthocoelous or biconcave, and there are numerous precaudal vertebrae. Portions of the notochord commonly persist in the intervertebral spaces. In the skull there is no sphenethmoid forming a ring encircling the anterior end of the brain, its place being in many cases partly taken by a pair of orbitosphenoids. There is no quadratojugal, and the quadrate is more or less ossified. The mandible has a distinct splenial, and the articular is ossified.

There is no definite tympanic cavity. The hyoid apparatus is throughout life connected to the quadrate by ligament, and a large basilingual plate does not occur. The ribs are short structures with bifurcated proximal ends. In the pelvis the pubis remains cartilaginous, and there is a bifid cartilaginous epipubis. The bones of the fore-arm and shin remain distinct, and the manus never has more than four digits.

Suborder (1). Ichthyoidea.

The vertebrae are amphicoelous, but the notochord remains but little constricted throughout the whole length of the vertebral column. Three or four branchial arches nearly always persist in the adult. The cartilages of the carpus and tarsus remain unossified.

The Ichthyoidea may be subdivided again into two groups:—

A. *Perennibranchiata*, whose chief distinguishing skeletal characters are that the skull is elongated, the premaxillae are not ankylosed, the maxillae are vestigial or absent; there are sometimes no nasals, and the palatines bear teeth;

e.g. *Siren, Proteus, Menobranchus.*

B. *Derotremata*, whose chief distinguishing skeletal characters are that there are large maxillae and nasals; teeth are borne by both maxillae and premaxillae; there are no palatines; and both pectoral and pelvic limbs are always present;

e.g. *Amphiuma, Megalobatrachus, Cryptobranchus.*

Suborder (2). Salamandrina.

The vertebrae are opisthocoelous. The skull is broad, and teeth are borne by both premaxillae and dentaries. Nasal bones are present. The remains of only two branchial arches are found in the adult. The carpus and tarsus are more or less ossified.

This suborder includes the Newts (*Molge*), Salamanders (*Salamandra*), and *Amblystoma*.

Order 2. Labyrinthodontia.

These are extinct Amphibia with a greatly developed dermal exoskeleton, which is generally limited to the ventral surface. The body and tail are long and in some cases limbs are absent. The teeth are pointed and often have the dentine remarkably folded. The vertebrae are amphicoelous, and are generally well ossified. The skull is very solid, and has a greatly-developed secondary roof which hides the true cranium and is very little broken up by fossae. Paired dermal supra-occipitals are found, and there is an interparietal foramen. The epi-otics and opisthotics form a pair of bones distinct from the exoccipitals. Four simple limbs of moderate length are generally present, and in some cases all four limbs are pentedactylate. Among the better known genera of Labyrinthodonts are *Mastodonsaurus, Nyrania*, and *Archegosaurus*.

Order 3. Gymnophiona.

These animals form a group of abnormal worm-like Amphibia having an exoskeleton in the form of subcutaneous scales arranged in rings. The vertebrae are biconcave and are very numerous; very few however belong to the tail. The skull has a complete secondary bony roof, the mandible bears teeth and has an enormous backward projection of the angular. The hyoid arch has very slender cornua and no distinct body, it is attached neither to the cranium nor to the suspensorium. The ribs are very long and there are no limbs or limb girdles.

Order 4. Anura.

These are tailless Amphibia, which except in a few instances, are devoid of an exoskeleton. The vertebrae are as a rule procoelous, and are very few in number. The post-sacral part of the spinal column ossifies continuously, forming an unsegmented cylindrical rod, the urostyle. Remains of the notochord persist, lying *vertebrally*, i.e. enclosed within the centra of the several vertebrae, and not as in Urodela lying between one vertebra and the next. The skull is very short and wide. The mandible is almost always, if not invariably, toothless.

The frontals and parietals on each side are united so as to form a pair of fronto-parietals, and a girdle-like sphenethmoid is present.

The quadrate is not generally ossified. A predentary or mento-meckelian bone is commonly present in the mandible, and a single bone represents the angular and splenial. The branchial arches are much reduced in the adult, and the distal ends of the cornua unite to form a flat basilingual plate of a comparatively large size.

Ribs are very little developed. Clavicles are present. The ilia are very greatly elongated. The anterior limb has four well-developed digits and a vestigial pollex, and is of moderate length; the radius and ulna have fused. The posterior limb is greatly elongated and is pentedactylate; the tibia and fibula are fused, while the calcaneum and astragalus are greatly elongated, and it is largely owing to them that the length of the limb is so great. The group includes the Frogs and Toads, the predominant Amphibia of the present time.

CHAPTER X.
THE SKELETON OF THE NEWT (Molge cristata).

I. EXOSKELETON.

The skin of the Newt is quite devoid of any exoskeletal structures. The only exoskeletal structures that the animal possesses are the teeth, and these are most conveniently described with the endoskeleton.

II. ENDOSKELETON.

The endoskeleton of the Newt, though ossified to a considerable extent, is more cartilaginous than is that of the frog. It is divisible into an **axial portion** including the vertebral column, skull, ribs, and sternum, and an **appendicular portion** including the skeleton of the limbs and their girdles.

1. The Axial Skeleton.

A. The Vertebral column.

This consists of about fifty vertebrae arranged in a regular continuous series. The first vertebra differs a good deal from any of the others; the seventeenth or sacral vertebra and the eighteenth or first caudal also present peculiarities of their own. The remaining vertebrae are divided by the sacrum into an anterior series of **trunk** vertebrae which bear fairly large ribs, and a posterior series of **caudal** vertebrae, all of which except the first few are ribless.

The trunk vertebrae.

Any vertebra from the second to the sixteenth may be taken as a type of the trunk vertebrae.

The general form is elongated and somewhat hour-glass shaped, and the **centra** are convex in front and concave behind; an opisthocoelous condition such as this is quite exceptional in Anura. The **notochord** may persist intervertebrally, but in the centre of each vertebra it becomes greatly constricted or altogether obliterated, and replaced by marrow. The superficial portion of the centrum is ossified, while the articular surfaces are cartilaginous. The **neural arches** are low and articulate together by means of **zygapophyses** borne on short diverging processes. The anterior zygapophyses look upwards, the posterior downwards. Each neural arch is drawn out dorsally into a very slight cartilaginous **neural spine**.

On each centrum, at a little behind the middle line, there arise a pair of short backwardly-directed **transverse processes**; each of which becomes divided into two slightly divergent portions, a dorsal portion which meets the tubercular process of the rib and is derived from the neural arch, and a ventral portion which meets the capitular process of the rib and is derived from the ventral or haemal arch. The division between these two parts of the transverse processes can be traced back as far as the sacrum.

The **first vertebra** as already mentioned differs much from all the others. It has no ribs, and presents anteriorly two slightly divergent concave surfaces which articulate with the occipital condyles of the skull. Between these surfaces the dorsal portion of the anterior face of the centrum is drawn out into a prominent **odontoid process**, the occurrence of which renders it probable that the first vertebra of the newt is really the axis, and that the atlas with the exception of the odontoid process has become fused with the skull. The sacral vertebra or **sacrum** differs from the vertebrae immediately in front of it only in the fact that its transverse processes are stouter and more obviously divided into dorsal and ventral portions.

The caudal vertebrae.

The **caudal vertebrae** are about twenty-four in number. The anterior ones have hour-glass shaped centra, and short backwardly-directed transverse processes. The middle and posterior ones have rather shorter centra, and are without transverse processes. The neural arches resemble those of the trunk vertebrae, but each is drawn out into a rather high cartilaginous neural spine abruptly truncated anteriorly. All the caudal vertebrae except the first have also a haemal arch, which is very similar to the neural arch, and is drawn out into a haemal spine quite similar to the neural spine. Both neural and haemal arches are ossified continuously with the centra.

B. The Skull.

The skull of the newt is divisible into three principal parts:—

(1) an axial part, the **cranium proper**, which encloses the brain and to which

(2) the **capsules** of the **auditory and olfactory sense organs** are fused;

(3) the skeleton of the **jaws and hyoid apparatus**. The skull is much flattened and expanded, though not so much as in the frog.

(1) The cranium proper.

The **cranium proper** or **brain case** is an unsegmented tube which remains partly cartilaginous, and is partly converted into cartilage bone, partly sheathed by membrane bone. The roof and floor of the cartilaginous cranium are, as is the case also in the frog, pierced by holes or fontanelles, and these are so large that the main part of the roof and floor comes to be formed by membrane bone.

Two pairs of large ossifications take place in the cranial walls. Of these the more posterior on each side represents the **exoccipital** and all three **periotic** bones. It bears a small convex patch of cartilage for articulation with the atlas, and with its fellow forms the boundary of the foramen magnum.

Two foramina pierce the exoccipital just in front of the occipital condyle and transmit respectively the glossopharyngeal and pneumogastric (fig. 21, X) nerves. Lying laterally to these nerve openings is seen a patch of cartilage, the **stapes**, which is homologous with the stapes or proximal element of the columellar chain in the frog. Further forward in front of the stapes is the small opening for the exit of the facial nerve, and seen in a lateral view close to the orbitosphenoid, that for the trigeminal (fig. 21, C, 5).

In front of these large bones the lateral parts of the cranial walls remain cartilaginous for a short distance, and then there follow two elongated bones, the **orbitosphenoids** (fig. 21, B and C, 11), pierced by the foramina for the exit of the optic nerves. These bones partly correspond to the sphenethmoid of the frog.

The *membrane bones* connected with the cranium are the *parietals, frontals* and *prefronto-lachrymals* on the dorsal surface, and the *parasphenoid* on the ventral surface.

The *parietals* (fig. 21, A and C, 6) roof over the posterior part of the great dorsal fontanelle and overlap the exoccipito-periotics. They meet one another along a sinuous suture in the middle line, as do also the *frontals* which overlap them in front. The *frontals* and *parietals* both extend for a short distance down the sides of the cranium and meet the orbitosphenoids. The *prefronto-lachrymals* (fig. 21, A and C, 7) connect the frontals with the maxillae.

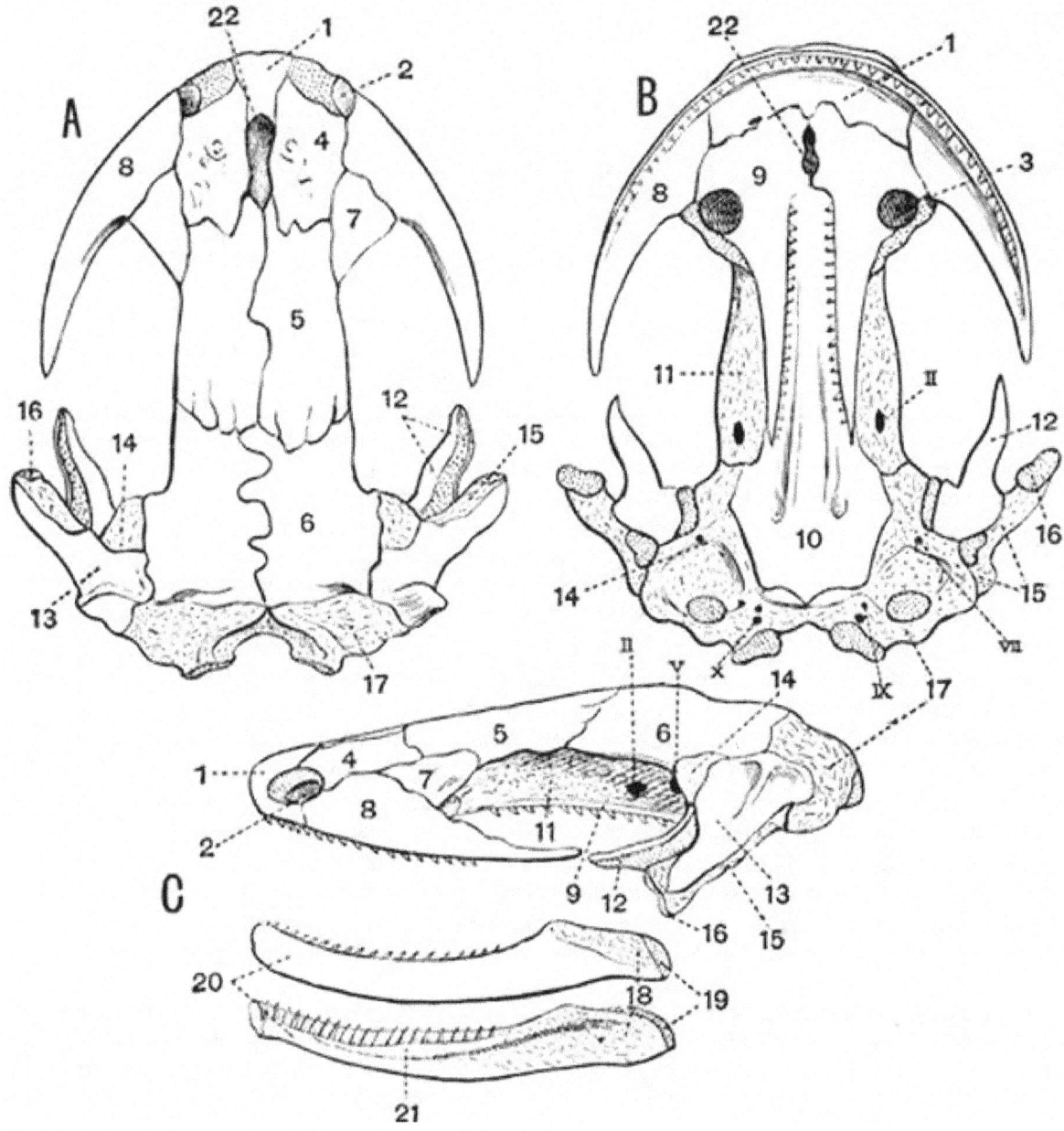

Fig. 21. A dorsal, B ventral, and C lateral views of the skull
of a Newt (*Molge cristata*) × 2½ (after Parker).
The cartilage is dotted, the cartilage bones are marked with dots and
dashes, the membrane bones are left white.

1. premaxillae.
2. anterior nares.
3. posterior nares.
4. nasal.
5. frontal.
6. parietal.
7. prefronto-lachrymal.
8. maxillae.
9. vomero-palatine.

14. pro-otic region of
exoccipito-periotic.
15. quadrate.
16. quadrate cartilage.
17. exoccipital region of
exoccipito-periotic.
18. articular.
19. articular cartilage.
20. dentary.

69

10. parasphenoid.

11. orbitosphenoid.

12. pterygoid.

13. squamosal.

21. splenial.

22. middle narial passage.

II. V. VII. IX. X. foramina for

the exit of cranial nerves.

On the ventral surface is the large *parasphenoid* (fig. 21, B, 10), which is widest behind and overlapped anteriorly by the vomero-palatines.

(2) The sense capsules.

The **auditory capsules** become almost completely ossified continuously with the exoccipitals; they have been already described.

The **nasal capsules** are large and quite unossified though they are overlain by membrane bone. They appear on the dorsal surface between the anterior nares and the nasal process of the premaxillae. They enclose the nasal organs, bound the inner side of the anterior narial opening, and are connected with one another posteriorly by a cartilaginous area.

Developed in connection with the nasal capsules are a pair of rather large *nasals* (fig. 21, A and C, 4), which lie on the dorsal surface immediately in front of the frontals. Each forms part of the posterior boundary of one of the anterior nares, and the two are separated from one another in the middle line by the nasal process of the premaxillae (fig. 21, A, 1), and the opening of the **middle narial passage** (fig. 21, A and B, 22), which passes right through the skull.

On the ventral surface of the skull and forming the greater part of the boundary of the posterior nares are two large bones, the *vomero-palatines* (fig. 21, B and C, 9). Each consists of a wide anterior portion, partly separated from its fellow in the middle line by the ventral opening of the middle narial passage, and of a long pointed posterior portion which is separated from its fellow by the *parasphenoid*, and bears a row of small pointed teeth formed of dentine capped with enamel.

(3) The jaws.

The **upper jaw** of the newt is a discontinuous structure divided into two parts, an anterior part which consists of membrane bones, the *maxillae* and *premaxillae*, and a posterior part which remains mainly cartilaginous.

The *premaxillae* are united, forming a single bone, which in a ventral view is seen to meet the maxillae and vomero-palatines, and in a dorsal view to send back a nasal process (fig. 21, A, 1) between the nasals.

The *maxillae* are large bones, each terminating in a point posteriorly. A single row of teeth similar to those on the vomero-palatines runs along the outer margin of the maxillae and premaxillae.

The posterior part of the upper jaw forms a mass of cartilage which extends forwards towards the maxillae as a long pointed process whose ventral surface and sides are overlapped by a membrane bone, the *pterygoid* (fig. 21, 12).

The suspensorial bones include the **quadrate** and *squamosal*. The **quadrate** (fig. 21, 15) which forms the true **suspensorium** is directed forwards and outwards, and is terminated by a patch of cartilage with which the mandible articulates.

The lower jaw or mandible remains partly cartilaginous, while its ossifications include two membrane bones and one cartilage bone. The cartilage bone is the **articular** (fig. 21, C, 18), it forms the posterior part of the ramus, extends forwards for some distance along its inner side, and is terminated posteriorly by a patch of cartilage which articulates with the quadrate. The *dentary* (fig. 21, C, 20) is a large bone which forms the anterior part and nearly all the outer half of each ramus, and bears teeth similar to those of the upper jaw. Attached to its inner face is a long slender *splenial* (fig. 21, C, 21).

The Hyoid apparatus.

This consists of the hyoid arch and part of the first two branchial arches.

The **hyoid arch** (fig. 29, A, 2) consists of a pair of **cornua**, each of which is divided into two halves. The dorsal half forming the **cerato-hyal** is mainly ossified though tipped with cartilage, and is connected by ligament with the suspensorium. The ventral half (**hypo-hyal**) is cartilaginous, and is connected with the basibranchial.

The **branchial arches** consist of a median piece, the **basibranchial**, which is ossified in the centre and cartilaginous at either end, and of two pairs of **cerato-branchials** which are attached to the cartilaginous part (fig. 29, A, 8) of the basibranchial. The first cerato-branchial is chiefly ossified, the second (fig. 29, A, 4) is a good deal smaller and is cartilaginous. Both are united dorsally to a single **epi-branchial**, which is terminated by a small cartilaginous area at the free end but is elsewhere well ossified.

C. The Ribs.

The ribs are short imperfectly ossified structures, bifid at their proximal end where they articulate with the transverse processes, and tipped both proximally and distally with cartilage. The dorsal portion of the proximal end corresponds to the **tuberculum** of the ribs of higher animals, and the ventral portion to the **capitulum**. Some of the anterior ribs have a step-like notch on their dorsal surfaces.

The second to twelfth ribs are fairly equal in size, but further back they decrease slightly. The ribs which connect the sacral vertebrae with the ilia are however large. The short ribs borne on the anterior caudal vertebrae are cartilaginous.

D. The Sternum.

The sternum (fig. 22, A, 6) is a rather broad plate of cartilage, drawn out posteriorly into a median process marked by a prominent ridge. On its antero-lateral margins it bears surfaces for articulation with the pectoral girdle.

2. The Appendicular Skeleton.

A. The Pectoral girdle.

This is of a very simple character, and remains throughout life in an imperfectly ossified condition. It consists of a dorsal **scapular portion**, and a ventral **coracoid portion** partially divided into an anterior part, the precoracoid, and a posterior part, the **coracoid**.

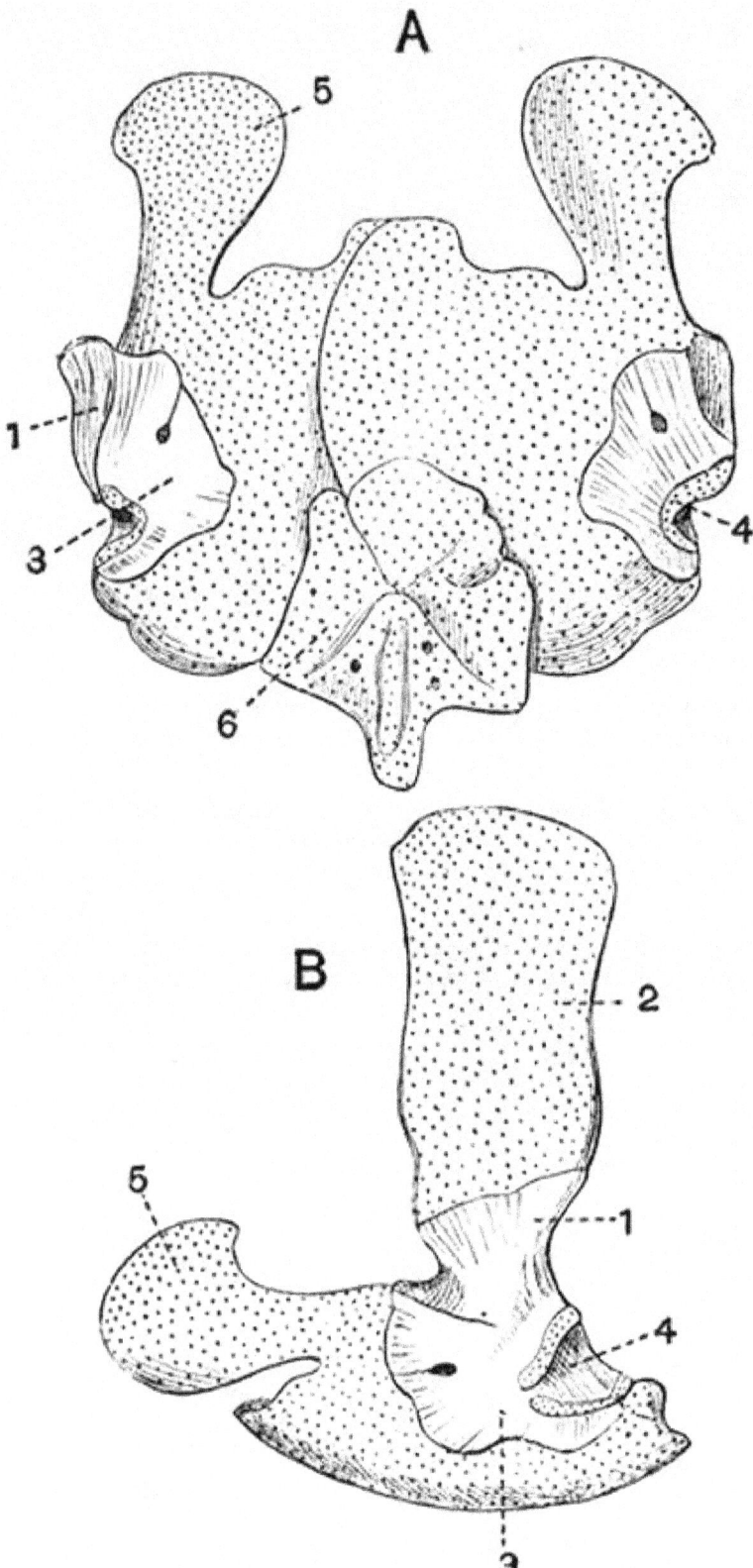

Fig. 22. A ventral, and B lateral view of the shoulder girdle and sternum of an old male Crested Newt (*Molge cristata*) × 3 (after Parker).

1. scapula.   4. glenoid cavity.

| 2. suprascapula. | 5. precoracoid. |
| 3. coracoid. | 6. sternum. |

The **scapular portion** is a slightly curved oblong plate; its proximal third the **scapula** (fig. 22, 1) is ossified and bounds part of the well-marked **glenoid cavity** (fig. 22, 4); its distal portion forms a large oblong cartilaginous plate, the suprascapula (fig. 22, 2).

The **precoracoid** (fig. 22, 5) forms a small forwardly-directed cartilaginous plate. The **coracoid** (fig. 22, 3) forms a much larger plate, the greater part of which is unossified and overlaps its fellow in the middle line, the two being overlapped by the sternum. Around the glenoid cavity is an area which is mainly ossified and is continuous with the scapula.

B. The Anterior limb.

This is divisible into three parts, the **upper arm** or **brachium**, the **fore-arm** or **antibrachium**, and the **manus**.

The **upper arm** includes a single bone, the **humerus**.

The **humerus** is a slender bone cylindrical in the middle and expanded at either end, the proximal part forms a rounded **head** which articulates with the glenoid cavity. Along the proximal part of the anterior or pre-axial surface runs a strong **deltoid ridge**. The proximal part of the postaxial surface also bears a small outgrowth.

The **fore-arm** contains two bones, the **radius** and **ulna**, both of which are small and imperfectly ossified at their terminations.

The **radius** (fig. 23, B, 11) or pre-axial bone is rather the larger of the two, and is considerably expanded at its proximal end. The **ulna** or postaxial bone is somewhat expanded distally, but is not drawn out proximally into an olecranon process.

The **manus** consists of two parts, a group of small bones forming the **carpus** or **wrist**, and the **hand**.

The **carpus** is in a very simple unmodified condition as compared with that of the Frog. It consists of a proximal row of two bones and a distal row of four, with one, the **centrale**, interposed between. All these bones are small and polygonal and are imbedded in a plate of cartilage.

The bones of the proximal row are a smaller pre-axial bone, the **radiale** (fig. 23, B, 13), and a larger postaxial bone, which represents the fused **ulnare** and **intermedium** of the very simple carpus described on pp. 26 and 27.

The four bones of the distal row are respectively **carpalia** 2, 3, 4 and 5.

The **hand** consists of four digits, that corresponding to the thumb of the human hand, judging from the analogy of the frog probably being the one that is absent.

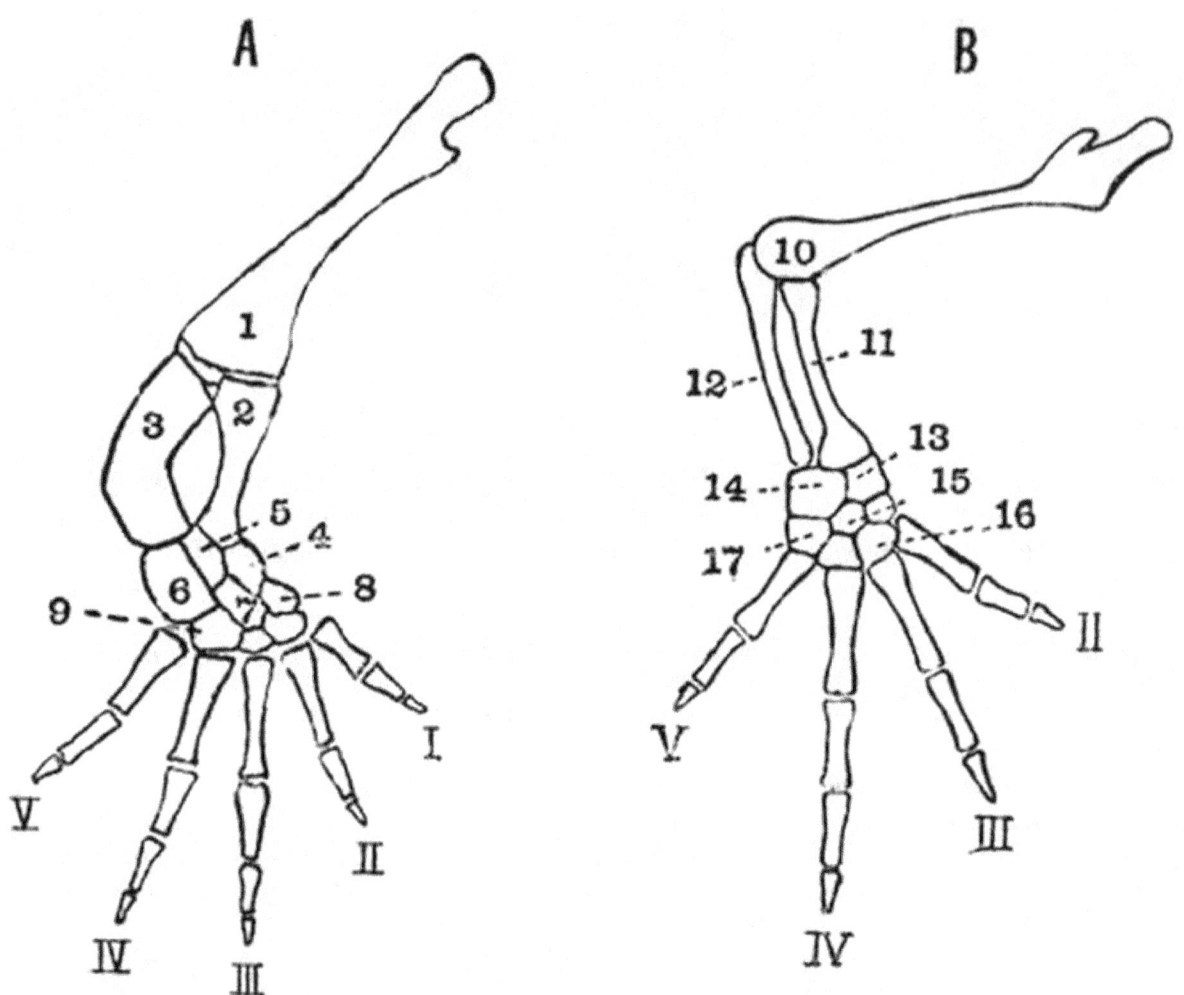

Fig. 23. A right posterior, and B right anterior limb of a Newt × 1½ (*Molge cristata*).

1. femur.	10. humerus.
2. tibia.	11. radius.
3. fibula.	12. ulna.
4. tibiale.	13. radiale.
5. intermedium.	14. intermedium and ulnare fused.
6. fibulare.	15. centrale of carpus, the pointing
7. centrale of tarsus.	line passes across carpale 2.
8. tarsale 1.	16. carpale 3.
9. tarsalia 4 and 5 fused.	17. carpale 5.

I. II. III. IV. V. digits.

Each digit consists of a somewhat elongated **metacarpal** and of two or three phalanges. The metacarpals are contracted in the middle and expanded at either end. They are connected with the carpus by cartilage, and the articulations between the several phalanges, and between the metacarpals and phalanges are also cartilaginous. The second, third, and fifth digits have two phalanges apiece, the fourth, which is the longest, has three. The second metacarpal in the specimen examined and figured articulates partly with carpale 2, partly with carpale 3.

C. The Pelvic girdle.

The pelvic girdle of the Newt is in a much less modified condition than is that of the Frog (see p. 165). It consists of a dorsal element, the **ilium**, a posterior ventral element, the **ischium**, and an anterior ventral element, the **pubis**, to which is attached an **epipubis**.

The **ilium** is a somewhat cylindrical bone which at its ventral end meets the ischium, and forms part of the **acetabulum**. It is then directed upwards and slightly backwards, and is attached to the ribs of the sacral vertebra.

The **ischia** are a pair of somewhat square bones which meet one another in the middle line; they form part of the acetabulum, and are united to the ilia above.

In front of the ischia is a narrow cartilaginous area which represents the **pubes**. Projecting forwards from it is a bifid cartilaginous **epipubis**.

D. The Posterior limb.

This is divisible into a proximal portion, the **thigh**, a middle portion, the **crus** or **shin**, and a distal portion, the **pes**.

The **thigh** consists of a single bone, the **femur** (fig. 23, A, 1), which has a thin shaft and expanded ends. The anterior part of the pre-axial border and posterior part of the postaxial border bear slight outgrowths.

The **crus** or **shin** includes two short bones, the **tibia** and **fibula**, which are nearly equal in length. The pre-axial bone or tibia is a straight bone thickest at its proximal end, the postaxial bone or **fibula** (fig. 23, A, 3) is a rather stouter curved bone of nearly equal diameter throughout.

The **pes** includes the **tarsus** or **ankle**, and the **foot**.

The **tarsus** consists of eight small bones arranged in a proximal row of three, the **tibiale**, **intermedium** and **fibulare**, and a distal row of four **tarsalia**, with one bone, the **centrale** (fig. 23, A, 7), interposed between the two rows. In the specimen examined, the **tibiale**, is a small bone articulating with the tibia, the **intermedium** (fig. 23, A, 5) is larger and articulates with both tibia and fibula, the **fibulare** is the largest of the three and articulates with the fibula.

The bones of the distal row are **tarsalia 1, 2, 3**, and a bone representing **4** and **5** fused. In the specimen examined tarsale 1 is pushed away dorsally (fig. 23, A, 8), so as to lie between the tibiale and tarsale 2. All the tarsal bones are small and somewhat polygonal, and are connected with one another, and with the tibia and fibula on the one hand, and with the metatarsals on the other by a thin layer of cartilage.

The five **digits** of the foot each consist of a **metatarsal** and of a certain number of **phalanges**. In the specimen examined, owing to the shifting of tarsale 1, the first metatarsal as well as the second articulates with tarsale 2, while the fifth metatarsal articulates partially with the bone representing the fused tarsalia 4 and 5, partially with the fibulare. All the bones of the digits except the distal phalanges are terminated at each end by cartilaginous epiphyses, the distal phalanx of each digit has a cartilaginous epiphysis only on its proximal end.

The first, second, and fifth digits have two phalanges apiece, the third and fourth have three.

Figure 31 B, showing a Newt's tarsus copied from Gegenbaur, has precisely the arrangement generally regarded as primitive for the higher vertebrates, except that tarsalia 4 and 5 are fused.

CHAPTER XI.
THE SKELETON OF THE FROG (Rana temporaria).

I. EXOSKELETON.

The skin of the frog is smooth and quite devoid of scales or other exoskeletal structures. The only exoskeletal structures met with in the frog are:—

1. The **teeth**, which are most conveniently described with the endoskeleton.

2. The horny covering of the calcar or prehallux (see p. 167).

II. ENDOSKELETON.

The endoskeleton of the adult frog consists partly of cartilage, partly of bone and each of these types of tissue occurs in two forms. The cartilage may be hyaline, as in the omosternum and xiphisternum, or may be more or less calcified as in part of the suprascapula and the epiphyses of the limb bones. The bone may be cartilage bone, or membrane bone.

The skeleton is divisible into an **axial portion** consisting of the skull, vertebral column, and sternum, and an **appendicular portion** consisting of the skeleton of the limbs and their girdles.

1. The Axial Skeleton.

A. The Vertebral column.

The vertebral column is a tube, formed of a series of ten bones which surround and protect the spinal cord. Of these ten bones nine are vertebrae, while the tenth is a straight rod, the **urostyle**, and is almost as long as all the vertebrae put together. The second to eighth vertebrae inclusive have a very similar structure, but the first and ninth differ from the others.

Any one of the second to eighth vertebrae forms a bony ring with a somewhat thickened floor, the **centrum** or body, which articulates with the centra of the immediately preceding and succeeding vertebrae. The articulating surfaces are covered with cartilage and are procoelous, or convex in front and concave behind. The eighth vertebra is however amphicoelous or biconcave. The centrum of each vertebra encloses an isolated vestige of the notochord. The **neural arch** forms the roof and sides of the neural canal, which is very spacious in the anterior vertebrae, but becomes more depressed in the posterior ones. The arch bears the **neural spine**, a low median ridge of variable character, and is drawn out in front and behind, forming the two pairs of articulating surfaces or **zygapophyses** by means of which the vertebrae are attached together. Of these the anterior articulating surfaces or **prezygapophyses** look upwards and slightly inwards, while the posterior articulating surfaces or **postzygapophyses** look downwards and slightly outwards. The sides of the neural arches are drawn out into a pair of prominent **transverse processes**. Those of the second vertebra look somewhat forwards, those of the third look directly outwards or somewhat forwards, while those of the fourth, fifth, and sixth are directed slightly backwards, and those of the seventh and eighth nearly straight outwards. All the transverse processes are terminated by very small cartilaginous **ribs**.

Special vertebrae.

The **first vertebra** is a ring-like structure with a much depressed centrum. It bears in front two oval concave surfaces for articulation with the condyles of the skull, while the centrum is terminated behind by a prominent convex surface. There are as a rule no transverse processes, and the postzygapophyses look downwards and outwards. Occasionally however transverse processes do occur. Projecting forwards from the centrum is a minute process better developed in the Newt. This resembles an odontoid process, and it has hence been supposed that the first vertebra is homologous with the axis of mammalia, and that the atlas of the frog is fused with the skull.

The **ninth vertebra** has very stout transverse processes directed backwards and somewhat upwards. They articulate with the pelvic girdle and hence this vertebra is regarded as the **sacrum**. The neural arch is much depressed, the centrum is convex in front and bears on its posterior surface two short rounded processes for articulation with the urostyle.

The **urostyle** is a long rod-like bone forming the posterior unsegmented continuation of the vertebral column. It is probably equivalent to three vertebrae, the tenth, eleventh, and twelfth fused together, and to an unsegmented rod of cartilage which lies ventral to the notochord. The anterior end is expanded and bears two concave articular surfaces by means of which it articulates with the sacrum. A prominent ridge runs along the dorsal surface, but gradually diminishes when traced back. The anterior portion contains a canal which is a continuation of the neural canal. At a point not far from the anterior end, this canal communicates with the exterior by a pair of minute holes which correspond with the intervertebral foramina.

B. The Skull.

The skull of the Frog consists of three principal parts:—

(1) an axial part, the **cranium proper**, which encloses the brain. To it are firmly fused

(2) the **capsules of the olfactory and auditory sense organs**,

(3) lastly there is the **hyoid apparatus** and the **skeleton of the jaws**.

The skull is by no means so completely ossified as is the vertebral column, but in addition to the cartilage bone, there is a great development of membrane bone in connection with it.

The skull has a peculiarly flattened and expanded form depending on the wide lateral separation of the jaws from the cranium.

(1) The Cranium proper or Brain case.

This is an unsegmented tube, which is widest behind. It remains to a considerable extent cartilaginous, but is partly converted into cartilage bone, partly sheathed in membrane bone. Its roof is imperfect, being pierced by three holes or **fontanelles**, one large anterior fontanelle (fig. 25, A, 9), and two smaller posterior fontanelles (fig. 25, A, 10).

The cartilage bones of the cranium proper are the two **exoccipitals** and the **sphenethmoid**.

The **exoccipitals** (figs. 24, 25, and 26, 6) are a pair of irregular bones bounding the foramen magnum at the posterior end of the skull. They almost completely surround the foramen magnum, and bear a pair of oval convex surfaces, the **occipital condyles**, with which the first vertebra articulates. The bones generally called the exoccipitals of the frog include the **epi-otic** and **opisthotic** elements of many skulls, in addition to the exoccipitals.

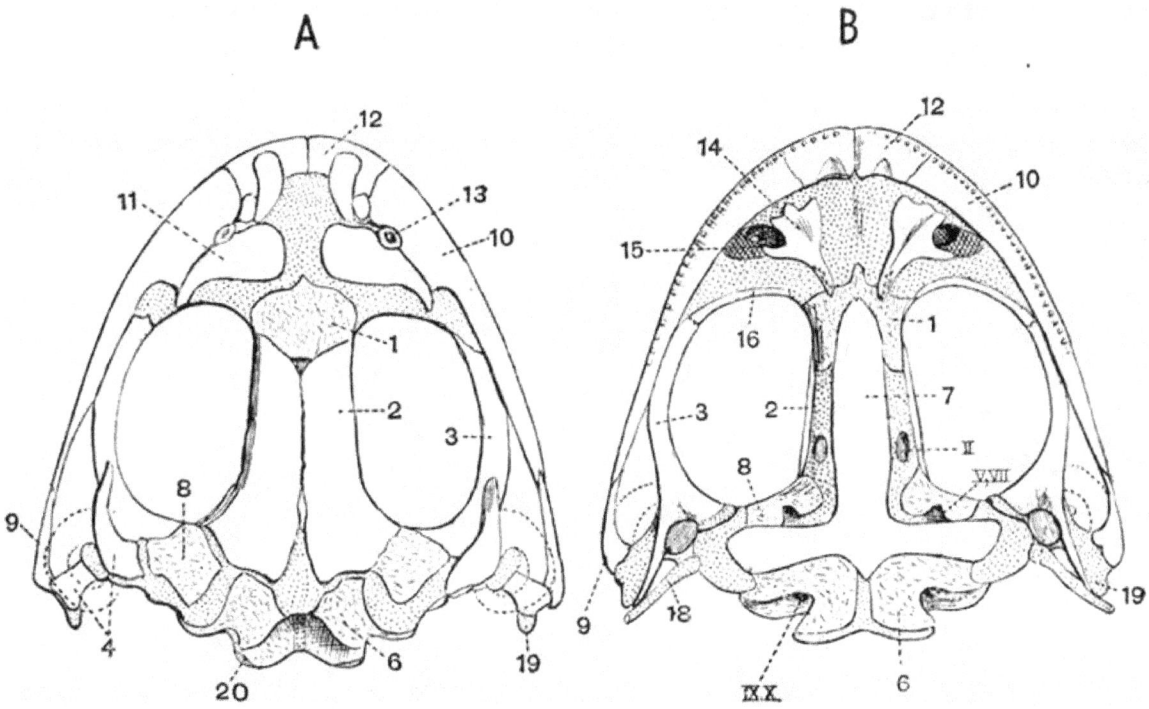

Fig. 24. A dorsal, and B ventral views of the cranium of a Common Frog (*Rana temporaria*) × 2 (after Parker). In this and the next two figs. cartilage is dotted, cartilage bones are marked with dots and dashes, membrane bones are left white.

1. sphenethmoid.	14. vomer.
2. fronto-parietal.	15. posterior nares.
3. pterygoid.	16. palatine.
4. squamosal.	18. columella.
6. exoccipital.	19. quadrate.
7. parasphenoid.	20. occipital condyle.
8. pro-otic.	II. optic foramen.
9. quadratojugal.	V. VII. foramen for exit of
10. maxillae.	trigeminal and facial nerves.
11. nasal.	IX. X. foramina for exit of
12. premaxillae.	glossopharyngeal and
13. anterior nares.	pneumogastric nerves.

The patch of unossified cartilage immediately external to the occipital condyle is pierced by two small foramina, through which the ninth and tenth nerves leave the cranial cavity. The ninth nerve passes through the more external of these foramina, the tenth through the one nearer the condyle. The foramina lie however very close together and are sometimes confluent. The cranial walls for a considerable distance in front of the occipitals are unossified, but the anterior end of the cranial cavity is encircled by another cartilage bone, the **sphenethmoid** (figs. 24 and 25, 1) or girdle bone. This partly corresponds to the orbitosphenoids of the Newt's skull. Anteriorly it is pierced by a pair of small foramina through which the ophthalmic branches of the trigeminal nerve pass out.

The anterior part of the cranial cavity is divided into two halves by a vertical plate, the **mesethmoid**. Some little distance behind the sphenethmoid the ventro-lateral walls of the cartilaginous cranium are pierced by a pair of rather prominent holes, the **optic foramina** (figs. 24 and 25, B, II), and at a similar distance further back, occupying a kind of notch in the pro-otic are the large **trigeminal foramina**, through which the fifth and seventh nerves leave the cranium. Between the trigeminal and optic foramina are the very small foramina for the sixth nerves (fig. 25, B, VI).

The *membrane bones* of the cranium proper include the *fronto-parietals* and the *parasphenoid*.

The *fronto-parietals* (figs. 24 and 26, A, 2) form a pair of long flat bones closely applied to one another in the middle line, the line of junction being the **sagittal suture**. They cover over the fontanelles and overlap the sphenethmoid in front.

77

The *parasphenoid* (figs. 24 and 26, B, 7) is a bone shaped like a dagger with a very short handle. It lies on the ventral surface of the cranium, the blade being directed forwards and underlying the sphenethmoid; its lateral processes underlie the auditory capsules.

(2) The sense capsules.

The sense capsules are cartilaginous or bony structures which surround the olfactory and auditory organs and are closely united to the cranium.

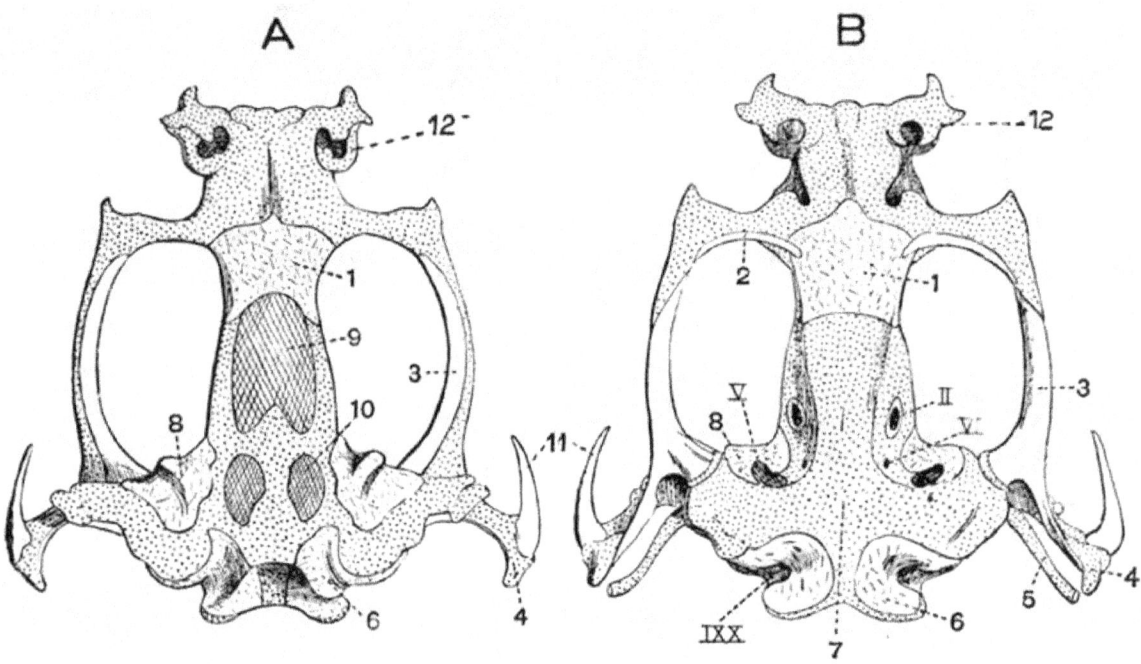

Fig. 25. A dorsal and B ventral view of the cranium of a common Frog (*Rana temporaria*) from which the membrane bones have mostly been removed. × 2 (after Parker).

1. sphenethmoid.	8. pro-otic.
2. palatine.	9. anterior fontanelle.
3. pterygoid.	10. right posterior fontanelle.
4. quadrate.	11. quadratojugal.
5. columella.	12. nasal capsule.
6. exoccipital.	II. V. VI. IX. X. foramina for
7. ventral cartilaginous wall of	exit of cranial nerves.
cranium.	

The **auditory capsules** are fused with the sides of the posterior end of the cranium just in front of the exoccipitals. They are largely cartilaginous, but include in their anterior walls a pair of irregular cartilage bones, the **pro-otics** (figs. 24 and 25, 8). The cartilaginous area lying ventral to the pro-otic and external to the exoccipital is pierced by a rather prominent hole, the **fenestra ovalis**, which forms a communication between the internal ear cavity, and a space the tympanic cavity, which lies at the side of the head, and is bounded externally by the tympanic membrane. The fenestra ovalis is occupied by a minute cartilaginous structure, the **stapes**, and articulated partly to this and partly to a slight recess in the pro-otic is the **columella** (fig. 25, B, 5), a rod in part bony and in part cartilaginous, whose outer end is attached to the tympanic membrane. The columella and stapes are together homologous with the mammalian auditory ossicles and with the hyomandibular of Elasmobranchs. Sometimes the term columella is used to include the whole ossicular chain,—the columella together with the stapes.

The **olfactory** or **nasal capsules** (fig. 25, B, 12) are fused with the anterior end of the cranium and differ from the auditory capsules in being to a great extent unossified. There are however two pairs of membrane bones developed in connection with them, the *vomers* and the *nasals*. They are drawn out into three pairs of cartilaginous processes, on the dorsal surface into the **prenasal** and **alinasal** processes which bound the external nares, and on the ventral surface towards the middle line into the forwardly-projecting **rhinal** processes.

The *nasals* (figs. 24 and 26, 11) form a pair of triangular bones lying dorsolaterally in front of the fronto-parietals. Their bases are turned towards one another and their apices are directed outwards and backwards. They correspond in position with the prefrontals of the reptilian skull as well as with the nasals.

The *vomers* are a pair of irregular bones lying on the ventral surface of the olfactory capsules. Each bears on its inner and posterior angle a group of minute pointed teeth, while its outer border is drawn out into three or four small slightly diverging processes, the two posterior of which form the inner boundary of the **posterior nares** (fig. 24, B, 15).

(3) The jaws.

The **upper jaw** consists of a rod of cartilage connected with the cranium near its two ends, but widely separated from it in the middle. It is almost completely overlain by membrane bone. With its posterior end the lower jaw articulates.

The membrane bones of the upper jaw include first the *premaxillae*, a small bone meeting its fellow in the middle line, and forming the extreme anterior end of the upper jaw. It gives off on its dorsal surface a backwardly-projecting process. It is connected behind with the *maxillae* (figs. 24 and 26, 10), a long flattened bone which forms the greater part of the margin of the upper jaw, and gives off near its anterior end a short process which projects upwards and meets the nasal.

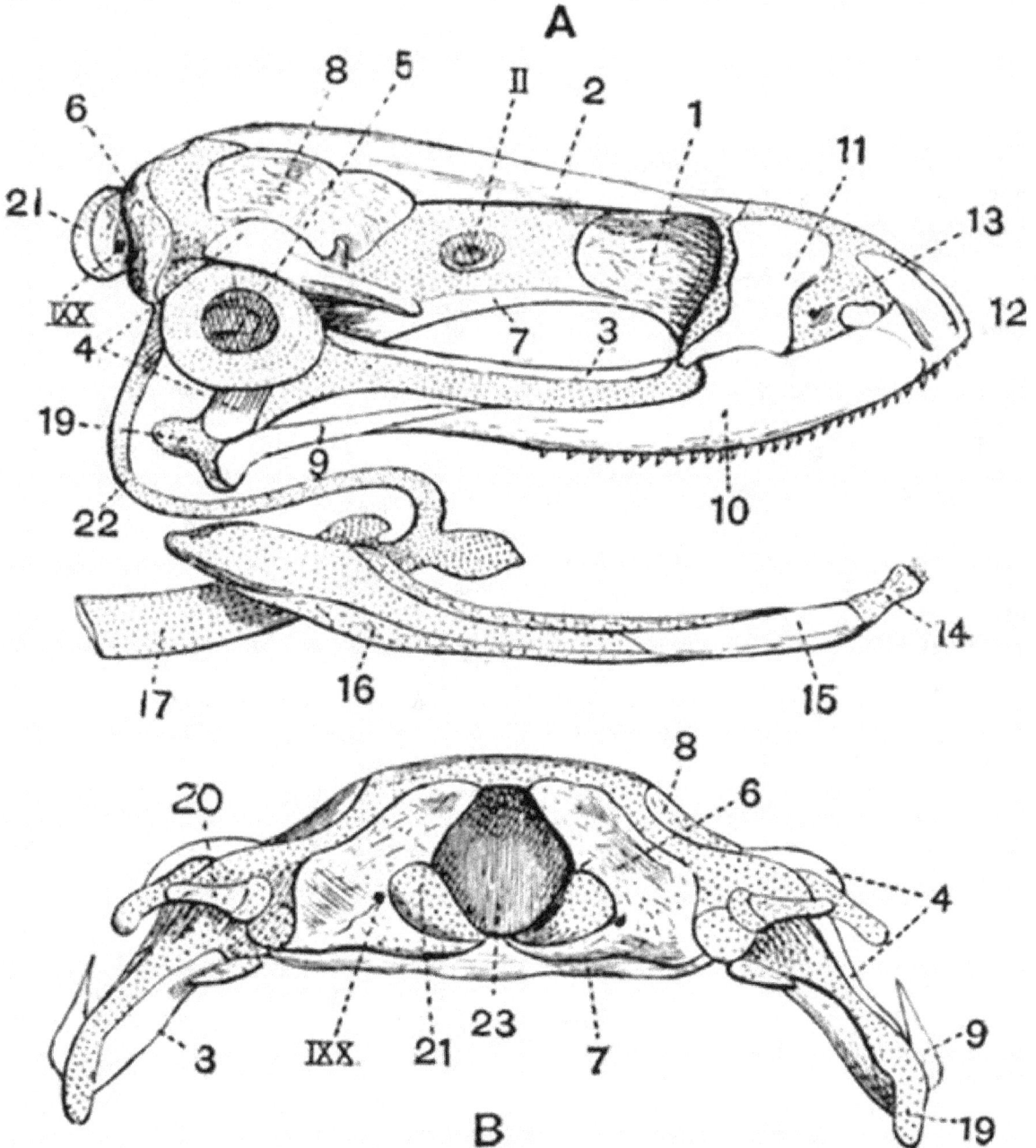

Fig. 26. A, Lateral view of the skull, B, Posterior view of the cranium of a Common Frog (*Rana temporaria*) × 2 (after

79

Parker).

1. sphenethmoid.
2. fronto-parietal.
3. pterygoid.
4. squamosal.
5. tympanic membrane.
6. exoccipital.
7. parasphenoid.
8. pro-otic.
9. quadratojugal.
10. maxillae.
11. nasal.
12. premaxillae.
13. anterior nares.

14. mento-meckelian.
15. dentary.
16. angulo-splenial.
17. basilingual plate.
19. quadrate.
20. columella.
21. occipital condyle.
22. anterior cornu of the hyoid
(cerato-hyal).
23. foramen magnum.
II. IX. X. foramina for the exit
of cranial nerves.

Both maxillae and premaxillae are grooved ventrally, and bear, attached to the outer more prominent margin of the groove, a row of minute conical teeth. These teeth are **pleurodont**, that is, are ankylosed by their bases and outer sides to the margin of the jaw. Each tooth is a hollow cone, the basal part of which is formed of bone, the apical part of dentine, capped by a very weak development of enamel.

The posterior end of the maxillae is overlapped by a small bone, the *quadratojugal* (figs. 24 and 26, 9), whose posterior end forms part of the articular surface for the lower jaw. Just behind the quadratojugal there is a small unossified area which lies at the angle of the mouth, and is connected by a narrow bar of cartilage with the cranium; this forms the **quadrate** (figs. 24 and 26, 19). A backwardly-directed outgrowth from the cartilaginous bar more or less completely surrounds the tympanic membrane, forming the tympanic ring. When followed back the maxillae and quadratojugal diverge further and further from the cranium, till the angle of the mouth comes to be separated from the foramen magnum by a space nearly double the width of the cranium. This space is bridged over to a considerable extent by two triradiate bones, the *pterygoid* and *squamosal*.

The *pterygoid* (figs. 24 and 26, 3) is a large bone, whose anterior limb runs forwards meeting the maxillae and palatine; while its inner limb meets the auditory capsule and parasphenoid, and its outer limb runs backwards and outwards to the angle of the mouth. The *palatine* is a small transversely-placed bone, which connects the pterygoid with the anterior part of the sphenethmoid. The *squamosal* (figs. 24 and 26, 4) is a T-shaped bone whose anterior arm is pointed and passes forwards to meet the pterygoid. The posterior upper arm is closely applied to the pro-otic, while the posterior lower arm meets the pterygoid and quadratojugal at the angle of the jaw, and surrounds the narrow cartilaginous bar of the quadrate which goes to join the cranium. The squamosal is probably homologous with the squamosal together with the pre-opercular of Bony Ganoids.

The quadrate and squamosal form the **suspensorium** by which the lower jaw is connected with the cranium.

The **lower jaw** or **mandible** consists of a pair of cartilaginous rods (**Meckel's cartilages**) in connection with each of which there are developed two membrane bones and one cartilage bone. The cartilage bone is the **mento-meckelian** (fig. 26, A, 14), a very small ossification at the extreme anterior end. The membrane bones are the *angulo-splenial* and the *dentary*. The *angulo-splenial* is a strong flat bone which forms the inner and lower part of the mandible for the greater part of its length. Its dorsal surface is produced into a slight **coronoid process**. The *dentary* (fig. 26, A, 15) is a flat plate which covers the outer surface of the anterior half of the mandible, as far forwards as the mento-meckelian. The lower jaw is devoid of teeth. The part of Meckel's cartilage which in most vertebrates ossifies, forming the articular bone, remains unossified in the Frog.

The Hyoid apparatus.

The **hyoid** of the adult Frog is formed of the modified hyoid and branchial arches of the tadpole. It consists of a broad thin plate of cartilage, the **basilingual plate** (fig. 29, B, 1), drawn out into two pairs of long processes, the **cornua**. The basilingual plate is broader in front than behind, and is formed from the fused ventral ends of the hyoid and branchial arches of the tadpole.

The **anterior cornua** (fig. 29, B, 2) form a pair of long slender cartilaginous rods which project from the body of the hyoid at first forwards, then backwards, and finally upwards and somewhat forwards again, to be united to the auditory capsules just below the fenestrae ovales. They are formed from the dorsal portion of the hyoid arch of the tadpole and are homologous with the cerato-hyals of the Dogfish.

The **posterior cornua** form a pair of straight bony rods diverging outwards from the posterior end of the basilingual plate. They are formed from the fourth branchial arches of the tadpole, and differ from the rest of the hyoid apparatus in being well ossified.

The **columellar chain**, which has been already described (p. 157), should be mentioned with the hyoid as it is homologous to the hyomandibular of fishes.

The **sternum** of the **Frog**, though regarded as part of the axial skeleton, is so intimately connected with the pectoral girdle, that it will be described with the appendicular skeleton.

2. The Appendicular Skeleton.

This consists of the skeleton of the two pairs of limbs and their respective girdles. It is at first entirely cartilaginous but the cartilage becomes later on mainly replaced by bone. The only bone developed in connection with the appendicular skeleton, which has no cartilaginous predecessor, is the *clavicle*.

A. The Pectoral girdle.

This consists originally of two half rings of cartilage encircling the sides of the body a short way behind the head. These two halves meet one another in the ventral middle line, and separate the anterior elements of the sternum from the posterior ones.

Each half-ring bears on the middle of its outer and posterior surface a prominent cup, the **glenoid cavity**, with which the proximal arm-bone articulates. This cup divides the half-arch into a dorsal **scapular** and a ventral **coracoid** portion.

The **scapular portion** consists of two parts, the **suprascapula** and the **scapula**.

The **suprascapula** (fig. 30, A, 2) is a wide, thin plate attached by its ventral and narrowest border to the scapula. Its proximal and anterior half is imperfectly ossified, its whole border or sometimes only its dorsal and posterior borders consist of unaltered hyaline cartilage, while the rest of it is composed of calcified cartilage. The **scapula** (fig. 30, A, 3) is a fairly stout rod of bone constricted in the middle, and forming the dorsal half of the glenoid cavity.

The **coracoid portion** consists of three parts, the **coracoid**, **precoracoid** and *clavicle*.

The largest and most posterior of these is the *coracoid* (fig. 30, A, 4) which like the scapula, is contracted in the middle and expanded at the ends, especially at the ventral end. It forms a large part of the glenoid cavity. The ventral ends of the coracoids which meet one another in the middle line are unossified, and form narrow strips of calcified cartilage, the **epicoracoids** (fig. 30, A, 5); these are often regarded as sternal elements.

The **precoracoid** forms a narrow strip of cartilage lying in front of the coracoid, from which it is separated by the wide **coracoid foramen** (fig. 30, A, 9). The dorsal end is continuous with an area of unossified cartilage which separates the coracoid and scapula and forms part of the glenoid cavity.

The *clavicle* is a narrow membrane bone closely attached to the anterior surface of the precoracoid, its dorsal end is expanded.

The Sternum.

The sternum consists of four parts arranged in two groups; two parts to each group. The anterior members are the episternum and omosternum.

The **episternum** (fig. 30, A, 10) is a thin almost circular plate of cartilage much of which remains hyaline.

The **omosternum** (fig. 30, A, 11) is a slender bony rod widest at its posterior end; it connects the episternum with the ventral ends of the precoracoids.

The **sternum proper** is a short rod of cartilage sheathed in bone; it is contracted in the middle and expanded at each end. It bears attached to its posterior end a broad somewhat bilobed plate of partially calcified cartilage, the **xiphisternum** (fig. 30, A, 13).

B. The Anterior limb.

This is divisible into three parts, the **upper arm** or **brachium**, the **fore-arm** or **antibrachium**, and the **manus**.

All the larger bones have their ends formed by prominent epiphyses which do not unite with the shaft till late in life. Their articulating surfaces are covered by hyaline cartilage.

In the **upper arm** there is a single bone, the **humerus**.

This has a more or less cylindrical shaft and articulates by a prominent rounded **head** with the glenoid cavity. The distal end shows a large rounded swelling on either side of which is a **condylar ridge**, the inner or postaxial one being the larger. A prominent **deltoid ridge** runs along the proximal half of the anterior surface, and in the male frog a second equally prominent ridge runs along the distal half of the posterior surface.

The **fore-arm** consists of two bones, the **radius** and **ulna**, united together and forming the **radio-ulna**. The two bones are quite fused at their proximal ends where they form a deep cup which articulates with the distal end of the humerus, and is drawn out into a rather prominent backwardly-projecting **olecranon process**, which ossifies from a centre distinct from that of the shaft. The distal end is distinctly divided by a groove into an anterior radial and a posterior ulnar portion.

The **manus** consists of two parts, the **wrist** or **carpus** and the **hand**.

The **carpus** consists of six small bones arranged in two rows. The three bones of the proximal row are the **ulnare**, **radiale** and **centrale**. The **ulnare** and **radiale** are about equal in size and articulate regularly with the radio-ulna. The **centrale** is pushed out of its normal position and lies partly on the pre-axial side, partly in front of the radiale. Of the three bones of the distal row the two pre-axial ones, **carpalia 1** and **2**, are small; carpale 2 articulates with the second metacarpal, carpale 1 with both the first and second. The third bone is large and articulates with the third, fourth and fifth metacarpals, it represents **carpalia 3-5**, with probably in addition the representative of a second centrale.

The **hand** consists of four complete digits, and a vestigial **pollex** reduced to a short metacarpal.

Each of the four complete digits consists of a **metacarpal** and a variable number of **phalanges**. The first digit, as just mentioned, has no phalanges, the second and third have two, and the fourth and fifth have three.

C. The Pelvic girdle.

The pelvic girdle of the Frog is much modified from the simple or general type found in the Newt (p. 149).

It is a V-shaped structure consisting of two halves which are fused together in the middle line posteriorly, while in front they are attached to the ends of the transverse processes of the sacral vertebra. Each half bears at its posterior end a deep cup, the **acetabulum**, with which the head of the femur articulates.

Each half of the pelvis ossifies from two centres. The anterior and upper half of the acetabulum, and the long laterally compressed bar extending forwards to meet the sacral vertebra ossify from a single centre and are generally called the **ilium**; it is probable however that they represent both the **ilium** and **pubis** of mammals. The posterior part of this bone meets its fellow in a median symphysis.

The posterior third of the acetabulum is formed by a small bone, the **ischium**, which likewise meets its fellow in a median symphysis.

The ventral portion of the pelvic girdle never ossifies, even in old animals being formed only of calcified cartilage. This is generally regarded as the pubis, but it perhaps corresponds to the **acetabular bone** of mammals.

D. The Posterior limb.

This corresponds closely to the anterior limb and, like it, is divisible into three parts, the **thigh**, the **shin** or **crus** and the **pes**.

As was the case with the anterior limb, all the long bones have their ends formed by prominent epiphyses which do not unite with the shaft till late in life.

In the **thigh** there is only a single bone, the **femur**.

The **femur** is a moderately long, slender bone with a well-ossified hollow shaft slightly curved in a sigmoid manner. Both ends are expanded, the proximal end is hemispherical and articulates with the acetabulum, the distal end is larger and more laterally expanded.

The **shin** likewise includes a single bone, the **tibio-fibula**, but this, as can be readily seen by the grooves at the proximal and distal ends of the shaft, is formed by the fusion of two distinct bones, the **tibia** and **fibula**. The tibio-fibula is longer and straighter than the femur.

The **pes** consists of two parts, the **ankle** or **tarsus** and the **foot**.

The **tarsus** consists of two rows of structures, very different in size. The proximal row consists of two long bones, the **tibiale** and **fibulare**, which are united by common epiphyses at the two ends, while in the middle they are widely separated. The tibiale lies on the tibial or pre-axial side, and the fibulare which is the larger of the two bones on the fibular or postaxial side. The distal row of tarsals consists of three very small pieces of calcified cartilage. The postaxial of these is the largest, it articulates with the second and third metatarsals and is probably homologous with tarsalia 2 and 3 fused. The middle one is very small, it articulates with the first metatarsal and is probably tarsale 1. The pre-axial one articulates with the metatarsal of the calcar, a structure to be described immediately, and has been regarded as a **centrale**.

The **foot** includes five complete digits and a supplemental toe as well. Each of the five digits consists of a long **metatarsal** with epiphyses at both ends, and of a variable number of phalanges. The first digit or **hallux** and the second have two phalanges, the third three, the fourth, which is the largest, four, and the fifth, three. The distal phalanges have epiphyses only at their proximal ends, the others at both ends.

On the pre-axial side of the hallux is the supplemental digit, the **prehallux** or **calcar**. It consists of a short metatarsal and one or two phalanges, and is terminated distally by a horny covering of epidermal origin.

CHAPTER XII.
GENERAL ACCOUNT OF THE SKELETON IN AMPHIBIA.

EXOSKELETON.

The exoskeleton, at any rate in most living forms, is very slightly developed in Amphibia. The only representatives of the epidermal exoskeleton are (1) the minute horny beaks found coating the premaxillae and dentaries in *Siren* and the tadpoles of most Anura, (2) the nails borne by the first three digits of the pes in *Xenopus* and by the Japanese Salamander *Onychodactylus*, (3) the horny covering of the calcar or prehallux of frogs. The Urodela and nearly all the Anura, which form the vast majority of living Amphibia, have naked skins. A few Anura belonging to the genera *Ceratophrys* and *Brachycephalus* have bony dermal plates developed in the skin of the back, and these plates become united with some of the underlying vertebrae.

In the Gymnophiona the integument bears small cycloid scales arranged in rings which are equal in number to the vertebrae. These scales contain calcareous concretions. Scales also occur between the successive rings.

In the Labyrinthodontia the dermal exoskeleton is in many genera greatly developed. It is generally limited to the ventral surface and consists principally of a buckler formed of three bony plates, one median and two lateral. These plates protect the anterior part of the thorax, and are closely connected with the adjacent endoskeleton. They probably represent the interclavicle and clavicles. Behind this buckler numerous scutes are generally developed, which often cover the whole ventral surface, and may cover the whole body.

Teeth.

In Amphibia teeth are generally present on the maxillae, premaxillae and vomers, and except in Anura on the dentaries; sometimes they occur on the palatines as in many Urodela, most Labyrinthodontia, and the Gymnophiona; less commonly on the pterygoids as in *Menobranchus*, *Siredon*, some Labyrinthodontia, and *Pelobates cultripes*, or on the splenials as in *Siren* and *Menobranchus*, or parasphenoid as in *Pelobates cultripes*, *Spelerpes belli* and *Batrachoseps*. In some Anura such as *Bufo* and *Pipa* the jaws are toothless.

In Gymnophiona, *Menobranchus*, and *Siredon*, the teeth are arranged in two concentric curved rows. The teeth of the outer row are borne on the premaxillae and maxillae if present, (the maxillae are absent in *Menobranchus*), the teeth of the second row on the vomers and pterygoids in *Menobranchus* and *Siredon*, and on the vomers and palatines in Gymnophiona. In some Gymnophiona there is a double row of mandibular teeth. The vomerine, palatine and parasphenoid teeth of all forms are numerous and are not arranged in rows.

The teeth of all living Amphibia are simple conical structures ankylosed to the bone, and consisting of dentine, coated or capped with a thin layer of enamel. In the Labyrinthodontia teeth of more than one size are sometimes present. The dentine of the basal part of the larger teeth is in some genera very greatly folded, causing the structure to be highly complicated. These folds, the intervals between which are filled with cement, radiate inwards from the exterior and outwards from the large pulp cavity. The basal part of the teeth of *Ceratophrys* (Anura) has a similar structure.

ENDOSKELETON.

Vertebral column.

Four regions of the vertebral column can generally be recognised in Amphibia, viz. the cervical, the trunk or thoraco-lumbar, the sacral and the caudal regions. In the limbless Gymnophiona, however, only three regions, the cervical, thoracic, and post-thoracic can be made out. The cervical region is limited to a single vertebra which generally differs from the others in having no transverse processes or indication of ribs. It is generally called the atlas, but it commonly bears a small process arising from the anterior face of the centrum which resembles the odontoid process of higher animals, and renders it probable that the first vertebra of Amphibia corresponds to the axis, not to the atlas. Amphibia generally have a single sacral vertebra.

Three elements go to make up the vertebral column in Amphibia, viz.

1. the notochord,

2. the long vertebral centra,

3. intervertebral cartilage which forms the joints between successive centra.

The relations which these three elements bear to one another are subject to much variation. The successive stages can be well traced in the Urodela.

1. The first stage is found in larval Urodeles in general and in adult Ichthyoidea, and some Salamandrina. In these forms the notochord persists and retains approximately the same diameter throughout the whole length of the vertebral column. Bony biconcave centra are present and constrict it to a certain extent vertebrally, while intervertebrally there is a development of cartilage. The connection between the bony vertebrae is effected mainly by the expanded notochord.

2. In the next stage, as seen in *Gyrinophilus porphyriticus*, the growth of intervertebral cartilage has caused the almost complete obliteration of the notochord intervertebrally, and its entire disappearance vertebrally, i.e. in the centre of each vertebra. The intervertebral cartilage now forms the main connection between successive vertebrae, and sometimes cases are found whose condition approaches that of definite articulations. Readily recognisable remains of the notochord are still found at each end of the intervertebral constriction.

3. In the third stage differentiation and absorption of the intervertebral cartilage has given rise to definitely articulating opisthocoelous vertebrae. These are found in most adult Salamandrina.

The transverse processes of the earlier trunk vertebrae are divided into two parts, a dorsal part which meets the tubercular process of the rib and is derived from the neural arch, and a ventral part which meets the capitular process of the rib, and is derived from the ventral or haemal arch. In the caudal vertebrae and often also in the posterior trunk vertebrae the two processes are fused.

Siren and *Proteus*, although they possess minute posterior limbs, have no sacral vertebrae, while *Cryptobranchus lateralis* has two. The caudal vertebrae, except the first, have haemal arches very similar to the neural arches.

In Labyrinthodontia the centra of the vertebrae are generally well ossified biconcave discs. In some forms however, like *Euchirosaurus*, the centra are imperfectly ossified, and consist of bony rings traversed by a wide notochordal canal. Each ring is formed of four pieces, a large well-ossified neural arch, a basal piece, and a pair of lateral pieces. Vertebrae of this type are called *rachitomous*.

In the tail region of other forms each vertebra consists of an anterior centrum bearing the neural arch, and a posterior intercentrum bearing chevron bones. Vertebrae of this type are called *embolomerous*. Haemal arches similar to the neural arches are often found as in Urodela. The transverse processes are sometimes well developed and are divided into tubercular and capitular portions.

In Gymnophiona the vertebrae are biconcave and are very numerous, they sometimes number about two hundred and thirty. Only quite the last few are ribless and so can be regarded as post-thoracic vertebrae. The first vertebra has nothing of the nature of an odontoid process.

In Anura the number of vertebrae is very greatly reduced, only nine and the urostyle being present. Of these, eight are presacral and one sacral. The urostyle is post-sacral and corresponds to three or more modified vertebrae. The first vertebra is without transverse processes, the remaining presacral vertebrae have the transverse processes fairly large, while the sacral vertebra has them very large, forming in some genera widely expanded plates. The urostyle is a long cylindrical rod which articulates with the sacrum generally by two facets. Ankylosed to its anterior end are the remains of two neural arches.

In Anura remains of the notochord are found in the centre of each vertebra, i.e. vertebrally, while in the Urodela they only occur intervertebrally.

The vertebrae in Anura are, as a rule, procoelous. The eighth vertebra is however generally amphicoelous, while the ninth commonly has one convexity in front, and two behind.

In some forms such as *Bombinator*, *Pipa*, *Discoglossus* and *Alytes* they are opisthocoelous; in others like *Pelobates* they are variable.

The Skull.

Cranium and mandible.

In the Amphibian skull there are as a rule far fewer bones than in the skull of bony fish. The primordial cartilaginous cranium often persists to a great extent. Only quite a few ossifications take place in it; namely in the occipital region—the exoccipitals, further forwards—the pro-otics, and at the boundary of the orbital and ethmoidal regions—the sphenethmoid. The basi-occipital and basisphenoid are never ossified. As in Mammalia there are two occipital condyles formed by the exoccipitals.

Large vacuities commonly occur in the cartilage of both floor and roof of the primordial cranium. These are roofed over to a greater or less extent by the development of membrane bone. Thus on the roof of the cranium there are paired parietals, frontals, and nasals, and on its floor are paired vomers, and a median unpaired parasphenoid.

In all living forms the parietals meet and there is no interparietal foramen, though this exists in Labyrinthodonts.

The palato-pterygo-quadrate bar is united at each end with the cranium, but elsewhere in most cases forms a wide arch standing away from it. The suspensorium is, as in Dipnoi and Holocephali, autostylic. The palato-pterygo-quadrate bar sometimes remains entirely cartilaginous, sometimes its posterior half is ossified forming the quadrate. In connection with it a number of membrane bones are generally developed, viz. the maxillae, premaxillae, palatines, pterygoids, quadratojugals, and squamosals. The pterygoids are, however, sometimes partially formed by the ossification of cartilage. The cartilage of the lower jaw and its investing membrane bones generally have much the same relations as in bony fishes.

Urodela. The skulls of the various Urodeles show an interesting series of modifications and differ much from one another, but all agree in the absence of the quadratojugals, in the fact that the palatines lie parallel to the axis of the cranium, and in the large size of the parasphenoid.

The lower types *Menobranchus*, *Siren*, *Proteus*, and *Amphiuma* have longer and narrower skulls than do the higher types.

Menobranchus has a very low type of skull which remains throughout life in much the same condition as that of a young tadpole or larval salamander. The roof and floor of the cranium internal to the membrane bones are formed of fibrous tissue, not of well-developed cartilage. The epi-otic regions of the skull are ossified, forming a pair of large bones which lie external to, and distinct from, the exoccipitals. *Proteus* and the Labyrinthodonts are the only other Amphibia which have these elements separately ossified. The parietals send a pair of long processes forwards along the sides of the frontals. Nasals and maxillae are absent, as is likewise the case in *Proteus*. Teeth are borne on the vomers, premaxillae, pterygoids, dentaries and angulo-splenials. The suspensorium is forwardly directed.

The skull of *Siren* resembles that of *Menobranchus* in several respects, as in the forward direction of the suspensorium and in the absence of maxillae, but differs in the possession of nasals, in the toothless condition of the premaxillae and dentaries, and in the fusion and dentigerous condition of the vomers and palatines.

Amphiuma has a skull which, though narrow and elongated, differs from those of *Menobranchus*, *Proteus*, and *Siren*, and resembles those of higher types in the following respects:—

(1) the suspensorium projects nearly at right angles to the cranium instead of being directed forwards, (2) the maxillae are well developed, and the premaxillae are completely ankylosed together, (3) there are no palatines.

The skulls of *Megalobatrachus*, *Cryptobranchus* and *Siredon* resemble those of the highest Urodeles the Salamanders in their wide form, in having the pro-otics distinct from the exoccipitals which are ossified continuously with the epi-otics and opisthotics, and in having no palatines, but differ in having the two premaxillae separate, and in the arrangement of the vomerine teeth which in *Megalobatrachus* and *Cryptobranchus* are placed along the anterior boundaries of the bones, these meeting in the middle line. In *Siredon* the vomers are separated by the very large parasphenoid.

The suspensorium in *Megalobatrachus* and *Cryptobranchus* projects at right angles to the cranium; in *Siredon* it projects somewhat downwards and forwards as in the Salamandrina.

Modifications of the vomers, pterygoids and palatines accompany the changes of the larval ichthyoid *Siredon* into the adult salamandroid *Amblystoma*, the vomers especially come to resemble to a much greater extent those of the Salamandrina.

The ossification of the skull in the Salamandrina is carried further than in the Ichthyoidea, though the supra-occipital and basi-occipital are not ossified. The skull differs from that in the Ichthyoidea in the size of the part of the vomero-palatines which lies in front of the teeth, in the frequent union of the two premaxillae and in the ossification of all the periotic bones continuously with the exoccipital.

The skull differs from that of Anura in the following respects:—

(1) the bones of the upper jaw do not form a complete arch standing away from the cranium, and the maxillae are not united to the quadrates by quadratojugals, (2) the long axis of the suspensorium passes obliquely downwards and forwards instead of downwards and backwards, (3) there is no sphenethmoid encircling the anterior end of the brain, its place being partly taken by a pair of orbitosphenoids, (4) there is no definite tympanic cavity.

85

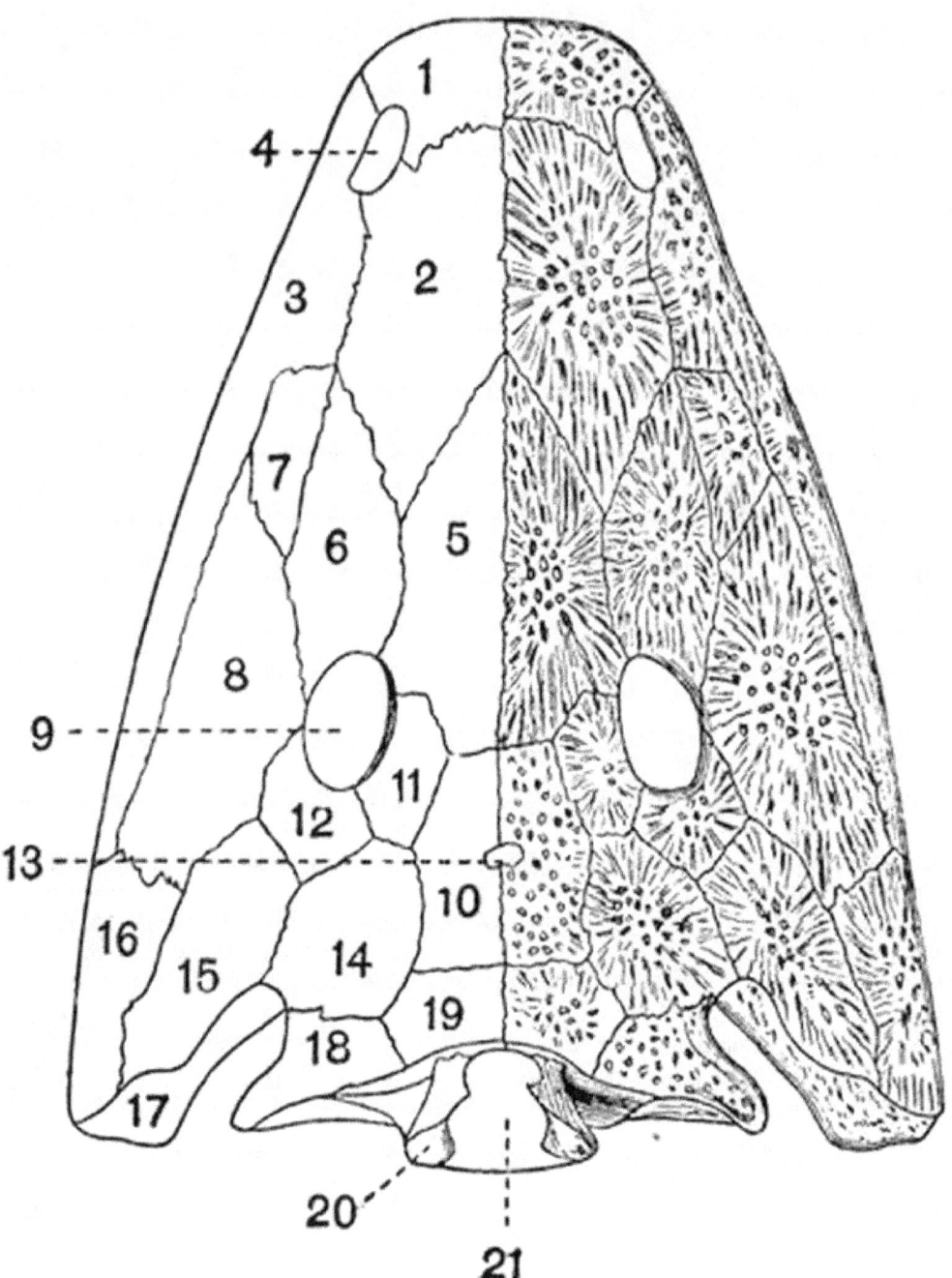

Fig. 27. Dorsal view of the skull of a Labyrinthodont (*Capitosaurus nasutus*) × 1/9 (from von Zittel).

1. premaxillae.

2. nasal.

3. maxillae.

4. anterior nares.

5. frontal.

6. prefrontal.

7. lachrymal.

8. jugal.

9. orbit.

12. postorbital.

13. interparietal foramen.

14. squamosal.

15. supratemporal.

16. quadratojugal.

17. quadrate.

18. epi-otic.

19. dermo-supra-occipital.

20. exoccipital.

10. parietal.

21. foramen magnum.

11. postfrontal.

Labyrinthodontia. The skull in Labyrinthodontia is remarkable for its extreme solidity, the large number of bones which are present, and the extent to which the roofing over of the temporal and other fossae has taken place. In many forms the surface of the bones is as in Crocodiles, strongly sculptured (fig. 27, right half) with ridges and grooves which probably lodged sensory organs. The bones forming the roof of the skull are generally very uniform in size, perhaps the most noticeable of them being the paired dermo-supra-occipitals (fig. 27, 19). Paired dermo-supra-occipitals occur also in certain Ganoids. The Labyrinthodont skull also bears resemblance to that of many fish in the development of a pair of long pointed epi-otics (fig. 27, 18), which remain permanently distinct from the surrounding bones. The parietals are small and enclose between them the interparietal foramen (fig. 27, 13). In some forms in which the head is protected with an armour of scutes, these do not roof over the interparietal foramen, and from this fact it has been inferred that the Labyrinthodonts had a functional pineal eye. Both supra- and infra-temporal fossae are partially or completely roofed over by the postorbitals and large supra-temporals (fig. 27, 15).

There is generally a ring of bones in the sclerotic coat of the eye. The pterygoids do not meet in the middle line, being separated by the parasphenoid. The palatines bear teeth, and in some genera (*Archegosaurus*) form long splints lying along the inner side of the maxillae and more or less surrounding the posterior nares. In others (*Nyrania*) they lie in the normal position near the middle line, one on each side of the parasphenoid. The vomers bear teeth and sometimes meet in the middle line; they are sometimes confluent with the parasphenoid. On the ventral surface of the cranium there are generally large palatal vacuities.

In the mandible there is often a well-marked postglenoid process, and the articular is generally completely ossified.

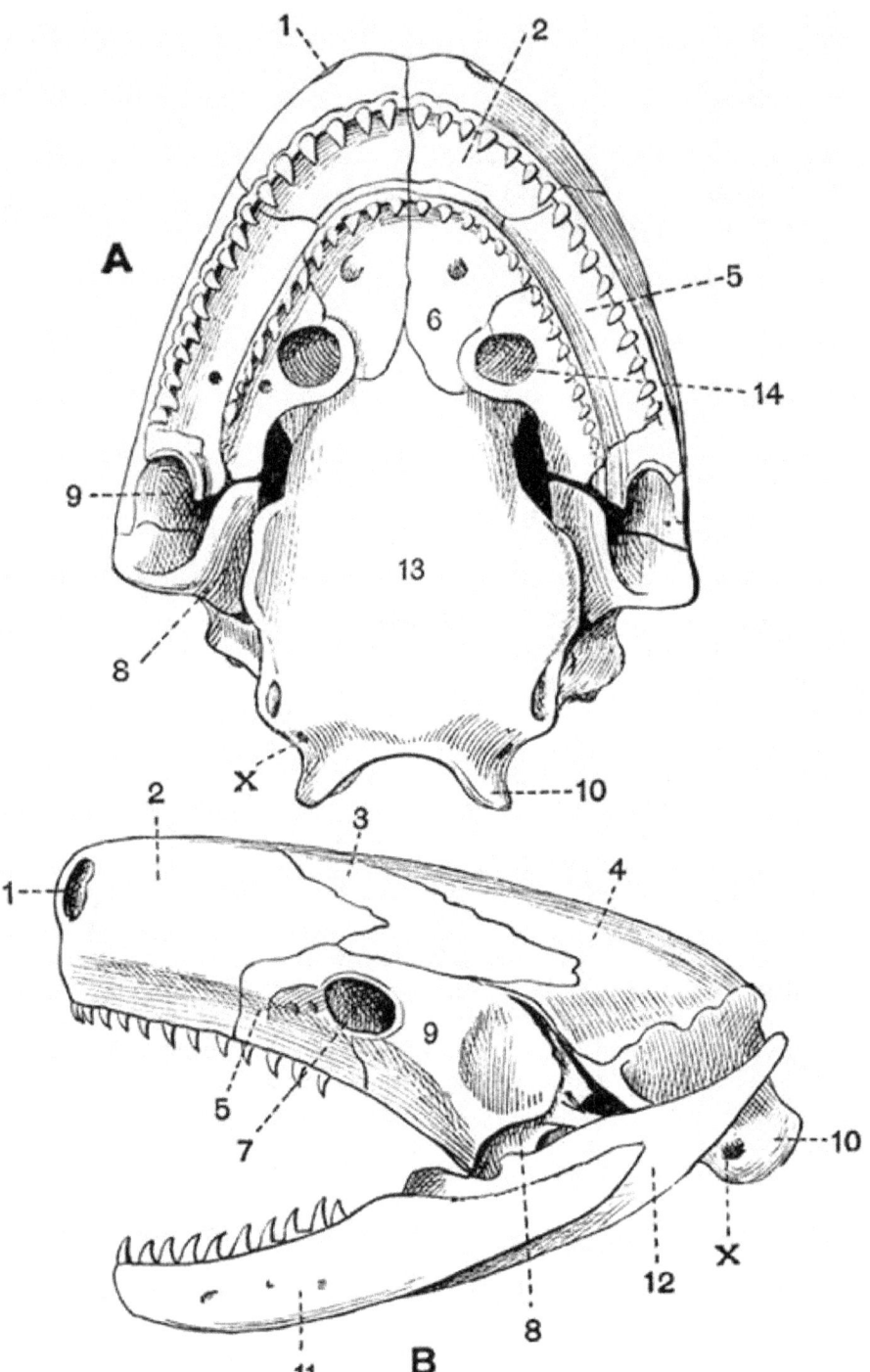

Fig. 28. A, ventral view of the cranium; B, lateral view of the cranium and mandible of *Siphonops annulatus* (after Wiedersheim).

1. anterior nares.
2. naso-premaxillae.
3. frontal.
4. parietal.
5. maxillae.
6. vomer.
9. squamosal.
10. exoccipital.
11. dentary.
12. angular.
13. basi-occipital and basisphenoid fused.

7. orbit.	14. posterior narial opening
8. quadrate united with the	surrounded by the palatine.
pterygoid in front.	X. pneumogastric foramen.

Gymnophiona. The skull bears a considerable resemblance to that of Labyrinthodonts, especially in the arrangement of the bones which bound the mouth cavity. The cranium is very hard, and is covered by a complete bony roof formed mainly of the exoccipitals, parietals, frontals, prefrontals, nasals and premaxillae. The nasals and premaxillae are sometimes ossified continuously. There is a median unpaired ethmoid whose dorsal end appears at the surface wedged in between the frontals and parietals. The bone generally regarded as the squamosal is very large, and it and the maxillae generally together surround the orbit, which, in *Epicrium*, has in it a ring of bones. The palatines form long tooth-bearing bones fused with the inner sides of the maxillae; they nearly surround the posterior nares.

The quadrate bears the knob, and the angular the cup for the articulation of the mandible,—a very primitive feature. The mandible is also noticeable for the enormous backward projection of the angular.

Anura. In Anura the skull is very short and wide owing to the transverse position of the suspensorium. There is often a small ossification representing the quadrate. Sometimes as in *Hyla* and *Alytes* there is a fronto-parietal fontanelle.

As compared with the skull in Urodela the chief characteristics of the skull of Anura are:—

1. the presence of a sphenethmoid,

2. the union of the frontals and parietals on each side,

3. the occasional occurrence of small supra- and basi-occipitals,

4. the backward growth of the maxillae and its connection with the suspensorium by means of the quadratojugal,

5. the dagger-like shape of the parasphenoid,

6. the occurrence of a definite tympanic cavity,

7. the frequent occurrence of a predentary or mento-meckelian ossification in the mandible.

The skull of *Pipa* is abnormal, being greatly flattened and containing little cartilage. The fronto-parietals are fused, and there is no sphenethmoid. The quadrates are well developed and the squamosals and parasphenoid differ much from those of other Anura.

Hyoid and branchial arches.

In larval Amphibia the hyoid and four branchial arches are generally present, and in adult Ichthyoidea they are frequently almost as well represented as in the larva, and are of use in strengthening the swallowing apparatus. They are very well seen in *Siredon*, and consist of a hyoid attached by ligaments to the suspensorium, followed by four branchial arches of which the first and second are united by a copula (fig. 29, D, 8), while the third and fourth are not. The hyoid is not always the largest and best preserved of the arches, for sometimes as in *Spelerpes* one of the branchials is far larger than the hyoid. Four branchial arches occur in *Siren* as in *Siredon*, but in *Proteus* there are only three.

In some larval Labyrinthodontia (*Branchiosaurus*) four branchial arches are known to occur, and their arrangement is almost precisely similar to that in *Siredon*.

In Gymnophiona the remains of only three branchial arches occur in addition to the hyoid. The four arches are all very similar to one another, each consists of a curved rod of uniform diameter throughout. The hyoid is united with the first branchial arch, but has no attachment to the cranium.

In larval Anura (fig. 29, C) the arrangement of the hyoid and branchial arches is much as in Urodela. In the adult, however, the ventral parts of all the arches unite, forming a compact structure, the *basilingual plate* (fig. 29, B, 1).

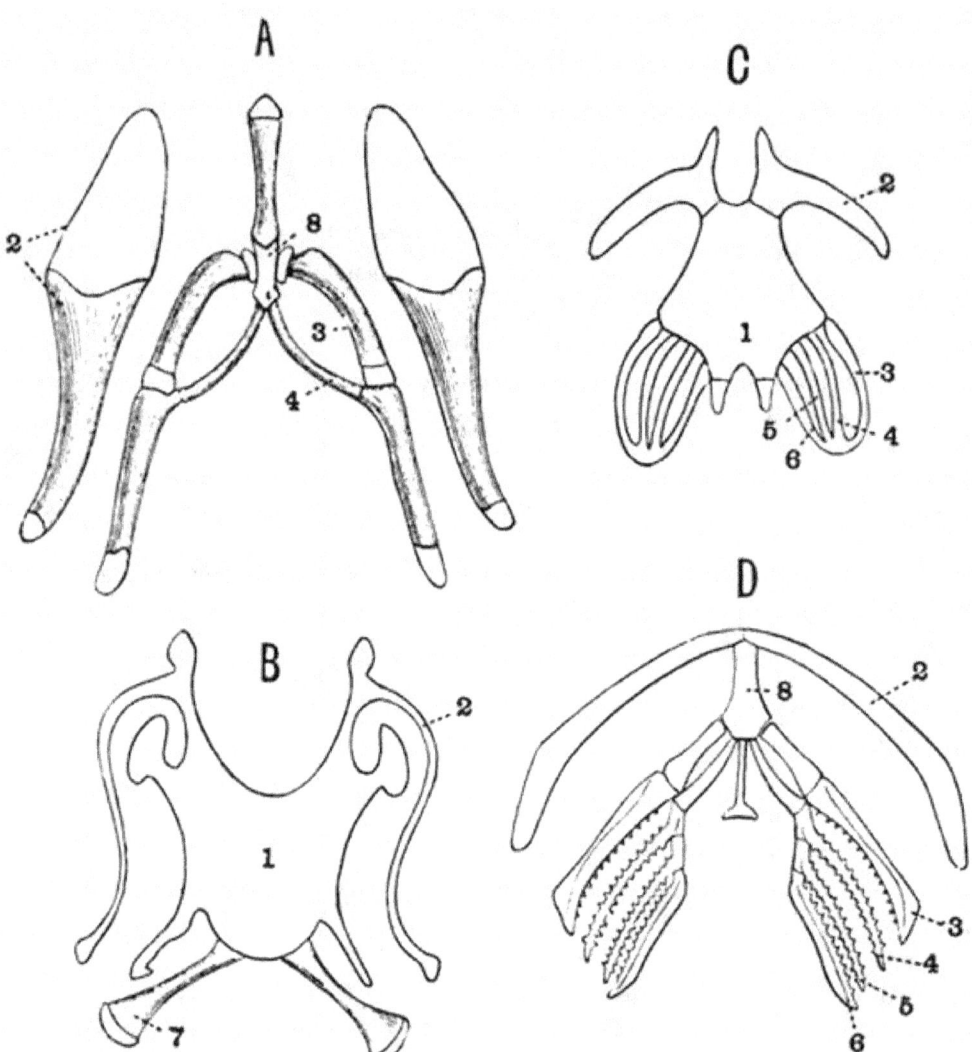

Fig. 29. Visceral arches of Amphibia.

A. *Molge cristata*   (after Parker).

B. *Rana temporaria*   adult (after Parker).

C. Tadpole of *Rana*   (after Martin St Ange).

D. *Siredon pisciformis*   (after Credner).

In each case the ossified portions are slightly shaded, while the cartilaginous portions are left white.

1. basilingual plate.   5. third branchial arch.

2. hyoid arch.   6. fourth   do.

3. first branchial arch.   7. thyro-hyal.

4. second   do.   8. copula.

The dorsal parts of the first three branchial arches disappear, but those of the fourth become ossified and form the short, stout thyro-hyals or posterior cornua. The dorsal parts of the hyoid arch in the adult form a pair of long bars, the anterior cornua, which are united to the periotic region of the skull in front of the fenestra ovalis either by short ligaments or by fusion as in *Bufo*. In *Pipa* and *Xenopus* the first and second branchial arches persist as well as the fourth (thyro-hyal), but in *Pipa* the hyoid is wanting.

Ribs.

Ribs are generally very poorly developed in Amphibia. In Anura they are in most cases absent; when present they generally form minute unossified appendages attached to the transverse processes, but in *Discoglossus* and *Xenopus* the anterior vertebrae are provided with distinct ribs. In Urodela and Labyrinthodontia they are generally short structures, each as a rule attached to the vertebra by a bifurcated proximal end. The number of rib-bearing vertebrae varies, but the first and the posterior caudal vertebrae are always ribless. The anterior caudal vertebrae too are generally ribless, but sometimes a few of them bear small ribs. In *Spelerpes* the last two trunk vertebrae are ribless, and hence may be regarded as lumbar vertebrae.

In Gymnophiona ribs are better developed than in any other Amphibia; they occur on all the vertebrae except the first and last few, and are attached to the transverse processes, sometimes by single, sometimes by double heads.

Sternal ribs are almost unknown in Amphibia, but traces of them occur in *Menobranchus.*

Sternum.

In Amphibia the sternum is not very well developed; sometimes as in Gymnophiona and *Proteus* no traces of it occur, and in the Urodela it is never ossified. It is always very intimately related to the pectoral girdle. In the Salamandrina it has the form of a broad thin plate of cartilage, grooved and overlapped by the coracoid.

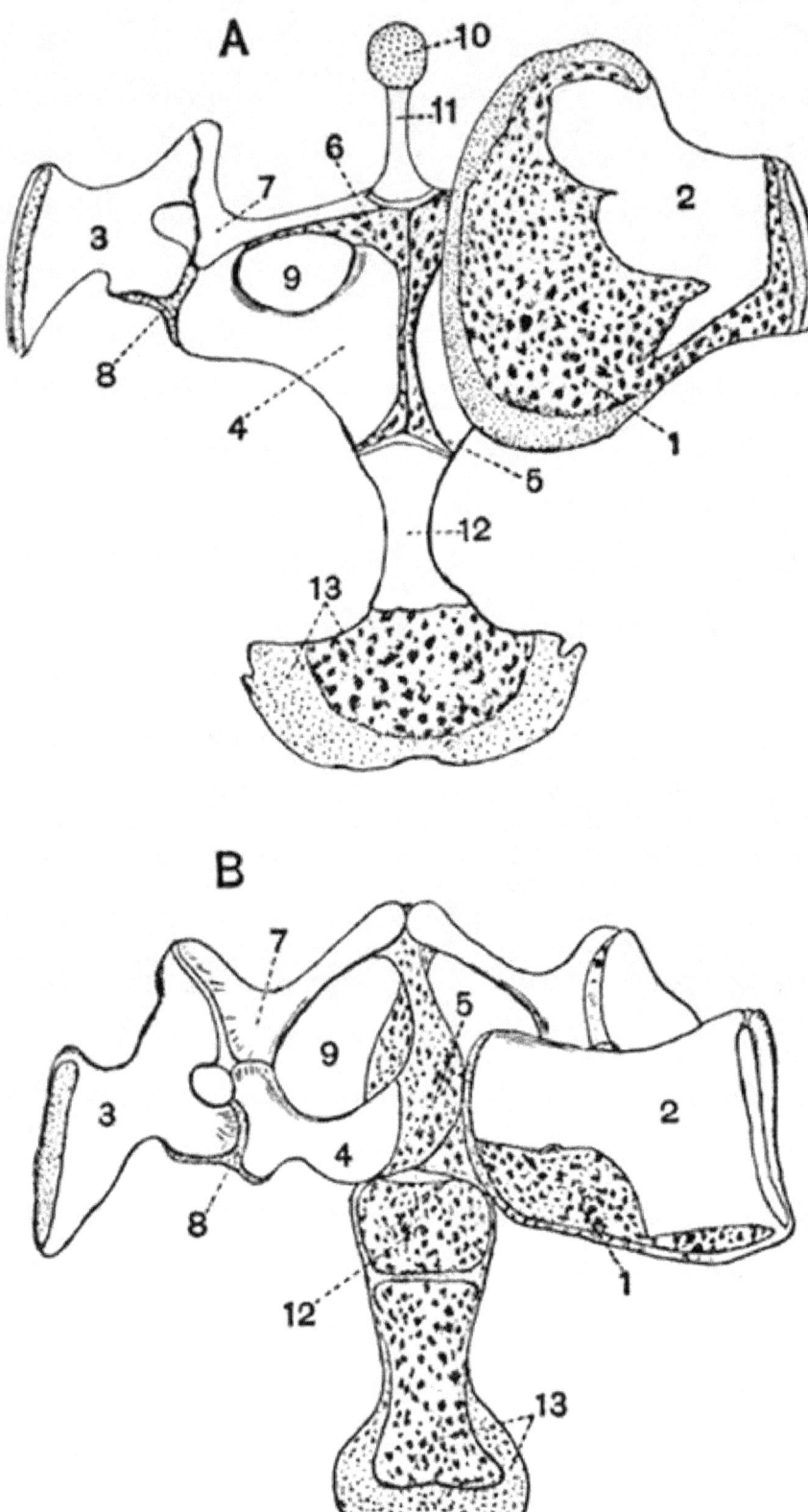

Fig. 30. Shoulder-girdle and sternum of
A. An old male common Frog (*Rana temporaria*).
B. An adult female *Docidophryne gigantea* (after Parker).

In both A and B the left suprascapula is removed. The parts left
unshaded are ossified; those marked with small dots consist of hyaline

cartilage, those marked with large dots of calcified cartilage.

1. calcified cartilage of suprascapula.
2. ossified portion of suprascapula.
3. scapula.
4. coracoid.
5. epicoracoid.
6. precoracoid.
7. clavicle.
8. glenoid cavity.
9. coracoid foramen.
10. episternum.
11. omosternum.
12. sternum.
13. xiphisternum.

In most Anura the sternum consists of a number of parts arranged in series. At the anterior end is a flat cartilaginous plate with a bony basal stalk. This plate is called the episternum, and its stalk the omosternum. The continuity of the sternum is now interrupted by a pair of cartilaginous structures, the epicoracoids, which are shoulder-girdle elements, and represent the unossified ventral ends of the coracoids. In some cases cartilaginous epiprecoracoids can also be distinguished. Further back is the long sternum proper, while last comes the xiphisternum, a broad expanded plate of cartilage.

In some Anura such as *Pipa* and *Hyla* the number of sternal elements is considerably reduced.

Appendicular Skeleton.

Pectoral girdle.

The most primitive Amphibian shoulder-girdle is found in the Urodela. It consists of a dorsal element, the scapula, a posterior ventral element, the coracoid, and an anterior ventral element, the precoracoid. The clavicle is not developed, and the two coracoids overlap in the middle line. The shoulder-girdle remains largely cartilaginous but the proximal end of the scapula is ossified, and the ossification may extend through part of the coracoid and precoracoid.

In Labyrinthodontia there is an exoskeletal ventral buckler formed of three plates, a median one, which probably represents an interclavicle, and two lateral ones, which are probably clavicles. Traces of endoskeletal structures, probably corresponding to the precoracoid and scapula, are also known in some cases. The Gymnophiona and some of the Labyrinthodontia have lost the pectoral girdle and limbs.

The ossification of the shoulder-girdle has gone on much further in Anura than it has in Urodela. Clavicles are present and the scapula and coracoid of each side are ossified from separate centres. The distal part of the scapula forms a large imperfectly ossified plate, the suprascapula.

The shoulder-girdle of Anura is however subject to considerable variations. In the Toads (Bufonidae) the epicoracoids or unossified ventral ends of the coracoids and precoracoids overlap in the middle line (fig. 30, B, 5). This arrangement is called *Arciferous*. In the Frogs,—Ranidae, and other forms belonging to the group *Firmisternia*,—the epicoracoids do not overlap but form a narrow cartilaginous bar separating the ventral ends of the coracoids (fig. 30, A, 5).

Anterior limb.

In many Amphibia and especially in the Urodela the anterior limb has a very simple unmodified arrangement. The humerus is straight and of moderate length, its ends are rounded for articulation on the one hand with the shoulder-girdle, and on the other hand with the radius and ulna. In the Urodela the radius and ulna are distinct. In the Anura they have fused, though the line of junction of the two is not obliterated. Their proximal ends are hollowed for articulation with the convex end of the humerus.

The manus in all recent Amphibia agrees in never having more than four complete digits, but is subject to considerable variation, this statement applying especially to the carpus.

In the larva of Salamandra (fig. 31, A), except that the pollex is absent, the manus retains completely the condition which is generally regarded as primitive for the higher Vertebrata. It consists of an anterior row of three elements, the ulnare, intermedium, and radiale, and a posterior row of four, the carpalia 2, 3, 4, and 5. Interposed between the two rows is a centrale. *Menobranchus* has a similar very simple carpus. In most other Amphibia this simplicity is lost. This loss may be due to:—

(*a*) fusion of certain structures, e.g. in the adult *Salamandra* the intermedium and ulnare have fused,

(*b*) displacement of structures, e.g. in *Bufo viridis*, the centrale has been pushed up till it comes to articulate with the radius,

(*c*) the development of supernumerary elements, especially of extra centralia. In *Megalobatrachus* two or even three centralia sometimes occur.

93

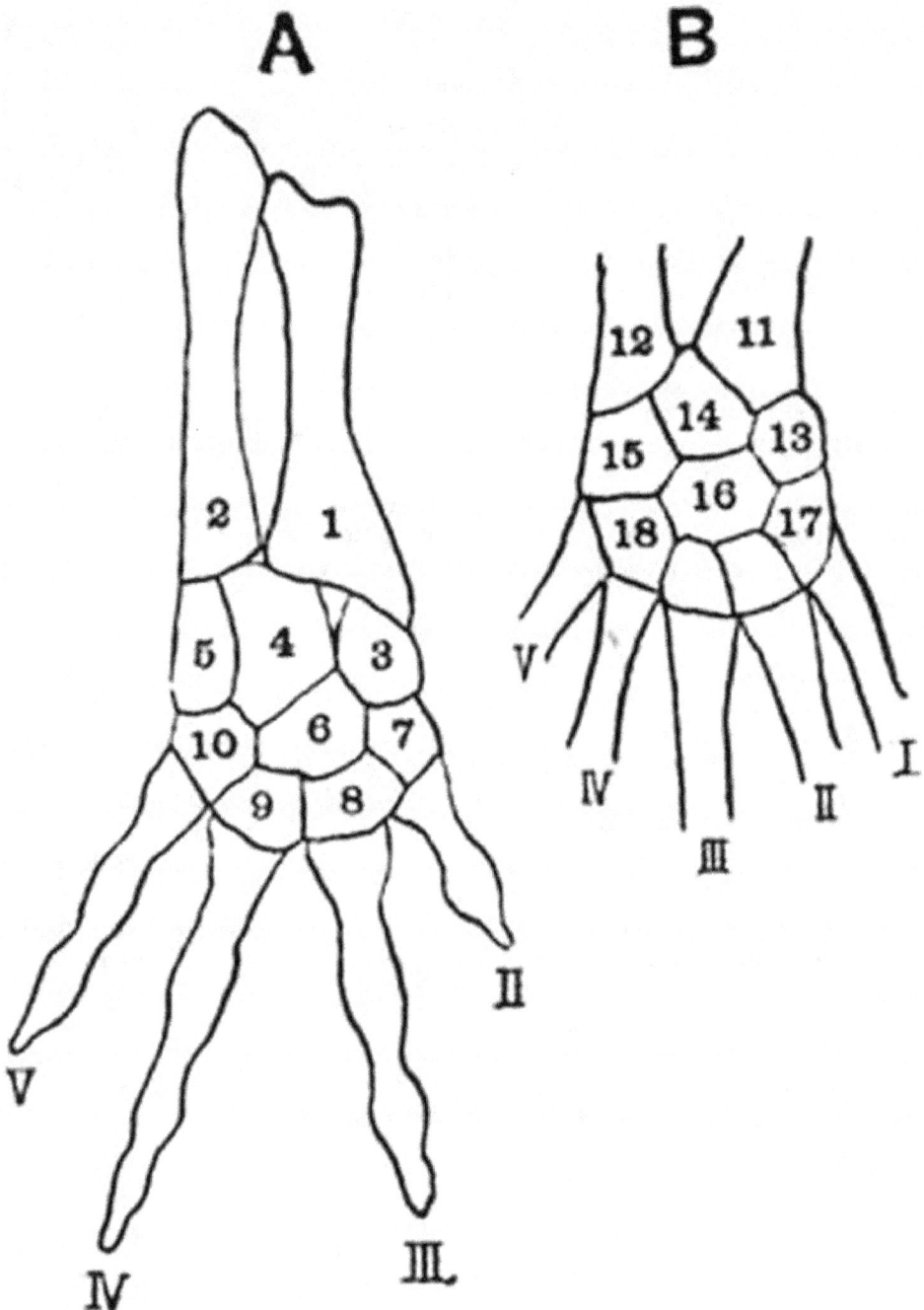

Fig. 31. A, Right Antibrachium and Manus of a larval Salamander (*Salamandra maculosa*) (after Gegenbaur).
B, Right Tarsus and adjoining bones of *Molge sp.* (after Gegenbaur).

1. radius.	11. tibia.
2. ulna.	12. fibula.
3. radiale.	13. tibiale.
4. intermedium.	14. intermedium.
5. ulnare.	15. fibulare.
6. centrale.	16. centrale.
7. carpale 2.	17. tarsale 1.
8. " 3.	18. tarsalia 4 and 5 fused.

9. " 4.   I. II. III. IV. V. digits.

10. " 5.

In the great majority of Amphibia while one digit, probably the first, is absent, the other four digits are well developed. In the forms however with degenerate limbs like *Amphiuma*, *Siren* and *Proteus* the number of digits is still further reduced. In *Siren* there are three or four, in *Proteus* three, and in *Amphiuma* two or three digits in the manus.

In Anura the pollex is represented only by a short metacarpal. There are sometimes traces of a prepollex. The carpus often has two centralia and the intermedium is absent.

In Labyrinthodontia the limbs are generally very simple and resemble those of Urodela. In some forms, however, the manus differs from that of all living Amphibia in possessing five well-developed digits.

Pelvic girdle.

The simplest Amphibian pelvis is that of some of the Labyrinthodontia; thus in *Mastodonsaurus* it consists dorsally of a short broad ilium placed vertically and attached to the sacrum, and ventrally of a small pubis and of a large ischium meeting its fellow in the middle line. In some Labyrinthodonts the pubes as well as the ischia meet in a ventral symphysis, and in many there are no obturator foramina. In *Siren*, Gymnophiona and some Labyrinthodontia the pelvic girdle and limbs are absent.

In Urodela the ventral element of the pelvis on each side forms a flat plate which meets its fellow of the opposite side. The anterior part of the plate, representing the pubis, generally remains cartilaginous throughout life; the posterior part representing the ischium is in almost every case well ossified. Attached to the anterior end of the pubes there is an unpaired bifid cartilaginous structure, the epipubis. The ilia are vertically placed.

In most Anura the pelvis is peculiarly modified in correlation with the habits of jumping. The long bone generally called the ilium is placed horizontally and is attached at its extreme anterior end to the sacrum. The ischium is ossified and distinct. Ventrally in front of the ischium there is a tract of unossified cartilage which is often regarded as the pubis. In *Xenopus*, however, the bone corresponding to the ilium of the Frog is seen to ossify from two centres, one forming the ilium, the other, which lies at the symphysis, being apparently the pubis. This makes it probable that the so-called ilium of the Frog is really to be regarded as an ilio-pubis, and renders the homology of the cartilaginous part uncertain, but it probably corresponds to the acetabular bone of mammals. In *Xenopus* also there is a minute epipubis similar to that of Urodeles.

Posterior limb.

In Urodela the posterior limb (fig. 31, B) closely resembles the anterior limb, but is even less removed from the primitive condition of the higher vertebrates in the fact that all five digits are commonly present. The tibia and fibula are short bones approximately equal in size. In some cases the number of digits is reduced. Thus in *Menobranchus* the pes has four digits, in *Proteus* it has two, and in *Amphiuma* two or three, while in *Siren* the posterior limbs have atrophied.

In correlation with their habits of jumping, the posterior limbs in Anura are much lengthened and considerably modified. The tibia and fibula are completely fused. The intermedium is absent, while the tibiale and fibulare are greatly elongated. Tarsalia 4 and 5 are absent. Five digits are always present, and there is a prehallux formed of two or more segments.

In general the posterior limbs in Labyrinthodontia bear the closest resemblance to the anterior limbs; in some cases three centralia are found.

In Ichthyoidea, and in most Labyrinthodontia, the cartilages of the carpus and tarsus remain unossified; in Salamandrina and in Anura they are generally ossified.

CHAPTER XIII.
SAUROPSIDA.

This great group includes the Reptiles and Birds and forms the second of the three into which the Gnathostomata may be divided. There is nearly always a strongly-developed epiblastic exoskeleton which has the form of scales or feathers, and in some cases a dermal exoskeleton is also well developed. In living forms the notochord never persists, being replaced by vertebrae, but in some extinct forms the centra are notochordal. The vertebral centra are ossified, and only in exceptionally rare cases have terminal epiphyses. The skull is well ossified and has membrane bones incorporated in its walls.

The occipital segment is completely ossified, and an interorbital septum or bony partition separating the two orbits is usually developed to a greater or less extent. The skull generally articulates with the vertebral column by a single occipital condyle into the composition of which the exoccipitals and basi-occipital enter in varying proportions. The pro-otic ossifies, and either remains distinct from the epi-otic and opisthotic throughout life, or unites with them only after they have fused with the adjacent bones. The hyoid and branchial arches are much reduced; and the representative of the hyomandibular is connected with the auditory apparatus, forming the auditory ossicles. Each ramus of the mandible always consists of a cartilage bone, the articular, and several membrane bones. The mandible articulates with the cranium by means of a quadrate. The ribs in Birds and some Reptiles bear *uncinate processes*, i.e. small, flat, bony or cartilaginous plates projecting backwards from their posterior borders. The sternum is not transversely segmented as in mammals, and there are commonly distinct cervical ribs. The ankle joint is intertarsal, or situated between the proximal and distal row of tarsal bones, not cruro-tarsal as in Mammalia.

Class I. Reptilia.

The axial skeleton is generally long, and that of the limbs frequently comparatively short, or sometimes absent.

The exoskeleton generally has the form of epidermal scales, which are often combined with underlying bony dermal plates or scutes and may sometimes form a continuous armour. Neither feathers nor true hairs are ever present. The vertebral column is generally divisible into the five usual regions. The centra of the vertebrae vary enormously, and may be amphicoelous, procoelous, opisthocoelous or flat, but they never have saddle-shaped articulating surfaces. The quadrate is always large, and is sometimes fixed, sometimes movable. A transpalatine bone uniting the pterygoid and maxillae is generally present.

Free ribs are often borne along almost the whole length of the trunk and tail, and often occur attached to the cervical vertebrae. The sacrum is generally composed of two vertebrae which are united with the ilia by means of expanded ribs. The sternum is rhomboidal, and may either be cartilaginous or formed of cartilage bone, but never of membrane bone; it differs from that of birds also in the fact that it does not ossify from two or more centres. An interclavicle is generally present. There are always more than three digits in the manus, and never less than three in the pes. In all living reptiles the ilia are prolonged further behind the acetabula than in front of them, and the bones of the pelvis remain as a rule, distinct from one another throughout life.

The pubes (pre-pubes) and ischia both commonly meet in ventral symphysis, and the acetabula are wholly or almost wholly ossified. The metatarsals are not ankylosed together.

Order 1. Theromorpha.

This order includes a number of mainly terrestrial, extinct reptiles, which differ much from one another, and show remarkable points of affinity on the one hand with the Labyrinthodont Amphibia, and on the other with the Mammalia. The vertebrae are nearly always amphicoelous and sometimes have notochordal centra. The skull is short and has the quadrate immovably fixed. There is an interparietal foramen, and generally large supratemporal fossae bounded by supratemporal arcades, but with no infratemporal arcades; *Elginia* however has the whole of the temporal region completely roofed over.

The teeth are placed in distinct sockets and are very variable in form, the dentition sometimes resembling the heterodont dentition of mammals. The humerus has distinct condyles and an ent-epicondylar foramen as in many mammals.

The pubis is fused with the ischium, and both pectoral and pelvic girdles are remarkably solid. The obturator foramen is remarkably small or even absent. The anterior ribs have two articulating surfaces, and each articulates by its tuberculum with the transverse process, and by its capitulum with the centrum as in mammals.

These reptiles occur chiefly in deposits of Triassic and Permian age. Some of the best known genera are *Dicynodon, Udenodon, Placodus, Pariasaurus* and *Galesaurus*. They will be noticed in the general account of the skeleton in reptiles.

Order 2. Sauropterygia.

This order includes a number of extinct marine reptiles, devoid of an exoskeleton. The tail is short, the trunk long, and the neck in the most typical forms extremely long. The vertebrae have slightly biconcave, or nearly flat centra. The skull is relatively small and has large supratemporal fossae. The teeth are placed in distinct sockets, and are generally confined to the margins of the jaws; they are sharp and curved and are coated with grooved enamel. The premaxillae are large, and there is an interparietal foramen. The quadrate is firmly united to the cranium. The anterior nares are separate and are placed somewhat close to the orbits. There is no ossified sclerotic ring. The palatines and pterygoids meet the vomers, and more or less completely close the palate, and in some forms, e.g. *Plesiosaurus*, there is a distinct parasphenoid. Thoracic ribs are strongly developed and each articulates with its vertebra by a single head. The cervical vertebrae have well-marked ribs, which articulate only with the centra, in this respect differing from those of Crocodiles. The caudal vertebrae bear both ribs and chevron bones, and abdominal splint-ribs are largely developed.

In the shoulder-girdle the coracoids are large and meet in a ventral symphysis; precoracoids and a sternum are apparently absent, but parts generally regarded as the clavicles and interclavicle are well developed. In the pelvis, the pubes and ischia meet in a long symphysis. The limbs are pentedactylate, and in the best known forms, the Plesiosauridae, form swimming paddles.

The Sauropterygia occur in beds of Secondary age, and some of the best known genera are *Plesiosaurus, Pliosaurus* and *Nothosaurus*.

Order 3. Chelonia.

In the Tortoises and Turtles the body is enclosed in a bony box, formed of the dorsal carapace, and a flat ventral buckler, the plastron. Except in *Dermochelys* the carapace is partly formed from the vertebral column and ribs, partly from dermal bones. Both carapace and plastron are, except in *Dermochelys, Trionyx* and their allies, covered with an epidermal exoskeleton of horny plates, which are regularly arranged, though their outlines do not coincide with those of the underlying bones. The thoracic vertebrae have no transverse processes, and are quite immovably fixed, but the cervical and caudal vertebrae are very freely movable. There are no lumbar vertebrae. The skull is extremely solid, and frequently has a very complete false roof. Teeth have been detected in embryos of *Trionyx* but with this exception the jaws are toothless, and are encased in horny beaks. The quadrate is firmly fixed. The facial part of the skull is very short, and the alisphenoidal and orbitosphenoidal regions are unossified. In living forms there are no separate nasal bones, while large prefrontals and postfrontals are developed. There is a comparatively complete bony palate chiefly formed of the palatines and pterygoids. The anterior nares are united and placed at the anterior end of the skull, and the premaxillae are very small. There is no transpalatine bone and the vomer is unpaired. The dentaries are generally fused together.

There are ten pairs of ribs, and each rib has only a single head and is partially attached to two vertebrae; there are no cervical or sternal ribs. There is no true sternum.

The three anterior elements of the plastron are respectively homologous with the interclavicle and two clavicles of other reptiles, while the remaining elements of the plastron are probably homologous with the abdominal ribs of Crocodiles. The pectoral girdle lies within the ribs, and the precoracoids and coracoids do not meet in ventral symphyses. The scapula and precoracoid are ossified continuously. The pubis probably corresponds with the pre-pubis of Dinosaurs. There are four limbs each with five digits.

The order includes three suborders:—

Suborder (1). Trionychia.

The carapace and plastron have a rough granular surface covered with skin and without any horny shields.

The plastron is imperfectly ossified, and marginal bones may be absent, or if present are confined to the posterior portion of the carapace. The pelvis is not united to the plastron. The cranium has not a complete false roof and the head can be drawn back under the carapace.

The first three digits of both manus and pes bear claws, and the fourth digit in each case has more than three phalanges. The most important genus is *Trionyx.*

Suborder (2). Cryptodira.

The carapace and plastron vary in the extent to which they are ossified, and except in *Dermochelys* and its allies are covered by horny plates. Marginal bones are always present. The head can generally be drawn back under the carapace. The pelvis is not firmly united to the plastron. The cranium often has a complete false roof, and in the mandibular articulation the cup is borne by the cranium, and the knob by the mandible. Among the more important genera are *Dermochelys, Chelone,* and *Testudo.*

Suborder (3). Pleurodira.

The carapace and plastron are strongly ossified, and firmly united to the pelvis. The head and neck can be folded laterally under the carapace, but cannot be drawn back under it. The cranium has a more or less complete false roof, and in the mandibular articulation the knob is borne by the cranium, and the cup by the mandible. *Chelys* is a well-known genus.

Order 4. Ichthyosauria.

The order includes a number of large extinct marine reptiles whose general shape is similar to that of the Cetacea. The skull is enormously large, and the neck short. The tail is very long, and is terminated by a large vertically-placed bilobed fin, the vertebral column running along the lower lobe. The very numerous vertebrae are short and deeply biconcave. The vertebral column can be divided into caudal and precaudal regions only, as the ribs which begin at the anterior part of the neck are continued to the posterior end of the trunk without being connected with any sternum or sacrum. The precaudal vertebrae bear two surfaces for the articulation of the ribs, while in the caudal vertebrae the two surfaces have coalesced. The caudal region is also distinguished by its chevron bones. The vertebrae have no transverse processes, and the neural arches are not firmly united to the centra, and have only traces of zygapophyses. The atlas and axis are similar to the other vertebrae, but there is a wedge-shaped intercentrum between the atlas and the skull, and another between the atlas and the axis. The skull is greatly elongated (fig. 32) and pointed, mainly owing to the length of the premaxillae. The orbits are enormous, and there is a ring of bones in the sclerotic (fig. 32, 15). The anterior nares are very small; and are placed far back just in front of the orbits. There is an interparietal foramen, and the supratemporal fossae (fig. 32, 9) are very large, while there are no infratemporal fossae. An epipterygoid occurs. The quadrate is firmly fixed to the cranium, and there is a large parasphenoid. There are large prefrontals, but the frontals are very small. The very numerous teeth are large and conical, and are placed in continuous grooves without being ankylosed to the bone. They are confined to the jaw-bones.

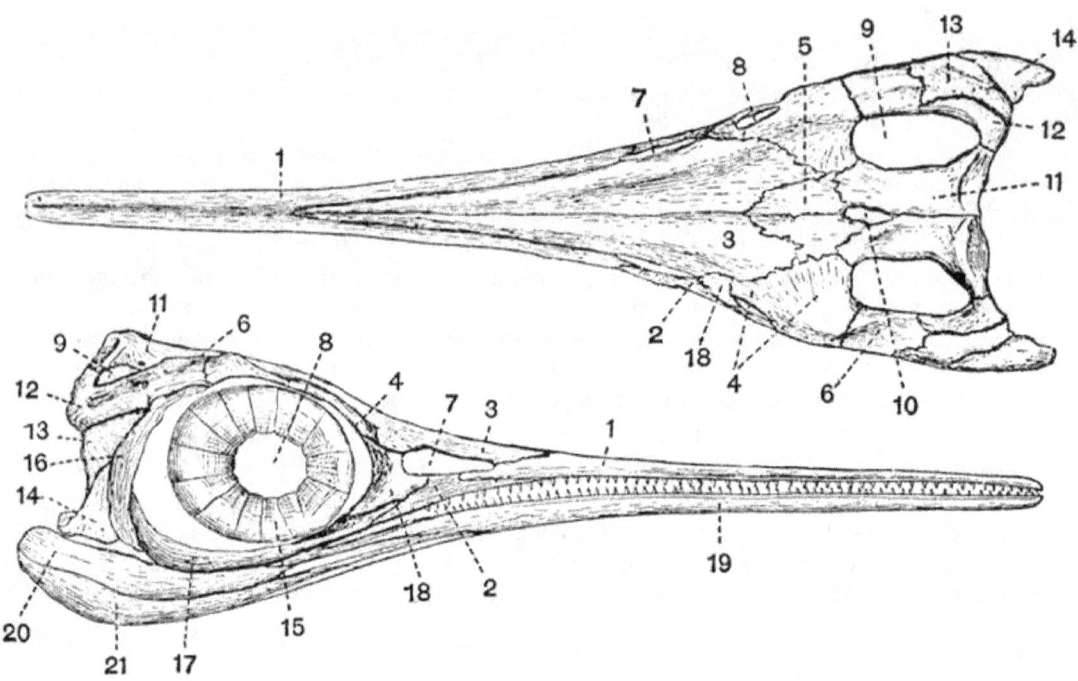

Fig. 32. Lateral (below) and dorsal (above) views of the skull of an *Ichthyosaurus*. (Modified from Deslongchamps.)

1. premaxillae.	12. squamosal.
2. maxillae.	13. supratemporal.
3. nasal.	14. quadratojugal.
4. prefrontal.	15. sclerotic ring.
5. frontal.	16. postorbital.
6. postfrontal.	17. jugal.
7. anterior nares.	18. lachrymal.
8. orbit.	19. dentary.
9. supratemporal fossa.	20. articular.
10. interparietal foramen.	21. angular.
11. parietal.	

The ribs are long, and the anterior ones have capitula and tubercula. There is no sternum, but the ventral body wall is strengthened by a complex system of abdominal splint ribs.

The pectoral girdle is strongly developed, the scapulae are narrow, the coracoids broad, and meet ventrally without overlapping. There are probably no precoracoids, but clavicles and a T-shaped interclavicle are well developed.

The limbs are very short, and completely modified into swimming paddles. The humerus and femur are both short, while the radius and ulna, tibia and fibula are generally still further reduced to the form of short polygonal bones.

The digits are formed of longitudinal series of very numerous small bones. The number of digits is five, but there sometimes appear to be more owing to the bifurcation of certain of them, or to the addition of marginal bones, either to the radial or ulnar side of the limb. The humerus has no foramen, and both humerus and femur are unique in that they are distally terminated by concave surfaces instead of by convex condyles. The pelvic limb is much smaller than the pectoral. The pelvis has no bony connection with the vertebral column, and all the component bones are small and rod-like.

The Ichthyosauria are confined to beds of Secondary age and by far the best known genus is *Ichthyosaurus*.

Order 5. Rhynchocephalia.

This order includes the living *Sphenodon* (*Hatteria*) and various extinct forms. The general shape of these animals is lizard-like and the tail is long.

The vertebrae are amphicoelous or sometimes nearly flat, and the notochord sometimes persists to some extent. *Proterosaurus* differs from the other members of the order in having opisthocoelous cervical vertebrae.

98

The sacrum is composed of two vertebrae. Ossified inter centra (interdorsalia) generally occur in the cervical and caudal regions, and sometimes throughout the whole vertebral column. In the skull the quadrate is immovably fixed and united to the pterygoid. The palate is well ossified, while the premaxillae which are often beak-like are never ankylosed together. The jaws may be toothless or may be provided with teeth which are usually acrodont (see p. 199). The palatines frequently bear teeth, and in *Proterosaurus* teeth occur also on the pterygoids and vomers. The rami of the mandible are united by ligament at the symphysis except in the Rhynchosauridae, in which the union is bony. Superior and inferior temporal arcades occur.

The ribs have capitula and tubercula, and often uncinate processes (see p. 190) as in birds. A pectoral girdle and sternum, with clavicles and a T-shaped interclavicle are developed, and abdominal ribs are always found. The precoracoid is however absent. The limbs are pentedactylate.

Sphenodon (Hatteria) now living in some of the islands of the New Zealand group, is certainly the most generalised of all living reptiles. Though lizard-like in form it differs from all living lizards in the possession of two temporal arcades, abdominal ribs and a fixed quadrate; and is often considered to be nearly allied in many respects to the type of reptile from which all the others took their origin.

Among the better known extinct forms are *Proterosaurus* of Permian and *Hyperodapedon* of Triassic age.

Order 6. Squamata.

This order includes the extinct Mosasaurians, and the lizards and snakes which form the vast majority of living reptiles. The trunk may be moderately elongated and provided with four short limbs as in lizards, or it may be limbless, extremely elongated, and passing imperceptibly into the tail. The surface is generally completely covered with overlapping horny epidermal scales, below which bony dermal scutes may be developed.

The vertebrae are procoelous, rarely amphicoelous. There are no inter centra, and the neural arches are firmly united to the centra. Additional articulating surfaces, the zygosphenes and zygantra, are often developed. The sacrum is formed of two or rarely three vertebrae, or may be wanting as in Ophidia. In the skull an infratemporal arcade forming the lower boundary of the infratemporal fossa is absent, and the quadrate, except in the Chamaeleons, is movably articulated to the squamosal. The palatal vacuities are large and the nares are separate. There is often a distinct parasphenoid. The teeth are either *acrodont* (i.e. ankylosed to the summit of the jaw), or *pleurodont*, i.e. ankylosed to the inner side of the jaw. The thoracic ribs each have a single head which articulates with the centrum of the vertebra; while uncinate processes and abdominal ribs never occur.

A pectoral girdle and sternum may be present, or may be completely absent as in snakes. Except in snakes there are generally four pentedactylate limbs which may either form paddles or be adapted for walking.

Suborder (1). Lacertilia.

The body is elongated, and as a rule four short pentedactylate limbs are present, but sometimes limbs are vestigial or absent. The exoskeleton generally has the form of horny plates, spines, or scales; while sometimes as in the Chamaeleons and Amphisbaenians it is absent. In other forms such as *Tiliqua* and *Scincus*, the body has a complete armour of bony scutes, whose shape corresponds with that of the overlying horny scales.

The vertebrae are procoelous, rarely as in the Geckos amphicoelous; they are usually without zygosphenes and zygantra, but these structures occur in the Iguanidae. The sacral vertebrae of living forms are not ankylosed together, and the caudal vertebrae usually have well-developed chevron bones.

In the skull the orbits are separated from one another, only by an imperfectly developed interorbital septum, the cranial cavity not extending forwards between them, while the alisphenoidal region is unossified. The premaxillae may be paired or united (Amphisbaenidae), and there is usually an interparietal foramen. There may be a complete supratemporal arcade bounding the lower margin of the supratemporal fossa, or the supratemporal fossa may be open below. The quadratojugal is not ossified, and the quadrate articulates with the exoccipital. There is no infratemporal arcade. There is commonly a rod-like epipterygoid (fig. 33, 14) connecting the pterygoid and parietal.

Teeth are always present, and may be confined to the jaws or may be developed also on the pterygoids and rarely on the palatines; they are either acrodont or pleurodont. The rami of the mandible are suturally united.

A pectoral girdle is always present, and generally also a sternum. Clavicles and a T-shaped interclavicle are commonly present, but are absent in the Chamaeleons.

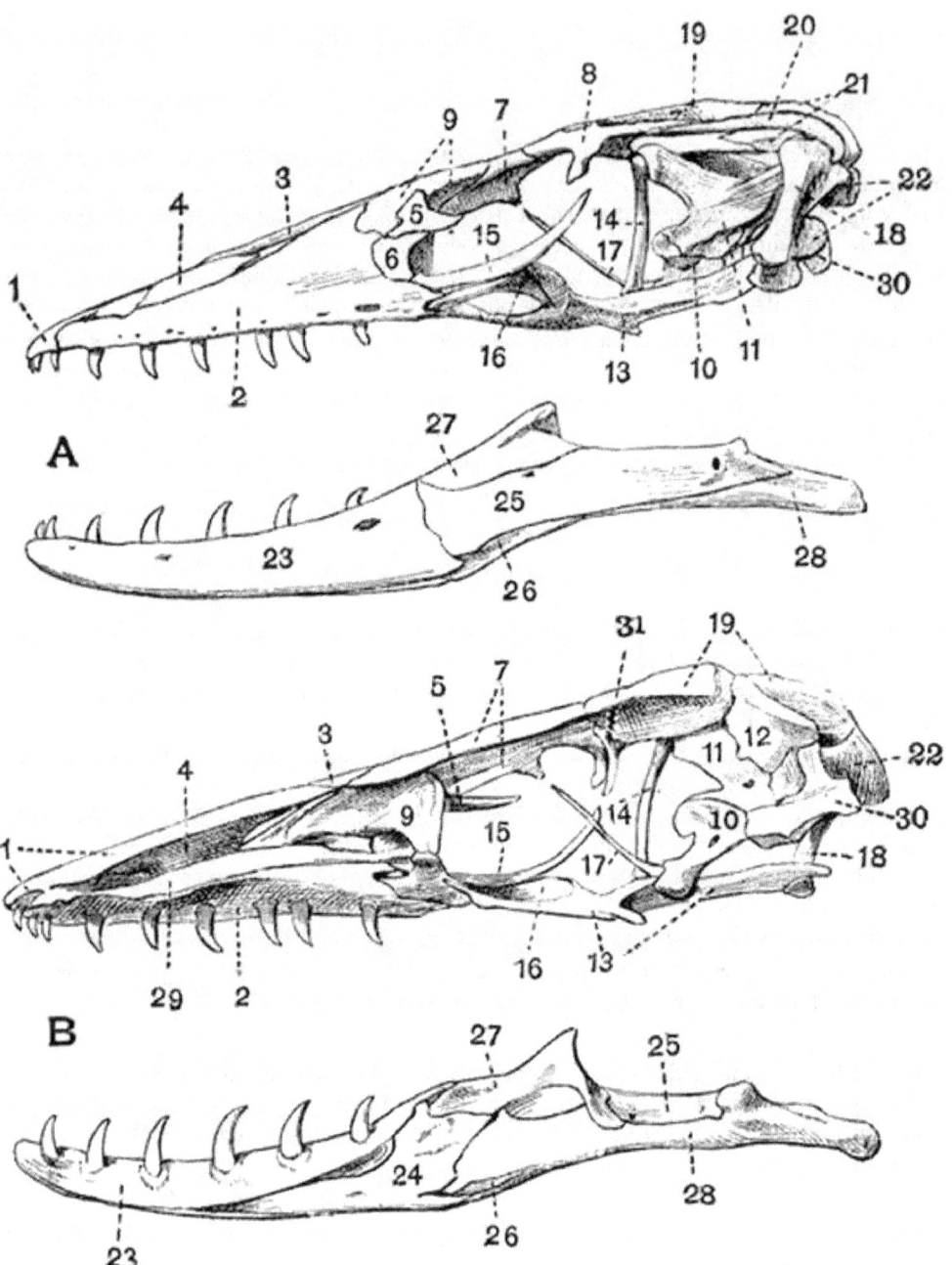

Fig. 33. A, Lateral view, and B, longitudinal section of the skull of a Lizard (*Varanus varius*). × 3/5. (Brit. Mus.)

1. premaxillae.
2. maxillae.
3. nasal.
4. lateral ethmoid.
5. supra-orbital.
6. lachrymal.
7. frontal.
8. postfrontal.
9. prefrontal.
10. basisphenoid.

16. transpalatine.
17. parasphenoid.
18. quadrate.
19. parietal.
20. squamosal.
21. supratemporal.
22. exoccipital.
23. dentary.
24. splenial.
25. supra-angular.

11. pro-otic.

12. epi-otic.

13. pterygoid.

14. epipterygoid (columella cranii).

15. jugal.

26. angular.

27. coronoid.

28. articular.

29. vomer.

30. basi-occipital.

31. orbitosphenoid.

There is no separate precoracoid but a precoracoidal process (fig. 34, 7) of the coracoid is generally prominent.

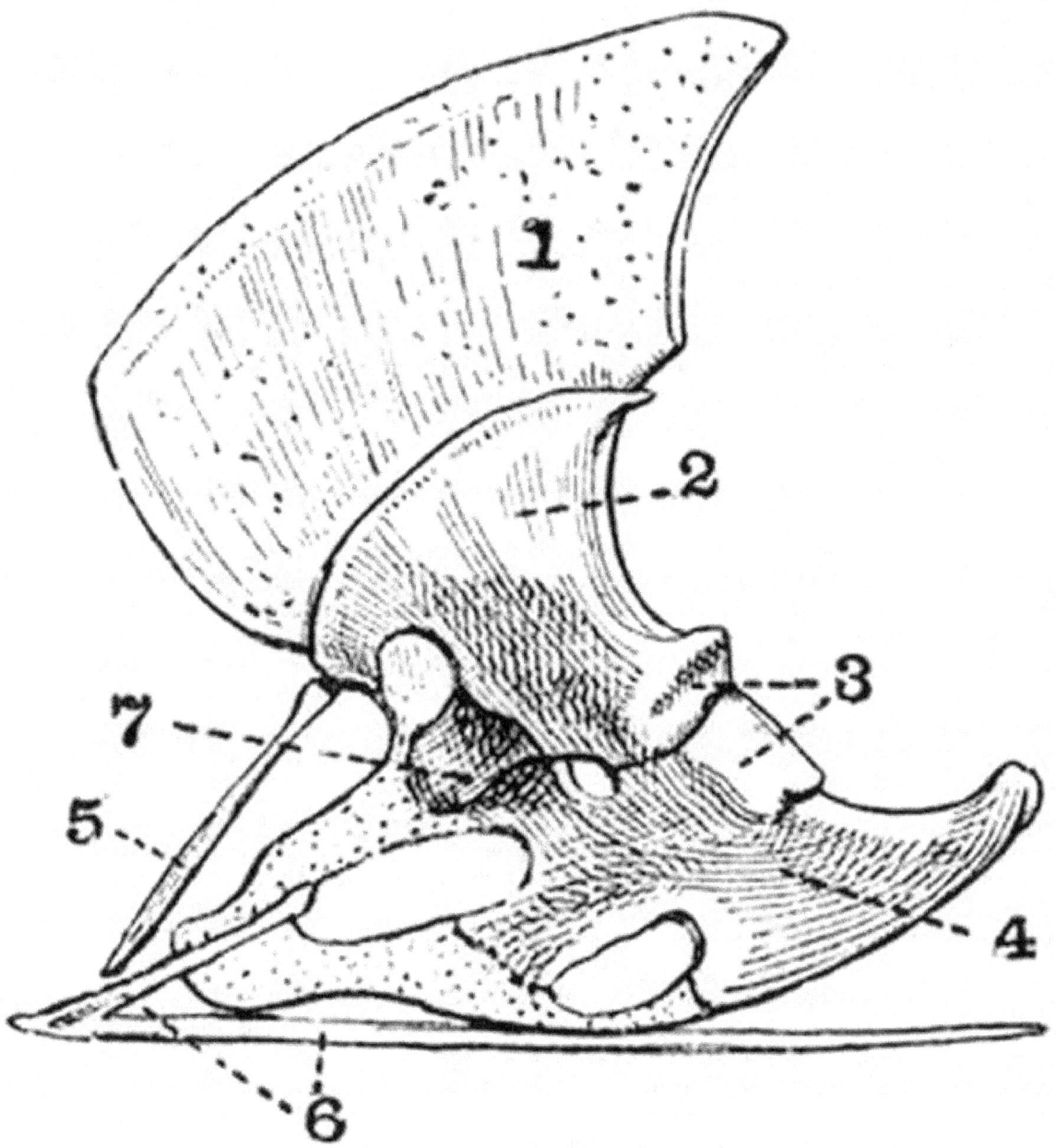

Fig. 34. Lateral view of the shoulder-girdle of *Varanus.* × 3/5. (Brit. Mus.).

1. suprascapula.

2. scapula.

3. glenoid cavity.

5. clavicle.

6. interclavicle.

7. precoracoidal process.

4. coracoid.

Sternal ribs are present in chamaeleons and scinks. The limbs are in the great majority of cases pentedactylate and the digits are clawed. The phalanges articulate by means of condyles. Sometimes one or both pairs of limbs are absent. When the posterior limbs are absent the pelvis is also wanting, though the loss of the anterior limbs does not lead to a corresponding loss of the pectoral girdle.

The pubis corresponds to the pre-pubis of Dinosaurs, and both pubes and ischia meet in ventral symphyses.

The suborder includes the Lizards, Chamaeleons and Amphisbaenians.

Suborder (2). Ophidia.

The Ophidia or snakes are characterised by their greatly elongated body and want of limbs. The body is covered with overlapping horny scales and bony dermal scutes are never present. The vertebrae are procoelous, and are distinguishable into two groups only, precaudal or rib-bearing, and caudal or ribless. The atlas vertebra is also ribless. The neural arches are always provided with zygosphenes and zygantra. Many of the vertebrae have strong hypapophyses, and the caudal vertebrae are without chevron bones.

In the skull the cranial cavity extends forwards between the orbits, and is closed in front by downgrowths from the frontals and parietals which meet the well-ossified alisphenoids and orbitosphenoids. The cranium is strongly ossified, and there are no parotic processes or interparietal foramen. There are no temporal arcades and no epipterygoid. The premaxillae if present are very small (fig. 51, 1) and usually toothless. The quadrates articulate with the squamosals, and do not as in Lacertilia meet the exoccipitals. The palatines do not unite directly with the vomers or with the base of the cranium, and the whole palato-maxillary apparatus is more loosely connected with the cranium than it is in Lacertilia. The pterygoids, and in most cases also the palatines, bear teeth. The dentition is acrodont, and the rami of the mandible are united only by an elastic ligament—an important point serving to distinguish the Ophidia from the Lacertilia. There is an imperfectly developed interorbital septum, the ventral part of which is formed by the parasphenoid. The postfrontal is generally well developed, while the jugals and quadratojugals are absent. There are never any traces of the anterior limbs or pectoral girdle, but occasionally there are vestiges of a pelvis and posterior limbs.

Suborder (3). Pythonomorpha.

This suborder includes *Mosasaurus* and its allies, a group of enormous extinct marine reptiles found in beds of Cretaceous age.

The skin is in most forms at any rate unprovided with dermal scutes. The vertebrae may be with or without zygosphenes and zygantra. The skull resembles that of lizards, having an interparietal foramen, and a cranial cavity open in front. The squamosal takes part in the formation of the cranial wall, and the quadrate articulates with the squamosal, not as in Lacertilia with the exoccipital. There are large supratemporal fossae, bounded below by supratemporal arcades. The teeth are large and acrodont, and occur on the pterygoids as well as on the jaws. The two rami of the mandible are united by ligament only. Pectoral and pelvic girdles are present, but clavicles are wanting, and the pelvis is not as a rule united to any sacrum.

The limbs are pentedactylate, and are adapted for swimming, while all the limb bones except the phalanges are relatively very short. The number of phalanges is not increased beyond the normal, and they articulate with one another by flat surfaces. The terminal phalanges are without claws.

Order 7. Dinosauria.

The extinct reptiles comprising this order were all terrestrial, and include the largest terrestrial animals known. They vary greatly in size and in the structure of the limbs, some approach close to the type of structure met with in birds, others are allied to crocodiles.

Passing to the more detailed characters:—there is sometimes a well-developed exoskeleton having the form of bony plates or spines. The vertebrae may be solid or their centra may be hollowed internally; their surfaces may be flat, biconcave or opisthocoelous. The sacrum is composed of from two to six vertebrae.

As regards the skull, the quadrate is large and fixed, and supratemporal and infratemporal fossae bounded by bone occur. The teeth are more or less laterally compressed, and often have serrated edges; they may be placed in distinct sockets or in a continuous groove. The ribs have capitula and tubercula, and sternal ribs often occur. The scapula is very large, the coracoid small, and there is no precoracoid, or T-shaped interclavicle. Clavicles are only known in a few cases. In the pelvis the ilium is elongated both in front of, and behind, the acetabulum, sometimes the pre-pubis, sometimes the post-pubis is the better developed. The anterior limbs are shorter than the posterior, and the long bones are sometimes solid, sometimes hollow.

There are three well-marked suborders of the Dinosauria.

Suborder (1). Sauropoda.

The reptiles belonging to this group were probably quadrupedal and herbivorous.

They have the cervical and anterior trunk vertebrae opisthocoelous, while the posterior vertebrae are biconcave; all the presacral, and sometimes the sacral vertebrae are hollowed internally. The teeth are spatulate and without serrated edges, they are always planted in distinct sockets, and some of them are borne by the premaxillae.

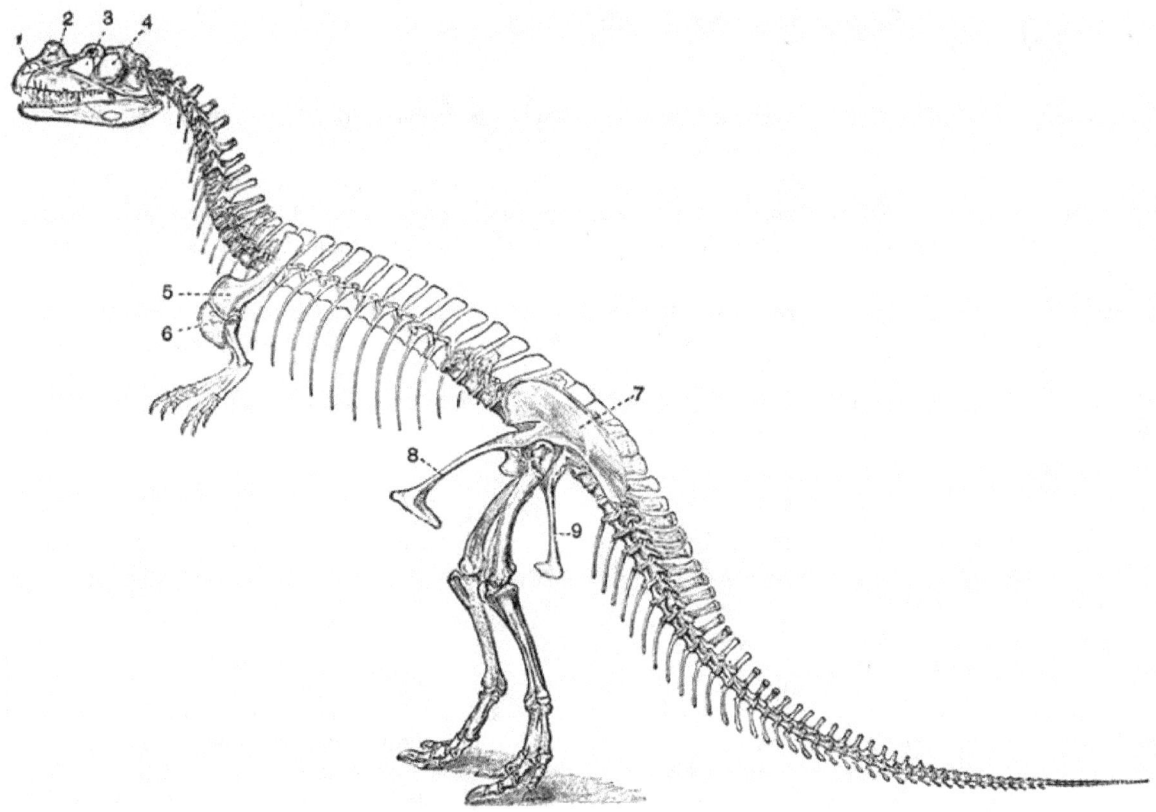

Fig. 35. Restored skeleton of *Ceratosaurus nasicornis*. × 1/30. (After Marsh.)

1. anterior nares.

2. prominence on the nasal

bones which probably

carried a horn.

3. pre-orbital vacuity.

4. orbit.

5. scapula.

6. coracoid.

7. ilium.

8. pubis (pre-pubis).

9. ischium.

The nares have the form of long slits and there are large pre-orbital vacuities.

The limb bones are solid, and the anterior limbs are not much shorter than the posterior ones. All the limbs are plantigrade and pentedactylate, and the digits of the pes are clawed. There is a large pre-pubis directed downwards and forwards, meeting its fellow in a ventral symphysis, but there is no post-pubis.

The Sauropoda are found in the secondary rocks of Europe and N. America and include the largest land animals that are known to have existed. Many of the best known forms such as *Brontosaurus* and *Morosaurus* are North American.

Suborder (2). Theropoda.

The members of this suborder were all carnivorous, and from the small comparative size of the anterior limbs many of them were probably bipedal.

The vertebrae are opisthocoelous or amphicoelous, their neural arches are provided with zygosphenes and zygantra, and their centra are frequently hollowed internally; the limb bones are also hollow, and in fact the whole skeleton is extremely light. The tail is of great length. The teeth are pointed and recurved, and have one or both borders serrated; they are always planted in distinct sockets, and some of them are borne by the premaxillae. There are large pre-orbital vacuities. The digits of both manus and pes are terminated by pointed ungual phalanges which must have borne claws. In the pelvis the pre-pubes and ischia are slender bones, the former meeting in a ventral symphysis. The ilia are very deep vertically and there are no post-pubes. The astragalus is closely applied to the tibia, in front of which it sends an ascending process, sometimes the two bones appear to have been ankylosed together, as in birds. The metatarsals are elongated and the feet digitigrade.

The Theropoda vary greatly in size, one of the best known genera *Compsognathus* was about as large as a cat, another, *Megalosaurus*, perhaps as large as an elephant. *Ceratosaurus* is the name of a well-known North American form regarded by many authorities as identical with *Megalosaurus*.

Suborder (3). Orthopoda.

This suborder includes the most specialised of the Dinosaurs, certain of which resemble the Theropoda in being bipedal. In some of them such as *Stegosaurus* the exoskeleton is strongly developed, in others such as *Iguanodon* it is absent.

The vertebrae are solid and may be opisthocoelous, biconcave, or flat. The teeth are compressed and serrated, often irregularly, and are frequently not set in distinct sockets. The anterior part of the premaxillae is without teeth, and a toothless predentary or mento-meckelian bone is present. The pre-orbital vacuities are small or absent, and the nares are large and placed far forwards.

The most characteristic features of the group are found in the pelvis which, except in the Ceratopsia, bears a striking resemblance to that of birds. The ischium and post-pubis are long slender bones directed backwards parallel to one another, and the pre-pubis is also well developed. The ischium has an obturator process. The limb bones are sometimes hollow, sometimes solid. The anterior limbs are much shorter than the posterior, pointing to a bipedal method of progression. The pes is digitigrade or plantigrade, and has three, rarely four, digits.

The suborder Orthopoda may be further subdivided into three sections:—

A. Stegosauria.

A dermal exoskeleton is strongly developed. The vertebral centra are flat or biconcave, and neither they nor the limb bones are hollowed out by internal cavities. The limbs are plantigrade, the anterior ones short, the posterior ones very large and strong. The post-pubis is well developed;

e.g. *Stegosaurus* from the Upper Jurassic of Colorado.

B. Ceratopsia.

There is sometimes a well-developed dermal exoskeleton formed of small granules and plates of bone. The bones are solid, and the vertebral centra flat. The cranium bears a pair of enormous pointed frontal horns, and the parietal is greatly expanded and elevated behind, forming with the squamosals a shield which overhangs the anterior cervical vertebrae. The premaxillae are united, and in front of them is a pointed beak-like bone which bites upon a toothless predentary ossification of the mandible. The teeth have two roots. The anterior limbs are but little shorter than the posterior ones. There is no post-pubis;

e.g. *Polyonax* from the uppermost Cretaceous of Montana.

C. Ornithopoda.

There is no dermal exoskeleton. The cervical vertebrae are opisthocoelous, and so are sometimes the thoracic. The limb bones are hollow and the anterior limbs are much shorter than the posterior ones. The feet are digitigrade and provided with long pointed claws. The post-pubis is long and slender and directed back parallel to the ischium;

e.g. *Iguanodon* from the European Cretaceous.

Order 8. Crocodilia.

This order includes the Crocodiles, Alligators and Garials and various extinct forms, some of which are closely allied to the early Dinosaurs.

There is always a more or less complete exoskeleton formed of bony scutes overlain by epidermal scales; these bony scutes are specially well developed on the dorsal surface but may occur also on the ventral. The vertebral column is divisible into the five regions commonly distinguishable. In all living forms the vertebrae, with the exception of the atlas and axis, the two sacrals, and first caudal, are procoelous, but in many extinct forms they are amphicoelous. The atlas (fig. 71) is remarkable, consisting of four pieces, and the first caudal is biconvex.

The teeth are, in the adult, planted in separate deep sockets. The skull is very dense and solid, and all the component bones including the quadrate are firmly united. The dorsal surface of the skull is generally characteristically sculptured. There is an interorbital septum, and the orbitosphenoidal and presphenoidal regions are imperfectly ossified. Supratemporal, infratemporal, and post-temporal fossae occur, but no interparietal foramen. In living genera there is a long secondary palate formed by the meeting in the middle line of the palatines, pterygoids and maxillae (fig. 43, A).

Cervical ribs (fig. 41, 8 and 9) are well developed, and articulate with rather prominent surfaces borne on the neural arches and centra respectively. The thoracic ribs articulate with the long transverse processes, and sternal ribs and abdominal splint ribs (fig. 46, 4) occur. The sternum is cartilaginous, and both it and the shoulder-girdle are very simple. The precoracoid is represented by merely a small process on the coracoid, while the clavicles are absent, except in the Parasuchia. In the pelvis (fig. 49) there is a large ilium, and an ischium meeting its fellow in a ventral symphysis; these two bones form almost the whole of the acetabulum. In front of the acetabulum, in the Eusuchia, projects a bone which is generally called the pubis, but is in reality rather an epipubis (fig. 49, 4), the true pubis being probably represented by a fourth element which remains cartilaginous for some time, and later on ossifies and attaches itself to the ischium. The limbs are small in proportion to the size of the body, and are adapted for swimming or for shuffling along the ground; they are plantigrade and the bones are all solid. In living forms the anterior limbs have five digits and the posterior four, the fifth being represented only by a short metatarsal. The first three digits in each case are clawed. The calcaneum has a large backwardly-projecting process.

The order Crocodilia may be subdivided into two suborders.

Suborder (1). Parasuchia.

The vertebral centra are flat or biconcave. The premaxillae are very large, and the nares are separated, and placed far back. The posterior narial openings lie comparatively far forward between the anterior extremities of the palatines.

The palatines and pterygoids do not form a secondary palate. The supratemporal fossae are small, and open posteriorly, the lateral temporal fossae are very large. The parietals and frontals are paired. Clavicles are present. The best known and most important genus of these extinct crocodiles is *Belodon*.

Suborder (2). Eusuchia.

The vertebrae are either biconcave or procoelous. The premaxillae are small, and the anterior nares are united and placed far forwards. The posterior nares lie far back, the palatines and in living genera the pterygoids, meeting in the middle line, and giving rise to a closed palate. The supratemporal fossae are surrounded by bone on all sides, and the parietals, and often also the frontals are united. There are no clavicles. The suborder includes the genera *Crocodilus*, *Alligator*, *Garialis* and others living and extinct.

Order 9. Pterosauria.

These animals, called also the pterodactyles or Ornithosauria, are a group of extinct reptiles, whose structure has been greatly modified from the ordinary reptilian type for the purpose of flight.

The skin was naked and they vary greatly in size and in the length of the tail. The vertebrae and limb bones are pneumatic just as in birds. The presacral vertebrae are procoelous and have their neural arches firmly united to the centra. The neck is long, the caudal vertebrae are amphicoelous, and from three to five vertebrae are fused together in the sacral region. The skull is large and somewhat bird-like, the facial portion being much drawn out anteriorly, and the sutures being obliterated. It resembles that of other reptiles in having large supratemporal fossae; large pre-orbital vacuities also occur. The jaws may be toothed or toothless, and the teeth, when present, are imbedded in separate sockets. The premaxillae are large, and the quadrate is firmly attached to the skull. The rami of the mandible are united at the symphysis, and there is an ossified ring in the sclerotic. The occurrence of a postfrontal and its union with the jugal behind the orbit, are characteristic reptilian features.

The ribs have capitula and tubercula, and sternal and abdominal ribs occur. The sternum has a well-developed keel, and the scapula and coracoid are large and bird-like. There are no clavicles or interclavicle.

The anterior limbs are modified to form wings by the great elongation of the fifth digit, to which a membrane was attached. The second, third and fourth digits are clawed and are not elongated in the way that they are in bats. The pollex, if present at all, is quite vestigial.

The pelvis is weak and small, and though the ilia are produced both in front of and behind the acetabula, in other features the pelvis is not bird-like. The ischia are short and wide, and the pubes are represented only by the pre-pubes. The posterior limbs are small and the fibula is much reduced. The pes is quite reptilian in type, and has five separate slender metatarsals. The two best known genera are *Pterodactylus*, in which the tail is short, and *Rhamphorhynchus*, in which it is long. The Pterosauria are found throughout the Jurassic and Cretaceous formations in both Europe and North America.

CHAPTER XIV.
THE SKELETON OF THE GREEN TURTLE. (Chelone midas.)

The most striking feature as regards the skeleton of the Turtle is that the trunk is enveloped in a bony box, the dorsal portion of which is called the **carapace**, while the ventral portion is the **plastron**.

I. EXOSKELETON.

a. The **epidermal exoskeleton** in the Green Turtle as in all other Chelonia except *Dermochelys*, *Trionyx* and their allies is strongly developed, its most important part consisting of a series of horny **shields** which cover over the bony plates of the carapace and plastron but do not at all correspond to them in size and arrangement.

The shields covering over the **carapace** consist of three rows of larger central shields,—five (**vertebral**) shields being included in the middle row and four (**costal**) in each lateral row,—and of a number of smaller **marginal** shields.

Of the marginal shields, that lying immediately in front of the first vertebral is termed the **nuchal**, while the two succeeding the last vertebral are called sometimes **pygal**, sometimes **supracaudal**; the remainder are the marginal shields proper.

The epidermal covering of the **plastron** consists principally of six pairs of symmetrically arranged shields, called respectively the **gular**, **humeral**, **pectoral**, **abdominal**, **femoral**, and **anal**, the gular being the most anterior. In front of the gular shields is an unpaired **intergular**, and the shields of the plastron are connected laterally with those of the carapace, by five or six pairs of rather irregular **infra-marginal** shields. Smaller horny plates occur on other parts of the body, especially on the limbs and head.

Two other sets of structures belong also to the epidermal exoskeleton, viz. (*a*) horny **beaks** with denticulated edges which ensheath both upper and lower jaws, (*b*) **claws**, which as a rule are borne only by the first digit of each limb. Sometimes in young individuals the second digit is also clawed.

b. The **dermal exoskeleton** is strongly developed, and is combined with endoskeletal structures derived from the ribs and vertebrae to form the carapace.

The **Carapace** (fig. 36) consists of a number of plates firmly united to one another by sutures. They have a very definite arrangement and include:

(*a*) the **nuchal** plate (fig. 36, 1), a wide plate forming the whole of the anterior margin of the carapace. It is succeeded by three series of plates, eight in each series, which together make up the main part of the carapace. Of these the small

(*b*) **neural plates** (fig. 36, A, 2) form the middle series. They are closely united with the neural arches of the underlying vertebrae;

(*c*) the **costal plates** (fig. 36, A, 3) are broad arched plates united to one another by long straight sutures. They are united at their inner extremities with the neural plates, but the boundaries of the two sets of plates do not regularly correspond. Each is united ventrally with a rib which projects beyond it laterally for some distance; (*d*) the **marginal plates** (fig. 36, 4) are twenty-three in number, eleven lying on each side, while an unpaired one lies in the middle line posteriorly. Many of them are marked by slight depressions into which the ends of the ribs fit; (e) the **pygal** plates (fig. 36, 5) are two unpaired plates lying immediately posterior to the last neural.

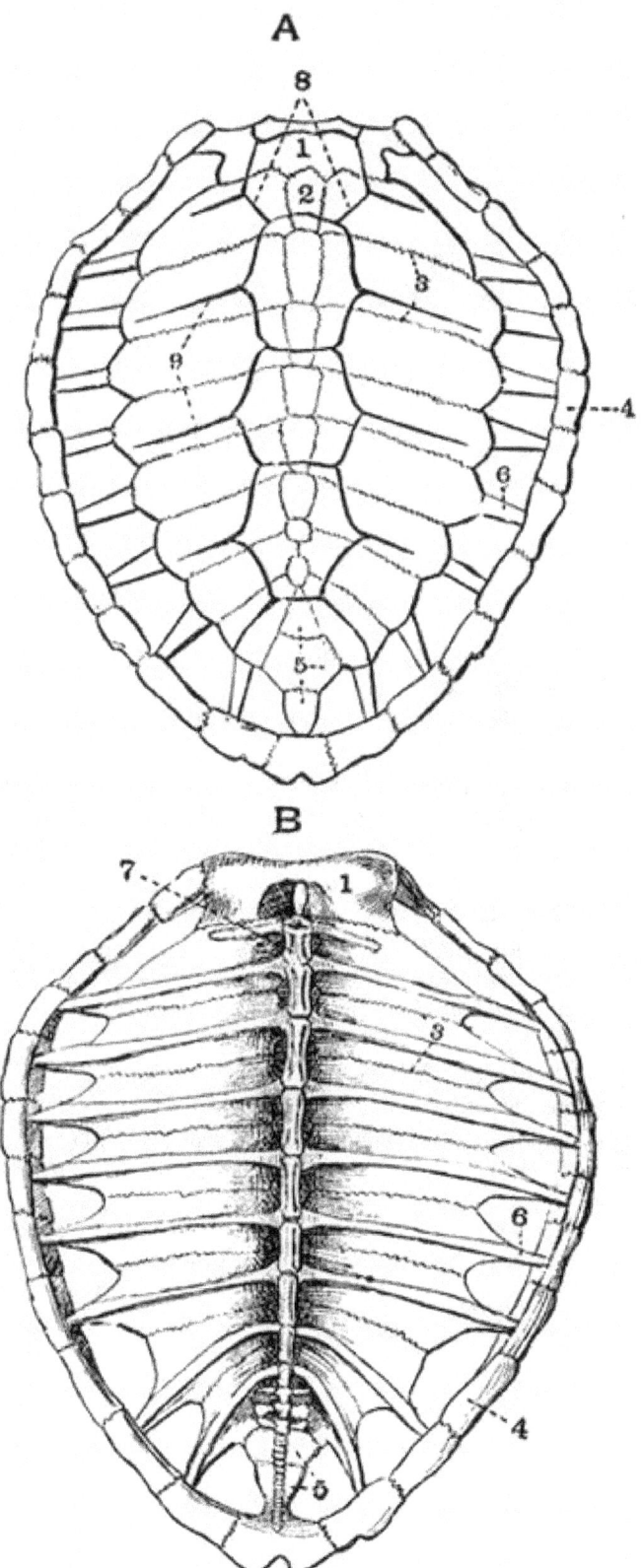

Fig. 36. A, dorsal and B, ventral view of the carapace of a Loggerhead Turtle (*Thalassochelys caretta*), (after Owen).

1. nuchal plate.   6. rib.

2. first neural plate.   7. thoracic vertebra.

107

3. second costal plate.

4. marginal plate.

5. pygal plate.

8. first vertebral shield.

9. costal shield.

The sculpturing due to the epidermal shields is very obvious on the carapace.

The **plastron** (fig. 37) consists of one unpaired ossification, the **entoplastron**, and four pairs of ossifications called respectively the **epiplastra, hyoplastra, hypoplastra**, and **xiphiplastra**.

The **epiplastra** (fig. 37, 1) are the most anterior, they are expanded and united to one another in the middle line in front, while behind each tapers to a point which lies external to a process projecting forwards from the hyoplastron. They are homologous with the *clavicles* of other vertebrates.

The **entoplastron** or **episternum** (fig. 37, 2) which is homologous with the *interclavicle* of other reptiles, is expanded at its anterior end and attached to the symphysis of the epiplastra, while behind it tapers to a point and ends freely.

The **hyoplastra** are large irregular bones each closely united posteriorly with the corresponding hypoplastron, and drawn out anteriorly into a process which lies internal to that projecting backwards from the epiplastron. Each gives off on its inner surface a slender process which nearly meets its fellow, while the anterior half of the outer surface is drawn out into several diverging processes.

The **hypoplastra** (fig. 37, 4) are flattened bones resembling the hyoplastra, with which they are united by long sutures; the posterior half of both outer and inner surfaces is drawn out into a number of pointed processes.

The **xiphiplastra** are small flattened elongated bones meeting one another in the middle line posteriorly. In front they are notched and each interlocks with a process from the hypoplastron of its side. The hyoplastra, hypoplastra and xiphiplastra are homologous with the abdominal ribs of Crocodiles.

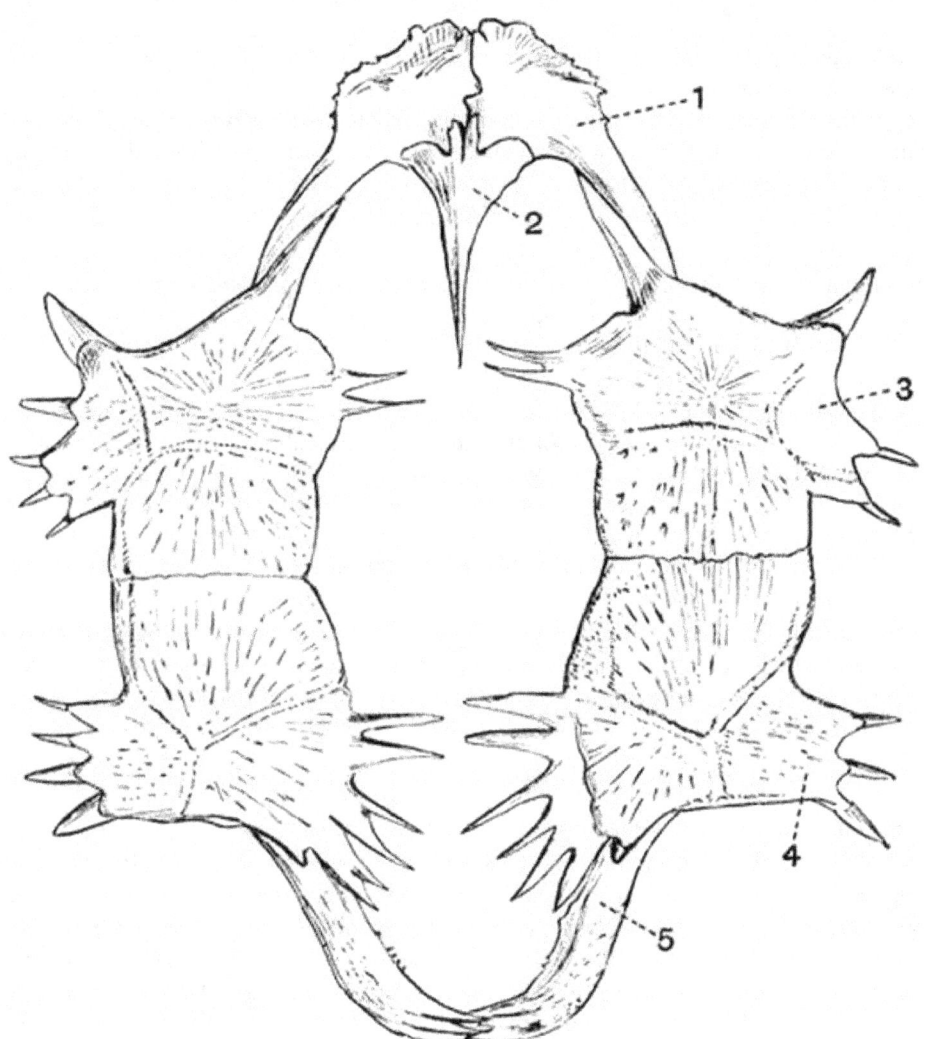

Fig. 37. The Plastron of a Green Turtle (*Chelone midas*). × 1/7. (Camb. Mus.)

1. epiplastron (clavicle).

2. entoplastron (interclavicle).

3. hyoplastron.

4. hypoplastron.

5. xiphiplastron.

II. ENDOSKELETON.

1. The Axial Skeleton.

The axial skeleton includes the vertebral column, the ribs, and the skull.

A. The Vertebral column and Ribs.

The number of vertebrae in the Green Turtle is thirty-eight, not a great number as compared with that in many reptiles, and of these eighteen are caudal.

The vertebral column is divisible into four regions only—**cervical**, **thoracic**, **sacral**, and **caudal**.

The Cervical vertebrae.

These are eight in number, and are chiefly remarkable for the great variety of articulating surfaces which their centra present, and for their mobility upon one another.

The first or **atlas** vertebra differs much from all the others and consists of the following parts:—

a. the **neural arch**, formed of two separate ossifications united in the mid-dorsal line;

b. the **inferior arch**;

c. the **centrum**, which is detached from the rest and forms the odontoid process of the second vertebra.

109

Each half of the **neural arch** consists of a ventral portion, the **pedicel**, which lies more or less vertically and is united ventrally to the inferior arch, and of a dorsal portion, the **lamina**, which lies more or less horizontally and meets its fellow in the middle line in front, partially roofing over the neural canal. Each pedicel bears a facet on its anterior surface, which, with a corresponding one on the inferior arch, articulates with the occipital condyle of the skull. Three similar facets occur also on the posterior surface of the pedicel and inferior arch, and articulate with the odontoid process. The laminae meet one another in front, but do not fuse, while behind they are separated by a wide triangular space. They bear a pair of small downwardly-directed facets, the **postzygapophyses**, for articulation with the prezygapophyses of the second vertebra.

The **inferior arch** is a short irregular bone bearing two converging facets for articulation with the occipital condyle and odontoid process respectively.

The **centrum** or **odontoid process** has a convex anterior surface for articulation with the neural and inferior arches, and a concave posterior surface by which it is united with the centrum of the second or axis vertebra. It bears posteriorly a small epiphysis which is really a detached portion of the inferior arch.

The second or **axis** and following five cervical vertebrae, though showing distinct differences, resemble one another considerably, each having a fairly elongated centrum with a keel-like **hypapophysis**, each having also a neural arch with prominent articulating surfaces, the anterior of which, or **prezygapophyses**, look upwards and inwards, while the posterior ones, the **postzygapophyses**, look downwards and outwards. They however, as was previously mentioned, differ very remarkably in the character of the articulating surfaces of the centra. Thus the second and third vertebrae are convex in front and concave behind, the fourth is biconvex, the fifth is concave in front and convex behind. The sixth is concave in front and attached to the seventh by a flat surface behind, the seventh has a flat anterior face and two slightly convex facets behind. The vertebrae all have short blunt transverse processes and the second has a prominent **neural spine**.

The **eighth cervical vertebra** is curiously modified, the centrum is very short, has a rather prominent hypapophysis, and is convex behind, while in front it articulates with the preceding centrum by two concave surfaces. The neural arch is deeply notched in front and bears two upwardly-directed prezygapophyses, while behind it is very massive and is drawn out far beyond the centrum, bearing a pair of flat postzygapophyses. The top of the neural arch almost or quite meets a blunt outgrowth from the nuchal plate.

The Thoracic vertebrae.

These are ten in number and are all firmly united with the ribs and elements forming the carapace.

The first thoracic vertebra differs from the others, the centrum is short and has a concave anterior surface articulating with the centrum of the last cervical vertebra, and a pair of prezygapophyses borne on long outgrowths. The neural spine arises only from the anterior half of the centrum, and is not fused to the carapace. Arising laterally from the anterior part of the centrum are a small pair of ribs each of which is connected with a process arising from the rib of the succeeding vertebra.

The next seven thoracic vertebrae are all very similar, each has a long cylindrical centrum, expanded at the ends, and firmly united to the preceding and succeeding vertebrae. The neural arches are flattened and expanded dorsally, and are united to one another and to the overlying neural plates; each arises only from the anterior half of its respective centrum, and overlaps the centrum of the vertebra in front of it. Between the base of the neural arch and its successor is a small foramen for the exit of the spinal nerve. There are no transverse processes or zygapophyses.

To each thoracic vertebra from the second to ninth inclusive, there corresponds a pair of **ribs** (fig. 36, 6) of a rather special character. Each is suturally united with the anterior half of the edge of its own vertebra, and overlaps on to the posterior half of the edge of the next preceding vertebra. The ribs are much flattened, and each is fused with the corresponding costal plate, beyond which it projects to fit into a pit in one of the marginal plates.

The tenth thoracic vertebra is smaller than the others, and its neural arch does not overlap the preceding vertebra, it bears a pair of small ribs which are without costal plates, but meet those of the ninth vertebra.

There are no **lumbar** vertebrae.

The Sacral vertebrae.

The **sacral vertebrae** are two in number, they are short and wide, their centra are ankylosed together, and their neural arches are not united to the carapace.

The first has the anterior face of the centrum concave and the posterior flat, while both faces of the second are flat. Each bears a pair of short ribs which meet the ilia, but are not completely ankylosed either with them or the centra.

The Caudal vertebrae.

The **caudal vertebrae** are eighteen in number. The centrum of the first is flat in front and is ankylosed to the second sacral; behind it is convex. The others are all very similar to one another, and decrease gradually in size when followed back. Each has a moderately long centrum, concave in front and convex behind, both terminations being formed by epiphyses. The neural arch arises only from the anterior half of the vertebra; it bears a blunt truncated neural spine and prominent pre- and post-zygapophyses. The first seven caudal vertebrae bear short ribs attached to their lateral margins, the similar outgrowths on the succeeding vertebrae do not ossify from distinct centres, and are transverse processes rather than ribs.

B. The Skull.

The skull of the Turtle is divisible into the following three parts:—

(1) the cranium;

(2) the lower jaw or mandible;

(3) the hyoid.

(1) The Cranium.

The **cranium** is a very compact bony box, containing a cavity in which the brain lies, and which is a direct continuation of the neural canal of the vertebrae.

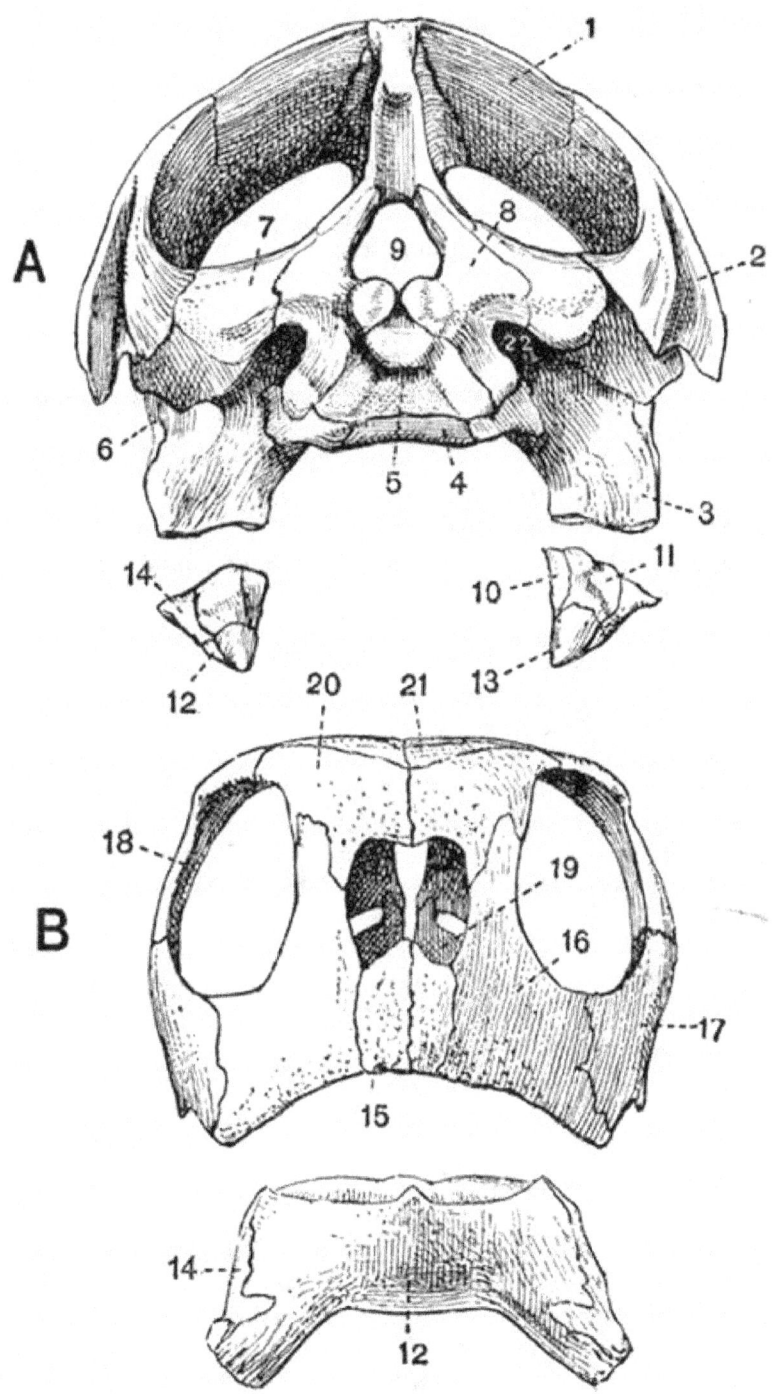

Fig. 38. The skull of the Green Turtle (*Chelone midas*). × ½. A, posterior half, B, anterior half. (Brit. Mus.)

1. parietal.   13. angular.

111

2. squamosal.

3. quadrate.

4. basisphenoid.

5. basi-occipital.

6. quadratojugal.

7. opisthotic.

8. exoccipital.

9. foramen magnum.

10. splenial.

11. articular.

12. dentary.

14. supra-angular.

15. premaxillae.

16. maxillae.

17. jugal.

18. postfrontal.

19. vomer.

20. prefrontal.

21. frontal.

22. external auditory meatus
leading into tympanic
cavity.

Like those of the skull as a whole its component bones may be subdivided into three sets:—

1. those forming the brain-case or **cranium proper**;

2. those developed in connection with the special sense organs;

3. those forming the upper jaw and suspensorial apparatus.

Both cartilage and membrane bones take part in the formation of the skull, and a considerable amount of cartilage remains unossified, especially in the ethmoidal and sphenoidal regions.

1. The Cranium proper or Brain-case.

The cartilage and membrane bones of the brain-case when taken together can be seen to be more or less arranged in three rings or segments, called respectively the occipital, parietal, and frontal segments.

The **occipital segment** is the most posterior of these, and consists of four cartilage bones, the **basi-occipital**, the two **exoccipitals** and the **supra-occipital**; these bound the **foramen magnum**.

The **basi-occipital** (figs. 38 and 39, 5) lies ventral to the foramen magnum and only bounds a very small part of it; it forms one-third of the **occipital condyle** by which the skull articulates with the atlas vertebra. It unites dorsally with the exoccipitals and anteriorly with the basisphenoid.

The **exoccipitals** are rather small bones, which form the sides and the greater part of the floor of the foramen magnum, and two-thirds of the occipital condyle. Laterally each is united with the pterygoid and opisthotic of its side. At the sides of the occipital condyle each exoccipital is pierced by a pair of foramina, the more dorsal and posterior of which transmits the hypoglossal nerve.

The **supra-occipital** (fig. 39, 14) is a larger bone than the others of the occipital segment. It forms the upper border of the foramen magnum and is drawn out dorsally into a large crest which extends back far beyond the occipital condyle. In the adult the supra-occipital is completely ankylosed with the epi-otics.

The **Parietal segment**.

The ventral portion of the parietal segment is formed by the **basisphenoid** (figs. 38 and 39, 4) which lies immediately in front of the basi-occipital. A triangular portion of it is seen in a ventral view of the skull, but it is quickly overlapped by the pterygoids. It gives off dorsally a pair of short processes which meet the pro-otics.

The alisphenoidal region is unossified and the only other constituents of the parietal segment are the *parietals* (fig. 39, 1). These are large bones which, after roofing over the cranial cavity, extend upwards and become expanded into a pair of broad plates which unite with the squamosal and bones of the frontal segment to form a wide, solid, false roof to the skull. Each also sends ventralwards a plate which meets an upgrowth from the pterygoid and acts as an alisphenoid.

The **Frontal segment**.

Of the frontal segment the basal or presphenoidal and lateral or orbitosphenoidal portions do not become ossified, the dorsal portion however includes three pairs of membrane bones, the *frontals*, *prefrontals* and *postfrontals*.

The *frontals* are a pair of small bones lying immediately in front of the parietals, and in front of them are the *prefrontals* (figs. 38 and 39, 20), a pair of similar but still smaller bones, which are produced ventrally to meet the vomer and palatines. They form also the dorsal boundary of the anterior nares. The *postfrontals* (figs. 38 and 39, 18) are larger bones, united dorsally to the frontals and parietals, posteriorly to the squamosals, and ventrally to the jugals and quadratojugals. All three pairs of frontal bones, especially the postfrontals, take part in the bounding of the orbits.

112

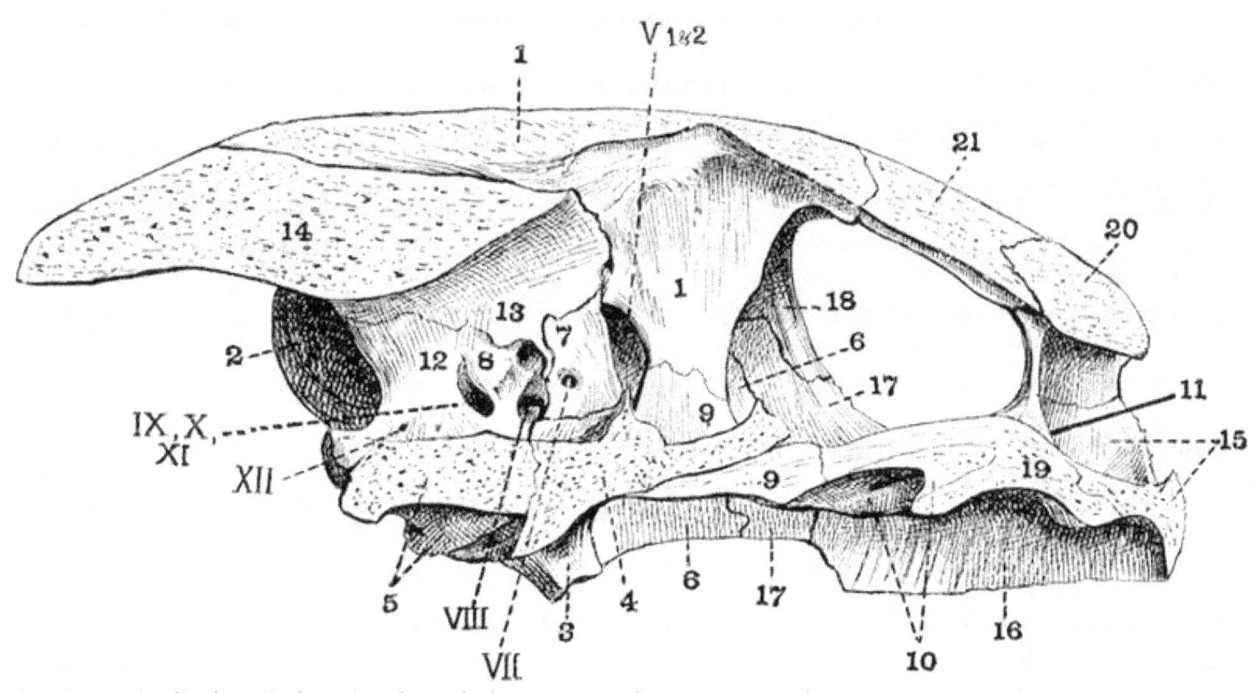

Fig. 39. Longitudinal vertical section through the Cranium of a Green Turtle (*Chelone midas*). × 2/3. (Camb. Mus.)

1. parietal.

2. squamosal.

3. quadrate.

4. basisphenoid.

5. basi-occipital.

6. quadratojugal.

7. pro-otic.

8. opisthotic.

9. pterygoid.

10. palatine.

11. rod passed into narial

passage.

12. exoccipital.

13. epi-otic fused to

supra-occipital.

14. supra-occipital.

15. premaxillae.

16. maxillae.

17. jugal.

18. postfrontal.

19. vomer.

20. prefrontal.

21. frontal.

V, VII, VIII, IX, X, XI, XII,

foramina for the exit of

cranial nerves.

2. The Sense capsules.

Skeletal structures occur in connection with each of the three special sense organs of hearing, sight, and smell.

The **Auditory capsules**.

The auditory or periotic capsule of the turtle is rather large and its walls are well ossified, epi-otic, pro-otic and opisthotic bones being present.

The **epi-otic** (fig. 39, 13) is the more dorsal of the three bones, and in the adult is completely ankylosed with the supra-occipital.

The **opisthotic** (fig. 39, 8) is the ventral posterior element. On its inner side it is united to the supra-occipital above, and to the exoccipital below; it sometimes becomes completely fused with the exoccipital. In front it meets the pro-otic, and on its outer side the squamosal and quadrate. Its anterior portion is hollowed out by the cavity in which the auditory organ lies, it gives off also a process which is separated from the exoccipital by an oval foramen through which the glossopharyngeal, pneumogastric, and spinal accessory nerves leave the cranial cavity.

The **pro-otic** is the anterior element; it meets the supra-occipital and opisthotic posteriorly, while anteriorly it is separated from the alisphenoidal plate of the parietal and pterygoid by a large oval foramen through which the maxillary and mandibular branches of the trigeminal nerve pass out (fig. 39, V 1 & 2). It is hollowed out posteriorly by the cavity in which the auditory organ lies, and its inner wall as seen in longitudinal section is pierced by a foramen through which the external

113

carotid artery and facial nerve leave the cranial cavity,—the nerve finally leaving the skull through a small oval foramen on the anterior face of the pro-otic near its junction with the quadrate.

Between the pro-otic and opisthotic as seen in a longitudinal section of the skull is a large opening constricted in the middle. This is the **internal auditory meatus** (fig. 39, VIII.). Through it the auditory nerve leaves the cranial cavity and enters the ear. The ramus vestibularis leaves through the dorsal part of the hole, the ramus cochlearis through the ventral.

The cavity of the auditory or periotic capsule communicates with the exterior by a fairly large hole, the **fenestra ovalis**, which lies between the opisthotic and pro-otic, and opens into a deep depression, the **tympanic cavity**, which is seen in a posterior view of the skull lying just external to the exoccipital. The cavity communicates with the exterior by a large opening, the **external auditory meatus** (fig. 38, 22).

Several other openings are seen in the tympanic cavity; through one at the extreme posterior end the pneumogastric and spinal accessory nerves leave the skull, and through another, a little further forwards, the glossopharyngeal.

The auditory ossicles consist of a long bony **columella**, whose inner end fits into the fenestra ovalis, while the outer end is attached to a small cartilaginous plate, the **extra-columella**, which is united to the tympanum.

The **Optic capsules**.

The skeletal structures developed in connection with the optic capsule do not become united to the skull. They consist of:—

(*a*) the **sclerotic**, a cartilaginous sheath investing the eye and bearing

(*b*) a ring of ten small bony scales.

There is no *lachrymal* bone.

The **Olfactory** or **Nasal capsules**.

The basicranial axis in front of the basisphenoid remains cartilaginous, neither presphenoid nor mesethmoid bones are developed, and the orbits in a dry skull communicate by a wide space through which the second, third, fourth, and sixth cranial nerves pass out. Separate *nasal* bones also do not occur, the large prefrontals extending over the area usually occupied by both nasals and lachrymals.

The only bone developed in connection with the nasal capsules is the *vomer* (fig. 39, 19), an unpaired bone lying ventral to the mesethmoid cartilage, and in contact laterally with the maxillae, premaxillae and palatines.

3. The Upper Jaw and suspensorial apparatus.

A number of pairs of bones are developed in connection with the upper jaw and suspensorial apparatus, one pair, the **quadrates**, being cartilage bones, while the rest are all membrane bones.

The *squamosals* (fig. 38, 2) are large bones which, lying external to the auditory bones, extend dorsalwards to meet the parietals and postfrontals, and form a large part of the false roof of the skull. They are united ventrally with the quadrates and quadratojugals.

Each **quadrate** (fig. 38, 3) forms the outer boundary of the tympanic cavity, and is firmly united on its inner side with the opisthotic, exoccipital, and pterygoid. Dorsally it is fixed to the squamosal and anteriorly to the quadratojugal. Its outer surface is marked by a deep recess, and it ends below in a strong condyle with which the mandible articulates. In front of the quadrates are a pair of thin plate-like bones, the *quadratojugals* which are united in front to the jugals or malars.

The *jugals* (fig. 38, 17) are also thin plate-like bones, and form part of the posterior boundary of the orbit. They are attached dorsally to the postfrontals, and anteriorly to the maxillae, while each also sends inwards a horizontal process which meets the pterygoid and palatine.

The *maxillae* (figs. 38 and 39, 16) are a pair of large vertically-placed bones, each drawn out ventrally into a straight, sharp, cutting edge. They form the lateral boundaries of the anterior nares, and each sends dorsalwards a process which meets the postfrontal. Each also sends inwards a horizontal **palatine process**, which meets the palatine and vomer, and also forms much of the floor of the narial passage.

The *premaxillae* (figs. 38 and 39, 15) are a pair of very small bones forming the floor of the anterior narial opening, they are wedged in between the two maxillae, and send back processes which meet the vomer and palatines.

The *palatines* (fig. 39, 10) are a pair of small bones firmly united with the pterygoids behind, with the maxillae and jugals externally, and with the vomer in the middle line. Each also gives off a palatine plate which unites with the expanded lower edge of the vomer, and forms the ventral boundary of the posterior nares. Anteriorly the palatines form the posterior boundary of a large foramen through which the ophthalmic branches of the fifth and seventh nerves pass to the olfactory organs.

The *pterygoids* (fig. 39, 9) are a pair of large bones which unite with one another by a long median suture. They are united also with the palatines in front, and with the quadrate, basisphenoid, basi-occipital, and exoccipitals behind. Each also sends dorsalwards a short **alisphenoid plate** which meets that from the parietal.

Piercing the posterior end of the *pterygoid* is the prominent opening of the carotid canal; a bristle passed into this hole emerges through a foramen lying between the pro-otic and the alisphenoid process of the pterygoid.

(2) The Lower Jaw or Mandible.

The **mandible** consists of one unpaired bone, formed by the fusion of the two *dentaries*, and five pairs of bones, called respectively the **articular**, *angular, supra-angular, splenial* and *coronoid*.

The fused *dentaries* (fig. 38, 12) form by far the largest of the bones; they constitute the flattened anterior part of the mandible, and extend back below the other bones almost to the end of the jaw.

The *coronoid* is the most anterior of the paired bones, it forms a prominent process to which the muscles for closing the jaw are attached.

The **articular** (fig. 38, 11) is expanded, and with the *supra-angular* forms the concave articulating surface for the quadrate.

The *splenial* (fig. 38, 10) is a thin plate applied to the inner surface of the posterior part of the mandible.

The *angular* (fig. 38, 13) is a slender plate of bone lying below the supra-angular and splenial.

(3) The Hyoid.

The hyoid apparatus is well developed, parts of the first two branchial arches being found, as well as of the hyoid proper. It consists of a more or less oblong flattened **basilingual plate** or **body of the hyoid** which represents the fused ventral ends of the hyoid and branchial arches of the embryo, and is drawn out into a point anteriorly. The greater part is formed of unossified cartilage, but at the posterior end it is bilobed, and a pair of ossified tracts occur. To its sides are attached three pairs of structures, which are portions of the hyoid and first and second branchial arches respectively.

The free part of the **hyoid** consists of a small piece of cartilage attached to the anterior part of the basilingual plate at its widest portion (fig. 53, 2).

The **anterior cornu** or free part of the **first branchial arch** is much the largest of the three structures. Its proximal portion adjoining the basilingual plate is cartilaginous, as is its distal end; the main part is however ossified.

The **posterior cornu** or free part of the **second branchial arch** (fig. 53, 4) consists of a short flattened cartilaginous bar arising from the bilobed posterior end of the basilingual plate.

The hyoid apparatus has no skeletal connection with the rest of the skull.

2. The Appendicular Skeleton.

This includes the skeleton of the two pairs of limbs and their girdles.

The Pectoral Girdle.

The pectoral girdle has an anomalous position, being situated internal or ventral to the ribs. It consists of three bones, a dorsal bone, the **scapula**, an anterior ventral bone, the **precoracoid**, and a posterior ventral bone, the **coracoid**.

The **scapula** is a small somewhat rod-shaped bone forming about two-thirds of the glenoid cavity. At its proximal end it is closely united with the precoracoid, the two bones ossifying continuously. It tapers away distally, and is directed dorsalwards towards the carapace.

The **precoracoid** forms an angle of about 130° with the scapula, with which it is completely fused at its proximal end. Its distal end is somewhat expanded and flattened, and is terminated by a fibrocartilaginous **epiprecoracoid** which meets its fellow. It takes no part in the formation of the glenoid cavity.

The **coracoid** is a large flattened blade-shaped bone forming about one-third of the glenoid cavity. It does not meet its fellow in a ventral symphysis, and is terminated by a cartilaginous **epicoracoid**. The glenoid articulating surfaces of both scapula and coracoid are lined by a thick pad of cartilage.

The Anterior Limb.

This is divisible into three portions, the **upper arm, fore-arm** and **manus**.

The **upper arm** contains a single bone, the **humerus**.

The **humerus** (fig. 40, A, 1) is a stout, nearly straight, somewhat flattened bone widely expanded at both ends. At the proximal end is the large hemispherical **head**, which articulates with the glenoid cavity. Behind the head the bone is drawn out into another large rounded process. Below the head the shaft bears a small outgrowth which is continuous with a larger one on the flexor surface (see p. 29). The bone is terminated distally by the **trochlea**, consisting of three partially distinct convex surfaces which articulate with the bones of the fore-arm.

The **fore-arm** includes two bones, the **radius** and **ulna**; both these are small bones, and are immovably fixed to one another proximally and distally.

The **radius** or pre-axial bone is the larger of the two, and is a rod-like bone terminated at either end by an epiphysis. It articulates at its proximal end with the humerus, and at its distal end with the radiale or scaphoid bone of the carpus.

The **ulna** (fig. 40, A, 3) or postaxial bone is shorter than the radius, and more expanded at its proximal end, where it articulates with the humerus. It articulates distally with the intermedium (lunar) and the ulnare (cuneiform) bones of the carpus. All three bones of the arm have their terminations formed by epiphyses which ossify from centres distinct from those forming the shafts.

115

The **Manus** consists of the **carpus** or **wrist** and the **hand** which includes the metacarpals and phalanges.

The **carpus** consists of a series of ten small bones, one of which, the **pisiform** (fig. 40, A, 10), differs from the others in being merely an ossification in the tendon of a muscle. The remaining nine bones are arranged in a proximal row of three, the **ulnare** (fig. 40, A, 6), **intermedium**, and **radiale**, and a distal row of five (carpalia 1-5), each of which supports one of the metacarpals. A ninth bone, the **centrale** (fig. 40, A, 7), is wedged in between the two rows. The ulnare, intermedium and pisiform are comparatively large flattened bones, the others are small and cubical.

The **hand**. This is composed of five digits, each of which consists of a metacarpal and of a varying number of phalanges.

The **metacarpals**. The first metacarpal (fig. 40, A, 11) is a short flattened bone, the others are all elongated and cylindrical, and are terminated proximally by slightly concave surfaces, and distally by slightly convex ones.

The **phalanges**. The first and fifth digits both have two phalanges, the second, third, and fourth have each three. The distal phalanx of the first digit is stout and curved, and bears a horny claw; those of the other digits are flattened and more or less pointed.

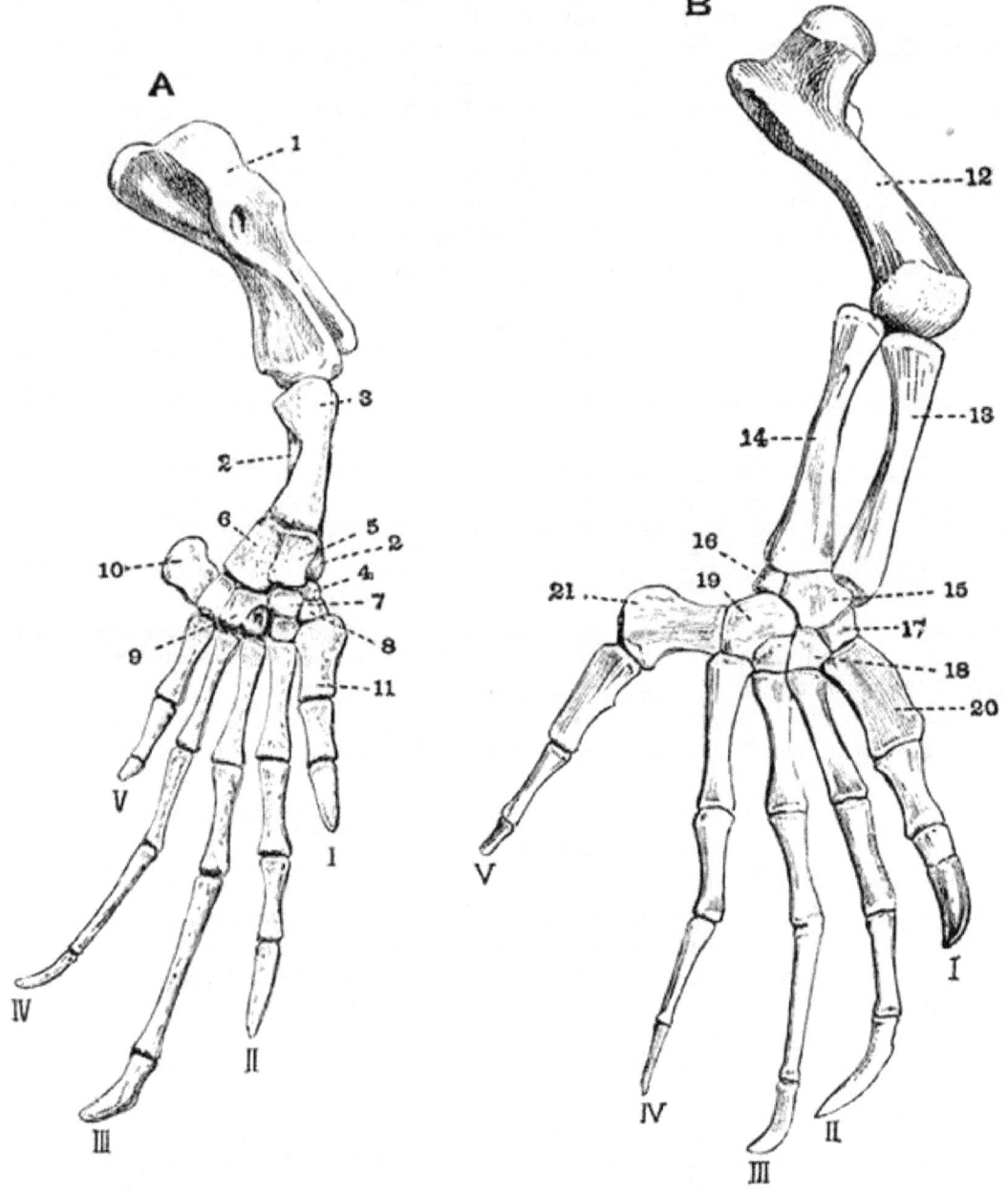

Fig. 40. A. Anterior limb of a young Hawksbill Turtle (*Chelone imbricata*) × ¼ (Brit. Mus.). B. Posterior limb of a large Green Turtle (*Chelone midas*) × 1/8 (Camb. Mus.).

1. humerus.   12. femur.

2. radius (almost hidden by the ulna).

3. ulna.

4. radiale.

5. intermedium.

6. ulnare.

7. centrale.

8. carpale I.

9. carpale IV.

10. pisiform.

11. first metacarpal.

13. tibia.

14. fibula.

15. tibiale intermedium and centrale fused.

16. fibulare.

17. tarsale 1.

18. tarsale 2.

19. tarsalia 4 and 5 fused.

20. first metatarsal.

21. fifth metatarsal.

I, II, III, IV, V, digits.

The Pelvic Girdle.

The pelvic girdle consists of three bones; a dorsal bone, the **ilium**, an anterior ventral bone, the **pubis**, and a posterior ventral bone, the **ischium**. All three bones contribute largely to the formation of the **acetabulum**, with which the head of the femur articulates.

The **ilium** is a small slightly curved bone, which unites ventrally with the pubis and ischium, and extends dorsalwards and backwards to meet the distal ends of the sacral ribs.

The **pubis** is the largest bone of the three; its distal end forms a wide bilobed plate, the inner lobe meeting its fellow in a median symphysis, while the other lobe or lateral process extends outwards. Attached to the symphysis in front is a cartilaginous **epipubis**, while behind, the two pubes are terminated by a wide rounded cartilaginous area.

The **ischium**, the smallest bone of the three, is flattened and like the pubis meets its fellow in a median symphysis. A narrow band of cartilage connects the symphysis pubis with the symphysis ischii, and separates the two **obturator foramina** from one another.

The Posterior Limb.

This is divisible into three portions, the **thigh**, the **crus** or **shin**, and the **pes**.

The **thigh** includes a single bone, the **femur**.

The **femur** (fig. 40, B, 12) is a short thick bone, with a prominent rounded **head** articulating with the acetabulum. Behind this head is a deep pit, beyond which is a roughened area corresponding with the great trochanter of mammals. The distal end is expanded and somewhat convex.

The bones of the **crus** or **shin** are the **tibia** and **fibula**. These are both straight rod-like bones with expanded terminations which closely approach one another, while elsewhere the bones diverge considerably.

The terminations of all three of the leg bones are formed by epiphyses.

The **Pes** consists of the **tarsus** or **ankle**, and the **foot**, which is made up of five digits.

The **tarsus**. The tarsal bones of the Turtle do not retain their primitive arrangement to such an extent as do the carpals. They are arranged in a proximal row of two and a distal row of four. Of the bones in the proximal row the postaxial one is much the smaller and is the **fibulare**; the larger pre-axial one (fig. 40, B, 15) represents the **tibiale**, **intermedium**, and **centrale** fused, and articulates with both tibia and fibula. The first three distal tarsalia are all small bones and are very similar in size, and each articulates regularly with the corresponding metatarsal. The fourth bone (fig. 40, B, 19) is much larger, and represents tarsalia 4 and 5 fused. The first two distal tarsalia articulate with the pre-axial tarsal of the proximal row, the third only with its neighbours the second, and the fused fourth and fifth. The latter articulates with both bones of the proximal row.

Each **digit** consists of a metatarsal and of a varying number of phalanges.

The **metatarsals**. The first metatarsal (fig. 40, B, 20) is broad and flattened, the second, third and fourth, are all elongated bones with nearly flat terminations formed by small epiphyses. The fifth is large and flattened, and the articular surface for the phalanx is situated somewhat laterally.

The **phalanges**. The first digit has two phalanges and is the stoutest of them all; its distal phalanx is sheathed in a large horny claw. The other digits, of which the third is the longest, have each three phalanges. The distal phalanges of the second and third digits are flattened and pointed and bear small horny claws.

CHAPTER XV.
THE SKELETON OF THE CROCODILE.

The species chosen for description is *C. palustris*, a form occurring throughout the Oriental region, but the description would apply almost equally well to any of the other species of the genus *Crocodilus*, and with comparatively unimportant modifications to any of the living Crocodilia.

I. EXOSKELETON.

The exoskeleton of the Crocodile is strongly developed and includes elements of both epidermal and dermal origin.

a. The **epidermal exoskeleton** is formed of a number of horny **scales** or plates of variable size covering the whole surface of the body. Those covering the dorsal and ventral surfaces are oblong in shape, and are arranged in regular rows running transversely across the body. The scales covering the limbs and head are mostly smaller and less regularly arranged, and are frequently raised into a more or less obvious keel. Those covering the dorsal surface of the tail are very prominently keeled.

The epidermal exoskeleton also includes the horny **claws** borne by the first three digits of both manus and pes.

b. The **dermal exoskeleton**. This has the form of bony **scutes** which underlie the epidermal scales along the dorsal surface of the trunk and anterior part of the tail. Except in very young individuals the epidermal scales are rubbed off from these scutes, which consequently come to project freely on the surface of the body. Each scute is a nearly square bony plate, deeply pitted or sculptured, and marked by a strong ridge on its dorsal surface, while its ventral surface is smooth. Contiguous scutes are united to one another by interlocking sutures.

The scutes are arranged in two distinct areas, viz. (1) a small anterior **nuchal shield** which lies just behind the head and is formed of six large scutes more or less firmly united together, and (2) a larger posterior **dorsal shield** covering the whole of the back and anterior part of the tail, and formed of smaller scutes, which are arranged in regular transverse rows, and progressively diminish in size when followed back.

The **teeth** are exoskeletal structures, partly of dermal, partly of epidermal origin. They lie along the margins of the jaws and are confined to the premaxillae, maxillae and dentaries. They are simple conical structures, without roots; each is in the adult placed in a separate socket, and is replaced by another which as it grows comes to occupy the pulp cavity of its predecessor. In the young animal the teeth are not placed in separate sockets but in a continuous groove. This feature is met with also in the Ichthyosauria. The groove gradually becomes converted into a series of sockets by the ingrowth of transverse bars of bone. The anterior teeth are sharply pointed and slightly recurved, the posterior ones are more blunt.

The upper jaw bears about nineteen pairs of teeth, the lower jaw about fifteen pairs. The largest tooth in the upper jaw is the tenth, and in the lower jaw the fourth.

The three living families of Crocodilia, the Crocodiles, Alligators and Garials, can be readily distinguished by the characters of the first and fourth lower teeth. In Alligators both first and fourth lower teeth bite into pits in the upper jaw; in Garials they both bite into notches or grooves in the upper jaw. In Crocodiles the first tooth bites into a pit, the fourth into a notch in the upper jaw.

II. ENDOSKELETON.

1. The Axial Skeleton.

This includes the vertebral column, the skull, and the ribs and sternum.

A. The Vertebral column.

The vertebral column is very long, consisting of some sixty vertebrae. It can be divided into the usual five regions, the cervical, thoracic, lumbar, sacral, and caudal regions.

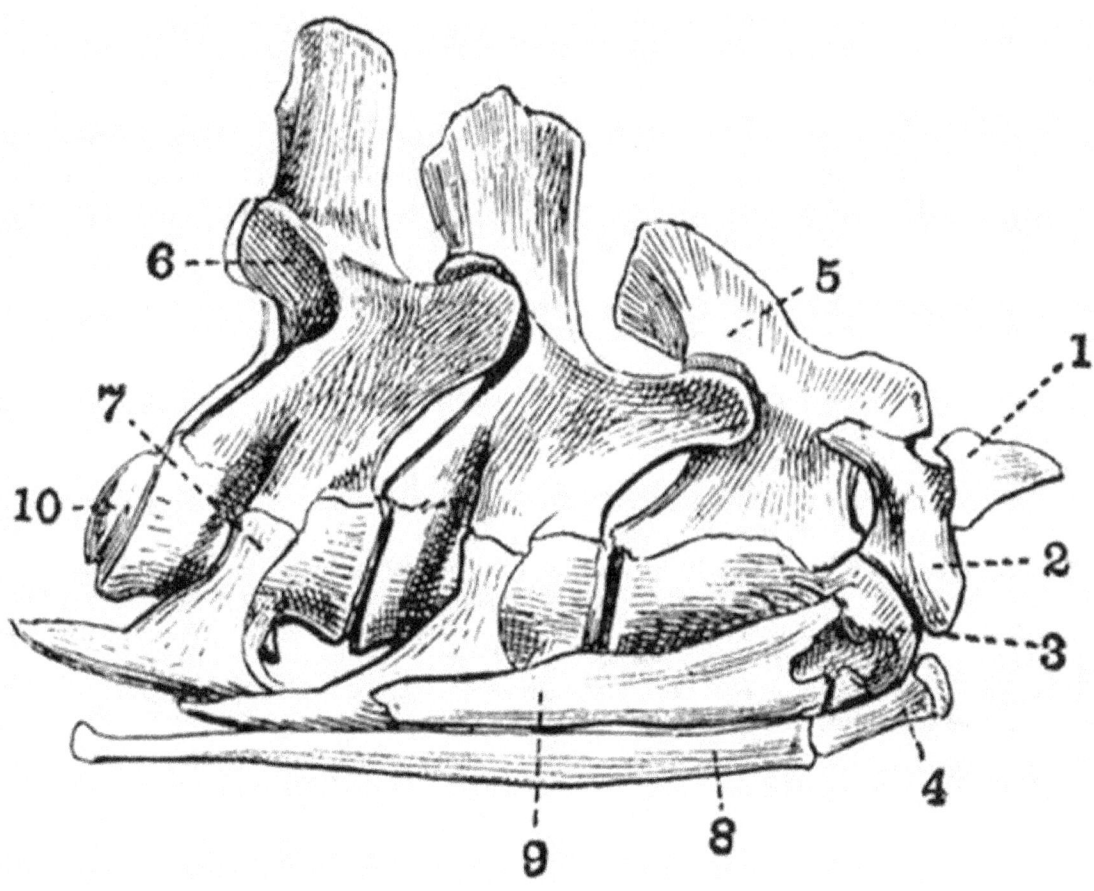

Fig. 41. First four cervical vertebrae of a Crocodile (*C. vulgaris*). (Partly after von Zittel.)

1. pro-atlas.
2. lateral portion of atlas.
3. odontoid process.
4. ventral portion of atlas.
5. neural spine of axis.
6. postzygapophysis of fourth vertebra.
7. tubercular portion of fourth cervical rib.
8. first cervical rib.
9. second cervical rib.
10. convex posterior surface of centrum of fourth vertebra.

The Cervical vertebrae.

Counting as cervical all those vertebrae which are anterior to the first one whose ribs meet the sternum, there are nine cervical vertebrae, all of which bear ribs.

As a type of the cervical vertebrae the fifth may be taken. It has a short cylindrical **centrum** deeply concave in front and convex behind. From the anterior part of the ventral surface of the centrum arises a short **hypapophysis**, and on each side is a facet with which the lower limb (**capitulum**) of the cervical rib articulates. The **neural arch** is strongly developed and drawn out dorsally into a long **neural spine**, in front of which are a pair of upstanding processes bearing the prominent upwardly and inwardly directed **prezygapophyses**. At the sides and slightly behind the neural spine are a corresponding pair of processes bearing the **postzygapophyses**, which look downwards and outwards. At the point where it joins the centrum the neural arch is drawn out into a short blunt **transverse process** with which the upper limb (**tuberculum**) of the cervical rib articulates. The sides of the neural arch are slightly notched behind for the exit of the spinal nerves.

The first or **atlas** vertebra differs much from any of the others, and consists of four quite detached portions, a ventral arch, with two lateral portions and one dorsal. The **ventral arch** (fig. 41, 4) is flat below and slightly concave in front, forming together with two flattened surfaces on the lateral portions a large articulating surface for the occipital condyle of the skull. Its posterior face is bevelled off and forms with a second pair of facets on the lateral portions a surface with which the odontoid process of the second vertebra articulates. The postero-lateral surfaces of the ventral arch also bear a pair of little facets with which the cervical ribs articulate. The lateral portions are somewhat flattened and expanded, and bear in addition

119

to those previously mentioned a pair of small downwardly directed facets, the postzygapophyses, which articulate with the prezygapophyses of the second vertebra. The dorsal portion (fig. 41, 1) is somewhat triangular in shape, and overhangs the occipital condyle. It is often regarded as the neural arch of a vertebra in front of the atlas and is called the *pro-atlas*; but as it is a membrane bone it is not properly a vertebral element.

The second or **axis vertebra** also differs a good deal from the other cervicals. The centrum is massive, and is terminated in front by a very large slightly concave articulating surface formed by the **odontoid process** (fig. 41, 3) which is united with the centrum by suture only, and is really the detached centrum of the first vertebra. The cervical rib (fig. 41, 9) articulates with two little irregularities on the odontoid process. The posterior surface of the centrum is convex. The neural arch is strongly developed and terminated dorsally by a long neural spine (fig. 41, 5), its sides are notched, slightly in front and more prominently behind for the exit of the spinal nerves. It is drawn out in front into two little processes bearing a pair of upwardly and outwardly directed prezygapophyses, while the postzygapophyses are similar to those of the other cervical vertebrae.

The last two cervical vertebrae resemble the succeeding thoracic vertebrae, in the increased length of the transverse processes, and the shifting dorsalwards of the facet with which the capitulum of the rib articulates.

The Thoracic vertebrae.

The thoracic vertebrae commence with the first of those that bears ribs reaching the sternum. They are ten in number, and the first eight are directly connected with the sternum by ribs.

The **third** of them may be taken as a type. It has a thick cylindrical centrum, concave in front and convex behind, there is a slight hypapophysis, and the centrum is suturally united with a strong neural arch enclosing a narrow neural canal. The neural arch is drawn out dorsally into a wide truncated neural spine, and laterally into two prominent transverse processes, with the ends of which the tubercula of the ribs articulate, while the capitulum articulates in each case with a step-like facet (fig. 42, A, 3) on the anterior face of the transverse process. The prezygapophyses (fig. 42, A, 2) are borne on outgrowths from the bases of the transverse processes, and the postzygapophyses on outgrowths at the base of the neural spine.

The thoracic vertebrae behind the third have no hypapophyses, and the capitular facets gradually come to be placed nearer and nearer the ends of the transverse processes, at the same time becoming less prominent; otherwise these vertebrae are just like the third.

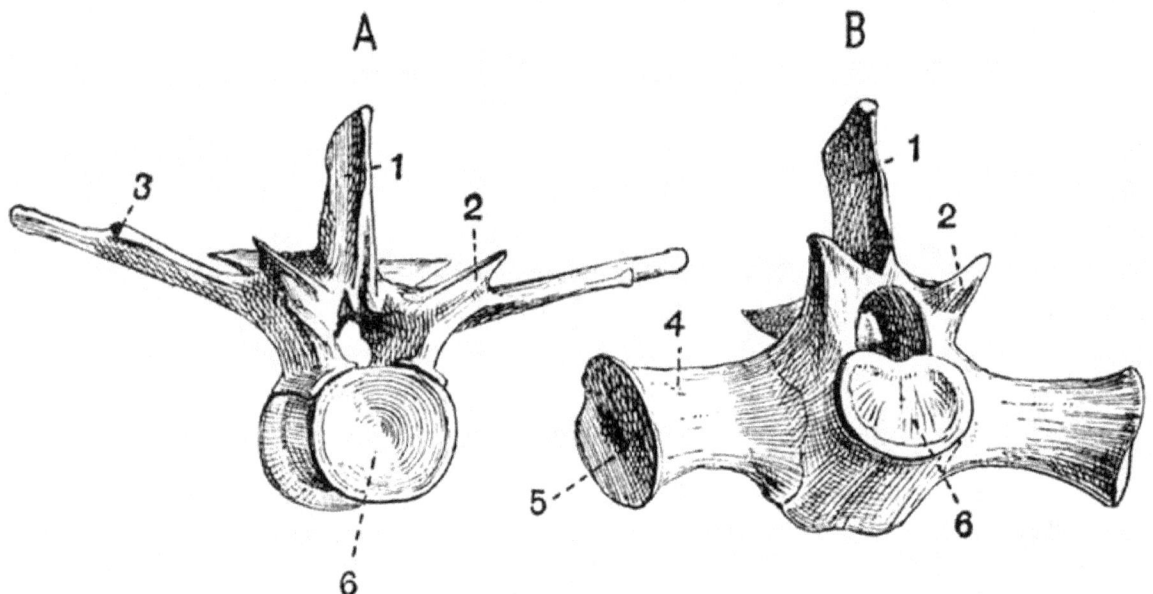

Fig. 42. Anterior view of A, a late thoracic and B, the first sacral vertebra of a young Crocodile (*C. palustris*). × 1/3.

1. neural spine.	4. sacral rib.
2. process bearing prezygapophysis.	5. surface which is united with the ilium.
3. facet for articulation with the capitulum of the rib.	6. concave anterior face of centrum.

In the first and second thoracic vertebrae the capitulum of the rib articulates, not with a facet on the transverse process, but with a little elevation borne at the line of junction of the centrum and neural arch.

The Lumbar vertebrae.

These are five in number, and are precisely like the posterior thoracic vertebrae, except in the fact that the transverse processes have no facets for the articulation of ribs.

The Sacral vertebrae.

These are two in number, and while the centrum of the first is concave in front (fig. 42, B, 6) and nearly flat behind, that of the second is flat in front and concave behind. Each has a pair of strong **ribs** (fig. 42, B, 4) firmly ankylosed in the adult with a wide surface furnished partly by the centrum, partly by the neural arch. The distal ends of these ribs are united with the ilia. The character of the neural spines and zygapophyses is the same as in the thoracic vertebrae.

The Caudal vertebrae.

These are very numerous, about thirty-four in number. The first differs from all the other vertebrae of the body in having a biconvex centrum. The succeeding ones are procoelous and are very much like the posterior thoracic and lumbar vertebrae, having high neural spines and prominent straight transverse processes. They differ however in having the neural spines less strongly truncated above, and the transverse processes arise from the centra and not from the neural arches. When followed further back the centra and neural spines gradually lengthen while the transverse processes become reduced, and after the twelfth vertebra disappear. Further back still the neural spines and zygapophyses gradually become reduced and disappear, as finally the neural arch does also, so that the last few vertebrae consist simply of cylindrical centra.

Each caudal vertebra, except the first and the last eleven or so, has a **V**-shaped **chevron bone** attached to the postero-ventral edge of its centrum. The anterior ones are the largest and they gradually decrease in size till they disappear.

B. The Skull.

The skull of the Crocodile is a massive depressed structure presenting a number of striking characteristics, some of the more important of which are:—

1. All the bones except the mandible, hyoid, and columella are firmly united by interlocking sutures. In spite of this, however, growth of the whole skull and of the component bones goes on continuously throughout life, this growth being especially marked in the case of the facial as opposed to the cranial part of the skull.

2. All the bones appearing on the dorsal surface are remarkable for their curious roughened and pitted character; this feature is prominent also in many Labyrinthodonts.

3. The size of the jaws and teeth is very great.

4. The mandibular condyle is carried back to some distance behind the occipital condyle.

5. The occipital plane (see p. 386) of the skull is vertical.

6. The length of the secondary palate is remarkably great, and the vomer takes no part in its formation.

7. The posterior nares are placed very far back, the nasal passages being as in mammals separated from the mouth by the long secondary palate.

8. There is a complicated system of Eustachian passages communicating at one end with the tympanic cavity and at the other end with the mouth cavity.

9. The interorbital septum is mainly cartilaginous, the presphenoidal and orbitosphenoidal regions remaining unossified.

The **skull** is divisible into three parts:—

(1) the cranium, (2) the lower jaw, (3) the hyoid.

The **cranium** may again for purposes of description be divided into:—

1. the cranium proper or brain case;

2. the bones connected with the several special sense organs;

3. the bones of the upper jaw, and suspensorial apparatus.

1. The Cranium proper or brain case.

121

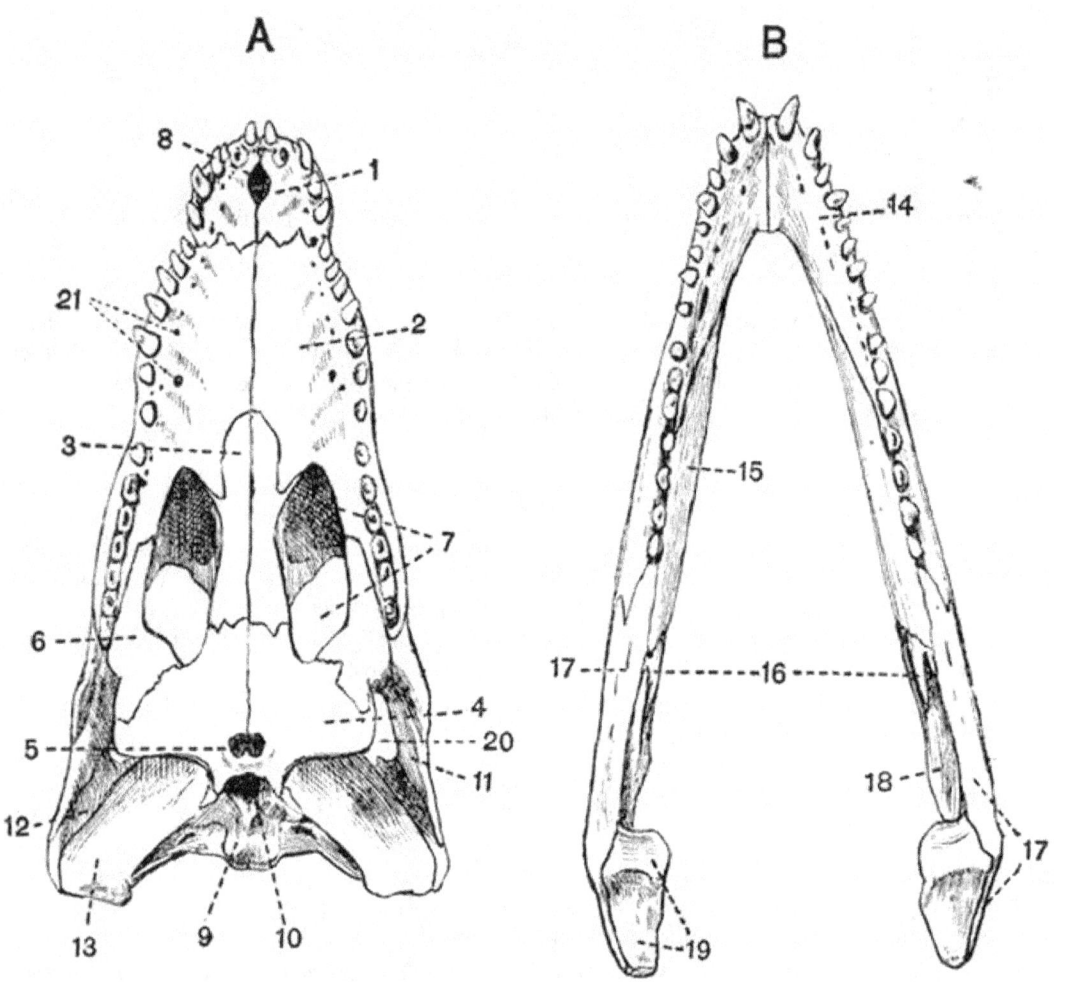

Fig. 43. Palatal aspect A, of the cranium, B, of the mandible of an Alligator (*Caiman latirostris*). × 1/3. (Brit. Mus.)

1. premaxillae.

2. maxillae.

3. palatine.

4. pterygoid.

5. posterior nares.

6. transpalatine.

7. posterior palatine vacuity.

8. anterior palatine vacuity.

9. basi-occipital.

10. opening of median

Eustachian canal.

11. jugal.

12. quadratojugal.

13. quadrate.

14. dentary.

15. splenial.

16. coronoid.

17. supra-angular.

18. angular.

19. articular.

20. lateral temporal fossa.

21. openings of vascular canals

leading into alveolar sinus.

The cartilage and membrane bones of the cranium proper when taken together can in most vertebrates be seen to be more or less arranged in three rings or segments called respectively the **occipital**, **parietal** and **frontal** segments; in the Crocodile however only the occipital and parietal segments are clearly seen.

The **occipital segment** consists of four cartilage bones, three of which together surround the **foramen magnum**.

The most ventral of these, the **basi-occipital** (figs. 43 and 45, 9), forms the single convex **occipital condyle** for articulation with the atlas, bounds the base of the foramen magnum, and is continuous laterally with two larger bones, the **exoccipitals** (fig. 45, 24), which meet one another dorsally and form the remainder of the boundary of the foramen magnum. Each is drawn out externally into a strong process, which is united below with the quadrate, and above with the squamosal by a surface seen in a disarticulated skull to be very rough and splintered. In a longitudinal section the anterior face of the exoccipital is seen to be closely united with the opisthotic.

The exoccipital is pierced by a number of foramina, four lying on the posterior surface. Just external to the foramen magnum is a small foramen for the exit of the hypoglossal nerve (figs. 44 and 45, XII). External to this is the foramen for the pneumogastric (fig. 44, X), while more ventrally still is the foramen (fig. 44, 15) through which the internal carotid artery enters the skull. Some distance further to the side, and more dorsally, is a larger foramen which gives passage to the facial nerve and certain blood-vessels.

In a median longitudinal section of the skull the hypoglossal foramen is seen, and just in front of it a small foramen for a vein. Further forwards the long slit-like opening between the exoccipital and opisthotic is the **internal auditory meatus** (fig. 45, VIII) through which the auditory nerve leaves the cranial cavity and enters the internal ear.

The **supra-occipital** (fig. 45, 5) is a small bone which takes no part in the formation of the foramen magnum, and is closely united in front with the epi-otic. It is characteristic of Crocodiles that all the bones of the occipital segment have their longer axes placed vertically, and that they scarcely if at all appear on the dorsal surface.

In front of the occipital segment is the **parietal segment**. The dorsal and ventral portions of the two segments are in contact with one another, but the lateral portions are widely separated by the interposition of the **auditory** and **suspensorial bones**.

The **basisphenoid** (fig. 45, 12) is an unpaired wedge-shaped bone, united along a deep vertical suture with the basi-occipital. The two bones are, however, partially separated in the mid-ventral line by a foramen, the opening of the **median Eustachian canal**, which leads into a complicated system of Eustachian passages ultimately communicating with the tympanic cavity.

The dorsal surface of the basisphenoid is well seen in a section of the skull, but owing to the way it tapers ventrally, it appears on the ventral surface only as a very narrow strip of bone wedged in between the basi-occipital and pterygoids. In a lateral view it is seen to be drawn out in front into an abruptly truncated process, the **rostrum**, which forms part of the **interorbital septum**. On the anterior part of the dorsal surface is a deep pit, the **pituitary fossa** or **sella turcica**, at the base of which are a pair of foramina, through which the carotid arteries pass. Dorsolaterally the basisphenoid articulates with the **alisphenoids**.

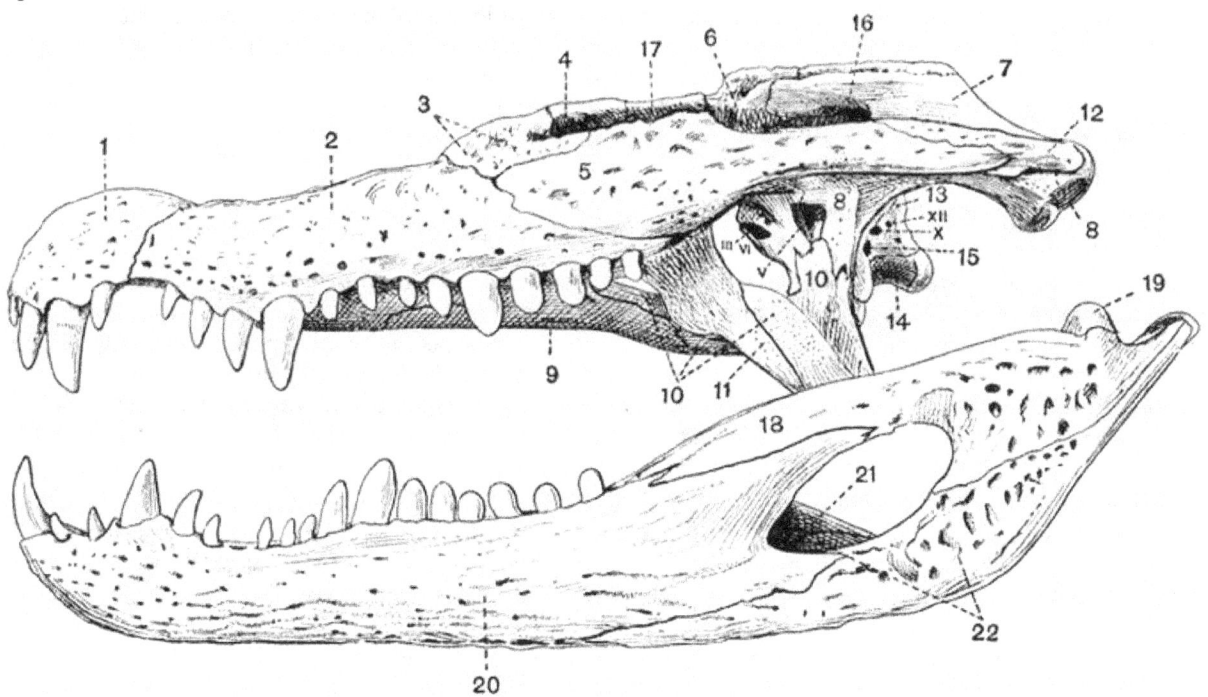

Fig. 44. Lateral view of the skull of an Alligator (*Caiman latirostris*). × 1/3. (Brit. Mus.)

1. premaxillae.	9. palatine.	16. external auditory	22. angular.
2. maxillae.	10. pterygoid.	meatus.	III, VI, opening for exit
3. lachrymal.	11. transpalatine.	17. frontal.	of oculomotor and
4. prefrontal.	12. quadratojugal.	18. supra-angular.	abducens nerves.
5. jugal.	13. exoccipital.	19. articular.	V, foramen ovale.

123

6. postfrontal.	14. basi-occipital.	20. dentary.	X, pneumogastric foramen.
7. squamosal.	15. foramen by which carotid	XII, hypoglossal foramen.	
8. quadrate.	artery enters skull.	21. coronoid.	

The **alisphenoids** (fig. 45, 13) are a pair of irregular bones which arise from the basisphenoid antero-laterally, and are united dorsally with the parietal, frontal, and postfrontals. They bound most of the anterior part of the brain case, and each presents on its inner face a deep concavity which lodges the cerebral hemisphere of its side. Viewed from the ventral side the two alisphenoids are seen to almost or quite meet one another immediately below the frontal, and then to diverge, forming an irregular opening—partially closed by cartilage in the fresh specimen,—through which the optic nerves leave the cranial cavity. Further back the alisphenoids meet one another for a narrow area, and then diverge again, so that between each and the rostrum of the basisphenoid there appears an opening (fig. 44, III, VI) through which the oculomotor and abducens nerves leave the cranium. Further back still each is united for a short space with the basisphenoid, pterygoid and quadrate, and then becomes separated from the quadrate by a large foramen, the **foramen ovale** (fig. 44, V), through which the whole of the trigeminal nerve passes out.

The dorsal portion of the parietal segment is formed by the *parietal* (fig. 45, 4), which though double in the embryo, early comes to form a single bone. It extends over the posterior part of the cranial cavity, and is continuous in front with the frontal, behind with the supra-occipital, and laterally with the postfrontals, squamosals, alisphenoids, pro-otics and epi-otics. It forms the inner boundary of a large rounded vacuity on the roof of the skull, the **supratemporal fossa**.

The **frontal segment** is very imperfectly ossified, there being no certain representatives of either the ventral member, the presphenoid, or the lateral members, the orbitosphenoids. On the dorsal side there is, however, a large development of membrane bones. There is a large *frontal* (fig. 45, 3), unpaired, except in the embryo, united behind with the parietal and postfrontal, and drawn out in front into a long process which is overlapped by the prefrontals and posterior part of the nasals. The frontal ends off freely below, owing to the orbitosphenoidal region being unossified, it forms a considerable part of the roof of the cranial cavity, but takes no part in the formation of the wall.

Each *prefrontal* (fig. 45, 14) forms part of the inner wall of the orbit and sends ventralwards a process which meets the palatine.

The *postfrontals* (fig. 44, 6) are small bones lying at the sides of the posterior part of the frontal. Each is united with a number of bones, on its inner side with the frontal and parietal, behind with the squamosal, and ventrally with the alisphenoid. It also unites by means of a strong descending process with an upgrowth from the jugal, and thus forms a **postorbital bar** separating the orbit from the lateral temporal fossa. The postfrontal forms also part of the outer boundary of the supratemporal fossa.

2. The Sense capsules.

Skeletal capsules occur in connection with each of the three special sense organs of sight, of hearing and of smell.

The **Auditory capsules** and associated bones.

Three bones, the **epi-otic**, **opisthotic** and **pro-otic**, together form the auditory or **periotic** capsule of each side. They are wedged in between the lateral portions of the occipital and parietal segments and complete the cranial wall in this region. Their relations to the surrounding structures are very complicated, and many points can be made out only in sections of the skull passing right through the periotic capsule. The relative position of the three bones is, however, well seen in a median longitudinal section. The **opisthotic** early becomes united with the exoccipital, while the **epi-otic** similarly becomes united with the supra-occipital, the **pro-otic** (fig. 45, 7),—seen in longitudinal section to be pierced by the prominent **trigeminal foramen**—alone remaining distinct throughout life. The three bones together surround the essential organ of hearing which communicates laterally with the deep tympanic cavity by the **fenestra ovalis**.

The **tympanic cavity**, leading to the exterior by the **external auditory meatus** (fig. 44, 16), is well seen in a side-view of the skull; it is bounded on its inner side by the periotic bones, posteriorly in part by the exoccipital, and elsewhere mainly by the quadrate. A large number of canals and passages open into it. On its inner side opening ventro-anteriorly is the **fenestra ovalis**, opening ventro-posteriorly the **internal auditory meatus** (fig. 45, VIII), while dorsally there is a wide opening which forms a communication through the roof of the brain-case with the tympanic cavity of the other side. On its posterior wall is the prominent foramen through which the facial nerve passes on its way to its final exit from the skull through the exoccipital, this foramen is bounded by the quadrate, squamosal, and exoccipital.

The opening of the fenestra ovalis is in the fresh skull occupied by the expanded end of the auditory ossicle, the **columella**, whose outer end articulates by a concave facet with a trifid **extra-columellar** cartilage which reaches the tympanic membrane. The lower process of this extra-columella passes into a cartilaginous rod which lies in a canal in the quadrate and is during life continuous with Meckel's cartilage within the articular bone of the mandible.

The columella and extra-columella are together homologous with the chain of mammalian auditory ossicles.

The **Optic capsules** and associated bones.

Two pairs of bones are associated with the optic capsules, viz. the *lachrymals* and the *supra-orbitals*. The *lachrymal* (fig. 44, 3) is a fairly large flattened bone lying wedged in between the maxillae, nasal, jugal, and prefrontal. It forms a considerable part of the anterior boundary of the orbit, and is pierced by two foramina. On the orbital edge is a large hole leading into a cavity within the bone which lodges the naso-lachrymal sac, and communicates with the narial passage by a wide second foramen near the anterior end of the bone. The *supra-orbital* is a very small loose bone lying in the eyelid close to the junction of the frontal and prefrontal.

The **Olfactory capsules** and associated bones.

Two pairs of membrane bones, the *vomers* and *nasals*, are developed in association with the olfactory organ, but the **mesethmoid** is not ossified.

The *vomers* form a pair of delicate bones, each consisting of a vertical plate (fig. 45, 15), which with its fellow separates the two narial passages, and of a horizontal plate which forms much of their roof. The vomers articulate with one another and with the pterygoids, palatines, and maxillae.

The *nasals* (fig. 45, 2) are very long narrow bones extending along the middle line from the frontal almost to the anterior nares. They are continuous laterally with the premaxillae, maxillae, lachrymals and prefrontals. They form the roof of the narial passages.

3. The Upper Jaw and suspensorial apparatus.

These are enormously developed in the Crocodile and are firmly united to the cranium. It will be most convenient to begin by describing the bones at the anterior end of the jaw and to work back thence towards the brain-case. The most anterior bones are the *premaxillae*. The *premaxillae* (figs. 44 and 45, 1) are small bones, each bearing five pairs of teeth, set in separate sockets in their alveolar borders. They constitute almost the whole of the boundary of the **anterior nares**, which are confluent with one another and form a large semicircular opening in the roof of the skull, leading into the wide narial passage. They are also partially separated from one another in the ventral middle line, by the small **anterior palatine vacuity** (fig. 43, A, 8). They form the anterior part of the broad **palate**. The alveolar border on each side between certain of the teeth is marked by pits which receive the points of the teeth of the other jaw. The first pair of these pits in the premaxillae are often so deep as to be converted into perforations. Pits of the same character occur between the maxillary and mandibular teeth.

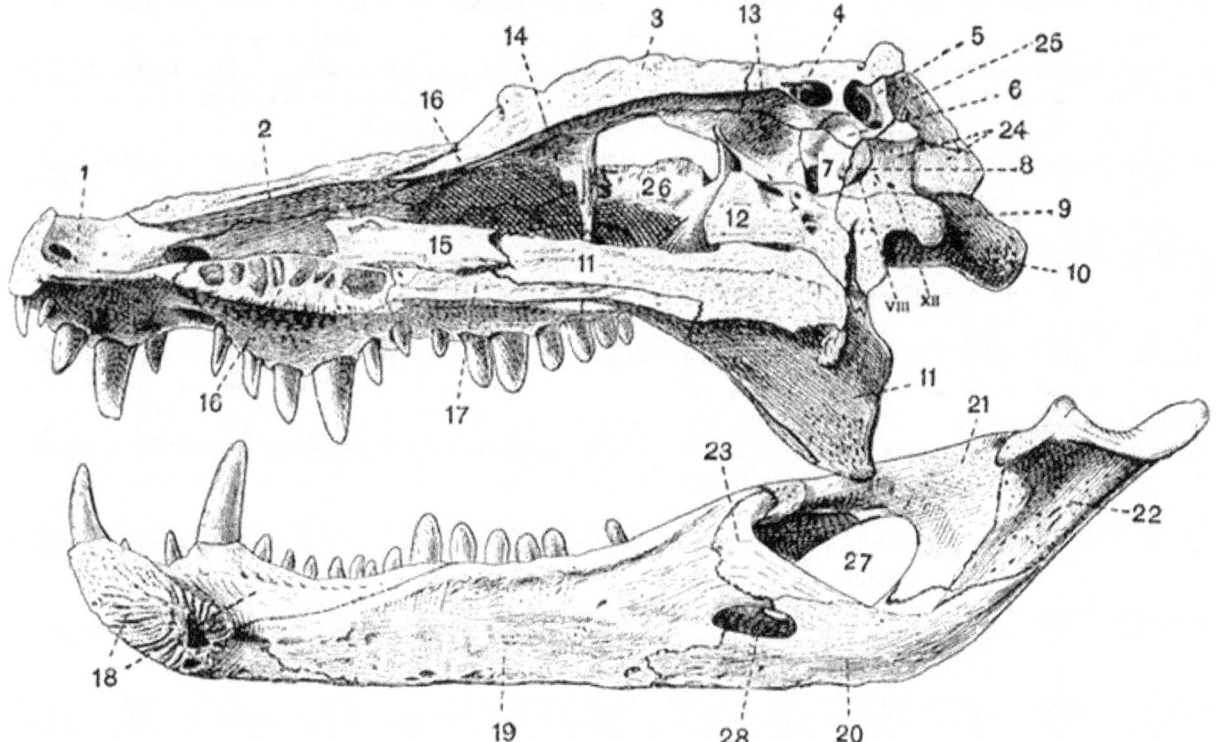

Fig. 45. Longitudinal section through the skull of an Alligator (*Caiman latirostris*). × 1/3. (Brit. Mus.)

1. premaxillae.		16. maxillae.	25. squamosal.
2. nasal.	for the trigeminal nerve.	17. palatine.	26. jugal.
	8. opisthotic.		

125

3. frontal.

4. parietal.

5. supra-occipital.

6. epi-otic.

7. pro-otic.

immediately in front of the

figure 7 is the prominent foramen

9. basi-occipital.

10. quadrate.

11. pterygoid.

12. basisphenoid.

13. alisphenoid.

14. prefrontal.

15. vomer.

18. dentary.

19. splenial.

20. angular.

21. supra-angular.

22. articular.

23. coronoid.

24. exoccipital.

27. external mandibular foramen.

28. internal mandibular foramen.

VIII. internal auditory meatus.

XII. hypoglossal foramen.

The *maxillae* (figs. 43, A, 2 and 44, 2) are a pair of very large bones and bear the remaining teeth of the upper jaw, set in sockets along their alveolar borders. On the dorsal side each maxillae is continuous with the premaxillae, nasal, lachrymal, and jugal, while ventrally it meets its fellow in a long straight suture and forms the greater part of the long bony palate. The maxillae are separated in the middle line posteriorly by processes from the palatines, while further back they meet the transpalatines. The internal or nasal surface, like that of the premaxillae, is excavated by a deep longitudinal groove, the **narial passage**. In a ventral view of the skull a number of small openings (fig. 43, A, 21) are seen close to the alveolar border, these are the openings of small vascular canals which lead into the **alveolar sinus**, a passage traversing the maxillae, and transmitting the superior maxillary branch of the trigeminal nerve and certain blood-vessels. This alveolar sinus opens posteriorly by the more external of the two large holes in the maxillae, which lie close to the anterior edge of the posterior palatine vacuity, to be described immediately. The more internal of these holes, on the other hand, leads into a cavity lodging the nasal sac. Behind the maxillae the completeness of the palate is broken up by the large oval **posterior palatine vacuities** (fig. 43, A, 7); these are separated from one another in the middle line by the palatines, and are bounded elsewhere by the maxillae, transpalatines, and pterygoids.

The *palatines* (fig. 43, A, 3) are long and rather narrow bones interposed between the maxillae in front and pterygoids behind. They meet one another in a long suture and form much of the posterior part of the palate, while the whole length of their dorsal surface contributes to the floor of the narial passage. The dorsal surface of each bone is also drawn out on its outer side into a prominent ridge which forms much of the side and roof of the narial passage, being in contact with the vomer and pterygoid, and at one point by means of a short ascending process with the descending process of the prefrontal.

The *pterygoids* (figs. 43, A, 4, and 45, 11) are a pair of large bones, each consisting of a median more or less vertical part, which becomes ankylosed to its fellow in the middle line early in life, and of a wide horizontal part which meets the transpalatine. They completely surround the posterior nares (fig. 43, A, 5) and their median portions form the whole boundary of the posterior part of the narial passage, and assist the palatines and vomers in bounding the middle part. The horizontal parts form the posterior part of the secondary palate, while the dorsal surface of each looks into the **pterygoid fossa**, a large cavity lying below the quadrate and quadratojugal at the side of the skull. The lateral margin adjoining the transpalatine is in the fresh skull terminated by a plate of cartilage against which the mandible plays. Dorsally the pterygoid articulates with the basisphenoid, quadrate, and alisphenoid.

The *transpalatines* (fig. 44, 11) connect the pterygoids with the jugals and maxillae, articulating with each of the three bones by a long pointed process. The jugal process meets also a down-growth from the postfrontal.

The *jugals* or *malars* (fig. 44, 5) are long somewhat flattened bones which are united to the lachrymals and maxillae in front, while passing backwards each is united behind to the *quadratojugal* (fig. 44, 12), the two forming the **infratemporal arcade** which constitutes the external boundary of the orbit and lateral temporal fossa. The jugal is united below to the transpalatine, and the two bones together form an outgrowth, which meeting that from the postfrontal forms the **postorbital bar**, and separates the orbit from the lateral temporal fossa. The quadratojugals are small bones and are united behind with the quadrates.

The **quadrate** (figs. 43, A, 13 and 44, 8) of each side is a large somewhat flattened bone firmly fixed in among the other bones of the skull. It is terminated posteriorly by an elongated slightly convex surface, coated with cartilage in the fresh skull, by which the mandible articulates with the cranium. The dorsal surface of the quadrate is flat behind, further forwards it becomes much roughened and articulates with the exoccipital and squamosal; further forwards still it becomes marked by a deep groove which forms the floor of the external auditory meatus and part of the tympanic cavity. The anterior boundary of the quadrate is extremely irregular, it is united dorsally with the postfrontal, pro-otic, and squamosal, and more ventrally with the alisphenoid. The smooth ventral surface looks into the pterygoid fossa. In front the quadrate forms the posterior boundary of the supratemporal fossa and foramen ovale, and is continuous with the alisphenoid, while it sends down a thin plate meeting the pterygoid and basisphenoid. On the inner side of the dorsal surface of the quadrate near the condyle, is a small foramen which leads into a tube communicating with the tympanic cavity, by a foramen lying in front of and ventral to that for the exit of the facial nerve. By this tube air can pass from the tympanic cavity into the articular bone of the mandible.

126

The *squamosal* (fig. 44, 7) meets the quadrate and exoccipital below, and forms part of the roof of the external auditory meatus, while above it forms part of the roof of the skull and has a pitted structure like that of the other bones of the roof. It is continuous with the postfrontal in front, forming with it the **supratemporal arcade** which constitutes the outer boundary of the supratemporal fossa. It meets also the parietal on its inner side, forming the **post-temporal bar**, the posterior boundary of the supratemporal fossa.

It may be useful to recapitulate the large vacuities in the surface of the Crocodile's cranium.

Dorsal surface.

1. **The Supratemporal fossae**. Each is bounded internally by the parietal, behind by the **post-temporal bar** formed by the parietal and squamosal, and externally by the **supratemporal arcade** formed by the squamosal and postfrontal. The postfrontal meets the parietal in front and forms the anterior boundary of the supratemporal fossa.

2. The **Lateral temporal** or **infratemporal fossae**. These lie below and to the outer side of the supratemporal fossae. Each is bounded dorso-internally by the supratemporal arcade; and behind by a continuation of the post-temporal bar formed by the quadrate and quadratojugal. The external boundary is the **infratemporal arcade** formed of the quadratojugal and jugal, while in front the fossa is separated from the orbit by the **postorbital bar** formed by the junction of outgrowths from the postfrontal and jugal.

3. The **Orbits**. Each is bounded behind by the postorbital bar, externally by the jugal forming a continuation of the infratemporal arcade, in front by the lachrymal, and internally by the frontal and prefrontal.

4. The **Anterior nares**. These form an unpaired opening bounded by the premaxillae.

Posterior surface.

5. The **Foramen magnum**. The exoccipitals form the chief part of its boundary, but part of the ventral boundary is formed by the basi-occipital.

6. The **Pterygoid fossae**. These form a pair of large cavities at the sides of the occipital region of the skull. The dorsal boundary is formed by the quadrate and quadratojugal, the ventral by the pterygoid, the internal chiefly by the quadrate, pterygoid, alisphenoid, and basisphenoid. The transpalatine forms a small part of the external boundary which is incomplete.

Ventral surface.

7. The **Posterior nares**. These form a median unpaired opening (fig. 43, A, 5) bounded by the pterygoids.

8. The **Posterior palatine vacuities**. Each is bounded by the maxillae in front, the maxillae and transpalatine externally, the transpalatine and pterygoid behind, and the palatine on the inner side (fig. 43, A, 7).

9. The **Anterior palatine vacuity**. This is unpaired and is bounded by the premaxillae (fig. 43, A, 8).

(*b*) The Lower Jaw or Mandible.

The mandible is a strong compact bony structure formed of two halves or **rami**, which are suturally united at the symphysis in the middle line in front. Each ramus is formed of six separate bones.

The most anterior and largest of these is the *dentary* (figs. 44, 20, and 45, 18), which forms the symphysis, and greater part of the anterior half of the jaw, and bears along the outer part of its dorsal border a number of sockets or **alveoli** in which the teeth are placed. Lying along the inner side of the dentary is a large splint-like bone, the *splenial* (fig. 45, 19), which does not extend so far forwards as the symphysis, and is separated from the dentary posteriorly by a large cavity. Forming the lower part of all the posterior half of the jaw is the large *angular* (figs. 44, 22, and 45, 20), which underlies the posterior part of the dentary in front and sends a long process below that bone to the splenial. On the inner side of the jaw there is an oval vacuity, the **internal mandibular foramen** (fig. 45, 28), between the angular and the splenial; through this pass blood-vessels and branches of the inferior dental nerve. Lying dorsal to the angular is another large bone, the *supra-angular* (figs. 44, 18, and 45, 21). It extends back as far as the posterior end of the jaw and forwards for some distance dorsal to the dentary and splenial. It forms part of the posterior margin of a large vacuity, the **external mandibular foramen**, which is bordered above and in front by the dentary and below by the angular; it gives passage to the cutaneous branch of the inferior dental nerve. The concave surface for articulation with the mandible and much of the posterior end of the jaw is formed by a short but solid bone, the **articular** (fig. 45, 22), which in young skulls rather readily becomes detached. The remaining mandibular bone is the *coronoid* (fig. 45, 23), a very small bone of irregular shape attached to the angular below, and to the supra-angular and splenial above.

(*c*) The Hyoid.

The hyoid of the Crocodile consists of a wide flattened plate of cartilage, the **basilingual plate** or **body of the hyoid**, and a pair of **cornua**.

The **basilingual plate** (fig. 53, 1) is rounded anteriorly and marked by a deep notch posteriorly. The **cornua** (fig. 53, 3), which are attached at a pair of notches near the middle of the outer border of the basilingual plate, are partly ossified, but their expanded ends are formed of cartilage. They pass at first backwards and then upwards and inwards. They are homologous with part of the first branchial arches of Selachians.

127

The columella and extra-columella have been already described (p. 251).

C. The Ribs and Sternum.

Thoracic ribs.

The Crocodile has ten pairs of **thoracic ribs**, all except the last one or two of which consist of three parts,—a vertebral rib, an intermediate rib and a sternal rib.

Of the **vertebral ribs** the third may be taken as a type, it consists of a curved bony rod which articulates proximally with the transverse process of the vertebra by two facets. The terminal one of these, the **capitulum** or **head**, articulates with a notch on the side of the transverse process; the other, the **tuberculum**, which lies on the dorsal surface a short distance behind the head, articulates with the end of the transverse process. From near the distal end an imperfectly ossified uncinate **process** (see p. 190) projects backwards.

The **intermediate ribs** are short and imperfectly ossified; they are united with the **sternal ribs** (fig. 46, 3), which are large, flattened, likewise imperfectly ossified structures, and articulate at their distal ends with a pair of long divergent **xiphisternal horns** (fig. 46, 5), which arise from the posterior end of the sternum proper. The last pair of sternal ribs are attached to the preceding pair, not to the xiphisternal horns.

The first and second vertebral ribs differ from the others in the fact that the tuberculum forms a fairly long outstanding process.

Cervical ribs.

Movable ribs are attached to all the cervical as well as to the thoracic vertebrae. Those borne by the atlas and axis are long, narrow structures attached by a fairly broad base, and tapering gradually. The ribs borne by the third to seventh cervical vertebrae are shaped like a **T** with a double base, one limb of which, corresponding to the tuberculum (fig. 41, 7), articulates with a short transverse process arising from the neural arch, while the other, corresponding to the capitulum, articulates with a surface on the centrum. The ribs attached to the eighth and ninth cervical vertebrae are intermediate in character between the **T**-shaped ribs and the ordinary thoracic ribs. The anterior limb of the **T** is shortened, the posterior one is drawn out, forming the shaft of the rib. The distal portion of the rib of the ninth cervical vertebra is unossified.

The **Sacral ribs** have been described in connection with the sacral vertebrae.

The Sternum.

The **sternum** of Crocodiles is a very simple structure, consisting of a plate of cartilage (fig. 46, 2) lying immediately dorsal to the interclavicle, and drawn out posteriorly into a pair of long **xiphisternal horns** (fig. 46, 5).

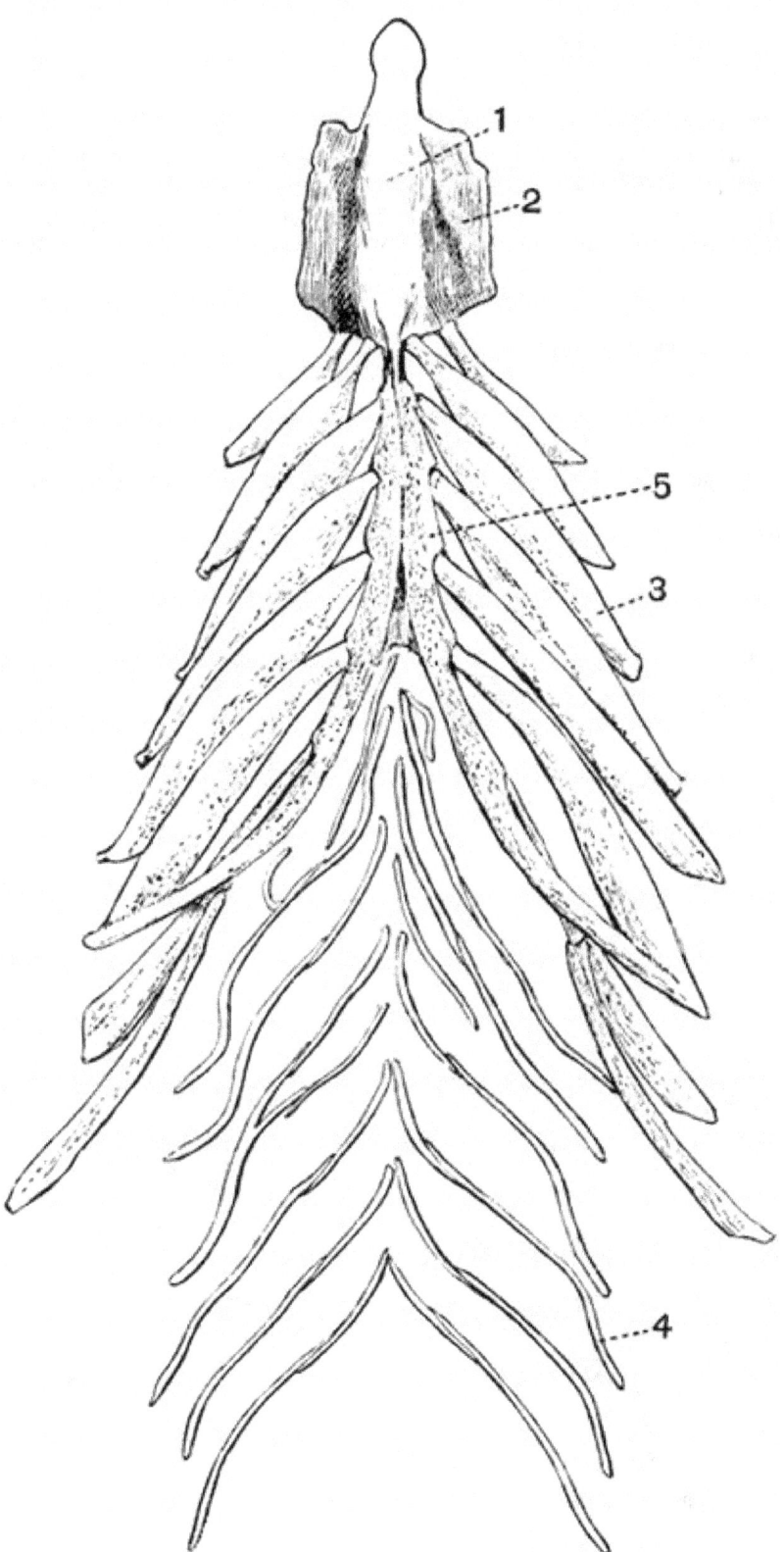

Fig. 46. Sternum and Associated Membrane bones of a Crocodile (*C. palustris*) × 1/3. (Brit. Mus.)
The last pair of abdominal ribs which are united with the epipubes by
a plate of cartilage have been omitted.

1. interclavicle.   4. abdominal splint rib.

129

2. sternum.   5. xiphisternal horn.

3. sternal rib.

The abdominal splint ribs.

Lying superficially to the recti muscles of the ventral body-wall, behind the sternal ribs, are seven or eight series of slender curved bones, the *abdominal ribs* (fig. 46, 4). Each series consists of four or more bones, arranged in a **V**-like form with the angle of the **V** directed forwards. They show a considerable amount of variability in number and character. They are really membrane bones, and are in no way homologous with true ribs, but correspond rather with the more posterior of the bones constituting the plastron of Chelonia.

2. The Appendicular Skeleton.

This includes the skeleton of the two pairs of limbs and their respective girdles.

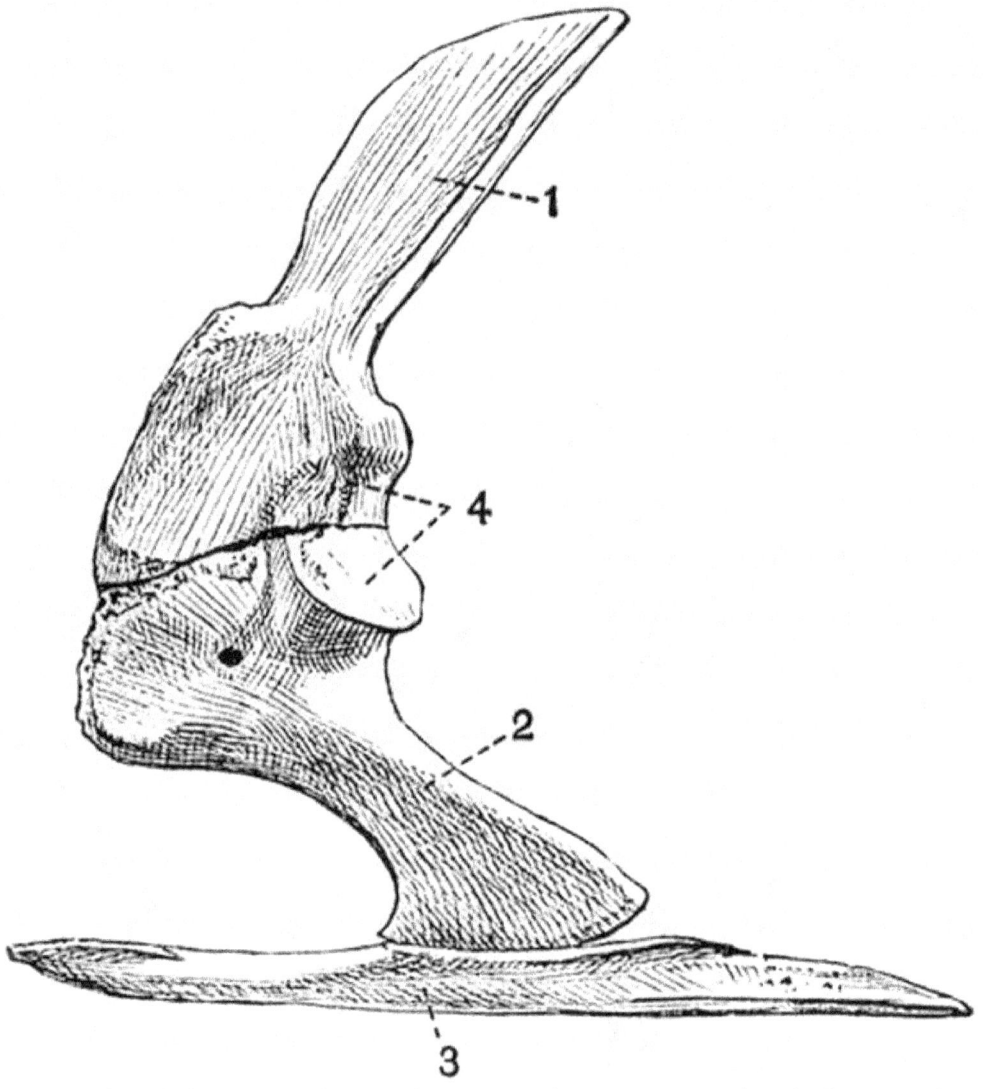

Fig. 47. Left half of the pectoral girdle of an Alligator (*Caiman latirostris*) × 2/3. (Brit. Mus.)

1. scapula.   3. interclavicle.

2. coracoid.   4. glenoid cavity.

The Pectoral girdle.

The pectoral girdle of the Crocodile is less complete than is that of most reptiles. It consists of a dorsal bone, the **scapula**, and a ventral bone, the **coracoid**, with a median unpaired element, the *interclavicle*; but there is no separate representative either of the clavicle or precoracoid.

The **scapula** (fig. 47, 1) is a large bone, flattened and expanded above where it is terminated by an unossified margin, the **suprascapula**, and thickened below where it meets the coracoid. The scapula forms about half the **glenoid cavity** (fig. 47, 4) for articulation with the humerus, and has the lower part of its anterior border drawn out into a roughened ridge.

The **coracoid** (fig. 47, 2) is a flattened bone, much expanded at either end; it bears on its upper posterior border a flattened surface which forms half the glenoid cavity, and is firmly united to the scapula at its dorsal end. Its ventral end meets the sternum.

The *interclavicle* (figs. 46, 1, and 47, 3) is a long narrow blade-shaped bone lying along the ventral side of the sternum; about a third of its length projects beyond the sternum in front.

The Anterior limb.

This is as usual divisible into three portions, the upper arm, fore-arm and manus.

The **upper arm** or **brachium** contains one bone, the **humerus.**

The **humerus** (fig. 48, A, 1) is a fairly long stout bone, considerably expanded at either end. The proximal end or head is evenly rounded and is formed by an epiphysis ossifying from a centre different from that forming the shaft. It articulates with the glenoid cavity. The shaft bears on the flexor surface, at some little distance behind the head, a prominent rounded protuberance, the **deltoid ridge.** The distal end or trochlea is also formed by an epiphysis and is partially divided by a groove into two convex surfaces; it articulates with the two bones of the fore-arm, the radius and ulna.

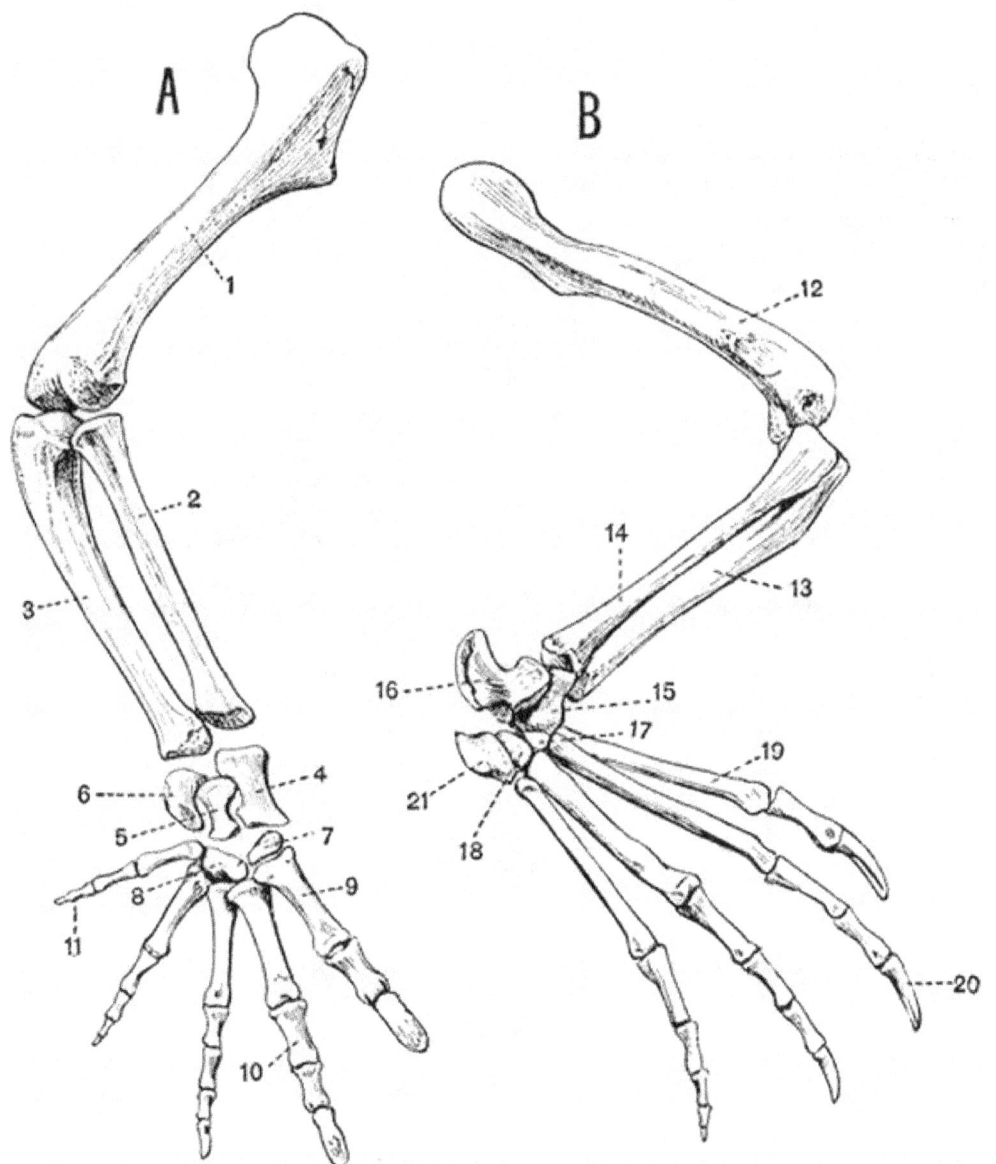

Fig. 48. A, right anterior, and B, right posterior limb of a young Alligator (*Caiman latirostris*). (Brit. Mus.) A × ½. B × about 1/3.

1. humerus.
2. radius.
3. ulna.
4. radiale.
5. ulnare.
6. pisiform.
7. patch of cartilage representing carpalia 1 and 2; between it and the radiale should be another flattened patch, the centrale.
8. carpalia 3, 4, and 5 (fused).
9. first metacarpal.
10. proximal phalanx of second digit.
11. second phalanx of fifth digit.
12. femur.
13. tibia.
14. fibula.
15. tibiale, intermedium and centrale (fused).
16. fibulare.
17. tarsalia 1, 2, and 3 (fused).
18. tarsalia 4 and 5 (fused).
19. first metatarsal.
20. ungual phalanx of second digit.
21. fifth metatarsal.

The **radius** and **ulna** are nearly equal in size and each consists of a long shaft terminated at either end by an epiphysis.

The **radius** (fig. 48, A, 2) or pre-axial bone is slightly the smaller of the two. It has a straight cylindrical shaft and is slightly and nearly evenly expanded at either end. The proximal end which articulates with the humerus is flat or slightly concave, the distal end which articulates with the carpus is slightly convex.

The **ulna** (fig. 48, A, 3) or postaxial bone is a curved bone rather larger than the radius. Its proximal end is large and convex, but is not drawn out into an olecranon process.

The **Manus** consists of the **carpus** or **wrist**, and the **hand**.

The **Carpus**. This differs considerably from the more primitive type met with in the Turtle. It consists of six elements arranged in a proximal row of three and a distal row of two, with one intervening. The bones of the proximal row are the radiale, the ulnare, and the pisiform. The **radiale** (fig. 48, A, 4) is the largest bone of the carpus: it is a somewhat hour-glass shaped bone, with its ends formed by flattened epiphyses. It articulates by its proximal end with the whole of the radius, and partly also with the ulna, and by its distal end with the centrale.

The **ulnare** (fig. 48, A, 5) is a smaller bone, also somewhat hour-glass shaped; it articulates proximally with the pisiform and radiale, not quite reaching the ulna. The third bone of the proximal row is the **pisiform** (fig. 48, A, 6), an irregular bone, articulating with the ulna, radiale, and fifth metacarpal. The **centrale** is a flattened cartilaginous element applied to the distal surface of the radiale.

The distal row of carpals consists of two small structures. The first of these forms a small cartilaginous patch, which is wedged in between the first and second metacarpals, the centrale and the bone representing carpalia 3, 4 and 5; this cartilaginous patch represents **carpalia 1 and 2** (fig. 48, A, 7). The bone representing **carpalia 3, 4 and 5** is a good deal larger, rounded, and well-ossified; it articulates with the ulnare, the pisiform, and the third, fourth, and fifth metacarpals.

The **hand**. Each of the five digits consists of an elongated metacarpal, terminated at each end by an epiphysis, and of a varying number of phalanges. The terminal phalanx of each digit has an epiphysis only at its proximal end, the others have them at both ends.

The first digit, or **pollex**, is the stoutest, and has two phalanges, the second has three, the third four, the fourth three, and the fifth two. The terminal phalanx of each of the first three digits is pointed and sheathed in a horny claw; and is also marked by a pair of prominent lateral grooves.

The Pelvic Girdle.

The pelvic girdle of the Crocodile consists of four parts, a dorsal element, the **ilium**, an anterior ventral element, the **pubis**, a posterior ventral element, the **ischium**, and an accessory anterior ventral element, the **epipubis**. All except the epipubis take part in the formation of the **acetabulum**, which is perforated by a prominent hole.

The **ilium** (fig. 49, 1) is a thick strong bone, firmly united on its inner side with the two sacral ribs. Its dorsal border is rounded, its ventral border bears posteriorly two irregular surfaces, completed by epiphyses, which are united respectively with the ischium and pubis.

The **ischium** (fig. 49, 2)—the largest bone of the pelvis, is somewhat contracted in the middle and expanded at either end. Its proximal end, which is formed by an epiphysis, bears two surfaces, one of which is united to the ilium, while the other forms part of the acetabulum. The anterior border is also drawn out dorsally into a strong process, which is terminated by a convex epiphysis, and is united to the pubis. The ventral end of the ischium forms a flattened blade, meeting its fellow in a median symphysis.

The **pubis** (fig. 49, 3) is much smaller than either the ilium or ischium; it forms a small patch of unossified cartilage, interposed between the anterior parts of the ilium and ischium.

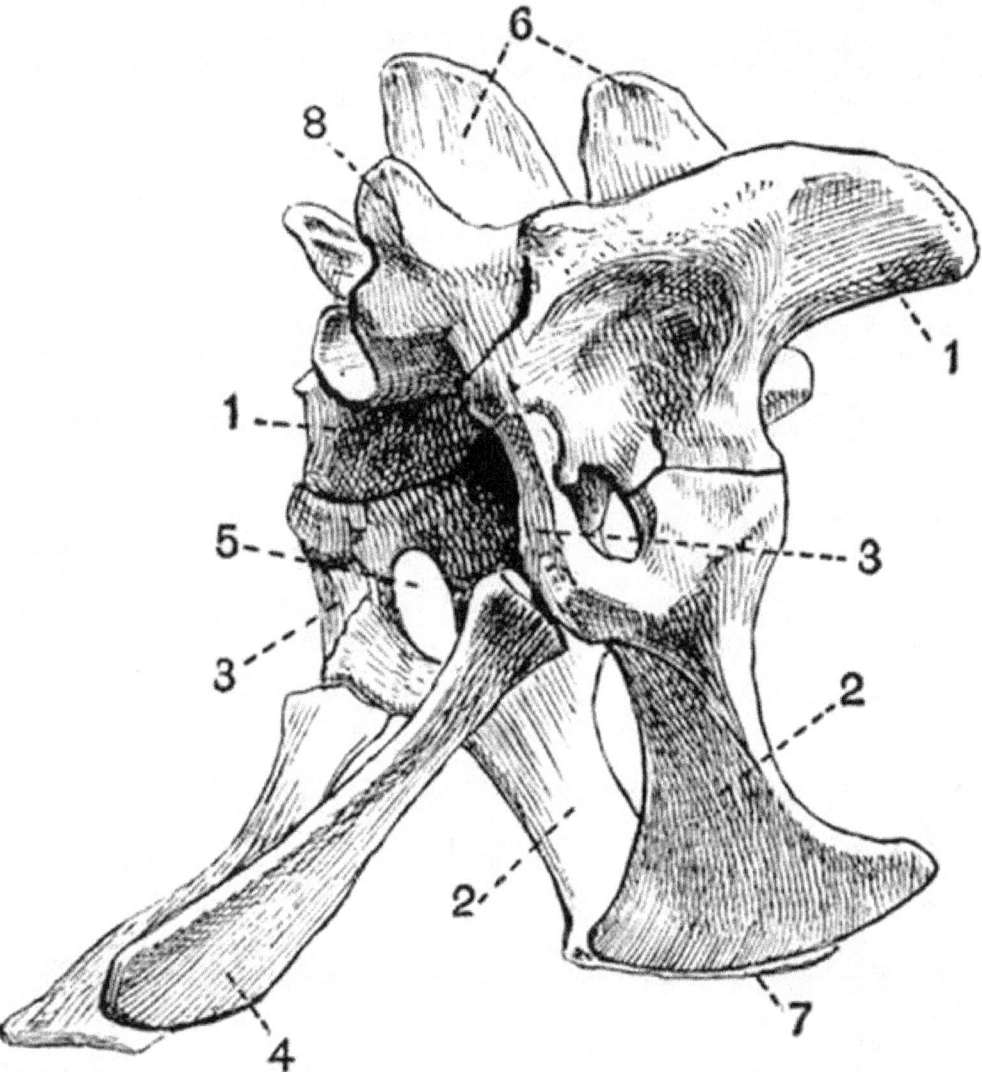

Fig. 49. Pelvis and sacrum of an Alligator (*Caiman latirostris*) × ½.
(Brit. Mus.)

1. ilium.
2. ischium.
3. true pubis.
4. epipubis (so-called pubis).
5. acetabular foramen.

6. neural spines of sacral
vertebrae.
7. symphysis ischii.
8. process bearing
prezygapophysis.

The **epipubis** (fig. 49, 4) is a large bone with a thickened proximal end, which is loosely articulated to the ischium, and a flattened expanded distal end, which is united with its fellow, and with the last pair of abdominal ribs by a large plate of cartilage. This bone is generally described as the pubis.

The Posterior limb.

This is as usual divisible into three portions, the **thigh**, the **crus** or **shin**, and the **pes**.

The **thigh** is formed by the **femur** (fig. 48, B, 12), a moderately long stout bone, not unlike the humerus; it articulates with the acetabulum by a fairly prominent rounded **head**. The distal end articulating with the tibia and fibula is also expanded, and

133

is partially divided into equal parts by anterior and posterior grooves. The flexor surface bears a fairly prominent trochanteric ridge. Each end of the femur is formed by an epiphysis.

The **crus** or **shin** includes two bones, the **tibia** and **fibula**. Both are well developed, but the tibia is considerably the larger of the two.

The **tibia** (fig. 48, B, 13) is a strong bone with a flattened expanded proximal end articulating with almost the whole of the end of the femur, and a similarly expanded distal end articulating with a bone representing the fused astragalus and centrale.

The **fibula** (fig. 48, B, 14) is flattened proximally, and articulates with only quite a small part of the femur, while distally it is more expanded, and articulates with the fibulare (calcaneum) and with a facet on the side of the fused astragalus and centrale.

The **Pes** consists of the **tarsus** or **ankle**, and the **foot**.

The **Tarsus**. This, like the carpus, is much reduced and modified from the primitive condition. It consists of only four bones, arranged in two rows of two each. The two bones of the proximal row are much larger than are those of the distal row. The pre-axial of them (fig. 48, B, 15) representing the fused **astragalus** (tibiale and intermedium) and **centrale**, articulates proximally with the tibia and fibula, and distally with the first metatarsal, and a small bone representing the first three tarsalia. The postaxial bone, the **calcaneum** (fibulare) (fig. 48, B, 16), is drawn out into a prominent posterior process forming a heel such as is almost unknown elsewhere except in mammals. It articulates with the fibula, the tibiale-centrale, and distally with a bone representing the fourth and fifth tarsalia, and with the fifth metatarsal.

The two bones forming the distal row of tarsals are both small and rounded; one represents the first three tarsalia fused together, the other tarsalia 4 and 5.

The **Foot**. The **foot** has five digits, but the fifth is much reduced, consisting only of a short metatarsal. The first four **metatarsals** are all long bones, slightly expanded at each end, and terminated by small epiphyses. The first digit has two phalanges, the second three, the third four, and the fourth five. The terminal or **ungual phalanx** in each instance is grooved and pointed, and in the case of the first three digits bears a horny claw. The ungual phalanx progressively decreases in size from the first to the fourth. The fifth digit consists only of a small, somewhat square metatarsal (fig. 48, B, 21), attached to the bone representing the fused fourth and fifth tarsalia.

CHAPTER XVI.
GENERAL ACCOUNT OF THE SKELETON IN REPTILES.

EXOSKELETON.

The exoskeleton both epidermal and dermal is exceedingly well developed in reptiles.

Epidermal Exoskeleton.

This generally has the form of overlapping horny **scales** which invest outgrowths of the dermis, and are found covering the whole body in most Rhynchocephalia, Ophidia, and Lacertilia, and many Crocodilia. In the Ophidia the ventral surface of the tail is commonly covered by a double row of broad scales, while the ventral surface of the precaudal part of the body is covered by a single row. In the burrowing snakes (Typhlopidae) and some sea snakes (Hydrophidae) these broad scales do not occur, the scales of the ventral surface being similar to those of the dorsal.

In the Chelonia with the exception of *Dermochelys*, *Trionyx* and their allies there is a well-developed system of horny shields having a regular arrangement which has been described in the account of the Turtle's skeleton.

The **rattle of the rattlesnake** is an epidermal structure formed of several loosely articulated horny rings, produced by the modification of the epidermal covering of the end of the tail, which instead of being cast off when the rest of the outer skin is shed is retained loosely interlocked with the adjoining ring or joint. New rings are thus periodically added to the base of the rattle, and in old animals the terminal ones wear away and are lost.

Horny claws occur on the ends of some or all of the digits in most living reptiles.

Owen's Chameleon bears three epidermal horns, one arising from the nasal and two from the frontal region.

In the Chelonia, some of the Theromorpha such as *Udenodon* and *Dicynodon*, probably also in the Pterosauria and *Polyonax* among the Dinosaurs, the jaws are more or less cased in horny beaks. The horny beaks of Chelonia are variable; sometimes they have cutting edges, sometimes they are denticulated, sometimes they are adapted for crushing.

Dermal Exoskeleton.

Nearly all Crocodilia, many Dinosauria, some Rhynchocephalia and Pythonomorpha, and some Lacertilia such as *Tiliqua*, *Scincus* and *Anguis* have a dermal exoskeleton of bony scutes, developed below and corresponding in shape to the epidermal scales. Sometimes as in *Caiman sclerops*, *Jacare* and *Teleosaurus*, the scutes completely invest the body, being so arranged as to form a dorsal and a ventral shield, and a continuous series of rings round the tail. In *Crocodilus* they are confined to the dorsal surface, and in *Alligator* to the dorsal and ventral surfaces. The scutes of some extinct forms articulate with one another by a peg and socket arrangement as in some Ganoid fish.

The **carapace** of most Chelonia is a compound structure, being partly endoskeletal and formed from the ribs and vertebrae, partly from plates derived from the dermal exoskeleton. The common arrangement is seen in fig. 36. All the surface plates are probably exoskeletal in origin, but united with the ventral surfaces of the costal and neural plates respectively are the expanded ribs and neural arches of the vertebrae.

The **plastron** in the common genus *Chelone* (fig. 37) includes nine plates of bone, one unpaired and four pairs; they will be referred to in connection with the ribs and pectoral girdle.

In the Leathery Turtle (*Dermochelys*) the carapace and plastron differ completely from those of any other living form. The carapace consists of a number of polygonal ossifications fitting closely together and altogether distinct from the vertebrae and ribs. The plastron is imperfectly ossified, and not united with the pelvis, and the whole surface of both carapace and plastron is covered with a tough leathery skin, without horny shields.

Some of the extinct Dinosauria have an enormously developed dermal exoskeleton. Thus in *Stegosaurus* and *Omosaurus* the dorsal surface is provided with flattened plates or with spines reaching a length of upwards of two feet. In *Polacanthus* the posterior part of the body is protected by a bony shield somewhat recalling that of the little armadillo *Chlamydophorus*. No exoskeleton is known in Ichthyosauria, Sauropterygia, Pterosauria, many Dinosauria and Theromorpha, and some Lacertilia, such as *Chamaeleon* and *Amphisbaena*.

Teeth.

The teeth of reptiles are generally well developed, and in the great majority of forms are simple conical structures, uniform in character, generally somewhat recurved, and often with serrated edges. Another common type of tooth is that with a laterally compressed triangular crown provided with a double cutting edge which may or may not be serrated. The teeth are mainly formed of dentine, with usually an external layer of enamel, and often a coating of cement on the root. Vasodentine is found below the dentine in *Iguanodon*. The teeth of reptiles never have the enamel deeply infolded, nor do they have double roots.

Teeth may be present not only on the jaw-bones, but as in many *Squamata*, also on the palatines, pterygoids or vomers. The method by which they are attached to the bones varies much. Sometimes as in *Iguana* and some other lizards, they are pleurodont, sometimes they are acrodont, as in the Rhynchocephalia, Pythonomorpha, Ophidia and some Lacertilia such as *Agama*. Again they may be set in a continuous groove as in the Ichthyosauria and young Crocodilia. Finally the teeth may be *thecodont* or placed in distinct sockets as in the Theromorpha, Sauropterygia, adult Crocodilia, Sauropoda and Theropoda. In *Iguanodon* the teeth are set in shallow sockets in a groove one side of which is higher than the other; the method of attachment thus shows points of resemblance to the thecodont condition, the pleurodont condition, and that met with in the Ichthyosauria.

In *Ichthyosaurus* the teeth are marked by a number of vertical furrows, and it is from a furrow of this nature greatly enlarged and converted into a tube that the channel down which flows the poison of venomous snakes is derived.

In most reptiles the dentition is more or less homodont. The only reptiles in which a definite heterodont dentition is known are the extinct Theromorpha, and in them the teeth vary greatly. Thus *Udenodon* is toothless, the jaws having been probably cased in a horny beak. In *Dicynodon* the jaws are likewise toothless with the exception of a pair of permanently growing tusks borne by the maxillae. *Dicynodon* is the only known reptile whose teeth have permanently growing pulps. In *Pariasaurus* the teeth are uniform and very numerous, and though placed in distinct sockets are ankylosed to the jaw. In *Galesaurus* and *Cynognathus* three kinds of teeth can be distinguished, slender conical incisor-like teeth, large canine-like teeth, and cheek teeth with two or three cusps. The teeth in *Galesaurus* are confined to the jaws, in *Placodus* and its allies, however, large flat crushing teeth are attached to the palatines as well as to the jaw-bones, and in *Pariasaurus* the vomer, palatine and pterygoid all bear teeth as well as the jaw bones. The upper jaw of *Sphenodon* and other Rhynchocephalia is provided with two parallel rows of teeth, one borne on the maxillae and one on the palatines, the mandibular teeth bite in a groove between these two rows. The bone of the jaws in *Sphenodon* is so hard that when the teeth get worn away, it can act as a substitute. In the young *Sphenodon* the vomers bear teeth, as they do also in *Proterosaurus*.

There is generally a continuous succession of teeth throughout life, the new tooth coming up below, or partly at the side of the one in use, and causing the absorption of part of its wall or base. In this way the new tooth comes to lie in the pulp cavity of the old one. This method of succession is well seen in the Crocodilia.

135

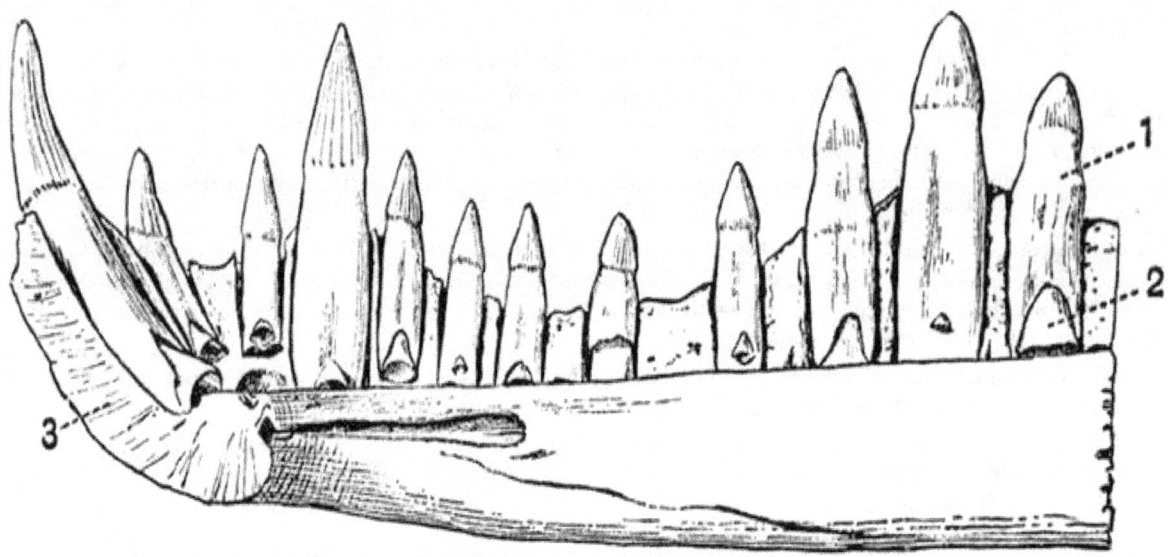

Fig. 50. Preparation of part of the right mandibular ramus of *Crocodilus palustris* × ½. (Brit. Mus.)

1. tooth in use.

2. fairly old germ of future
tooth.

3. symphysial surface of the
mandible.

Teeth have been detected in embryos of *Trionyx*, but otherwise no teeth are known to occur in Chelonia, or in *Pteranodon* (Pterosauria), while the anterior part of the jaw is edentulous in *Iguanodon*, *Polyonax* and some other Dinosaurs, and in *Rhamphorhynchus*.

ENDOSKELETON.

Vertebral column.

The vertebral column is commonly divisible into the usual five regions, but in the Ophidia, Ichthyosauria, and Amphisbaenidae among Lacertilia, only into caudal and precaudal regions. In the Chelonia there are no lumbar vertebrae.

The form of the vertebral centra is very variable. A large proportion of extinct reptiles,—several entire orders,—and the earlier and more primitive forms in some of the other groups have amphicoelous vertebrae. Vertebrae of this type occur in the Theromorpha, Ichthyosauria, most Sauropterygia and Rhynchocephalia, and many Dinosauria, also in some of the early Crocodilia such as *Belodon*, *Teleosaurus* and *Goniopholis*, and the Geckonidae among Lacertilia.

The majority of living reptiles have procoelous vertebrae. Thus they occur in the Lacertilia (excluding the Geckos), the Ophidia, and the Crocodilia, also among extinct forms in the Pterosauria and many Dinosauria. On the other hand some Dinosauria such as *Iguanodon* have opisthocoelous cervical vertebrae, while others have opisthocoelous thoracic vertebrae. The vertebrae of the Ceratopsidae and some Sauropterygia, the thoracic vertebrae of *Iguanodon*, and the sacral vertebrae of Crocodilia have flat centra. The first caudal vertebra of modern Crocodilia is biconvex, and in the Chelonia all types of vertebral centra are found. The cervical vertebrae of *Sphenodon* are noticeable for the occurrence of a small pro-atlas, which may represent the neural arch of a vertebra in front of the atlas.

In most reptiles the vertebrae are fully ossified, but in some of the more primitive forms the notochord persists in the centre of the vertebra (i.e. intervertebrally), this is the case for instance in many of the Theromorpha and Rhynchocephalia, and also in the Geckos. In other reptiles it persists longest intravertebrally.

The centrum of each of the caudal vertebrae of most Lacertilia is traversed by an unossified septum along which it readily breaks.

Chevron bones occur below the caudal vertebrae in Lacertilia, Chelonia, Ichthyosauria, many Dinosauria, and *Sphenodon*, articulating with quite the posterior part of the centrum which bears them. In Lacertilia and Crocodilia (fig. 41, 3) the axis has a well-marked odontoid process. The ventral portions of the intervertebral discs are sometimes ossified, forming wedge-shaped inter centra, as in Geckos, and the cervical vertebrae of *Sphenodon*.

In snakes, Theropod Dinosaurs, and the iguanas among lizards, the neural arches are provided with *zygosphenes*, and *zygantra*.

The neural arches are usually firmly ankylosed to the centra, but in the Crocodilia and some Chelonia, Sauropterygia, and Dinosauria, the suture between the centrum and neural arch persists at any rate till late in life. In the Ichthyosauria the neural arches were united to the centra by cartilage only.

The thoracic vertebrae of some of the Theromorpha (*Dimetrodon*) are remarkable for the extraordinary development of the neural spine, and those of Chelonia for the absence of transverse processes.

In living reptiles the number of sacral vertebrae is nearly always two, but in the Theromorpha, Dinosauria, and Pterosauria, as many as five or six bones may be ankylosed together in the sacral region. In Crocodiles the two halves of the pelvis sometimes articulate with different vertebrae. The vertebrae of some of the great Sauropoda are remarkably hollowed out, having a large vacuity on each side of the centrum communicating with a series of internal cavities. The whole structure of these vertebrae shows a combination of great strength with lightness.

The Skull.

The reptilian skull is well ossified and the bones are noticeable for their density. The true cranium is often largely concealed by a secondary or false roof of membrane bones, which is best seen in the Ichthyosauria and some of the Chelonia. In other reptiles the false roof is more or less broken up by vacuities exposing the true cranial walls. The ethmoidal region is the only one in which much of the primordial cartilaginous cranium remains. The lateral parts of the sphenoidal region are also as a rule not well ossified.

In some reptiles, such as most Lacertilia and Chelonia, the orbits are separated only by the imperfect interorbital septum, while in others, such as the Ophidia, Crocodilia and Amphisbaenidae, the cranial cavity extends forwards between the orbits.

In the occipital region all four bones are ossified. The great majority of reptiles have a single convex occipital condyle, but some of the Theromorpha such as *Cynognathus* have two distinct condyles as in mammals. Sometimes, as in Chelonia, Ophidia and Lacertilia, the exoccipitals, as well as the basi-occipital, take part in the formation of the single condyle; sometimes, as in Crocodiles, it is formed by the basi-occipital alone, as in birds. The relations of the bones to the foramen magnum vary considerably, in Chelonia the basi-occipital generally takes no part in bounding it, and in the Theromorpha, Crocodilia, and Ophidia, the supra-occipital is excluded. The parietals are paired in Geckos and Chelonia alone among living forms, and in the extinct Ichthyosauria and some Theromorpha; in all other reptiles they are united.

The frontals are paired in Ichthyosauria (fig. 32, 5), Chelonia, Ophidia, *Sphenodon* (fig. 52, B, 4) and some extinct crocodiles, such as *Belodon*. They are completely fused in living Crocodilia and some Lacertilia and Dinosauria. In the gigantic *Polyonax* they are drawn out into a pair of enormous horns, and the parietals and squamosals are greatly expanded behind.

An interparietal foramen occurs in the Theromorpha, the Ichthyosauria (fig. 32, 10), *Sphenodon*, the Sauropterygia and most Lacertilia. The posterior part of the skull is curiously modified in some Chamaeleons, the parietals and supra-occipitals being drawn out into a backwardly-projecting sagittal crest which unites with the two prolongations from the squamosals. In other Chamaeleons (*C. bifidus*) prolongations of the prefrontals and maxillae form large forwardly-projecting bony processes.

The roof of the skull is characterised by the development of prefrontals and postfrontals, which lie respectively near the anterior and posterior extremity of the orbit. In Theromorpha, Squamata, Crocodilia, and some Dinosauria lachrymals are developed. There is a ring of bones in the sclerotic in the Ichthyosauria (fig. 32, 15), the Metriorhynchidae among Crocodiles and some Rhynchocephalia, Dinosauria, and Pterosauria.

The pro-otic lies in front of the exoccipital and together with the opisthotic forms the hind border of the fenestra ovalis. In Chelonia the opisthotic remains separate, in all other living reptiles it fuses with the exoccipital. The epi-otic fuses with the supra-occipital.

The parasphenoid, so important in Ichthyopsids, has very often disappeared completely; it is present, however, in the Ichthyosauria, the Plesiosauridae, and a number of Squamata, in many Ophidia its anterior part forming the base of the interorbital septum.

In the Plesiosauridae and most Lacertilia, but not in the Amphisbaenidae, a slender bone, the epipterygoid, occurs uniting the parietal or the anterior end of the pro-otic with the pterygoid. A homologous arrangement occurs in the Ichthyosauria and some Chelonia.

In most reptiles a transpalatine occurs, connecting the maxillae with the pterygoid, but this is absent in the Chelonia, and some Dinosauria, and in the Typhlopidae among snakes.

The quadrate is always well developed, and except in the Squamata is firmly fixed to the surrounding bones. The Chamaeleons also, among the Squamata, have a fixed quadrate, and in them too the quadratojugal is absent. Separate nasal bones do not occur in any living Chelonia.

The vomers are generally paired as in Squamata, sometimes unpaired as in Chelonia.

137

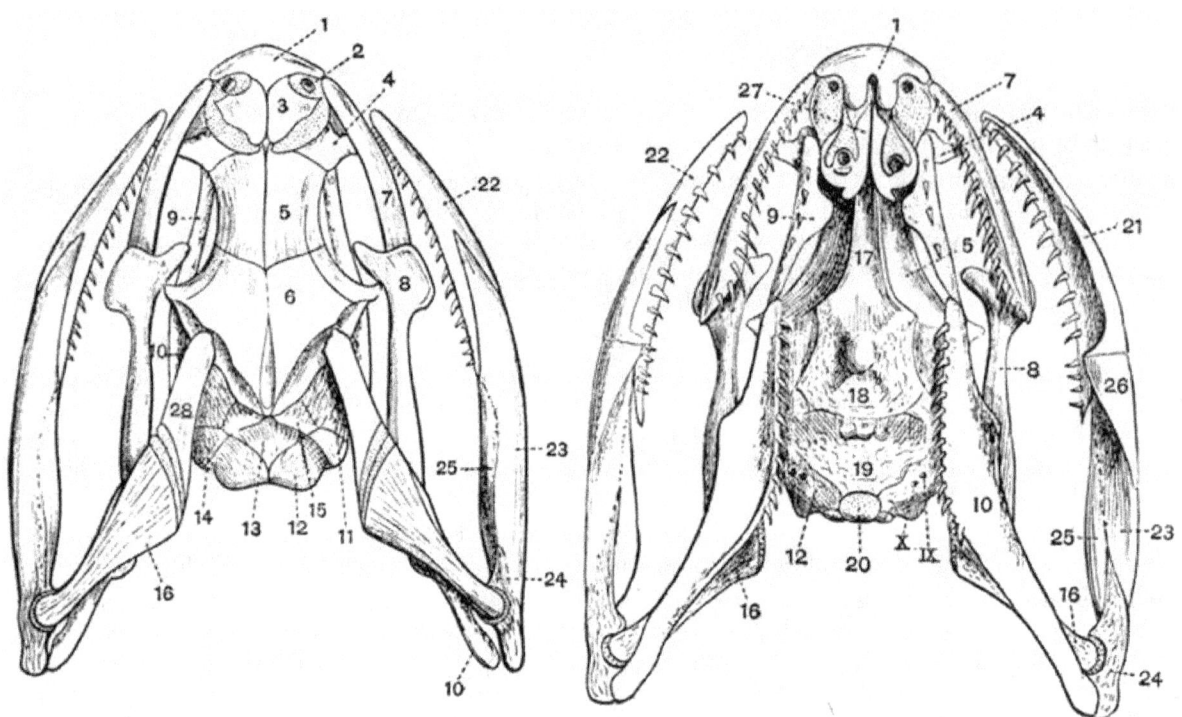

Fig. 51. Dorsal (to the left) and ventral (to the right) views of the skull of the Common Snake (_Tropidonotus_ natrix). (After Parker.)

1. premaxillae (fused).	16. quadrate.
2. anterior nares.	17. parasphenoid.
3. nasal.	18. basisphenoid.
4. prefrontal.	19. basi-occipital.
5. frontal.	20. occipital condyle.
6. parietal.	21. splenial.
7. maxillae.	22. dentary.
8. transpalatine.	23. angular.
9. palatine.	24. articular.
10. pterygoid.	25. supra-angular.
11. pro-otic.	26. coronoid.
12. exoccipital.	27. vomer.
13. supra-occipital.	28. squamosal.
14. opisthotic.	IX, X foramina for the ninth
15. epi-otic.	and tenth cranial nerves.

The disposition of the bones of the jaws is subject to much modification in the Ophidia in order to adapt them for swallowing very large prey. The arrangements again differ greatly in the venomous and non-venomous snakes. In the non-venomous snakes, such as _Python_ and _Tropidonotus_, the palatine is large and is fixed to the pterygoid which extends outwards (fig. 51, 10) so as to be united to the quadrate, and is at the same time firmly connected by the transpalatine with the maxillae. The quadrate is united to the squamosal, which is loosely attached to the cranium. The premaxillae is moderately developed and bears teeth, and the maxillae forms a long bar loosely connected with the rest of the skull. The rami of the mandible are united only by an extremely elastic ligament. It is as regards the maxillae and premaxillae that the skulls of venomous and non-venomous snakes differ most. In the rattlesnake (_Crotalus_) and other venomous snakes the premaxillae is extremely small and toothless. The maxillae is small and subcylindrical, and is movably articulated to the lachrymal, which also is capable of a certain amount of motion on the frontal. The maxillae is connected by means of the transpalatine with the pterygoid, which in its turn is united to the quadrate. When the mouth is shut the quadrate is directed backwards, and carrying back the pterygoid and transpalatine pulls at the maxillae and causes its palatal face, to which the poison teeth are attached, to lie back along the roof of the mouth. When the mouth opens the distal end of the quadrate is thrust forward, and this necessitates the pushing forward of the pterygoid and transpalatine, causing the tooth-bearing surface of the maxillae to look downwards and the tooth to come into the position for striking.

The Ophidian skull is also noticeable for the absence of the jugals and quadratojugals. In poisonous snakes the place of the jugal is taken by the zygomatic ligament which connects the quadrate and maxillae.

The extent to which the palate is closed in reptiles varies much. In many reptiles, such as the Squamata and Ichthyosauria, the palate is not complete, both palatines and pterygoids being widely separated in the middle line. In others, such as the Crocodilia, Sauropterygia, and Chelonia, there is a more or less complete bony palate. In many Chelonia this is chiefly formed of the vomer, palatines, and pterygoids, the posterior nares being mainly bounded by the palatines. In living Crocodilia, however, outgrowths are formed from the pterygoids and palatines which arch round and meet one another ventrally, forming a secondary palate (fig. 43, A), which completely shuts off the true sphenoidal floor of the skull, and causes the posterior nares which are bounded by the pterygoids to open very far back. Though this feature is common to all postsecondary crocodiles, it is interesting to notice that it is not found in the earlier forms, but that its gradual evolution can be traced. In the Triassic *Belodon*, for instance, the posterior nares open far forwards, and are not surrounded by either the palatines or pterygoids. In the Jurassic crocodile, *Teleosaurus*, the posterior nares lie further back, being surrounded by the palatines, but the pterygoids do not meet them. Finally, in the Tertiary forms the arrangements are as in living crocodiles.

A short secondary hard palate is found also in the Theriodontia. The palatines of *Ichthyosaurus* are noticeable for their transverse position, which recalls that in the Frog.

The various **fossae** or **vacuities** in the false roof of the skull are important, and their relations may best be understood by a description of their mode of occurrence in *Sphenodon*, a form in which they are very completely developed.

In *Sphenodon*, then, on the dorsal surface of the skull, are the large **supratemporal fossae** (fig. 52, 20). Their inner margins are separated from one another by the parietal walls of the cranium, while externally each is bounded by a bony arch, the **supratemporal arcade**, formed of the postfrontal, postorbital, and squamosal. Posteriorly the boundary is formed by a **post-temporal bar**, formed by the parietal and squamosal. Below the supratemporal arcade is another large vacuity, the **infratemporal** or **lateral temporal fossa** (fig. 52, 21). This is bounded above by the supratemporal arcade, and is separated from the orbit in front by the **postorbital bar**, formed by the union of outgrowths from the jugals and postorbitals. Behind it is bounded by a continuation of the post-temporal bar formed of the squamosal and quadratojugal, and below by an **infratemporal arcade**, which is chiefly composed of the quadratojugal and jugal.

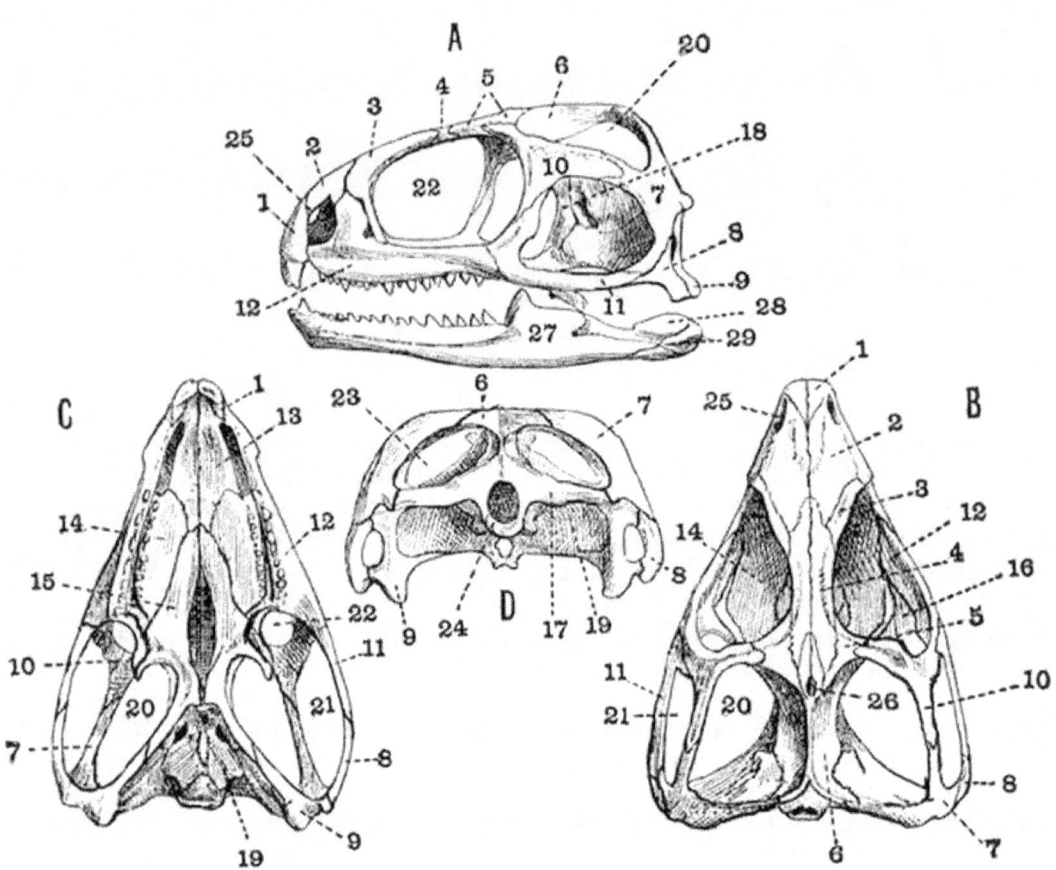

Fig. 52. Skull of Hatteria. (*Sphenodon punctatus*). A, lateral; B, dorsal; C, ventral; D, posterior. (After von Zittel.)

1. premaxillae.
2. nasal.
3. prefrontal.
4. frontal.
5. postfrontal.
6. parietal.
7. squamosal.
8. quadratojugal.
9. quadrate.
10. postorbital.
11. jugal.
12. maxillae.
13. vomer.
14. palatine.
15. pterygoid.
16. transpalatine.
17. exoccipital.
18. epipterygoid.
19. basisphenoid.
20. supratemporal fossa.
21. infratemporal or lateral temporal fossa.
22. orbit.
23. post-temporal fossa.
24. foramen magnum.
25. anterior nares.
26. interparietal foramen.
27. dentary.
28. supra-angular.
29. articular.

Below the post-temporal bar is a third vacuity, the **post-temporal fossa** (fig. 52, D, 23), bounded above by the post-temporal bar and below by the exoccipital and opisthotic.

Sphenodon and the Crocodilia are the only living reptiles with complete supratemporal and infratemporal arcades, but they are both present in the extinct Pterosauria and some Dinosauria.

Supratemporal fossae, bounded below by supratemporal arcades, occur in all reptiles except some Chelonia, the Ophidia, the Geckonidae among Lacertilia, and the Pariasauria and others among Theromorpha; they are specially large in

Nothosaurus among the Sauropterygia, *Dicynodon* among the Theromorpha, and many Crocodilia and Pterosauria. In some Dinosaurs, such as *Ceratosaurus*, they are very small, while the infratemporal fossae are correspondingly large.

In *Elginia* (Theromorpha) and some Chelonia, such as *Chelone*, there are no fossae on the surface of the skull, a complete false roof being developed; in other Chelonia, such as *Trionyx*, the true cranium is freely visible, the only part of the false roof developed being the infratemporal arcade.

In many reptiles large **pre-orbital vacuities** occur; they are specially large in the Pterosauria and in some of the Crocodilia and Dinosauria (fig. 35, 3). In some Pterosauria they are confluent with the orbits.

The premaxillae are usually separate, but sometimes, as in some Ophidia (fig. 51, 1), Chelonia, Lacertilia (Agamidae), and Dinosaurs (Ceratopsia) they are united. In the Dinosaur *Hadrosaurus* they are exceedingly large and spatulate. In the Rhynchocephalian *Hyperodapedon* they are drawn out into a strongly curved beak.

As regards the mandible, sometimes, as in most Rhynchocephalia, Ophidia and Pythonomorpha, the rami have only a ligamental union; sometimes, as in Crocodilia, the Rhynchosauridae and the majority of Lacertilia, they are suturally united. In Chelonia (fig. 28, B, 12), and apparently in Pterosauria, the two dentaries are completely fused together. The sutures between the various bones of the lower jaw usually persist, but in Ophidia those between the angular, supra-angular, articular and coronoid are obliterated. There are sometimes large vacuities in the mandible, as in Theromorpha, Crocodilia, and some Dinosauria. In *Iguanodon, Polyonax, Hypsilophodon* and *Hadrosaurus* among Dinosaurs the mandible has a predentary or mento-meckelian bone which, in some cases at any rate, was probably sheathed in a horny beak.

The principal part of the auditory ossicular chain is formed by a rod-like columella. The development of the hyoid apparatus varies, and it often happens that the first branchial arch is better developed than is the hyoid arch. In the Crocodilia and Chelonia there is a large basilingual plate or body of the hyoid (fig. 53, 1); but while in the Crocodilia the first branchial forms the only well-developed arch, in the Chelonia the first and second branchials are both strongly developed, and the hyoid is often fairly large.

The Ribs.

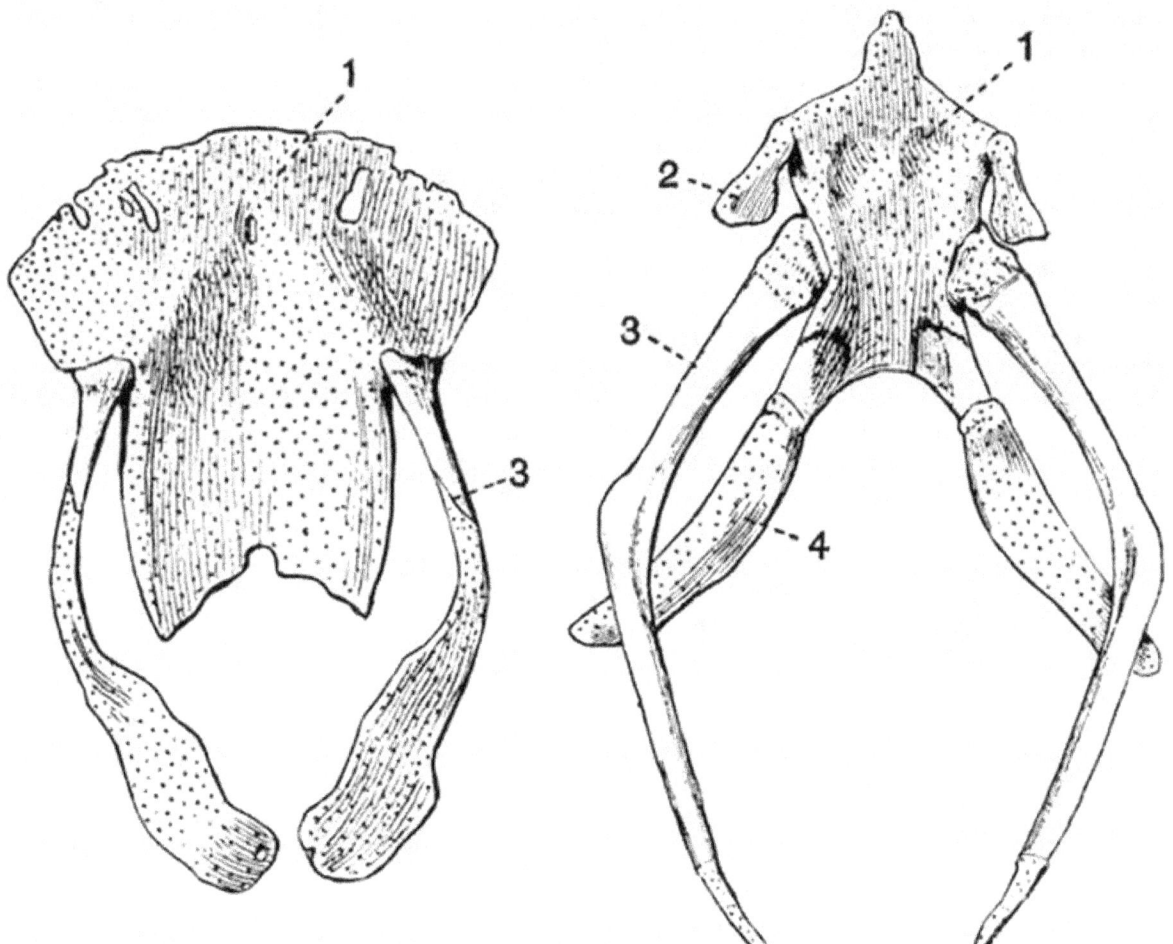

Fig. 53. Hyoids of an Alligator (*Caiman latirostris*) (to the left) and of a Green Turtle (*Chelone midas*) (to the right) × 5/8.

(Brit. Mus.)

The cartilaginous portions are dotted.

1. basilingual plate or body of the hyoid.

2. hyoid arch.

3. first branchial arch (anterior cornu).

4. second branchial arch (posterior cornu).

Ribs are always present, and may be attached to any of the precaudal vertebrae. In most reptiles the posterior cervical vertebrae bear ribs, while the atlas and axis are ribless; in Crocodiles and Geckos, however, ribs are borne even by the atlas and axis. On the other hand, in the Chelonia none of the cervical vertebrae bear obvious ribs. In the following groups the thoracic ribs have both capitula and tubercula—Theromorpha, Ichthyosauria, Crocodilia, Dinosauria, Pterosauria. In the other groups each rib articulates by a single head, and the position of the facet is subject to a considerable amount of variation, thus in the Squamata it lies on the centrum, and in the Sauropterygia on the neural arch, while in the Chelonia the rib articulates with the contiguous parts of two centra instead of directly with one.

In most reptiles a greater or smaller number of ribs are united ventrally with a sternum; but in snakes a continuous series of similar ribs, all articulating freely with the vertebral column, extends from the third cervical vertebra to the end of the trunk. The number of ribs connected with the sternum varies from three or four in Lizards to eight or nine in Crocodiles. Those which reach the sternum are nearly always divided into vertebral, sternal, and intermediate portions, and as a rule only the vertebral portion is completely ossified. In Crocodiles a number of sternal ribs are connected with a cartilaginous arch, which is attached to the hind end of the sternum, and represents the xiphisternum. The sacral ribs connecting the vertebral column with the ilia are very distinct in Crocodiles; in these animals and *Sphenodon* the vertebral ribs have backwardly-projecting uncinate processes as in birds.

In the curious arboreal lizard, *Draco volans*, the posterior ribs are long and straight, and support a parachute-like expansion of the integument used in its long flight-like leaps. In Chelonia the ribs are generally combined with the carapace.

In Ichthyosauria, Sauropterygia, Crocodilia and *Sphenodon*, abdominal splint ribs occur; and probably all except the first of the paired ossifications forming the plastron of Chelonia are of similar character. Abdominal ribs have quite a different origin from true ribs, for while true ribs are cartilage bones, abdominal ribs have no cartilaginous precursors, but are simply the ossified tendons of the rectus abdominalis muscle.

The Sternum.

A sternum occurs in the following groups of reptiles: Rhynchocephalia, nearly all Lacertilia, Pythonomorpha, Crocodilia, and Pterosauria, and is generally more or less rhomboidal or shield-shaped. In Pterosauria it is keeled and bears some resemblance to that of birds. It may have been replaced by membrane bone.

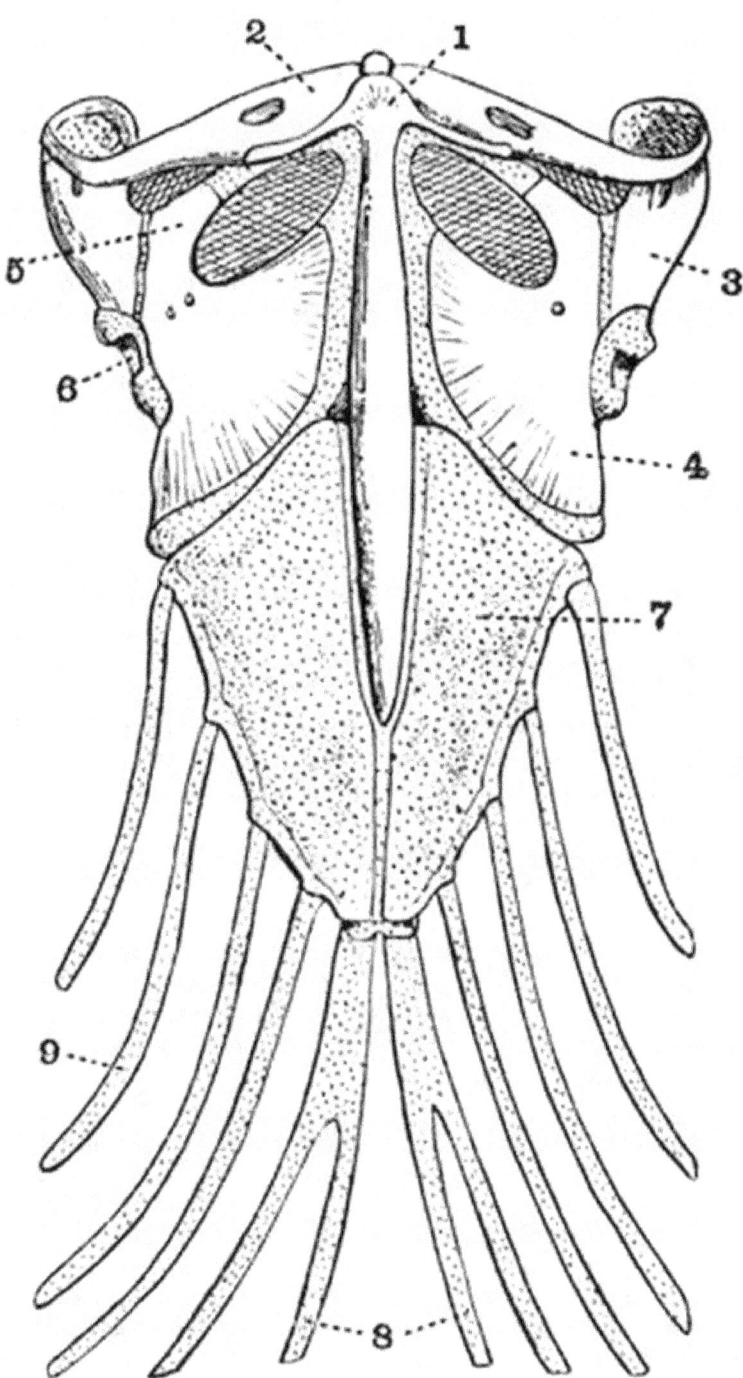

Fig. 54. Ventral view of the shoulder-girdle and sternum of a Lizard (*Loemanctus longipes*) × 2. (After Parker.)

1. interclavicle.

2. clavicle.

3. scapula.

4. coracoid.

5. precoracoidal process.

6. glenoid cavity.

7. sternum.

8. xiphisternum.

9. sternal rib.

The sternum is absent in Sauropterygia, Ichthyosauria, Chelonia, Ophidia, and most of the snake-like Amphisbaenidae among Lacertilia; while it is not well known in Theromorpha and Dinosauria. In the Sauropod *Brontosaurus*, however, two rounded bones occur near the base of the coracoids, and these probably represent ossified patches in a sternum, which was mainly cartilaginous; similar structures occur in *Iguanodon*.

The sternum frequently remains wholly cartilaginous, especially in Lacertilia; sometimes it becomes calcified, but true ossification does not as a rule take place.

APPENDICULAR SKELETON.

The Pectoral Girdle.

The pectoral girdle is well developed in all groups of reptiles except the Ophidia, occurring even in the limbless Amphisbaenidae. It is very solid in the Theromorpha. As a rule all three cartilage bones, scapula, coracoid, and precoracoid are represented, and frequently also the membrane bones,—clavicles, and interclavicle.

The coracoids are generally flat expanded bones, which sometimes, as in Sauropterygia and Ichthyosauria, meet in a ventral symphysis; sometimes, as in Lacertilia, are united with the sides of the sternum. In Chelonia neither the coracoids nor precoracoids meet one another, but their free ends are connected by fibrocartilaginous bands. In Lacertilia the coracoids are pierced by fenestrae.

The precoracoid is generally represented, but the Theromorpha are the only reptiles in which it is separately ossified; it forms a well-marked process on the coracoid in Lacertilia (fig. 54, 5). It is absent in Ichthyosauria, and Dinosauria, and probably in Sauropterygia. In some Lacertilia and Chelonia the sternal ends of the coracoids are unossified and form epicoracoids; in some Chelonia there are also epiprecoracoids; but neither these nor the epicoracoids overlap their fellows of the opposite side as they do in arciferous Anura (see p. 185). In some Lacertilia with degenerate limbs the pectoral girdle is also much reduced, in *Ophisaurus apus* the ventral borders of the coracoids are widely separated.

A scapula is always present, and is generally expanded distally, but in the Chelonia the distal end is cylindrical. In the Theromorpha it has an acromial process with which the precoracoid articulates, and it is very large in Dinosauria. In the Chelonia the scapula and precoracoid are ossified continuously. Among the Pterosauria, *Pteranodon* has an unique pectoral girdle; the scapula and coracoid are ankylosed and the scapula articulates with the neural spines of several ankylosed vertebrae.

Clavicles occur in some Theromorpha such as *Pariasaurus*, and also in the Ichthyosauria, Sauropterygia, Rhynchocephalia, and most Lacertilia. They are absent in the Pterosauria, the Chamaeleons among Lacertilia, the Ophidia and the Crocodilia. They are wanting too in the Chelonia, unless the first pair of ossifications in the plastron are to be regarded as clavicles. In the Sauropterygia bones regarded as the clavicles and interclavicle are generally well developed. The unpaired ossification in the plastron of Chelonia is an interclavicle, and a representative of the same bone occurs arising from the sternum in Pterosauria. A well developed **T**-shaped interclavicle is found in Ichthyosauria, Rhynchocephalia, Lacertilia, and some Theromorpha, such as *Pariasaurus*.

The Limbs.

In most reptiles there are two pairs of pentedactylate limbs provided with claws, but in nearly all Ophidia and some Lacertilia (*Amphisbaena, Lialis, Anguis*) the limbs have entirely disappeared. In a few Ophidia such as *Python* traces of the posterior limbs occur, and in *Chirotes* among the Amphisbaenidae there are minute anterior limbs. The Lacertilians, *Chalcides* (*Seps*) and *Ophisaurus* (*Bipes, Pseudopus*) have very small posterior limbs.

The limbs are as a rule adapted for walking, but in Ichthyosauria, Sauropterygia, Pythonomorpha and some Chelonia, they have the form of swimming paddles, the relative size of the manus and pes being increased, while that of the proximal and middle portions of the limbs is reduced. This reduction is carried to its furthest extent in the Ichthyosauria in which radius and ulna, tibia and fibula, have the form of short polygonal bones similar to those constituting the manus and pes. In the Pythonomorpha the reduction of the limb bones is not quite so marked, in the Sauropterygia it is less, and still less in the Chelonia. In the earlier Ichthyosauria too, the limb bones are not so short as they are in the later forms. The Ichthyosaurian limb is also remarkable, firstly for the fact that both humerus and femur are terminated by concave articulating surfaces instead of by convex condyles, and secondly for the great multiplication of the phalangeal bones, each digit being sometimes composed of a series of over twenty. Sometimes too the number of series is increased, either by the bifurcation of some of the digits or by the development of marginal bones. In the Sauropterygia the phalanges are likewise increased above the normal but not so much as in Ichthyosauria. The humerus and femur of Sauropterygia are noticeable for the enormous size of the terminal epiphyses which form in each case by far the greater part of the bone.

The Anterior Limb.

The anterior limb is usually approximately equal in length to the posterior, but in many Dinosauria it is considerably the shorter of the two. The humerus is generally without distinct condyles, but they are well developed in the Theromorpha, the Lacertilia and *Sphenodon*.

In the Theromorpha, some Rhynchocephalia, and some Sauropterygia, such as *Mesosaurus*, the humerus has an ent-epicondylar foramen; in Lacertilia, Chelonia and some Dinosauria there is an ect-epicondylar foramen or groove; *Sphenodon* possesses both ent- and ect-epicondylar foramina. The radius and ulna are always separate. In some Chelonia, such as *Chelydra*, the carpus has a very simple arrangement, namely, a proximal row of three bones, the radiale, intermedium and ulnare, and a distal row of five carpalia, with one bone, the centrale, between the two rows. Many reptiles have a carpus only slightly different from this. Thus the carpus in *Sphenodon* differs mainly in having two centralia, that of most Lacertilia, in having the centrale and intermedium fused.

Crocodiles have a much reduced carpus with the radiale and ulnare considerably elongated. The manus in Chamaeleons is curiously modified, having the first three digits arranged in one group and turned inwards, and the fourth and fifth in another group turned outwards; carpalia 3 and 4 are united.

In the Pterosauria the anterior limbs form wings, the phalanges of the fifth digit being very greatly elongated to support the wing membrane. The first digit is vestigial and the second, third, and fourth are clawed.

The Pelvic Girdle.

The pelvic girdle is well developed in all reptiles which have posterior limbs, but is absent or quite vestigial in Ophidia and those Lacertilia which have no posterior limbs. The ilium and ischium agree in their general characters throughout all the various groups of reptiles, but that is not the case with the pubis.

In many reptiles such as Chelonia, Ichthyosauria and Lacertilia the ilia are small, more or less cylindrical bones either directed backwards, or vertically placed as in the Chamaeleons. In the Crocodilia they are larger and more expanded, while in Dinosauria and Pterosauria they are greatly elongated both in front of, and behind, the acetabulum. The ischia are generally strongly developed somewhat square bones meeting in a ventral symphysis. In Dinosauria the ischium (fig. 35, 9) is a much elongated and backwardly-directed bone, bearing a forwardly projecting obturator process. In Pterosauria the ischium is fused with the ilium, and in both pterosaurs and crocodiles the ilium and ischium are the only bones taking part in the formation of the acetabulum. In most Lacertilia there is an unpaired structure, the *hypo-ischium* or *os cloacae* projecting back from the symphysis ischii, which is usually separated from the symphysis pubis by a large space, the *foramen cordiforme*. In some Lacertilia and Chelonia there is a cartilaginous bar dividing the foramen cordiforme into two obturator foramina; in many Chelonia this bar is ossified. Among *Ophidia, Python, Tortrix, Typhlops* and their allies have a structure representing a vestigial ischio-pubis: but in most Ophidia there is no trace of the pelvis. In some Theromorpha all the bones of the pelvis are completely fused, forming an os innominatum as in mammals; the pubes and ischia are so completely fused that sometimes as in *Pariasaurus* even the obturator foramina are closed.

Concerning the reptilian pubis there are considerable difficulties. Sometimes there is only a single pubic structure present, sometimes there are two. The reptilian pubis is best understood by comparing the arrangements met with in the various other groups with that in the Orthopod Dinosaurs such as *Iguanodon*. In *Iguanodon* the pubis consists of two portions, viz. of a moderately broad pre-pubis directed downwards and forwards, and of a narrow greatly elongated post-pubis directed backwards parallel to the ischium. The pubis is united to both ilium and ischium, the acetabulum has a large unossified space, and neither pre-pubes nor post-pubes meet in ventral symphyses. The arrangement bears a great resemblance to that of Ratite birds. In Lacertilia, Chelonia, Rhynchocephalia and Ichthyosauria together with Theropod and Sauropod Dinosaurs the pubis corresponds to the pre-pubis of *Iguanodon* and is a more or less cylindrical bone expanded at both ends, meeting its fellow in a ventral symphysis. In Chelonia and Lacertilia the pubis bears a lateral process which is homologous with the post-pubis of Iguanodon. In Lacertilia and sometimes in Chelonia there is a cartilaginous epipubis attached to the anterior border of the pubic symphysis; this is well developed in the Chamaeleons and Geckos. In Crocodilia there is, forming the anterior and ventral portion of the acetabulum, a patch of cartilage (fig. 49, 3) which is probably the true pubis homologous with that of lizards and with the pre-pubis of *Iguanodon*. The large bone generally called the pubis in Crocodiles is probably an epipubis.

The Posterior Limb.

The posterior limb is entirely absent in some Lacertilia and in most Ophidia, though traces occur in *Python, Tortrix* and *Typhlops*. In the Ichthyosauria, Sauropterygia and Pythonomorpha the posterior limbs form swimming paddles and have been already referred to.

The arrangement of the proximal and middle segments of the limb is fairly constant in all reptiles with limbs adapted for walking, and the tibia and fibula are always separate. The pes is however subject to a considerable amount of variation, especially as regards the tarsus. In some Chelonia the tarsus like the carpus has an extremely simple arrangement, consisting of a proximal row of three bones, the tibiale, intermedium and fibulare, a centrale, and a distal row of five tarsalia. In most living reptiles, however, the tibiale and intermedium are as in mammals united, forming the astragalus. In Crocodiles (fig. 48, B, 15) the centrale is also united with the tibiale while the distal tarsalia are very slightly developed. The calcaneum in Crocodiles is drawn out into a long process forming a heel in a manner almost unique among Sauropsida. In *Sphenodon* and Lacertilia the tibia and fibula articulate with a single large bone representing the whole proximal row of tarsalia.

The pes is generally pentedactylate, but in some Crocodiles the fifth digit is vestigial (fig. 48, B), and in some Dinosauria (fig. 35) there are only three digits. The North American Dinosaurs present a continuous series ranging from a pentedactylate plantigrade form like *Morosaurus*, to such a form as *Hallopus* with a highly digitigrade and specialised pes reduced to three functional digits, and a vestigial fifth metatarsal. The second, third and fourth metatarsals in this form are nearly two-thirds as long as the femur, and the calcaneum is drawn out into a heel much as it is in most mammals.

In Lacertilia, Orthopoda and many Chelonia, the ankle joint comes to lie between the proximal and distal row of tarsals as in birds.

145

CHAPTER XVII.
CLASS. AVES.

Birds form a large and extremely homogeneous class of the vertebrata, and are readily distinguished from all other animals by the possession of an epidermal exoskeleton having the form of feathers. Feathers differ from hairs in the fact that they grow from papillae formed of both the horny and the Malpighian layer of the epidermis, which papillae at first project from the surface, and only subsequently become imbedded in pits of the dermis. A dermal exoskeleton does not occur in birds.

The endoskeleton is characterised by its lightness, the large bones being generally hollow; but the pneumaticity does not vary in proportion to the power of flight. The cervical part of the vertebral column is very long and flexible, while the post-cervical portion is generally very rigid, owing to the fusion of many of the vertebrae, especially in the lumbar and sacral regions. The vertebrae are generally without epiphyses to their centra. The cervical vertebrae in living forms have saddle-shaped articulating surfaces, and many of them bear ribs. The thoracic ribs in almost all birds have large uncinate processes. The sternum is very large, and the ribs are always attached to its sides, not as in many reptiles to any backwardly-projecting process or processes. The sternum ossifies from two or more centres.

The skull is extremely light, and its component bones show a great tendency to fuse together completely. The facial part of the skull is prolonged into a beak, chiefly formed of the premaxillae; this beak is in all modern birds devoid of teeth, and is coated externally with a horny epidermal sheath. The quadrate is large and freely movable. The supratemporal arcade is imperfect, while the infratemporal arcade is complete. There are no postorbital or postfrontal bones. Neither parotic processes nor an interparietal foramen occur. There are commonly large pre-orbital vacuities. The palatines and pterygoids never form a secondary bony palate as in Crocodiles. Part of the floor of the skull is formed by a wide *basitemporal* (paired in the embryo) which is continued in front as a long slender *rostrum*; these structures have replaced the parasphenoid of Ichthyopsids. Cartilage or bone is always developed in the sclerotic. The first branchial arch is well developed, the hyoid arch but slightly. The coracoids are large, and the clavicles are nearly always united forming the *furcula*. There is no separate interclavicle and hardly any trace of a precoracoid.

The anterior limbs form wings, and the manus is in the adult always much modified, never having more than three digits. The three bones of the pelvis are, except in Archaeornithes, always ankylosed together in the adult, and the ilium is greatly prolonged in front of the acetabulum, which is perforated. The ilia are not connected with the sacrum by ossified sacral ribs. The pubes and ischia are directed backwards parallel to one another, and except in a very few forms never meet their fellows in ventral symphyses. The fibula is generally much reduced. The proximal tarsal bones are always ankylosed to the tibia, and the distal tarsals to the metatarsals, so that the ankle joint is *intertarsal*. The first metatarsal is nearly always free. The pes never has more than four digits in the adult.

The class Aves is most conveniently divided into two subclasses: 1. Archaeornithes. 2. Neornithes.

Subclass I. Archaeornithes.

The only form referred to this subclass of extinct birds is *Archaeopteryx*, the earliest known bird. In this animal the skeleton does not seem to be pneumatic. The cervical and trunk vertebrae are generally thought to be flat, certainly their articulating surfaces are not saddle-shaped. There is no long compound sacrum as in modern birds. The tail is longer than the whole body, the caudal vertebrae are twenty in number, they gradually taper as traced away from the trunk, and each bears a pair of feathers. The posterior caudal vertebrae are not united together to form a *pygostyle*. The upper jaw bears thirteen pairs of conical teeth, planted in distinct sockets in the maxillae and premaxillae, but the mandible has only three pairs. The presence of these teeth forms the most essential difference between the skull of *Archaeopteryx* and that of modern birds, and the fact that they occur on the premaxillae renders it improbable that a horny beak was present. There is a ring of ossifications in the sclerotic. The ribs do not show uncinate processes, and articulate with the vertebrae by single heads not divided into capitula and tubercula. Abdominal ribs appear to have been present. The furcula is large, and the scapula has a well developed acromion. The sternum is unknown. The radius and ulna are approximately equal in size. In the manus the first, second and third digits are present, each terminated by a claw. The second digit is considerably the longest, while the third includes four phalanges. The three bones of the pelvis probably remained distinct throughout life. The tarsals are ankylosed respectively to the tibia and metatarsals as in other birds. The metatarsals are ankylosed together, and the pes has four digits.

Subclass II. Neornithes.

To this subclass may be referred all known birds except *Archaeopteryx*. They all agree in having a short tail whose component vertebrae are commonly ankylosed together forming a pygostyle. The three metacarpals do not all remain distinct.

The bones of the pelvis are ankylosed together, and to a large though variable number of vertebrae. There are three orders, the Ratitae, Odontolcae, and Carinatae.

Order 1. Ratitae.

The Ratitae differ from *Archaeopteryx* and the great majority of Carinatae in being flightless. The bones are generally not pneumatic, containing marrow instead of air, in the Ostrich however they are very pneumatic. The tail is short and the posterior caudal vertebrae are generally ankylosed together forming a pygostyle. The pectoral girdle has comparatively a much smaller size than in Carinatae, clavicles are small or absent, and the scapula and coracoid lie nearly in the same straight line. The ilium and ischium do not as in Carinatae unite posteriorly, and enclose a foramen except in very old Rheas and Emeus. The quadrate articulates with the cranium by a single head. The vomers unite and form a broad plate, separating the palatines, pterygoids and basisphenoidal rostrum.

The anterior limbs are greatly reduced in size or even absent, while the posterior limbs are greatly developed and adapted for running. The tibia and fibula are quite distinct.

Many ornithologists agree that the various forms grouped together as Ratitae are not all very closely allied to one another, that they resemble one another mainly in having lost the power of flight, and do not form a natural group.

The Ratitae include the following groups:—

Æpyornithes, huge extinct birds from Madagascar.

Apteryges, including the Apteryx of New Zealand.

Dinornithes, the Moas, huge extinct birds from New Zealand, and some of the neighbouring islands.

Megistanes, including the Cassowaries (*Casuarius*) of Australia, New Guinea, and some of the neighbouring islands; and the Emeus (*Dromaeus*) of Australia.

Rheornithes, including the Rheas of S. America.

Struthiornithes, including the Ostriches (*Struthio*) now living in Africa, and found fossil in N. India and Samos.

Order 2. Odontolcae.

This order includes only an extinct N. American bird *Hesperornis*. The jaws are provided with a series of sharp teeth placed in continuous grooves, but the premaxillae are toothless, and were probably sheathed in a horny beak. The rami of the mandible are not ankylosed together in front. The skeleton is not pneumatic. The cervical vertebrae have saddle-shaped articulating surfaces as in ordinary birds, and the thoracic vertebrae are not ankylosed together. The tail is comparatively long, and formed of twelve vertebrae with only slight indications of a pygostyle. The ribs have uncinate processes. The anterior limb is quite vestigial, being reduced to a slender humerus. The posterior limb is very powerful and adapted for swimming.

Order 3. Carinatae.

This order includes the vast majority of living birds. The cervical vertebrae have saddle-shaped articulating surfaces (except in the Ichthyornithiformes). The posterior caudal vertebrae are ankylosed forming a pygostyle. The quadrate articulates with the cranium by a double head. In all except the Tinamidae the vomers are narrow behind and not interposed between the palatines, pterygoids and basisphenoidal rostrum. The sternum has a median keel, and the anterior limbs are in the great majority of cases adapted for flight. Clavicles are well developed, and the scapula and coracoid are nearly at right angles to one another. The various groups into which the Carinatae are divisible are shown in the table on pp. 40-42. Their special characters will not be dealt with.

Fig. 55. *Gallus bankiva* var. *domesticus*. The left half of the Skeleton. The skull, vertebral column, and sternum are bisected in the median plane. (After Marshall and Hurst.)

A, acetabulum. B, cerebral fossa. CB, cerebellar fossa. CL, clavicle. CO, coracoid. CR, cervical rib. C 1 = one, first cervical vertebra. FE, femur. HC, ventral end of clavicle. HU, humerus. HY, hyoid. IF, ilio-sciatic foramen. IL, ilium. IS, ischium. L, lachrymal. MC 3, postaxial metacarpal. MN, mandible. MS, xiphoid processes. MT, tarso-metatarsus. MT 1, first metatarsal. N, nasal. OP, optic foramen. P, premaxillae. PB, pubis. PL, palatine. PY, pygostyle. R, radius. RC, radial carpal. S, keel of sternum. SC, scapula. T, tibio-tarsus. TH 4, fourth thoracic vertebra. U, ulna. UC, ulnar carpal. UP, uncinate process. Z, infra-orbital bar. 1, 2, 3, 4, first, second, third and fourth digits of pes. 3, pre-axial, 4, middle, and 5, postaxial digit of manus.

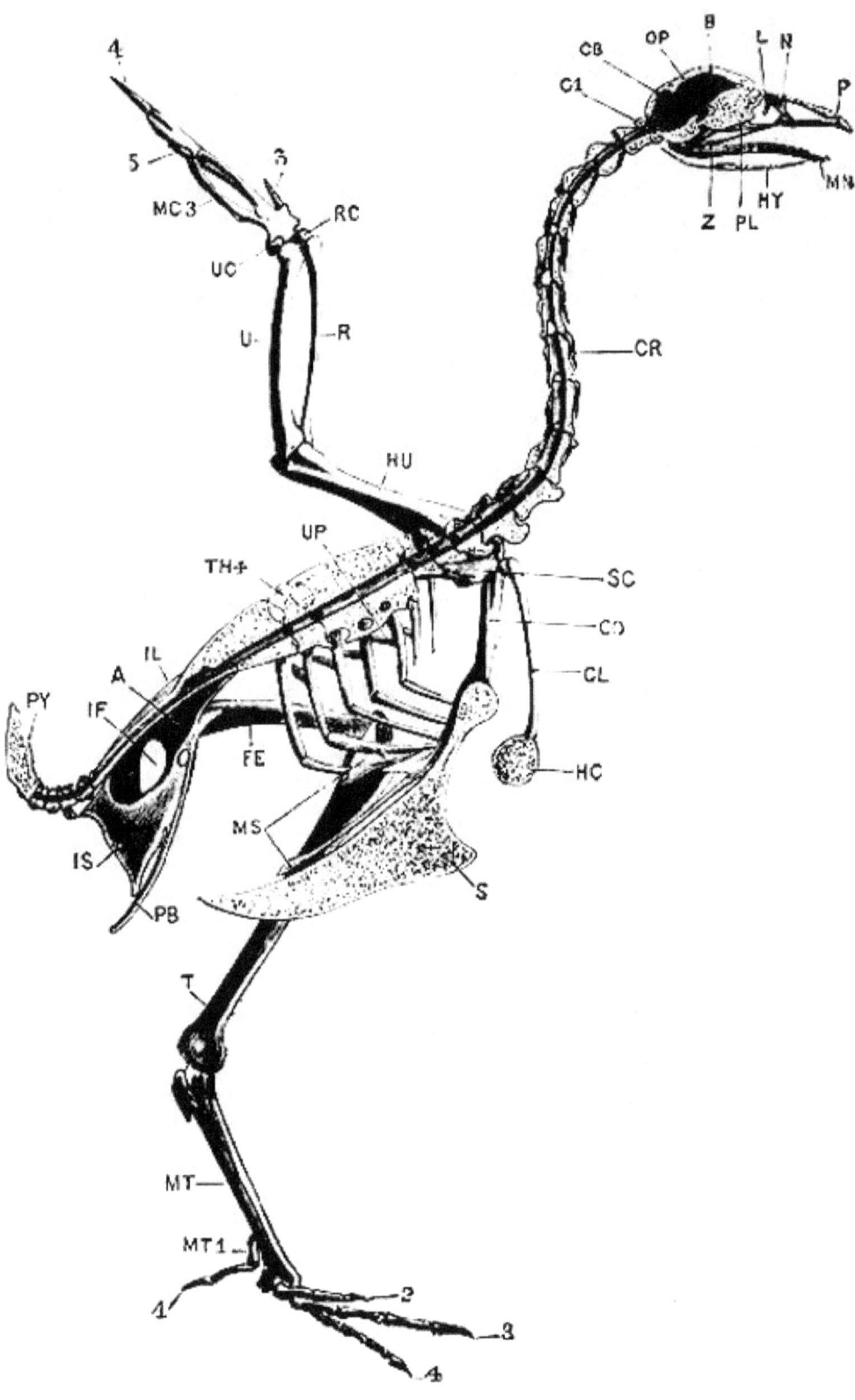

Fig. 55.

CHAPTER XVIII.
THE SKELETON OF THE WILD DUCK (Anas boschas).

I. EXOSKELETON.

The exoskeleton of the Duck and indeed of all birds is entirely epidermal in origin. Its most important part consists of **feathers**, but it includes also the following horny structures:—

(*a*) **scales**, which cover the toes and tarso-metatarsus;

(*b*) **claws**, which are attached to the distal phalanges of the toes and of the pollex;

(*c*) the wide **beak**, which sheaths both upper and lower jaws, and whose edges are raised into lamellae, which act as strainers.

Feathers.

A well developed feather, such as one of the large quill feathers of the wing or tail, consists of the following parts: A main stem, the **scapus**, which forms the axis running along the whole length of the feather, and is divided into (1) a proximal hollow cylindrical portion, the **calamus** or **quill**, and (2) a distal solid portion, the **rachis** or **shaft**, which is square in section, flexible and grooved along its ventral surface, and bears a number of lateral processes, the **barbs**. The **calamus** which is partly imbedded in a pit in the dermis, bears two holes: one, the **inferior umbilicus**, is at its proximal end, and into it enters a vascular outgrowth from the dermis; the other, the **superior umbilicus**, lies on the ventral surface at the junction of the calamus and scapus.

The **barbs** are a series of narrow elastic plates, attached by their bases to the rachis, and with their edges looking upwards and downwards. The barbs are connected together by a number of smaller processes, the **barbules**, which interlock with one another by means of hooklets, and bear the same relation to the barbs that the barbs do to the rachis. The barbs and barbules, together with the rachis, constitute the **vexillum** or **vane** of the feather. Any feather having the above type of structure is called a **penna** or a **contour feather**, from the fact that it helps to produce the contour of the body.

Varieties of feathers.

1. **Pennae.** There are two kinds of pennae or contour feathers.

(*a*) The **quills**. These form the large feathers of the wing and tail. They are divided into two groups, the **remiges**, or wing quills, and the **rectrices**, or tail quills.

The **remiges** include three sets of feathers, the **primaries** or **metacarpo-digitals**, which are attached to the bones of the manus, the **secondaries** or **cubitals**, which are attached to the ulna, and the **humerals**, which are attached to the humerus.

The **primaries** differ from all the other quill feathers in having the posterior half of the vane much wider than the anterior half. They are ten in number, and of these six, the **metacarpal** quills (fig. 57, 14), are attached to the second and third metacarpals, one, the **ad-digital** (fig. 57, 15), to the phalanx of the third digit, two, the **mid-digitals** (fig. 57, 16), to the first phalanx of the second digit, and two, the **pre-digitals** (fig. 51, 17), to the second phalanx of the second digit. One of the pre-digitals is very small, and is called the **remicle** (fig. 57, 11).

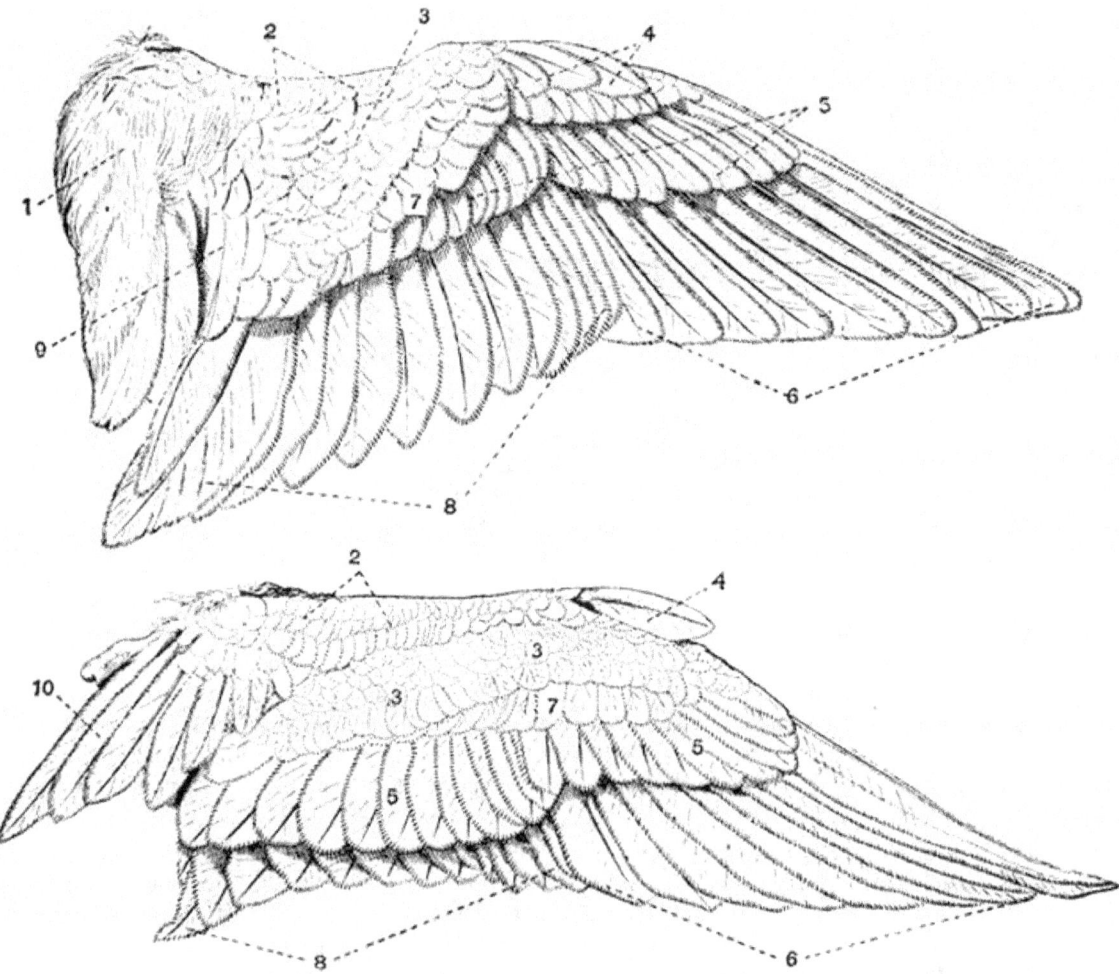

Fig. 56. The wing of a Wild Duck (*Anas boschas*).
The upper figure shows the dorsal side of a right wing, the lower figure
the ventral side of a left wing. × 1/3. (Brit. Mus.)

1. scapulars.

2. tectrices marginales.

3. tectrices minores.

4. bastard wing.

5. tectrices majores.

6. metacarpo-digitals or primaries.

7. tectrices mediae.

8. cubitals or secondaries.

9. pennae humerales.

10. pennae axillares.

In addition, a group of three quill feathers is attached to the first digit, constituting the **bastard wing** or **ala spuria** (fig. 56, 4).

The **secondaries** or **cubitals** (fig. 56, 8) form a group of seventeen feathers, attached to the ulna; they are shorter than the primaries, and do not have the posterior half of the vane much wider than the anterior half.

The **humerals** (figs. 56, 9 and 57, 12) form a group of eight small feathers, of varying length, attached to the anterior half of the humerus.

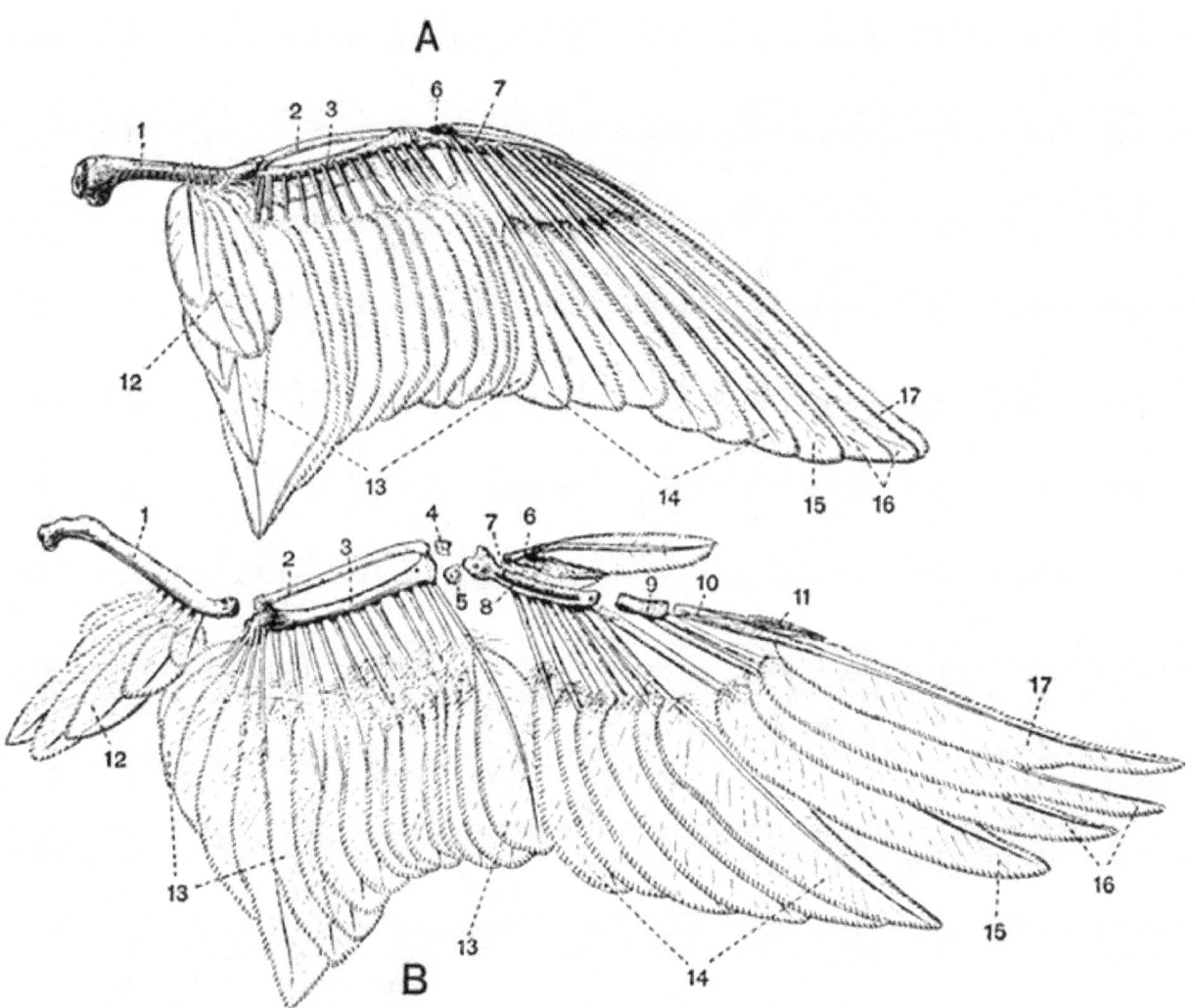

Fig. 57. Wings of a Wild Duck with the coverts removed (*Anas boschas*). × 1/3.
A. Right wing seen from the dorsal side. B. Left wing disarticulated
and seen from the ventral side. (Brit. Mus.)

1. humerus.
2. radius.
3. ulna.
4. radial carpal.
5. ulnar carpal.
6. first phalanx of first digit.
7. second metacarpal.
8. third metacarpal.
9. first phalanx of second digit.
10. second phalanx of second digit
11. remicle.
12. pennae humerales.
13. cubitals or secondaries.
14. metacarpal quills.
15. ad-digital.
16. mid-digitals.
17. pre-digital.

(b) The **tectrices** or **coverts** are short feathers, which cover over the quills of the rectrices and remiges, and clothe the body generally. Their barbules are less developed than is the case with the quill feathers, so that the barbs separate readily from one another, especially at the base of the vane. The nomenclature of the various patches of coverts on the wings is seen in fig. 56. A small patch of backwardly-directed feathers surrounding the external auditory opening are known as the **auriculars**.

2. The **filoplumes** are rudimentary feathers, consisting of a minute stem and slightly developed vane. They are left in the skin after the other feathers have been removed.

3. The **plumulae**, or down feathers, have the stem very slightly developed, while the barbs are soft and free from one another. They are distributed all over the body, not only among the contour feathers, but also over the spaces (*apteria*) which bear no contour feathers.

In the young bird the rudiments of the new feathers are formed at the bases of the embryonic down feathers, and as they grow they push them out from the skin. The embryonic down feathers however remain attached to the apices of the new feathers till these have reached a length of about an inch; they are then shed.

II. ENDOSKELETON.

As compared with that of the Turtle or Crocodile, the endoskeleton of the Duck is characterised by:

1. The great lightness of the bones, many of which contain air cavities.

2. The tendency to become ankylosed together shown by many of the bones.

3. The modification of the anterior limbs and girdle for the purpose of flight.

1. The Axial Skeleton.

This, as in other vertebrates, is divisible into—

A. The vertebral column. B. The skull. C. The ribs and sternum.

A. The Vertebral Column.

The vertebral column of the duck, like that of the great majority of birds, presents a number of well-marked characteristics, contrasting strongly with those of the generality of higher vertebrates. The centra are always without epiphyses. The neck is exceedingly long, about as long as all the rest of the vertebral column put together, and is remarkable for its flexibility. The trunk portion of the vertebral column on the other hand is characterised by extreme rigidity, and the marked tendency shown by the component vertebrae to fuse together into one almost continuous mass. The most rigid part of the vertebral column is that to which the pelvis is united, as no less than seventeen vertebrae take part in the union. The tail of the duck, like that of all living birds, is very short, and the posterior caudal vertebrae are united together, forming the **pygostyle**. The vertebral column may be divided into cervical, thoracic, lumbar, sacral, and caudal regions, but the boundaries between the several regions are ill-defined.

The Cervical Vertebrae.

All the vertebrae anterior to the first one that bears a rib meeting the sternum are regarded as cervical vertebrae. There are therefore sixteen cervical vertebrae, the last two of which bear well developed ribs. All are freely movable on one another.

As a typical cervical vertebrae, any one from the fifth to the ninth may be taken. The vertebra is rather elongated, and is very lightly and strongly made, its most characteristic feature being the shape of the articulating surfaces of the centra, which are generally described as saddle-shaped. The anterior surface is convex from above downward, and concave from side to side, while the posterior and more prominent surface is concave from above downwards and convex from side to side. The neural arch is low, and is drawn out into a slight blade-like **neural spine**. Its base is deeply notched on both sides posteriorly for the exit of the spinal nerves. Above these notches it is drawn out into two rather prominent diverging processes, which bear the **postzygapophyses**,—two flattened surfaces which look downwards and outwards. The **transverse processes** form irregular outgrowths from the anterior two-thirds of the sides of the vertebra; each projects for a short distance downwards and outwards, and is terminated posteriorly by a short backwardly-projecting spine. The transverse processes are shown by development to ossify from separate centres, and are therefore to be regarded as cervical ribs, and each is perforated at its base by a canal for the passage of the vertebral artery. Above the anterior end of the vertebrarterial canal are a pair of thickened outgrowths, which bear upwardly and inwardly directed **prezygapophyses**. Each transverse process is perforated near its middle by a prominent foramen through which passes a vein which is connected with the jugular vein.

The third and fourth cervical vertebrae resemble the succeeding ones in most respects, but have small **hypapophyses**, and the neural spines are less blade-like. The posterior cervical vertebrae (tenth to sixteenth) differ somewhat from the middle ones. They are shorter and more massive, the neural arch is much shorter, being deeply notched in the middle line in front and behind. The transverse processes arise from the anterior half of the vertebra only, and in the eleventh vertebra each is drawn out below into a pair of rather prominent downwardly and inwardly directed processes. In the twelfth vertebra these processes have almost coalesced, and in the thirteenth vertebra they have coalesced completely, forming a prominent **hypapophysis**. In the succeeding vertebrae this hypapophysis rapidly decreases in size.

The fifteenth and sixteenth cervical vertebrae resemble the succeeding thoracic vertebrae, having short thick centra and prominent squarely truncated neural spines; the sides of the neural arches are very deeply notched. The fifteenth vertebra has a short transverse process, perforated by a wide vertebrarterial foramen, but this foramen is absent in the sixteenth. The transverse processes of the fifteenth vertebra bear two facets for the articulation of the capitulum and tuberculum of the rib. The sixteenth vertebra has its tubercular facet on the transverse process, but the capitular facet is borne on the centrum.

The second or **axis** vertebra is small, and has the centrum drawn out into a comparatively very large hypapophysis. The posterior articulating surface of the centrum is saddle-shaped, the anterior nearly flat: above it the centrum is prolonged into the prominent **odontoid process**, which is shown by development to be the detached centrum of the atlas. The neural arch is deeply notched in the middle line in front, and at the sides behind. It is drawn out posteriorly into a wide massive outgrowth, which overhangs the third vertebra and bears the downwardly-directed postzygapophyses. The prezygapophyses are situated at the sides of the anterior end of the neural arch, and look directly outwards. The transverse processes are very slightly developed, and are pierced by the vertebrarterial canals.

The **atlas** vertebra is a very slight ring-like structure, thickened ventrally and bearing in front a prominent concave cavity for articulation with the occipital condyle of the skull. Posteriorly it bears a more or less flattened surface for articulation with the centrum of the axis. It surrounds a large cavity partially divided into a larger dorsal portion, which is the neural canal, and a smaller ventral portion which lodges the odontoid process. The sides of the atlas are pierced by the vertebrarterial canals, above which there are two slight backwardly-projecting outgrowths bearing the postzygapophyses on their inner faces.

The Thoracic Vertebrae.

The thoracic region includes all the vertebrae bearing free ribs, except the first two, viz. those whose ribs do not reach the sternum. There are seven thoracic vertebrae. The first four have centra with saddle-shaped articulating surfaces, but are more or less firmly united together by their neural spines; the last two are completely ankylosed by their centra to the lumbar vertebrae.

Each of the first five vertebrae has a prominent, vertical, abruptly terminated neural spine, and straight transverse processes. The zygapophyses and articulating surfaces at the ends of the centra are well developed. The third, fourth, fifth, and sixth vertebrae have very prominent hypapophyses. The articular facets for the ribs are well marked, those for the tubercula lying at the free ends of the transverse processes, and those for the capitula at the sides of the anterior ends of the centra. The sixth and seventh thoracic vertebrae are firmly fused by their centra and neural arches to one another and to the lumbar vertebrae behind, and by their transverse processes to the ilia. The sixth has its centrum terminated in front by a saddle-shaped articulating surface, and bears a pair of prominent prezygapophyses. Its transverse processes and centrum bear facets for the tubercula and capitula of the ribs respectively. In the seventh vertebra the tubercular facet is wanting.

153

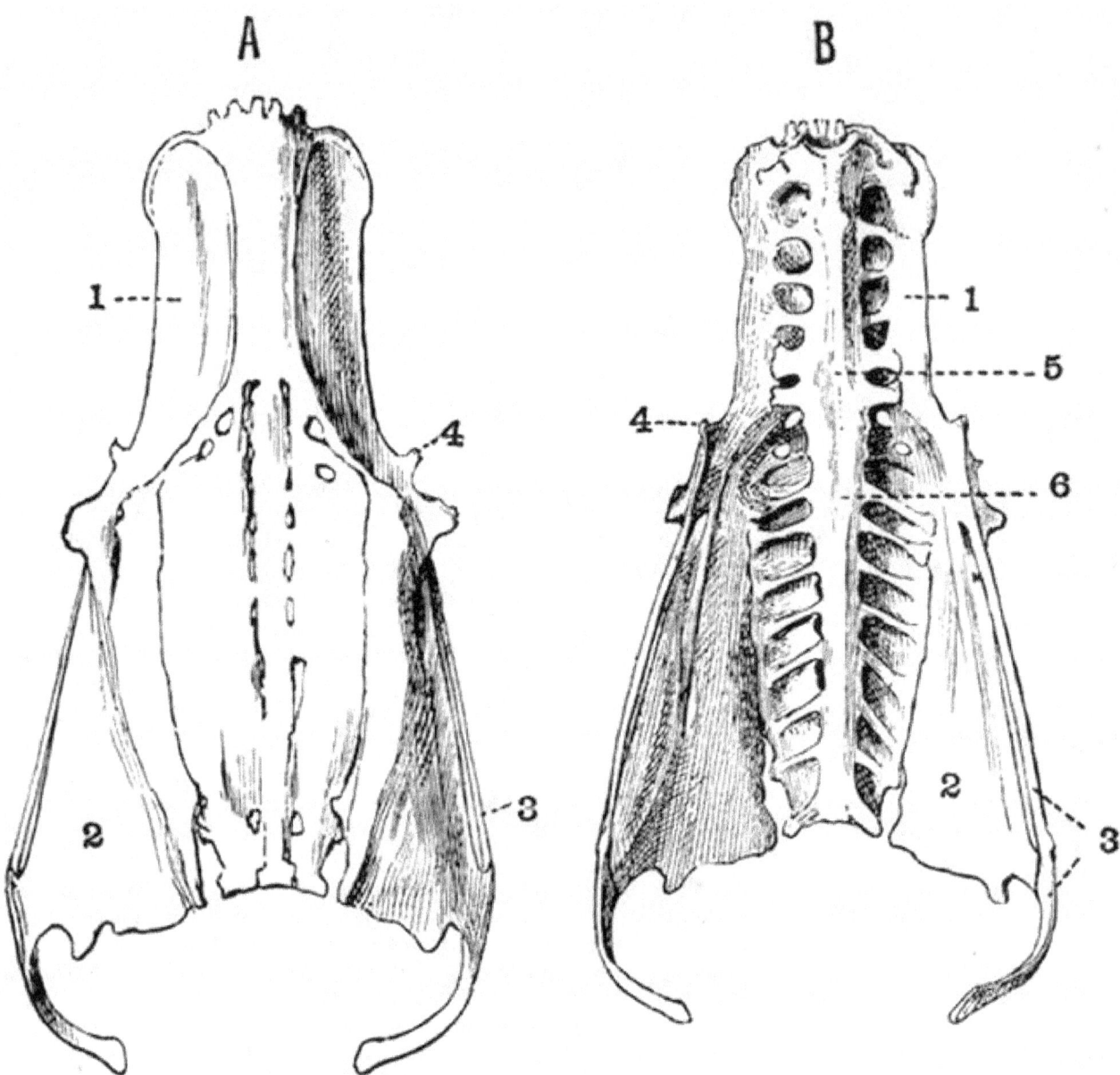

Fig. 58. A, dorsal and B, ventral view of the pelvis and sacrum of a Duck (*Anas boschas*).

1. ilium.	4. pectineal process.
2. ischium.	5. lumbar vertebrae.
3. pubis.	6. true sacral vertebrae.

The Sacrum.

The **sacrum** generally consists of seventeen vertebrae fused with one another and with the ilia. Their number may be reckoned from the number of foramina for the exit of spinal nerves. The two most anterior of these vertebrae bear ribs and have been already described with the other thoracic vertebrae. Their neural spines and those of the four succeeding vertebrae are fused together, forming a continuous crest of bone completely united laterally with the ilia. The transverse processes of all these six vertebrae are well developed, but those of the posterior two (fig. 58, B, 5) are much the stoutest. The next three vertebrae have broad centra, but their transverse processes are very slightly developed and have no ventral elements. These seven vertebrae belong to the **lumbar** series. The remaining eight vertebrae have well-developed transverse processes, which in the case of the first three or four are divisible into dorsal and ventral elements. All the dorsal elements are united to form a pair of flattened plates, partially separated by a series of foramina from a median plate formed by the united neural arches. Laterally they are continuous with the ischia. The first two of this series of vertebrae are shown by their relation to the nerves to be the true **sacrals** (fig. 58, B, 6), the remaining six belonging to the **caudal** series.

154

Behind them come the six free caudal vertebrae, succeeded by a terminal piece, the **pygostyle**, formed of a number of vertebrae fused together; this bears the rectrices or tail quills.

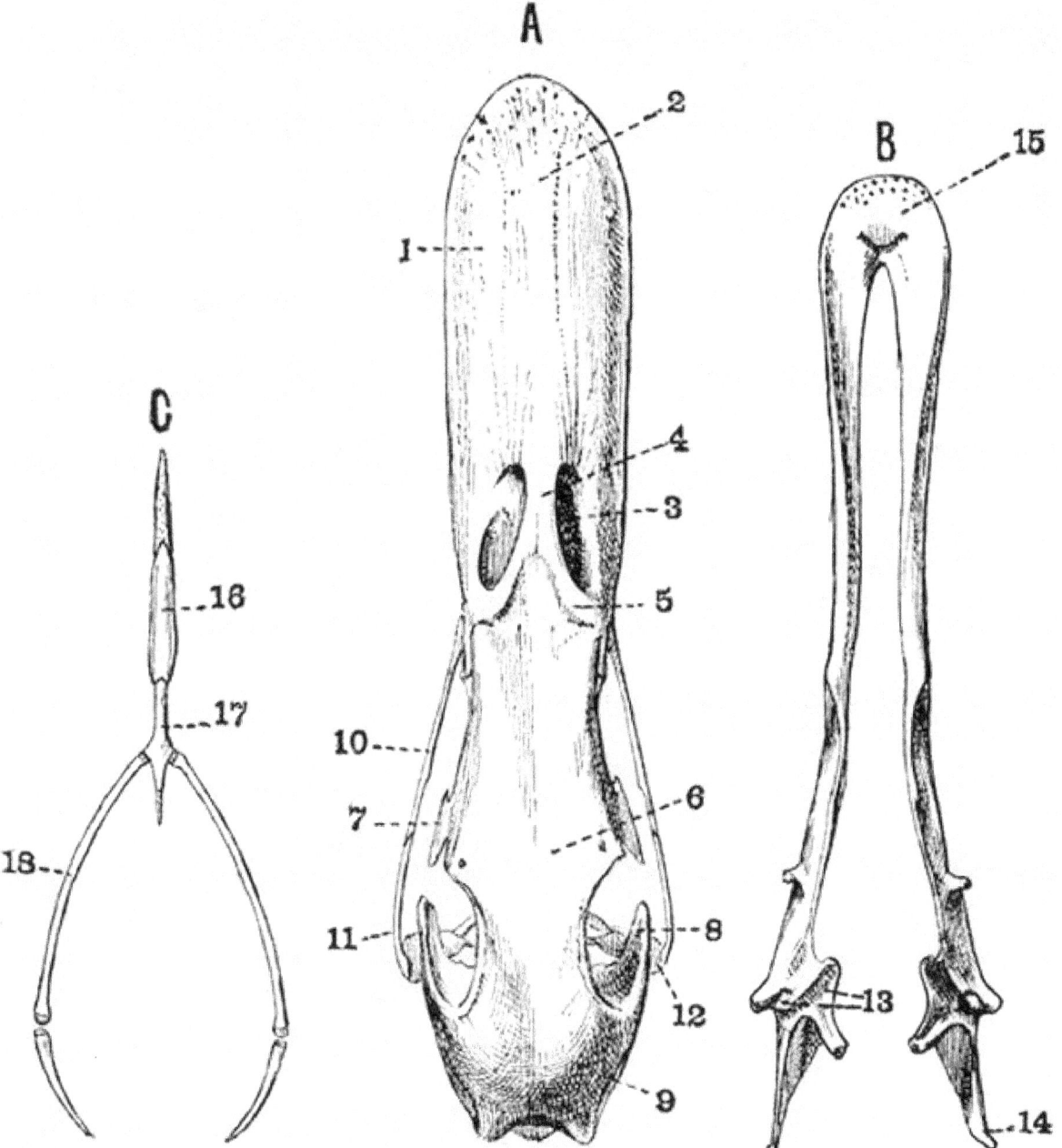

Fig. 59. Skull of a Duck (*Anas boschas*). × 1.
A. Dorsal view of the cranium. B. Palatal view of the mandible.
C. The Hyoid.
For numbers see Fig. 60.
B. The Skull.

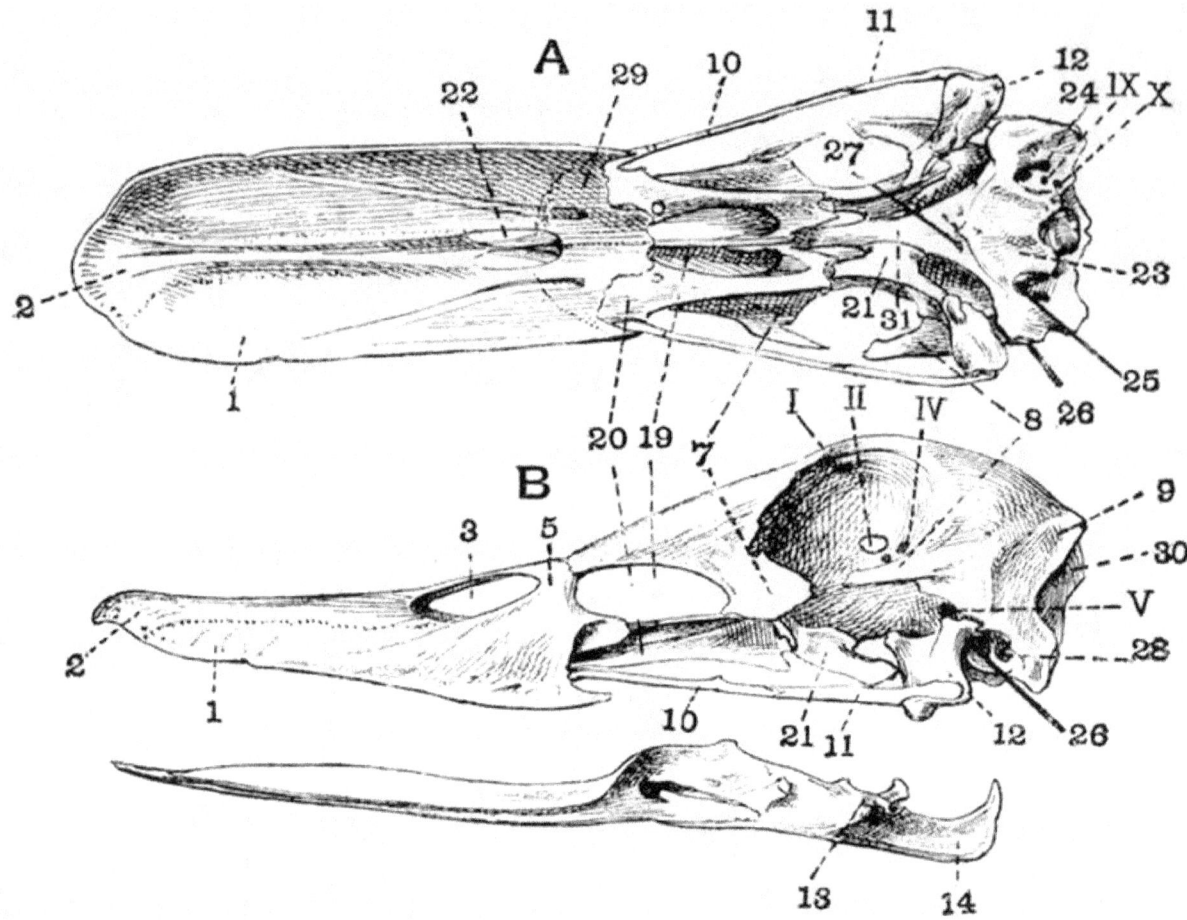

Fig. 60. A. Ventral view of the cranium of a Duck (*Anas boschas*).
B. Cranium and mandible seen from the left side. × 1.

1. maxillae.

2. premaxillae.

3. anterior nares.

4. nasal process of premaxillae
(fig. 59).

5. nasal.

6. frontal (fig. 59).

7. lachrymal.

8. postfrontal process.

9. parietal (fig. 59).

10. jugal.

11. quadratojugal.

12. quadrate.

13. condyle of mandible.

14. posterior articular process.

15. dentary at symphysis. }

16. basi-hyal.} (fig.

17. uro-hyal.} 59).

22. anterior palatine foramen.

23. basitemporal.

24. foramen leading into tympanic
cavity.

25. bristle inserted into posterior
opening of carotid canal.

26. bristle inserted into posterior
opening of Eustachian
canal.

27. bristle emerging through
anterior opening of carotid
canal. Close by is seen the
bristle emerging through
the anterior opening of the
Eustachian canal.

28. fenestral recess.

29. maxillo-palatine.

30. lambdoidal crest.

18. basibranchial.}

19. vomer.

20. palatine.

21. pterygoid.

31. rostrum.

I. II. IV. V. IX. X. nerve

foramina.

The skull of the duck, like that of birds in general, is characterised (1) by its lightness, (2) by the contrast between the bones of the cranium proper and those forming the rest of the skull, for the bones forming the cranium proper are closely fused together, the sutures between them being nearly all completely obliterated in the adult, while the bones forming the face are loosely connected with the cranium proper; (3) by the prolongation of the face into a long toothless beak; (4) by the size of the orbits, and their position entirely in front of the cranium, so that they are separated from one another only by a thin interorbital septum.

For purposes of description the skull may be divided into

(1) The cranial portion.

(2) The facial portion.

(3) The mandible.

(4) The hyoid.

(1) The Cranial portion.

This is a rounded box expanded dorsally and posteriorly, but tapering antero-ventrally. In the young skull the divisional lines between the several bones can be easily seen, but in the adult they are quite obliterated.

(*a*) The *dorsal surface* is rounded, expanded in front and behind, but encroached upon in the middle by the cavities of the orbits. There is a prominent divisional line in front, separating it from the facial part of the skull. It is formed mainly by the *frontal* (fig. 59, A, 6) and *parietal* bones, but the frontals diverge a little anteriorly and enclose between them the ends of the *nasal processes* (fig. 59, A, 4) of the *premaxillae*. Just in front of the orbit the outer margins of the frontals are either notched or pierced by a pair of foramina.

(*b*) At the *posterior end* of the cranium the most prominent feature is the large, almost circular **foramen magnum**, through which the spinal cord and brain communicate; this in young birds is seen to be bounded by four distinct bones, dorsally by the **supra-occipital**, ventrally by the **basi-occipital**, and laterally by the **exoccipitals**.

The **basi-occipital** forms the main part of a prominent convex knob, the **occipital condyle**, with which the atlas articulates. The occipital condyle is slightly notched above, and the ventral surface of the cranium is deeply pitted just in front of it; the exoccipitals also contribute slightly to its formation. Slightly in front of and ventral to the foramen magnum is a small foramen through which the hypoglossal nerve leaves the cranial cavity.

The **supra-occipital** is separated from the parietal by a suture line along which run a pair of prominent ridges, the **lambdoidal crests** (fig. 60, B, 30). There are often a pair of prominent vacuities in the supra-occipital dorsal to the foramen magnum. The **epi-otics** and **opisthotics** become completely fused with the bones of the occipital segment at a very early stage.

(*c*) The *ventral surface* of the cranium is wide behind, where it is formed by a broad transverse membrane bone, the *basitemporal* (fig. 60, A, 23), the sides of which are fused with the auditory capsules. Slightly in front of and an eighth of an inch external to the hypoglossal foramen the cranial wall is pierced by a pair of foramina through which the tenth or pneumogastric nerves leave (fig. 60, A, X). At the sides of the basitemporal are a pair of depressions, the **tympanic recesses**, in each of which are three holes. Straight lines joining these holes would form an isosceles triangle with its apex directed forwards. Of the two holes at the base of the triangle, the one nearer the middle line and leading into the cranial cavity, is for the exit of the ninth or glossopharyngeal nerve (fig. 60, A, IX), it lies just in front of the pneumogastric foramen. The more external leads into the tympanic cavity, while the more anterior at the apex of the triangle is the **posterior opening of the carotid canal** (fig. 60, A, 25), which traverses the base of the cranium, and during life lodges the carotid artery.

The anterior end of the basitemporal is pierced near the middle line by a pair of holes, the **anterior openings of the Eustachian canals**; while just in front of these and a little further removed from the middle line are the anterior openings of the **carotid canals**. Bristles passed in through the posterior openings of the carotid canals will emerge here (fig. 60, A, 27). In front of the basitemporal the base of the cranium is formed by the **rostrum** (fig. 60, A, 31), or thickened basal portion of the interorbital septum; this bears two prominent surfaces with which the pterygoids articulate. In some kinds of duck these surfaces are borne by well-marked basi-pterygoid processes.

(*d*) *The side of the cranium.* At the base of the posterior end is seen the deep **tympanic cavity**. The dorsal part of this is divided by a vertical partition into two halves; of these the more anterior is the larger, and forms a deep funnel-shaped cavity, the **posterior opening of the Eustachian canal** (fig. 60, B, 26). A bristle passed into this opening emerges through the anterior opening of the Eustachian canal. The more posterior of the two is the **fenestral recess** (fig. 60, B, 28), and is in its turn divided by a slender horizontal bar into a dorsal hole, the **fenestra ovalis**, and a ventral hole, the **fenestra rotunda**.

During life the fenestra ovalis lodges the proximal end of the **columellar** chain. Lying at the outer side and slightly dorsal to the tympanic cavity is a deep depression, the **lateral tympanic recess**, and immediately in front of this is the articular surface for the quadrate. The tympanic cavity is bounded below by the basitemporal, posteriorly by the exoccipital, and above by the *squamosal*, a membrane bone, which roofs over a good deal of the side of the cranium, and bears ventrally a prominent surface with which the quadrate articulates. Just in front of this is a large round hole, the **trigeminal foramen** (fig. 60, B, V), behind which the squamosal is drawn out into a short process.

In front of the squamosal there is a prominent forwardly-projecting **postfrontal process** (fig. 60, 8), which ossifies from a different centre from that forming the squamosal, but in the adult is completely fused with it.

The **orbit** forms a large more or less hemispherical cavity which lodges the eyeball. It is separated from its fellow of the opposite side by an imperfect partition, the **interorbital septum**. In the young skull it is seen to be bounded above by the frontal, with which the *lachrymal* (fig. 60, 7) is fused anteriorly, forming a large backwardly-projecting process; while behind it is bounded by the **alisphenoid**. The interorbital septum is formed by the ossification and coalescence of the **mesethmoid** in front, with the **orbitosphenoid** behind, and the **rostrum** below. The boundary of the orbit below is very imperfect, the zygomatic arch being incomplete.

The interorbital septum is pierced by the very prominent **optic foramen** (fig. 60, B, 2), just behind which are the two much smaller foramina for the exit of the oculomotor and pathetic (fig. 60, B, IV) nerves, the more anterior being that for the oculomotor.

Above and slightly in front of the optic foramen is a median opening, the **olfactory foramen.** This leads into the cranial cavity behind, and in front is continued forwards as a groove between the interorbital septum and the frontal.

(2) The Facial part of the Skull.

This includes the olfactory capsule and associated bones, and the upper jaw.

The bones associated with the olfactory capsules are the *nasals* and *vomer*. The *nasals* (figs. 59 and 60, 5) lie on the dorsal surface immediately in front of the cranium, and are separated from one another by the nasal processes of the premaxillae. Each is completely fused in the adult with the corresponding maxillae and premaxillae, the three bones together forming the boundary of the **anterior nares.** The *vomer* (fig. 60, 19) is unpaired and forms a small median vertical plate lying ventral to the anterior continuation of the interorbital septum.

The bones of the upper jaw consist on each side of two slender arcades which in front converge and are attached to the large beak, while behind they diverge but are united by the **quadrate.**

The **inner arcade** is formed by the pterygoid and palatine. The *pterygoid* (fig. 60, 21) is a short flattened bone, which articulates behind with the quadrate, and on its inner side with a large flattened surface borne by the rostrum, in front it meets the palatine, or sometimes ends freely with a long antero-dorsally directed point.

The *palatine* (fig. 60, 20) is a slender irregular bone flattened dorso-ventrally at its anterior end where it articulates with the beak, and laterally behind. It gives off at its posterior end a process, which is sometimes united with the vomer, sometimes projects forwards, and meets its fellow dorsal to the vomer. In the large space between it and the vomer is the opening of the **posterior nares.**

The *premaxillae* (figs. 59 and 60, 2) are very large, and form nearly a third of the big shovel-shaped beak. They constitute the inner, and part of the front boundary of the anterior nares, and send back a pair of *nasal processes* which partially separate the nasals from one another.

The **outer arcade** forms the slender **suborbital bar**, and consists mainly of two rod-like bones, which in the adult are completely fused together. The posterior of these is the *quadratojugal* (figs. 59 and 60, 11) which articulates with the quadrate, the anterior is the small and slender *jugal* or *malar* (figs. 59 and 60, 10). The extreme anterior part of the bar is formed by the *maxillae*. The main part of the maxillae however lies anterior to the suborbital bar, and extends forwards along the side of the premaxillae forming all the lateral part of the beak (figs. 59 and 60, 1); it also sends inwards a plate, the **maxillo-palatine** (fig. 60, A, 29), which completely fuses with its fellow in the middle line, and forms the posterior boundary of the anterior palatine foramen. The term **desmognathous** describes the condition of the skull in which the maxillo-palatines fuse with one another in the middle line in this way.

The **quadrate** (fig. 60, 12), which unites the two arcades behind, is a stout irregular four-cornered bone forming the **suspensorium**. It articulates by its dorso-posterior corner with the squamosal, and by its antero-internal corner with the pterygoid. The middle of its ventral surface forms a hemispherical knob with which the mandible articulates, while its dorso-anterior border is drawn out into a long point which extends towards the interorbital septum.

(3) The Mandible.

The **mandible** or lower jaw consists of two **rami** which are flattened and fused together in the middle line in front, while behind they diverge from one another and articulate with the quadrates.

Each ramus is composed of five bones fused together, one being a cartilage bone, and the other four membrane bones. The **articular** is the only cartilage bone of the mandible, it bears the double condyle (figs. 59 and 60, 13) or concave articular surface for the quadrate, and is drawn out behind into a large hooked **posterior articular process**. The articular is also drawn out into a prominent process on each side of the articular surface for the quadrate, and is marked by a deep pit opening posteriorly. The articular is continuous in front with **Meckel's cartilage** which forms the original cartilaginous bar of the lower jaw, and is ensheathed by the membrane bones. Of these the *supra-angular* forms the upper part of the mandible in front of the articular, its dorsal surface is drawn out into a small **coronoid process**, its outer surface also bearing a prominent process. The *angular* is a small bone which underlies the articular and supra-angular on the inner side of the jaw. The *dentary* (fig. 59, 15) forms the anterior half of each ramus, and is the largest bone of the mandible; it is fused with its fellow at the symphysis in front, and extends back below the supra-angular. The *splenial* is a small bone lying along the middle half of the inner side of each ramus of the mandible.

(4) The Hyoid.

With the hyoid apparatus is included the **columella**. This forms a minute rod of bone, one end of which is expanded and fits into the fenestra ovalis, while the other end, terminated by a triradiate piece of cartilage, is attached to the tympanic membrane. The structure is as a whole homologous with the auditory ossicles of mammals and the hyomandibular of fish.

The **hyoid** consists of a median unpaired portion, formed of two pieces of bone, the **basi-hyal** (fig. 59, C, 16) in front, and the **uro-hyal** (fig. 59, C, 17) behind, the two being placed end to end and terminated anteriorly by an unpaired cartilaginous plate, the **os entoglossum**. At the posterior end there come off a pair of long **posterior cornua**, each of which consists of two pieces, a longer **basibranchial** (fig. 59, C, 18), and a shorter **cerato-branchial**. For the homology of these parts see p. 336.

The Ribs and Sternum.

The last two cervical vertebrae bear long movable ribs which articulate by distinct capitular and tubercular processes, but do not meet the sternum. The thoracic ribs are eight in number, and each is divisible into a **vertebral** and a **sternal** portion. The first five thoracic ribs are flattened curved bars of bone, which articulate by a prominent **capitulum** with the centrum of the corresponding vertebra, and by a **tuberculum** with the transverse process. Projecting backwards from each is a large hooked **uncinate process.** The last three ribs which are without uncinate processes, become progressively more slender, and in the eighth the tubercular processes are lost.

The sternal portions of the ribs are imperfectly ossified pieces, short and comparatively thick in the case of the anterior ribs, longer and more slender in the case of the posterior ribs.

The Sternum.

The **sternum** or breast bone is exceedingly large in the Duck, as in all birds, and projects back far beyond the thorax over much of the anterior part of the abdomen. It is an irregularly oblong plate of bone, abruptly truncated behind, somewhat concave dorsally, and drawn out ventrally into a prominent keel, the **carina**, which projects for some distance forwards beyond the body of the sternum, and tapers off gradually behind. The point where the carina joins the body of the sternum is at the anterior end drawn out into a small process, the **rostrum**. Just dorsolateral to this are a pair of deep grooves, the **coracoid grooves**, with which the coracoids articulate.

The sides of the sternum are drawn out in front into a pair of short blunt **costal processes;** and just behind these are a series of seven surfaces with which the ends of the sternal ribs articulate. Immediately behind these surfaces the sides are produced into a pair of long backwardly-projecting **xiphoid processes** which nearly meet processes from the posterior end of the sternum.

2. The Appendicular Skeleton.

This consists of the skeleton of the anterior and posterior limbs and of their respective girdles.

A. The Pectoral Girdle.

The pectoral girdle in almost all birds is strongly constructed and firmly united to the sternum. It consists of three bones, a dorsal element, the **scapula**, a posterior ventral element, the **coracoid**, and an anterior ventral element, the *clavicle*.

The **scapula** forms a long curved flattened bone expanded at its anterior end, where it meets the coracoid, and lying across the ribs at its tapering posterior end. It helps to form the imperfect **glenoid cavity**, with which the humerus articulates. The **coracoid**, a shorter but stouter bone than the scapula, has its upper end or **head** thickened and bears on its posterior border an irregular surface, with part of which the scapula articulates, while the rest forms part of the glenoid cavity. The inner border of the coracoid adjoining the articular facet for the scapula is produced into a strong process which helps to complete the **foramen triosseum**, a space lying between the adjoining ends of the scapula and coracoid, through which the tendon of the second pectoral muscle passes. The lower part of the coracoid, which is much flattened and expanded, and abruptly truncated posteriorly, articulates with the coracoid groove of the sternum. The *clavicle* is a thickened curved membrane bone, which is fused with its fellow in the middle line below, the two forming the *furcula* or merrythought. Its dorsal end is drawn out into a process which articulates with the coracoid.

The Anterior Limb or Wing.

This consists of three parts, a proximal part, the upper arm or **brachium**, a middle part, the fore-arm or **antibrachium**, and a distal part, the **manus**. When extended for flight the parts lie almost in the same straight line, but when at rest they are

159

folded on one another in the form of a Z, the brachium and manus pointing backwards, and the antibrachium forwards. When extended for flight the surfaces and borders of the wing correspond in position with those of the primitive vertebrate limb, the pre-axial border being directed forwards and the postaxial backwards, while the dorsal and ventral surfaces look respectively upwards and downwards. But when the wing is at rest, the humerus as it extends backwards becomes slightly rotated, so that its dorsal surface looks more inwards than upwards, while the dorsal surface of the antibrachium looks partially outwards and upwards, and that of the manus mainly outwards.

The **brachium** or **upper arm** contains only a single bone, the **humerus** (fig. 57, 1). This is a large nearly straight bone expanded at both ends. The proximal end is specially expanded, forming two **tuberosities**, and a large convex **head** articulating with the glenoid cavity. The **pre-axial tuberosity** is the smaller of the two, but is continued by a prominent **deltoid ridge**, which extends for a very short distance down the shaft. The **postaxial tuberosity** is the larger, and below it there is a very deep pit, the **pneumatic foramen**, which leads into an air cavity in the shaft of the bone. The shaft is long and straight, and at the distal end of the bone is the **trochlea** with two convex surfaces, one pre-axial with which the radius articulates, the other postaxial for the ulna.

The **fore-arm** or **antibrachium** consists of two bones, the **radius** and **ulna**. These are of nearly equal length, and are separated from one another by a considerable space except at their terminations.

The **radius** (fig. 57, 2), the pre-axial and smaller bone, is straight and fairly stout; its proximal end articulates with the humerus by a slightly cupped surface, while its distal end, which articulates with the carpus, is convex and somewhat expanded.

The **ulna** (fig. 57, 3) is longer, stouter, and slightly curved. Its proximal end is expanded, forming two surfaces which articulate with the trochlea of the humerus; behind them it is drawn out into a short blunt **olecranon process**. Its distal end is less expanded, and articulates with the carpus and also with the radius.

The **Manus**. This includes the carpus or wrist, and the hand.

The **Carpus**. While in the embryo the carpus consists of five distinct elements arranged in a proximal row of two and a distal row of three, in the adult only the proximal bones can be clearly distinguished, the distal ones having become completely ankylosed with the metacarpals to form the **carpo-metacarpus**.

The two distinct carpal bones are the radial carpal and the ulnar carpal. The radial carpal (fig. 57, 4) is a small somewhat cubical bone, wedged in between the manus and the radius and ulna. The ulnar carpal (fig. 57, 5) is a somewhat larger, more irregular bone, lying adjacent to the end of the ulna. It is deeply notched to receive the carpo-metacarpus.

The hand. In the adult bird the hand is in a much modified condition; only the first three digits are represented, and the metacarpals are all fused with one another and with the distal carpalia to form the **carpo-metacarpus**.

The most prominent part of the carpo-metacarpus is formed by the **second metacarpal** (fig. 57, 7), a stout, straight bone expanded at both ends. The **third metacarpal** (fig. 57, 8) is a more slender curved bone fused at both ends with the second metacarpal. The **first metacarpal** forms simply a small projection on the radial side of the proximal end of the second metacarpal.

The **phalanges**. The first digit or **pollex** includes two phalanges, the distal one being very small and bearing a claw.

The second digit includes three phalanges, the proximal one being somewhat flattened. The third digit has a single small phalanx.

The Pelvic Girdle.

The bones constituting the pelvic girdle are not only as in other higher vertebrates ankylosed together forming the innominate bones, but are also ankylosed with a series of some seventeen sacral and pseudosacral vertebrae. The **acetabulum** (fig. 61, 5) with which the head of the femur articulates is incompletely ossified.

The **ilium** (figs. 58 and 61, 1) is the largest bone of the pelvis. It forms a long flattened plate extending for a considerable distance both in front of and behind the acetabulum, and is fused along its whole length with the transverse processes and neural spines of the sacral and pseudosacral vertebrae. It forms more than half the acetabulum, above and behind which it is produced to form a process, the **antitrochanter** (fig 61, 8), with which the great trochanter of the femur articulates.

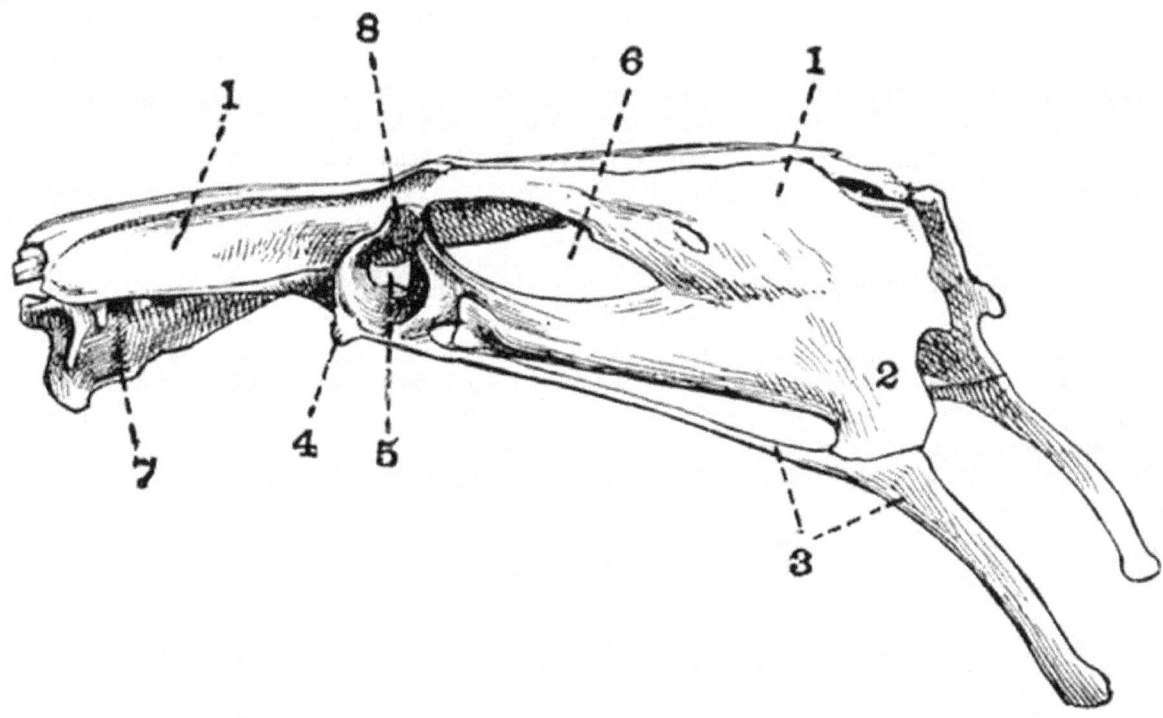

Fig. 61. Lateral view of the pelvis and sacrum of a Duck (*Anas boschas*) × 2/3.

1. ilium.	5. acetabulum.
2. ischium.	6. ilio-sciatic foramen.
3. pubis.	7. fused vertebrae.
4. pectineal process.	8. antitrochanter.

The **ischium** (figs. 58 and 61, 2) is a flattened bone which forms about one-third of the acetabulum, and lies ventral to the posterior part of the ilium. Its anterior portion is separated from the ilium by the large oval **ilio-sciatic foramen** (fig. 61, 6), while behind this the two bones are completely fused.

The **pubis** (figs. 58 and 61, 3) is a very long slender bar of bone which forms only a very small part of the acetabulum and runs back parallel to the ventral surface of the ischium with which it is loosely connected at its posterior end. For the greater part of their length the two bones are separated by the long narrow **obturator foramen**. Behind the ischium the pubis is produced into a long curved downwardly-projecting process, and in front of the acetabulum it bears a short blunt **pectineal** or **pre-pubic process** (fig. 61, 4) probably homologous with the pre-pubis of Orthopod Dinosaurs. The remainder of the pubis is homologous with the post-pubis of Orthopod Dinosaurs.

The Posterior Limb.

The leg of the bird is somewhat differently constructed from that of other vertebrates owing to the fact that there is no free tarsus, the proximal tarsals having fused with the tibia, and the distal with the metatarsals.

The **thigh** consists of a single bone, the femur. The **femur** is a comparatively short bone with a straight shaft and expanded ends. The proximal end bears on its inner side a rounded **head**, which articulates with the acetabulum. On its outer side is an irregular outgrowth, the **great trochanter**, while between the two is the surface which meets the antitrochanter of the ilium. The posterior end also is expanded and marked by a wide groove which lodges the **patella**. On each side of the groove is a strong **condylar ridge** for articulation with the tibia. The external condyle is deeply grooved behind for articulation with the fibula.

The **crus** or **shin** consists of two separate bones, (1) the **tibio-tarsus**, formed by the fusion of the tibia with the proximal row of tarsals, and (2) the **fibula**.

The **tibio-tarsus** is a thick straight bone nearly twice as long as the femur. Both ends of the bone are considerably expanded. The proximal end bears two slight depressions which articulate with the condyles of the femur, and a third depression which partly lodges the patella. The proximal end of the anterior or extensor surface is drawn out into a very prominent **cnemial crest** which bends over towards the postaxial side of the bone; a slight ridge is continued from it all the way down the shaft. The proximal part of the shaft of the tibio-tarsus bears a roughened ridge with which the fibula is closely

161

connected. The distal end is expanded and rotated outwardly and forms a prominent pulley-like surface which articulates with the tarso-metatarsus.

The **fibula** is reduced to the proximal portion only, which is expanded and articulates with a depression behind the external condyle of the femur. The fibula further extends about a third of the way down the shaft of the tibio-tarsus. The **patella** or **knee-cap** is a sesamoid bone due to an ossification in the tendon of the extensor muscles of the leg.

The **ankle joint** lies between the proximal and distal tarsals which as previously mentioned fuse respectively with the tibia and metatarsus.

The **Pes**. The pes includes four digits, and consists of the tarso-metatarsus and the phalanges. The proximal tarsals which are fused with the tibia also really belong to the pes.

The **tarso-metatarsus** is a strong straight bone nearly as long as the femur, and is formed by the fusion of the distal tarsals with the second, third and fourth metatarsals. The proximal end of the bone is expanded and bears two facets for articulation with the tibio-tarsus, and near them on the posterior surface is a large roughened projection. The lines of junction between the several metatarsals are marked along the shaft by slight ridges. At the distal end of the bone the three metatarsals diverge from one another and each bears a prominent convex pulley-like surface. The **first metatarsal** is reduced to the distal end, which tapers to a point proximally, and is attached by ligaments near the distal end of the tarso-metatarsus.

The **digits**. Four digits are present, each consisting of a metatarsal (already described) and a certain number of phalanges, the terminal one being in each case clawed. The first digit or **hallux** has two phalanges, the second three, the third four, and the fourth five.

CHAPTER XIX.
GENERAL ACCOUNT OF THE SKELETON IN BIRDS.

EXOSKELETON.

The epidermal exoskeleton of birds is very greatly developed, feathers constituting its most important part.

Three kinds of feathers are found, viz. (*a*) *pennae* including quills and coverts, (*b*) down feathers or *plumulae*, and (*c*) *filoplumes* which are rudimentary feathers. The structure of the different kinds of feathers is described on pp. 303-306.

Sometimes a fourth class of feathers, the *semiplumae*, is recognised. They have the stems of pennae, and the downy barbs and barbules of plumulae.

In most birds the pennae are not uniformly distributed over the whole surface of the body, but are confined to certain tracts, the **pterylae**; while the intervening spaces or **apteria** are either bare or covered only with down feathers. In some birds, however, such as the Ratitae and the Penguins, pennae are evenly distributed over the whole body.

In many birds the calamus or quill bears two vexilla or vanes, the second of which, called the **aftershaft** or **hyporachis**, is generally much the smaller, and is attached to the under surface of the main vexillum. In the Moas, Emeu and Cassowary the two vexilla in the adult bird are nearly equal in size; though in the nestling Emeu one is much longer than the other. The aftershaft is very small in most Passeres and gallinaceous birds, but is comparatively large in Parrots, Gulls, Herons and most birds of prey. It is absent or extremely small in the Ostrich, *Apteryx*, *Rhea*, Pigeons, Owls, Anseres, and others.

The quill feathers include two groups, the **remiges** or wing quills, and the **rectrices** or tail quills. In most birds the primary remiges, or those which are attached to the bones of the manus, are ten or eleven in number, and are set in grooves in the bones, being firmly attached to them. In the Ostrich however the primaries are little specialised in character and are as many as sixteen in number. They are also less definitely attached to the bones; as their ends do not lie in grooves in the bones, but project beyond them.

The secondary quills or those attached to the ulna vary much in number according to the length of the bone. The large dark quills in the wings of Cassowaries are the secondaries.

The wing of Penguins is very little differentiated. It is covered at the margin by overlapping scales which gradually merge into scale-like feathers at the proximal end. The wing of the Penguin has nothing comparable to the remiges of other birds.

In some birds, such as Herons (*Ardea*), there occur in places plumulae of a peculiar kind, which grow persistently and whose summits break off into fine powder as fast as they are formed. These feathers are known as *powder-down* feathers. They occur also in some Parrots and are then scattered indiscriminately all over the body.

Other exoskeletal structures besides feathers are commonly well developed. Thus the extremities of the jaws are sheathed in horny **beaks** whose form varies enormously according to the special mode of life.

In ducks and geese the beak with the exception of the anterior end is soft, and its edges are raised into lamellae, while in the Mergansers these lamellae become pointed processes supported by bony outgrowths. These lamellae act as strainers. In Parrots and Hawks, on the other hand, nearly the whole of the beak is hard.

The toes and tarso-metatarsus are usually featherless and are covered either with granular structures or with well-formed scales. The toes are nearly always provided with **claws**, and these vary in correlation with the character of the beak. Claws also sometimes occur on the manus. Thus *Archaeopteryx* and some Ostriches and Rheas have claws on all three digits. Most Ostriches and Rheas, and many Anseres and birds of prey, have them on the first two digits, while the Secretary bird (*Gypogeranus*) and many fowls, ducks, and birds of prey, especially kestrels, have a claw only on the pollex. In the Cassowary, Emeu, Apteryx and some Ostriches and Rheas only the second digit is clawed.

Claws should not be confounded with **spurs**, which are conical horny structures developed on bony outgrowths of the radial side of the carpus, metacarpus, or metatarsus. They occur in a number of birds, but are most commonly developed in gallinaceous birds, by which they are used for fighting. A single spur occurs on the metacarpus in *Megapodius*, in *Palamedea*, in *Parra jacana* and in *Hoplopterus spinosus*, the Spur-winged plover. The Derbian Screamer, *Chauna derbiana*, has two metacarpal spurs, borne on the first and second metacarpals. The Spur-winged goose, *Plectropterus gambensis*, has a carpal spur borne on the radial carpal. Metatarsal spurs are quite common.

The male Solitaire (*Pezophaps*) has large bony excrescences on the wrist which may, like spurs, have been sheathed in horn and used for fighting.

Teeth do not occur in any living birds, but conical teeth imbedded in separate sockets are present in *Archaeopteryx* and *Ichthyornis*, while in *Hesperornis* similar teeth occur implanted in continuous grooves in the mandibles and maxillae, the premaxillae being toothless.

Except that teeth are partly dermal in origin, a dermal exoskeleton is quite unrepresented in birds.

ENDOSKELETON.

Perhaps the most striking feature of the endoskeleton of birds is its pneumaticity. In the embryo all the bones contain marrow, but as growth proceeds this becomes replaced by air to a variable extent in different forms. In all birds some part of the skeleton is pneumatic. Many small birds and *Apteryx* and Penguins among larger ones have air only in the skull; in Pigeons air is present in all the bones except the caudal vertebrae, the leg bones, and those of the antibrachium and manus; in Hornbills every bone contains air.

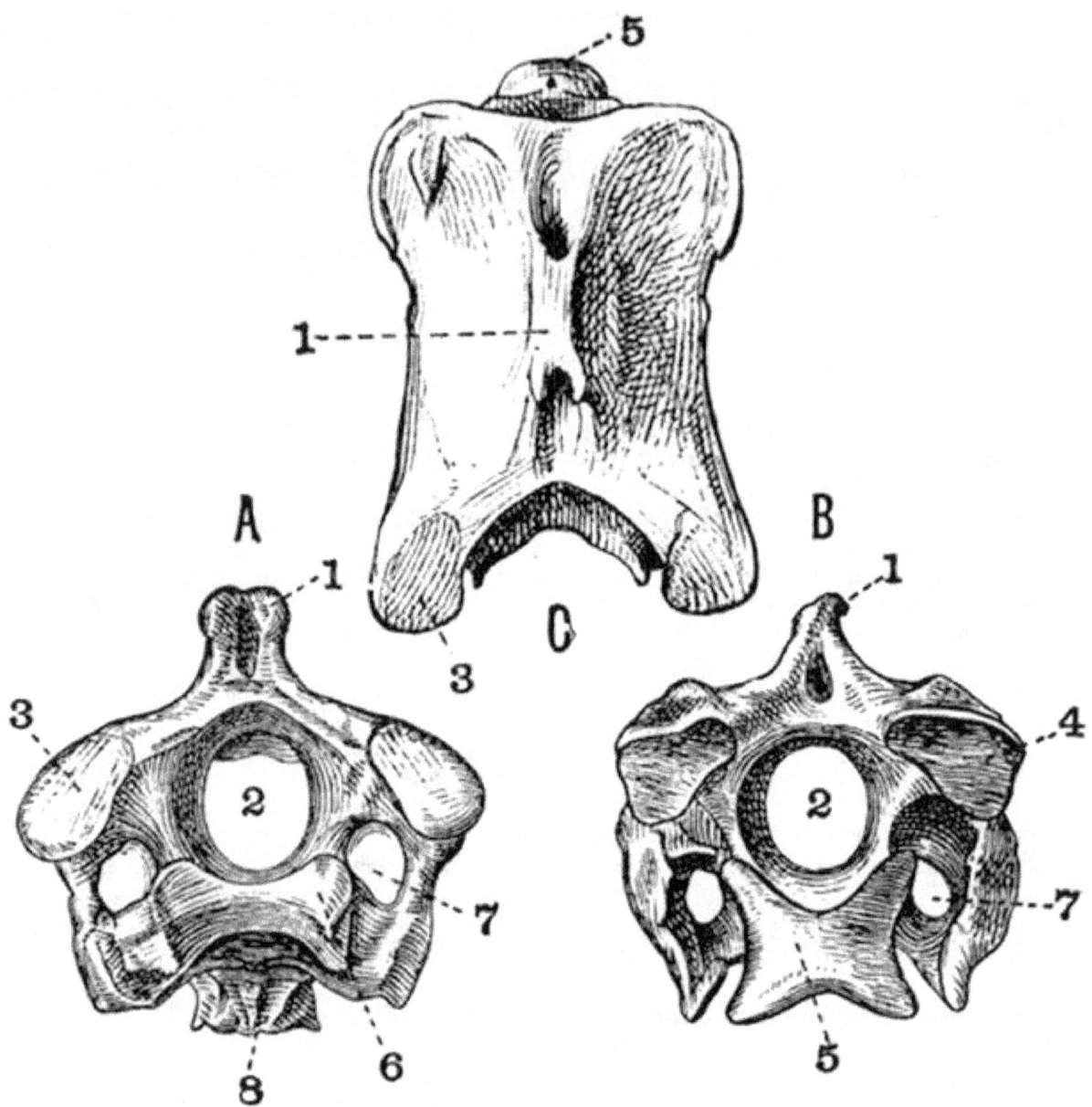

Fig. 62. Third Cervical Vertebra of an Ostrich (*Struthio camelus*).
× 1. A anterior, B posterior, C dorsal view (A and B after Mivart).

1. neural spine.

2. neural canal.

3. prezygapophysis.

4. postzygapophysis.

5. posterior articular surface of
centrum.

6. anterior articular surface of
centrum.

7. vertebrarterial canal.

8. hypapophysis.

Vertebral Column.

The vertebral column of birds is readily divisible into a very mobile cervical region, and an extremely rigid post-cervical region. In most birds the vertebral centra are without terminal epiphyses, but these structures are found in Parrots. The cervical vertebrae are generally large and vary in number from eight or nine to twenty-three in Swans. Except in some extinct forms, such as *Ichthyornis* and *Apatornis*, in which they are biconcave, the centra are characterised by having saddle-shaped articulating surfaces, which in front are concave from side to side and slightly convex from above downwards, while posteriorly they are convex from side to side and concave from above downwards. The atlas is small and ring-like, and its centrum is fused with the axis forming the odontoid process. Cervical ribs are often well developed, and in some of the Ratitae they remain for a long time distinct from the vertebrae.

The thoracic vertebrae are distinguished from the cervical by the fact that their true ribs are united to the sternum by means of sternal ribs. This distinction, however, though convenient, is somewhat arbitrary, as it has been shown that in the fowl and gannet, two pairs of ribs which in the adult are free from the sternum, are connected with it in the embryo. When, as in the Swans, the thoracic vertebrae are not all fused together, they generally have saddle-shaped articulating surfaces, but sometimes, as in the Penguins, Auks and Plovers, the centra are convex in front and concave behind. The trunk vertebrae generally have well-marked neural spines, while in the Divers the anterior ones have peculiar bifurcating hypapophyses.

The trunk vertebrae are not readily divisible into thoracic and lumbar. There are two true sacral vertebrae, but as development proceeds a number of other vertebrae become fused with the true sacrals, the whole forming a large compound sacrum. These pseudosacral vertebrae generally include the lumbar, and some of the thoracic and caudal vertebrae. Sixteen to twenty vertebrae or even more may be included in the compound sacrum, and sometimes the whole of the trunk vertebrae are fused together. In *Archaeopteryx* however but five vertebrae take part in the formation of the sacrum.

In *Archaeopteryx* there are twenty long caudal vertebrae, of which the last sixteen carry a pair of feathers apiece, but in all other birds the tail is short and in the great majority of cases the posterior vertebrae are fused together, forming the pygostyle. In the Ratitae and Tinamidae a pygostyle is rarely or imperfectly developed. In *Hesperornis* there are twelve caudal vertebrae, six or seven of which are united by their centra only, forming an imperfect pygostyle.

The free caudal vertebrae are generally amphicoelous.

The Skull.

The skull of all birds from *Archaeopteryx* onwards is essentially similar, differing from the skull of reptiles mainly in the extent to which the cranium is arched, and its greater size in proportion to the jaws.

Most of the bones of the cranium are pneumatic, and all show a marked tendency to fuse together, and have their outlines obliterated by the disappearance of the sutures. The several bones remain longest distinguishable in the Ratitae and to a less extent in the Penguins. The orbits are very large and lie almost entirely in front of the cranium; they are separated by an interorbital septum which is sometimes, as in *Chauna* and *Scythrops*, very complete, sometimes, as in Hornbills and the Common Heron, very slightly developed. As a general rule the sclerotic is cartilaginous.

The anterior nares are almost always situated far back at the base of the beak near the orbits, but in *Apteryx* they are placed right at its extremity. In *Phororhacos* they are placed very high up on the enormous beak and are not separated by any bony partition.

The skull of Parrots has some peculiarities. In some Parrots the lachrymal sends back a process which meets the postorbital process of the frontal and completes the orbit. In most birds the upper beak is immovably fixed, but in some it is attached to the cranium, only by the nasals and by flexible processes of the premaxillae, so that by this means a kind of elastic joint is established and the beak is able to be moved on the cranium. In the Parrots and *Opisthocomus* there is a regular highly movable joint.

In Cassowaries the fronto-nasal region of the skull is produced into an enormous bony crest, and in Hornbills a somewhat similar structure occurs. Although true teeth do not occur in any known bird except *Archaeopteryx*, *Hesperornis*, and *Ichthyornis*, another extinct bird, *Odontopteryx*, has the margins of both jaws provided with forwardly-directed tooth-like serrations, formed of part of the actual jawbone: a living hawk, *Harpagus*, too, has a deeply notched bill, to which correspond serrations in the premaxillae.

A basi-pterygoid process of the basisphenoid abuts against the pterygoid in Ratitae and in Tinamous, plovers, fowls, pigeons, ducks and geese among Carinatae, recalling the arrangement met with in many reptiles. The squamosal is sometimes, as in the fowl, united with the postorbital process of the frontal. In the Carinatae the quadrate articulates with the cranium by a double convex surface, in the Ratitae by a single one. The premaxillae are always comparatively large bones, the maxillae on the contrary are small, but give rise to important inwardly-projecting maxillo-palatine processes.

The relations of the palatines, pterygoids, maxillae, and vomers vary considerably, and on them Huxley has based a classification of birds. In the Ratitae and the Tinamous (Tinamidae), among Carinatae the vomers unite and form a large broad bone, separating the palatines and the pterygoids from the rostrum. Huxley uses the term **Dromaeognathous** to describe this condition. In all other Carinatae the vomers are narrow behind, and the palatines and pterygoids converge posteriorly and articulate largely with the rostrum. Three modifications of this condition are distinguished by Huxley, and termed **Schizognathous**, **Ægithognathous**, and **Desmognathous**.

In the **Schizognathae** the vomers coalesce and form a narrow elongated bone, pointed in front, separating the maxillo-palatine processes of the premaxillae. Waders, fowls, penguins, gulls, some falcons and eagles, American vultures, some herons and many owls have the Schizognathous arrangement. In pigeons and sandgrouse there is no vomer, but the other bones have the Schizognathous arrangement.

In the Ægithognathae the arrangement is the same as in the Schizognathae, except that the vomers are truncated in front. Passeres, swifts, woodpeckers, humming birds, rollers, hoopoes have this arrangement.

In the **Desmognathae** (fig. 60, A) the maxillo-palatine processes approach one another in the middle line, and either unite with the vomers, or unite with one another, hiding the vomers. Thus a more or less complete bony roof is formed across the palate. The vomers in Desmognathae are small or sometimes absent. Ducks, storks, most herons, most birds of prey and owls, pelicans, cormorants, parrots, and flamingoes are Desmognathous.

The mandible, as in other Sauropsids, consists of a cartilage bone, the articular, and a series of membrane bones, the dentary, splenial, coronoid, angular, and supra-angular, developed round the unossified Meckel's cartilage. The dentaries of the two rami are nearly always fused together, but in *Ichthyornis* and *Archaeopteryx* the two rami are but loosely united. There is often a fontanelle between the dentary and the posterior bones, while the angle is sometimes, as in the fowl, drawn out into a long curved process.

The hyoid apparatus (fig. 59, C) consists of a median portion, and a pair of cornua. The median portion is composed of three pieces placed end to end, and called respectively the os entoglossum, the basi-hyal, and the uro-hyal. The os entoglossum is shown by development to be formed by the union of paired structures and is probably homologous with the hyoid arch of fishes. The basi-hyal and the long cornua, each of which is composed of two or three pieces placed end to end, are homologous with the first branchial arch of fishes, while the uro-hyal is probably homologous with the second branchial arch of fishes. In Woodpeckers the cornua are enormously long, and curve over the skull, extending as far forwards as the anterior nares.

Ribs and Sternum.

Well-developed ribs are attached to the posterior cervical vertebrae as well as to the thoracic vertebrae. The ribs generally have uncinate processes and separate capitula and tubercula, but uncinate processes are absent in *Chauna Palamedea* and apparently in *Archaeopteryx*.

The sternum (fig. 63) is greatly developed in all birds. In the embryo it is seen to be derived from the union of right and left plates of cartilage, formed by the fusion of the ventral ends of the ribs. In the Ratitae and a few Carinatae, such as *Stringops*, it is flat, but in the great majority of birds it is keeled, though the development of the keel varies greatly. It is large in the flightless Penguins, which use their wings for swimming. Traces of an interclavicle may occur in the embryo.

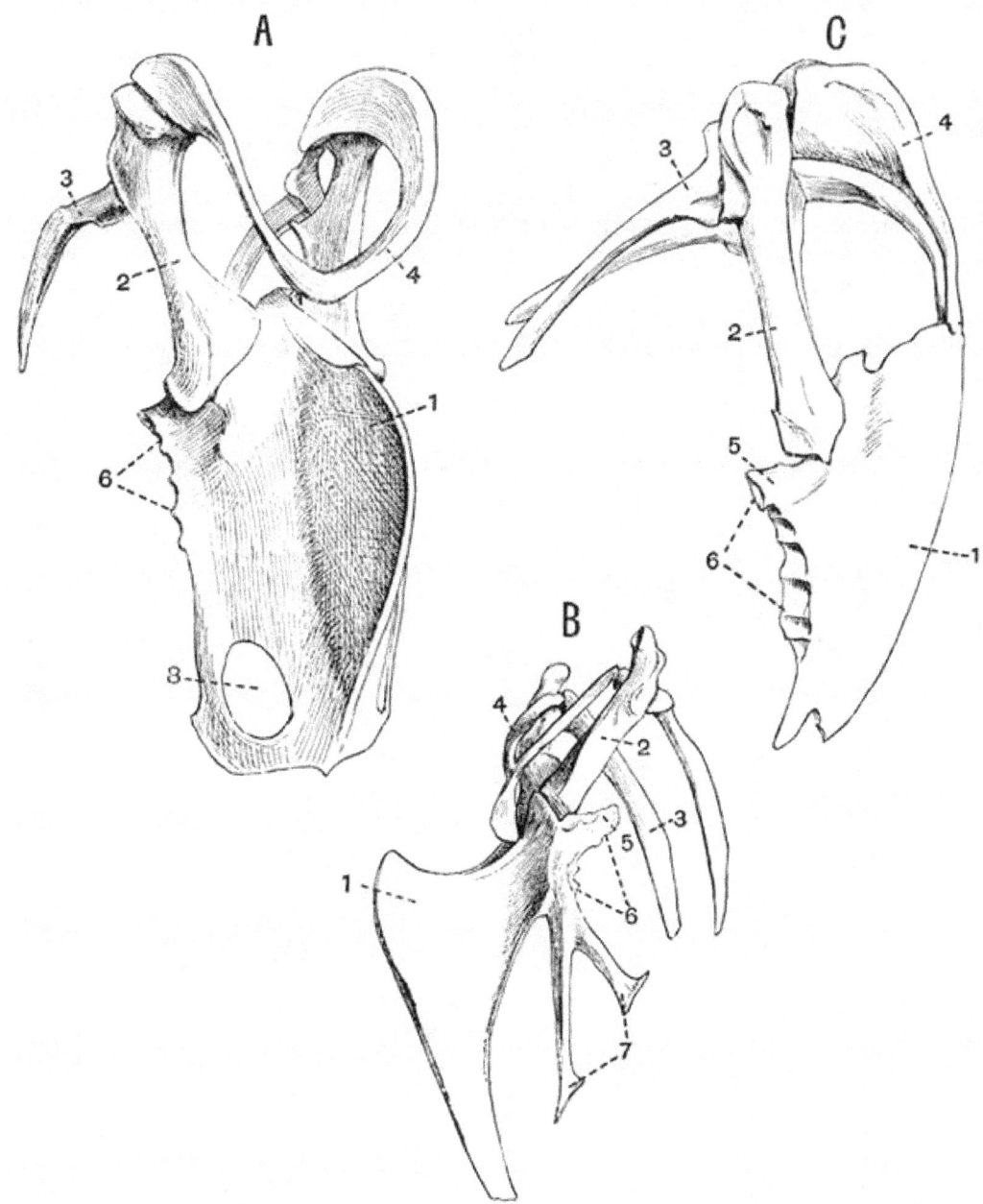

Fig. 63. Shoulder-girdle and sternum of

A. Black Vulture (*Vultur cinereus*) × 1/3.

B. Peacock (*Pavo cristatus*) × 3/8.

C. Pelican (*Pelicanus conspicillatus*) × 1/3. (All Camb. Mus.)

1. carina of the sternum.

2. coracoid.

3. scapula.

4. clavicle.

5. costal process.

6. surfaces for articulation with the sternal ribs.

7. xiphoid processes.

8. fontanelle.

Pectoral Girdle.

The pectoral girdle is also strongly developed in all Carinatae, but is much reduced in Ratitae. In some Moas the sternum has no facet for the articulation of the coracoid, and the pectoral girdle appears to have been entirely absent; it is extremely small also in *Apteryx*. Clavicles are generally well developed in the Carinatae, and small ones are found also in *Hesperornis*, and in Emeus and Cassowaries. In the other living Ratitae and in *Stringops* they are absent. In some Parrots, Owls and Toucans they do not meet one another ventrally. Clavicles are especially stout in some of the birds of prey. They do not generally touch the sternum, but sometimes, as in the Pelican (fig. 63, C), Adjutant and Frigate bird, they are fused with it.

In all Ratitae the scapula and coracoid lie almost in the same straight line with one another, in the Carinatae they are nearly at right angles to one another.

Anterior Limb.

In the wing of nearly all birds the ulna is thicker than the radius, but in *Archaeopteryx* the two bones are equal in size. In the wing of *Archaeopteryx* there are three long digits with distinct metacarpals. In all other birds the digits are modified, the metacarpals being commonly fused and the phalanges reduced in number. In *Palamedea* and some other birds the metacarpus bears a bony outgrowth, which when sheathed in horn forms a spur.

In most of the Ratitae and in the extinct Dodo (*Didus*) and Solitaire (*Pezophaps*) the wing is very small, but the usual parts are recognisable. In *Hesperornis* apparently only the humerus is present; in some Moas, in which the wing is imperfectly known, the presence of the humerus is indicated by traces of a glenoid cavity. In most Moas the wing is apparently completely absent. As compared with those in other Ratitae, the wings of the Ostrich and Rhea are well developed. In the Ostrich (fig. 64, B) and Rhea, as in nearly all Carinatae, the manus has three digits, but in *Apteryx* there is only a single digit, the second. The Penguins (fig. 64, A) too among Carinatae have only two digits, but in their case it is the pollex which is missing. In the Ostrich the third digit has two phalanges, in all other living birds it has only one phalanx.

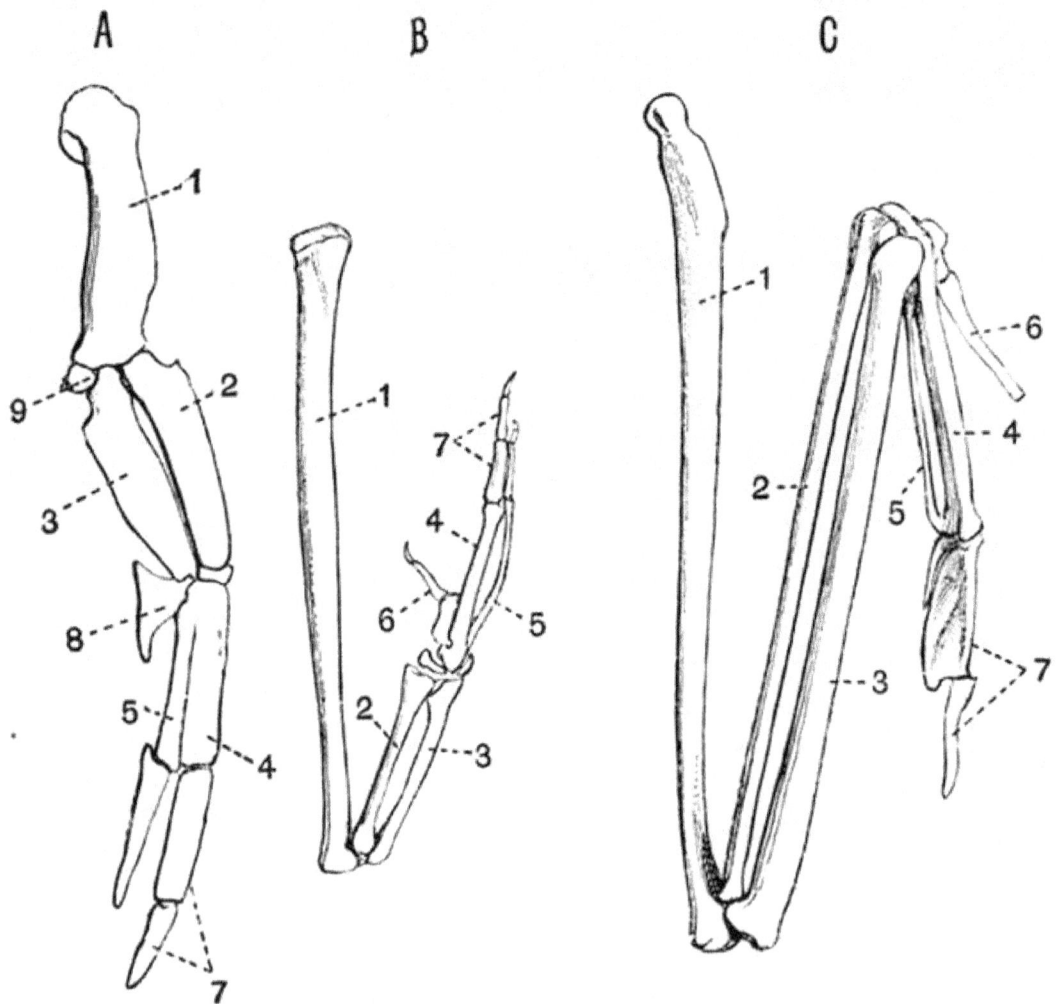

Fig. 64. Bones of the right wing of

A. A Penguin × 1/3. (Camb. Mus.)

B. Ostrich (*Struthio camelus*) × 1/7. (Partly after Parker.)

C. Gannet (*Sula alba*) × 1/3. (Camb. Mus.)

In C the distal phalanges of the pollex and second digit have been omitted.

168

1. humerus.

2. radius.

3. ulna.

4. second metacarpal.

5. third metacarpal.

6. pollex.

7. second digit.

8. cuneiform.

9. sesamoid bone.

Pelvic Girdle.

Birds have a very large pelvis and its characters are constant throughout almost the whole group. The ilium is very large, and is united along its whole length with the sacral and pseudosacral vertebrae. The ischium is broad and extends back parallel to the ilium with which in most birds it fuses posteriorly, further forward the ilio-sciatic foramen separates the two bones. In *Tinamus*, *Hesperornis*, *Apteryx* (fig. 65, B, 2), and *Struthio*, the ischia are separate from the ilia along their whole length except at the acetabulum; in *Phororhacos*, on the other hand, the two bones are fused along almost their whole length. The bone usually called the pubis in birds corresponds to the post-pubis of Dinosaurs and forms a long slender rod (fig. 65, 3) lying parallel to the ischium. In many birds the ischia and pubes are united at their distal ends. This is the case in the Ostrich (fig. 65, D), in which the ilia and ischia are widely separated. In many birds the pubis is drawn out in front into the pectineal process, this is specially large in *Apteryx* (fig. 65, B, 5), and in the embryos of many birds. It is probably homologous with the pre-pubis of Dinosaurs but in some birds is formed in part by the ilium. The acetabulum in birds is always perforate.

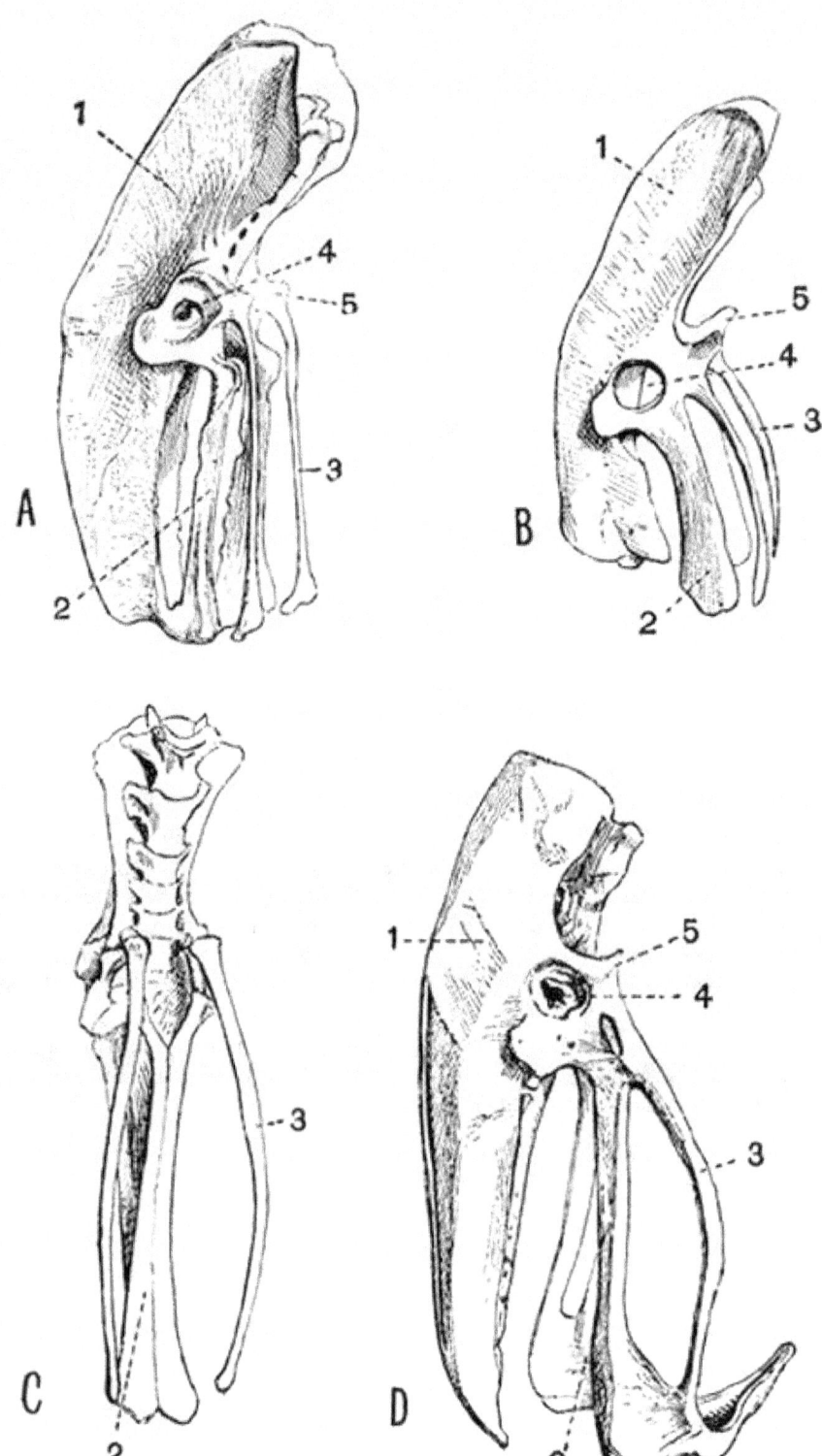

Fig. 65. Pelvic girdle and sacrum of
A. Cassowary (*Casuarius galeatus*) × 1/8.
B. Owen's Apteryx (*A. oweni*) × ½.
C. Broad billed Rhea (*R. macrorhyncha*) × 1/6.
D. Ostrich (*Struthio camelus*) × 1/10. (All Camb. Mus.)

1. ilium.   4. acetabulum.

2. ischium.   5. pectineal process.

3. pubis.

In *Rhea* (fig. 65, C, 2) and probably in *Archaeopteryx* a symphysis ischii occurs, and in the ostrich alone among birds there is a symphysis pubis. In *Archaeopteryx* all three bones of the pelvis are distinct, but they are imperfectly known. In Ichthyornis they are also distinct, in all other known birds they are fused together to a greater or less extent.

Posterior Limb.

The tibia is always well developed and has a very strong cnemial crest. The proximal tarsals are fused with its distal end, the whole forming a compound bone, the tibio-tarsus. There is frequently an oblique bar of bone crossing the anterior face of the tibio-tarsus at the distal end, just above the articular surface of the tarso-metatarsus, this is absent in Ostriches and *Æpyornis*. The fibula though in the embryo and in *Archaeopteryx* equal in length to the tibia, is in the adult of other birds always imperfect, its proximal end is often fused with the tibia, and its distal end is commonly atrophied. In the Penguins however the distal end is complete. The distal tarsals fuse with the second, third and fourth metatarsals, forming a compound bone, the tarso-metatarsus. The first metatarsal is nearly always free but occasionally as in *Phaëthon* it is fused with the others. No adult bird has more than four digits in the pes. In the Penguins the metatarsals are separate, and in many birds larger or smaller gaps exist between the fused metatarsals. In most birds the third metatarsal is curved so as not to lie in the same plane as the others, but in the Penguins they all three lie in the same plane. The metatarsals are clearly separated in *Archaeopteryx*. In Gallinaceous birds the tarso-metatarsus bears a bony outgrowth which is sheathed in horn and forms a spur.

In most birds the first four toes are present while the fifth is always absent. The first toe commonly has two phalanges, the second three, the third four, and the fourth five. In Swifts the third and fourth toes have only three phalanges. Many birds, such as all Ratitae except *Apteryx*, have only three toes, the hallux being absent; in the Ostrich the second toe is also gone with the exception of a small metatarsal, so that the foot retains only the third and fourth digits, the third being much the larger of the two and bearing a claw, while the fourth is clawless.

In the Swifts, Cormorants, and Penguins, all four toes are directed forwards. In most birds the hallux is directed backwards, and the other toes forwards. In the Owls the fourth toe can be directed backwards as well as the hallux, while in Parrots, Cuckoos, Woodpeckers, and Toucans the fourth toe is permanently reversed. In Trogons the second toe is reversed in addition to the hallux, but not the fourth.

CHAPTER XX.
CLASS MAMMALIA.

The skeleton of the members of this class, the highest of the vertebrata, has the following characteristics:—

Some part of the integument at some period of life is always provided with hairs; these are epidermal structures arising from short papillae of the Malpighian layer of the epidermis, which at once grow inwards and become imbedded in pits of the dermis. Sometimes scales or spines occur, and epidermal exoskeletal structures in the form of hoofs, nails, claws and horns are also characteristic. As regards the endoskeleton, the vertebral centra have terminal epiphyses except in the Ornithodelphia and some Sirenia. In the skull the cranial region is greatly developed as compared with that in lower vertebrates, and whereas in many reptiles the true cranium is largely concealed by a false roof, in mammals the only relic of this secondary roof is found in the zygomatic arch, and postorbital bar. In the adult all the bones except the mandible, hyoid, and auditory ossicles are firmly united together. The basisphenoid is well ossified, and there is no parasphenoid. The pro-otic ossifies, and unites with the epi-otic and opisthotic before they coalesce with any other bones.

The skull articulates with the vertebral column by means of two convex occipital condyles formed mainly by the exoccipitals, and the mandible articulates with the squamosal without the intervention of the quadrate. The latter is much reduced, and is converted into the tympanic ring, while the hyomandibular of fish is represented by the auditory ossicles.

The teeth are always attached to the maxillae, premaxillae and mandibles, never to any of the other bones. They are nearly always implanted in distinct sockets, and are hardly ever ankylosed to the bone. The teeth of mammals are generally markedly heterodont, four forms, incisors, canines, premolars, and molars, being commonly distinguishable. Some mammals are *monophyodont*, having only a single set of teeth, but the great majority are *diphyodont*, having two sets, a deciduous or milk dentition, and a permanent dentition.

The *incisors*, the front teeth, are simple, one-rooted, adapted for cutting, and are nearly always borne by the premaxillae. Next come the *canines*, one on each side in each jaw. They are generally large teeth adapted for tearing or holding, and get

171

their name from the fact that they are largely developed in the dog. The remaining teeth form the grinding series, the more posterior of them being the *molars*, which are not preceded by milk teeth. Between the molars and the canines are the *premolars*, which do as a rule have milk or deciduous predecessors, though very frequently the first of them is without a milk predecessor.

In describing the dentition of any mammal, for the sake of brevity a formula is generally made use of. Thus, the typical mammalian dentition is expressed by the formula

$i\ ^3/_3\ c\ ^1/_1\ pm\ ^4/_4\ m\ ^3/_3 = {}^{11}/_{11}$,

giving twenty-two teeth on each side, or forty-four altogether. The incisors are represented by *i*, the canines by *c*, the premolars by *p* or *pm*, and the molars by *m*. The numbers above the lines represent the teeth in the upper jaw, those below the lines the teeth in the lower jaw. The milk dentition is expressed by a similar formula with *d* (deciduous) prefixed to the letter expressing the nature of the tooth.

The following terms are of frequent use as characterising certain forms of the grinding surfaces of teeth, and it will be well to define them at once.

Bunodont is a term applied to teeth with broad crowns raised into rounded tubercles, e.g. the grinding teeth of Pigs and Hippopotami;

Bilophodont to teeth marked by a simple pair of transverse ridges, with or without a third ridge running along the outer border of the tooth at right angles to the other two, e.g. the grinding teeth of *Lophiodon*, Kangaroo, Manatee, Tapir, *Dinotherium*;

Selenodont to teeth marked by crescentic ridges running from the anterior towards the posterior end of the tooth, e.g. the grinding teeth of the Ox and Sheep.

Teeth whose crowns are low so that their whole structure is visible from the grinding surface are called *brachydont*, while those with higher crowns, in which the bases of the infoldings of enamel are invisible from the grinding surface are said to be *hypsodont*. Bunodont teeth are brachydont, the teeth of the Horse and Ox are hypsodont.

Passing now to the appendicular skeleton—the shoulder girdle differs markedly from that of Sauropsids in the fact that the coracoid, except in the Ornithodelphia, is greatly reduced, generally forming only a small process on the scapula. In the pelvis the pubes meet in a ventral symphysis, except in some Insectivora and Chiroptera. In many mammals a fourth pelvic element, the *acetabular bone*, is distinguishable. The ankle joint is *cruro-tarsal*, or situated between the proximal tarsal bones and the tibia and fibula. Carpalia 4 and 5 are united forming the *unciform*; and the ulnar sesamoid bone or *pisiform* is generally well developed. In the proximal row of tarsal elements there are only two bones, the calcaneum and astragalus. Of these the calcaneum is the fibulare, and the astragalus is generally regarded as the tibiale and intermedium fused.

Subclass I. Ornithodelphia or Prototheria.

This subclass contains only a single order, the Monotremata, and the following characteristics are equally applicable to the subclass and to the order. The vertebral centra have no epiphyses, and the odontoid process remains for a long time free from the centrum of the second vertebra. With the exception of the atlas of *Echidna* the cervical vertebrae are without zygapophyses. The cranial walls are smooth and rounded, and the sutures between the several bones early become completely obliterated as in birds. The mandible is a very slight structure, with no ascending ramus, and with the coronoid process (see p. 398) and angle rudimentary. The auditory ossicles show a low state of development. The tubercula of the ribs articulate with the sides of the centra of the thoracic vertebrae, not with the transverse processes. Some of the cervical ribs remain for a long time separate from the vertebrae. Well ossified sternal ribs occur. No true teeth are present in the adult. The young *Ornithorhynchus* has functional molar teeth, but in the adult their place is taken by horny plates. In the Echidnidae neither teeth nor horny plates occur.

The coracoid (fig. 66, 3) is complete and well developed, and articulates with the sternum. A precoracoid (epicoracoid) occurs in front of the coracoid, and there is a large interclavicle (fig. 66, 6). The ridge on the scapula, corresponding to the spine of other mammals, is situated on the anterior border instead of in the middle of the outer surface. Epipubic bones are present. In the Echidnidae, but not in *Ornithorhynchus*, the central portion of the acetabulum is unossified as in birds. The humerus has a prominent deltoid crest; its ends are much expanded, and the distal end is pierced by an ent-epicondylar foramen. The fibula has a broad proximal process resembling an olecranon. The limbs and their girdles bear a striking resemblance to those of some Theromorphous reptiles.

172

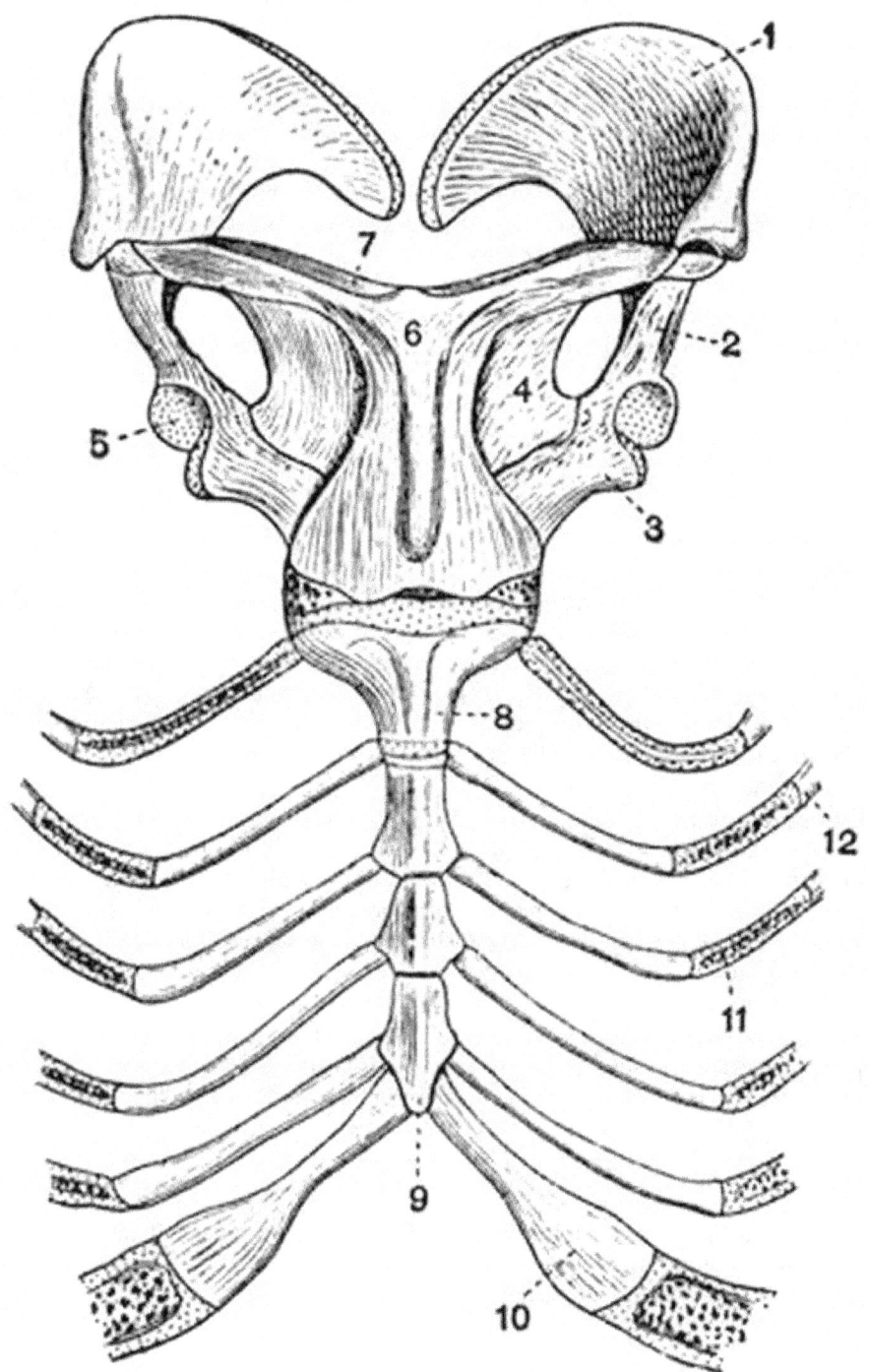

Fig. 66. Ventral view of the shoulder-girdle and sternum of a Duckbill (*Ornithorhynchus paradoxus*) × ¾ (after Parker).

1 and 2. scapula.

3. coracoid.

4. precoracoid (epicoracoid).

5. glenoid cavity.

6. interclavicle.

7. clavicle.

8. presternum.

9. third segment of

mesosternum.

10. sternal rib.

11. intermediate rib.

12. vertebral rib.

The order Monotremata includes only two living families, the Echidnidae and Ornithorhynchidae.

173

Mesozoic Mammalia.

It will be well here to briefly refer to certain mammals of small size, the remains of which have been found in deposits of Mesozoic age. In the great majority of cases they are known only by the lower jaw, or sometimes only by isolated teeth. A large number of them are commonly grouped together as the Multituberculata, and are sometimes, partly owing to the resemblance of their teeth to those of *Ornithorhynchus*, placed with the Prototheria, sometimes between the Prototheria and the Metatheria. They are characterised by having a single pair of large incisors in the lower jaw, and one large with one or two smaller incisors in each premaxillae. The lower canines are very small or altogether wanting. The incisors are separated by a diastema from the grinding teeth, which are sometimes (*Tritylodon*) characterised by the possession of longitudinal rows of little tubercles separated by grooves, sometimes by having the premolars provided with high cutting edges, whose surfaces are obliquely grooved. Some of the Mesozoic mammals found associated with the Multituberculata, have however a dentition of an altogether different type, with at least three lower incisors, well developed canines and premolars, and numerous molars with peculiar three-cusped or tritubercular grinding surfaces. These mammals, one of the best known of which is *Phascolotherium*, are commonly separated from the Multituberculata, and are divided by Osborn into two groups, one allied to the Marsupials, and one to the Insectivores. The group showing Marsupial affinities is further subdivided into carnivorous, omnivorous, and herbivorous subgroups. The members of both groups commonly have four premolars, and six to eight molars in each mandibular ramus.

Subclass II. Didelphia or Metatheria.

This subclass, like the previous one, contains only a single order, viz. the Marsupialia; but the forms referable to it are far more numerous than in the case of the Monotremata.

The integument is always furry, and the teeth are always differentiated into incisors, canines, premolars and molars. Except in *Phascolomys*, the number of incisors in the upper and lower jaws is never equal, and the number in the upper usually exceeds that in the lower jaw. There is no such regular succession and displacement of teeth as in most mammals. Sometimes the anterior teeth are diphyodont, and as a general rule the tooth commonly regarded as the last premolar is preceded by a milk tooth. The majority of the permanent teeth of most Marsupials are regarded as belonging to the milk series for two reasons, (1) they are developed from the more superficial tissues of the jaws, (2) a second set, the permanent teeth, begin to develop as outgrowths from them, but afterwards become aborted.

The odontoid process at an early stage becomes fused with the centrum of the second cervical vertebra, and the number of thoraco-lumbar vertebrae is always nineteen. The skull has several characteristic features. The tympanic bone remains permanently distinct, and the anterior boundary of the tympanic cavity is formed by the alisphenoid. The carotid canal perforates the basisphenoid, and the lachrymal canal opens either outside the orbit or at its margin. There are generally large vacuities in the palate. The angle of the mandible is (except in *Tarsipes*) more or less inflected; and as a rule the jugal furnishes part of the articular surface for the mandible. There is no precoracoid (epicoracoid) or interclavicle, and the coracoid is reduced to form a mere process of the scapula, not coming near the sternum.

Epipubic, or so-called marsupial bones, nearly always occur, and a fourth pelvic element, the acetabular bone, is frequently developed. The fibula is always complete at its distal end, sometimes it is fused with the tibia, but often it is not only free but is capable of a rotatory movement on the tibia. This is the case in the families Phascolomyidae, Didelphyidae, and Phalangeridae.

The Marsupialia can be subdivided into two main groups, according to the character of the teeth:—

1. Polyprotodontia.

In this group the incisors are small, subequal and numerous, not less than 4/3. The canines are larger than the incisors, and the molars have sharp cusps. The members of this group are all more or less carnivorous or insectivorous. The group includes the families Didelphyidae, Dasyuridae, Peramelidae, and Notoryctidae.

2. Diprotodontia.

In this group the incisors do not exceed 3/3, and are usually 3/1, occasionally 1/1. The first upper and lower incisors are large and cutting. The lower canines are always small or absent, and so in most cases are the upper canines. The molars have bluntly tuberculated, or transversely ridged crowns. The group includes the families Phascolomyidae, Phalangeridae, Macropodidae, and Epanorthidae.

Subclass III. Monodelphia or Eutheria.

This great group includes all the Mammalia except the orders Monotremata and Marsupialia. Coming to their general characteristics—as in the Didelphia the odontoid process and cervical ribs early become fused with the centra which bear them, while the coracoid is reduced so as to form a mere process on the scapula, and there is no precoracoid (epicoracoid), such as is found in Ornithodelphia. Clavicles may be present or absent; when fully developed they articulate with the sternum, usually directly, but occasionally, as in some Rodents and Insectivores, through the remains of the sternal end of the precoracoid. There is never any interclavicle in the adult, though sometimes traces of it occur during development. In the pelvis the acetabula are imperforate; and well-developed epipubic bones are never found in the adult, though traces of them occur in some Carnivores and foetal Ungulates.

Order 1. Edentata.

Teeth are not, as the name of the order seems to imply, always wanting; and sometimes they are very numerous. They are, however, always imperfect, and, with very few exceptions, are homodont and monophyodont. They have persistent pulps, and so grow indefinitely and are never rooted. In all living forms they are without enamel, consisting merely of dentine and cement, and are never found in the front part of the mouth in the situation occupied by the incisors of other mammals. These characters derived from the teeth are the only ones common to the various members of the order, which includes the living sloths, ant-eaters, armadillos, pangolins and aard varks, together with various extinct forms, chiefly found in beds of late tertiary age in both North and South America, the best known being the Megatheridae and Glyptodonts.

Order 2. Sirenia.

The skeleton of these animals has a general fish-like form, in correlation with their purely aquatic habits. The fore limbs have the form of paddles, but the number of phalanges is not increased beyond the normal. There are no external traces of hind limbs.

The whole skeleton and especially the skull and ribs is remarkably massive and heavy. The dentition varies; in the two living genera *Manatus* and *Halicore*, incisor and molar teeth are present, in one extinct genus, *Rhytina*, teeth are entirely absent, while in another, *Halitherium*, the dentition is more decidedly heterodont than in living forms. In the two living genera the dentition is monophyodont, but in *Halitherium* the anterior grinding teeth are preceded by milk teeth. The tongue and anterior part of the palate and lower jaw are covered with roughened horny plates. The skull is noticeable for the size and backward position of the anterior nares, also for the absence or small size of the nasal bones. There is no union of certain of the vertebrae to form a sacrum, and in living forms the centra are not terminated by well-formed epiphyses.

The cervical vertebrae are much compressed, but they are never ankylosed together. In *Manatus* there are only six cervical vertebrae. The caudal vertebrae have well-developed chevron bones. The humerus is distinctly articulated to the radius and ulna, and these two bones are about equally developed, and are often fused together. There are no clavicles, and the pelvis is vestigial, consisting of a pair of somewhat cylindrical bones suspended at some distance from the vertebral column. In living forms there is no trace of a posterior limb, but in *Halitherium* there is a vestigial femur connected with each half of the pelvis.

Order 3. Cetacea.

In these mammals the general form is more fish-like than is the case even in the Sirenia. The skin is generally almost completely naked, but hairs are sometimes present in the neighbourhood of the mouth, especially in the foetus. In some Odontoceti vestiges of dermal ossicles have been described, and in *Zeuglodon* the back was probably protected by dermal plates. The anterior limbs have the form of flattened paddles, showing no trace of nails, the posterior limb bones are quite vestigial or absent, and there is never any external sign of the limb. Teeth are always present at some period of the life history, but in the whalebone whales they are only present during foetal life, their place in the adult animal being taken by horny plates of baleen. In all living forms the teeth are simple and uniform structures without enamel; they have single roots, and the alveoli in which they are imbedded are often incompletely separated from one another. As in some forms traces of a replacing dentition have been described, it has been concluded that the functional teeth of Cetacea belong to the milk dentition.

The texture of the bones is spongy. The cervical vertebrae are very short, and though originally seven in number, are in many forms completely fused, forming one solid mass (fig. 67). The odontoid process of the axis is short and blunt, or may be completely wanting. The lumbar and caudal vertebrae are large and numerous, and as zygapophyses are absent, are very freely movable on one another; zygapophyses are also absent from the posterior thoracic vertebrae. The lumbar vertebrae are sometimes more numerous than the thoracic. The epiphyses are very distinct, and do not unite with the centra till the animal is quite adult. None of the vertebrae are united to form a sacrum, but the caudal vertebrae have large chevron bones.

175

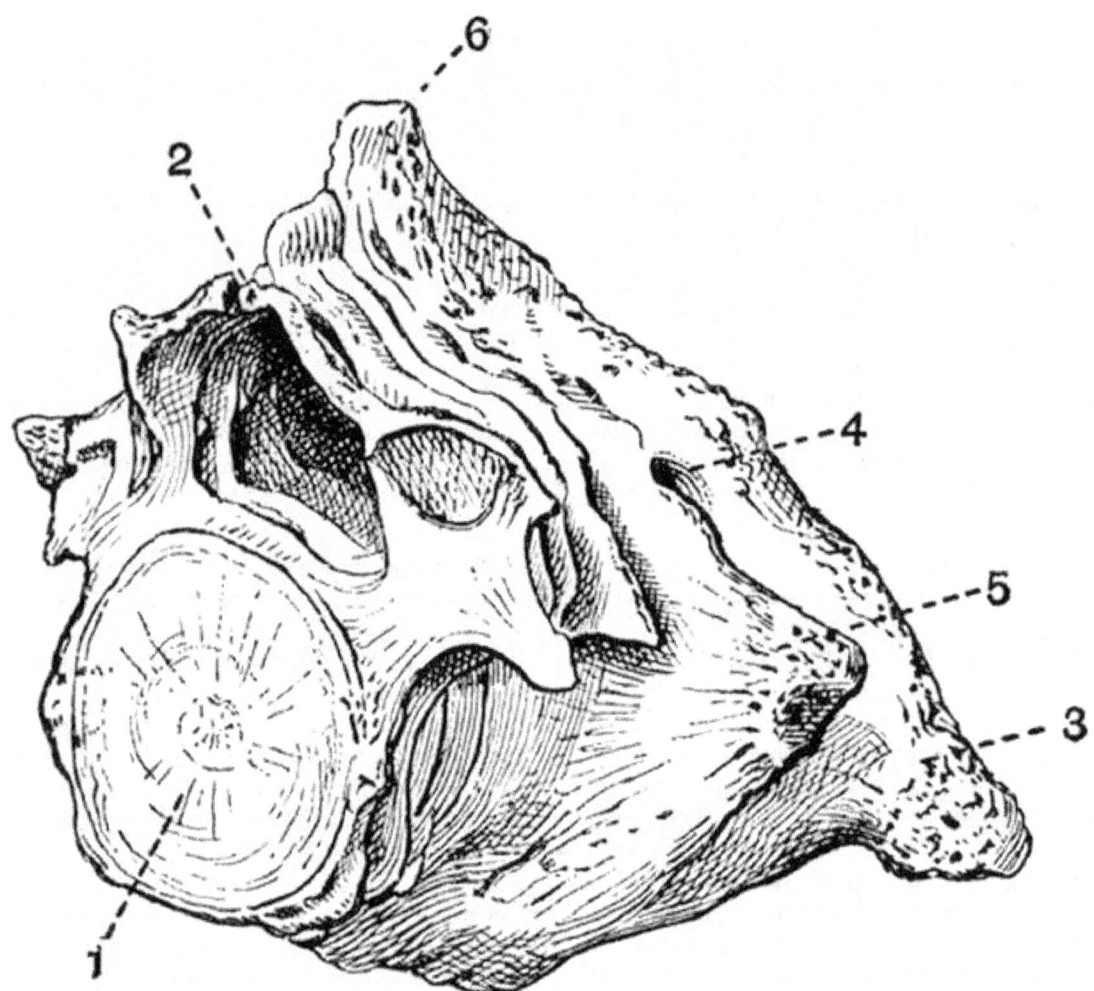

Fig. 67. Cervical vertebrae of a Ca'ing Whale (*Globicephalus melas*) × ¼. (Camb. Mus.)

1. centrum of seventh cervical vertebra.

2. neural arch of seventh cervical vertebra.

3. transverse process of atlas.

4. foramen for exit of first spinal nerve.

5. transverse process of axis.

6. fused neural spines of atlas and axis.

The skull is peculiarly modified; the bones forming the occipital segment show a specially strong development, and the cranial cavity is short, high, and almost spherical. The supra-occipital is very large and rises up to meet the frontals, thus with the interparietal completely separating the parietals from one another.

The frontals are expanded, forming large bony plates, which roof over the orbits. The zygomatic process of the squamosal is extremely large and extends forwards to meet the supra-orbital process of the frontal; the zygomatic process of the jugal is on the contrary very slender. The face is drawn out into a long rostrum, formed of the maxillae and premaxillae surrounding the vomer and the mesethmoid cartilage. The maxillae are specially large, and extend backwards so as to partially overlap the frontals. The nasals are always small, and the anterior nares open upwards between the cranium and rostrum. The periotics are loosely connected with the other bones of the skull and the tympanics are commonly large and dense. The mandible has hardly any coronoid process, and the condyles are at its posterior end.

There are no clavicles, but the scapula and humerus are well developed. The humerus moves freely in the glenoid cavity, but all the other articulations of the anterior limb are imperfect; the various bones have flattened ends, and are connected with one another by fibrous tissue, which allows of hardly any movement. Frequently the carpus is imperfectly ossified.

The number of digits in the manus is generally five, sometimes four, and when there are four digits it is the third and not the first that is suppressed. The number of phalanges in the second and third digits almost always exceeds that which is normal in mammals, and the phalanges are also remarkable for having epiphyses at both ends. The pelvis is represented by two small bones which lie suspended horizontally at some distance below the vertebral column; in some cases vestiges of the skeleton of the hind limb are attached to them.

The Cetacea are divided into three suborders.

Suborder (1). Archaeoceti.

The members of this group are extinct; they differ from all living Cetacea in having the dentition heterodont and in the fact that the back was probably protected by dermal plates. The skull is elongated and depressed, and the brain cavity is very small. The temporal fossae are large, and there is a strong sagittal crest. The nasals and premaxillae are a good deal larger than they are in living Cetacea, and the anterior nares are usually far forward. The cervical vertebrae are not fused with one another, and the lumbar vertebrae are unusually elongated.

The limbs are very imperfectly known, but while the humerus is much longer than in modern Cetaceans, it is nevertheless flattened distally, indicating that the limb was paddle-like, and that there was scarcely any free movement between the fore-arm and upper arm.

The best known genus is *Zeuglodon*, which is found in beds of Eocene age in various parts of Europe, and in Alabama.

Suborder (2). Mystacoceti or Balaenoidea.

These are the Whalebone Whales or True Whales.

Calcified teeth representing the milk dentition occur in the foetus, but the teeth are never functional, and always disappear before the close of foetal life. There is a definite though small olfactory fossa. The palate is provided with plates of baleen or whalebone. The skull is symmetrical, and is extremely large in proportion to the body. The nasals are moderately well developed, and the maxillae do not overlap the orbital processes of the frontals. The lachrymals are small and distinct from the jugals. The tympanics are ankylosed to the periotics, and the rami of the mandible do not meet in a true symphysis. The ribs articulate only with the transverse processes, and the capitula are absent or imperfectly developed. Only one pair of ribs meets the sternum, which is composed of a single piece.

The group includes among others the Right whale (*Balaena*), Humpbacked whale (*Megaptera*), and Rorqual (*Balaenoptera*).

Suborder (3). Odontoceti.

Teeth always exist after birth and baleen is never present. The teeth are generally numerous, but are sometimes few and deciduous; the dentition is homodont (except in *Squalodon*). The dorsal surface of the skull is somewhat asymmetrical, there is no trace of an olfactory fossa, the nasals are quite rudimentary, and the hind ends of the maxillae cover part of the frontals; in all these respects the skull differs from that of the Mystacoceti. The lachrymal may either be united to the jugal or may be large and distinct. The tympanic is not ankylosed to the periotic. The rami of the mandible are nearly straight and become united in a long symphysis. Some of the ribs have well developed capitula articulating with the vertebral centra. The sternum is almost always composed of several pieces as in other mammals, and several pairs of ribs are connected with it. There are always five digits to the manus, though the first and fifth are usually very little developed.

The suborder includes the Sperm Whale (*Physeter*), Narwhal (*Monodon*), Dolphin (*Delphinus*), Porpoise (*Phocoena*), and many other living forms as well as the extinct *Squalodon* which differs from the other members of the suborder in its heterodont dentition.

Order 4. Ungulata.

This order includes a great and somewhat heterogeneous group of animals, a large proportion of which are extinct. They all (except certain extinct forms) agree in having the ends of the digits either encased in hoofs or provided with broad flat nails. The teeth are markedly heterodont and diphyodont, and the molars have broad crowns with tuberculated or ridged surfaces. Clavicles are never present in the adult except in a few generalised extinct forms such as *Typotherium*, and it is only recently that vestigial clavicles have been discovered in the embryo. The scaphoid and lunar are always distinct.

The order Ungulata may be subdivided into two main groups, Ungulata vera and Subungulata.

Section I. Ungulata vera.

The cervical vertebrae except the atlas are generally opisthocoelous. The feet are never *plantigrade*. In all the living and the great majority of the extinct forms the digits do not exceed four, the first being suppressed. In the carpus the os magnum articulates freely with the scaphoid, and is separated from the cuneiform by the lunar and unciform. In the tarsus the cuboid articulates with the astragalus as well as with the calcaneum, and the proximal surface of the astragalus is marked by a pulley-like groove. All the bones of the carpus and tarsus strongly interlock. These characters with regard to the carpus and tarsus do not hold in *Macrauchenia* and its allies. The humerus never has an ent-epicondylar foramen.

The group is divided into two very distinct suborders:—

Suborder (1). Artiodactyla.

177

The Artiodactyla have a number of well marked characters, one of the most obvious being the fact that many of the most characteristic forms have large paired outgrowths on the frontal bones. These may be (1) solid deciduous bony *antlers*, or (2) more or less hollow bony outgrowths which are sheathed with permanently growing horn.

The premolar and molar teeth are usually dissimilar, the premolars being one-lobed and the molars two-lobed; the last lower molar of both the milk and permanent dentitions is almost always three-lobed.

The grinding surfaces of the molar teeth have a tendency to assume one of two forms. In the Pigs and their allies the crowns are bunodont, while in the more highly specialised Ruminants the crowns are selenodont. The nasals are not expanded posteriorly, and there is no alisphenoid canal. The thoraco-lumbar vertebrae are always nineteen. The symphysis of the ischia and pubes is very elongated, and the femur has no third trochanter. The limbs never have more than four digits, and are symmetrical about a line drawn between the third and fourth digits; the digits, on the other hand, are never symmetrical in themselves. The astragalus has pulley-like surfaces both proximally and distally, and articulates with the navicular and cuboid by two nearly equal facets. The calcaneum articulates with the lower end of the fibula when that bone is fully developed.

In the Artiodactyla are included the following living groups:—

a. Suina. Pigs and Hippopotami.

b. Tylopoda. Camels and Llamas.

c. Tragulina. Chevrotains.

d. Ruminantia or Pecora. Deer, giraffes, oxen, sheep and antelopes.

Suborder (2). Perissodactyla.

In this group there are never any bony outgrowths from the frontals. The grinding teeth form a continuous series, the posterior premolars resembling the molars in complexity, and the last lower molar generally has no third lobe. The cervical vertebrae with the exception of the atlas almost always have markedly opisthocoelous centra, but in *Macrauchenia* they are flat. The nasals are expanded posteriorly, and an alisphenoid canal is present. The thoraco-lumbar vertebrae are never less than twenty-two in number and are usually twenty-three. The femur has a third trochanter (except in *Chalicotherium*). The third digit of the manus and pes is symmetrical in itself, and larger than the others, and in some cases the other digits are quite vestigial. The number of the digits of the pes is always odd. The astragalus is abruptly truncated distally, and the facet by which it articulates with the cuboid, is much smaller than that by which it articulates with the navicular. The calcaneum does not articulate with the fibula, except in *Macrauchenia*. The group includes many extinct forms, and the living families of the Tapirs, Horses and Asses, and Rhinoceroses.

Section II. Subungulata.

In this group is placed a heterogeneous collection of animals, the great majority of which are extinct. There is really no characteristic which is common to them all, and which serves to distinguish them as a group from the Ungulata vera. But the most distinctive character common to the greatest number of them is to be found in the carpus, whose bones in most cases retain their primitive relation to one another, the os magnum articulating with the lunar and sometimes just meeting the cuneiform, but in living forms at any rate not articulating with the scaphoid. The feet frequently have five functional digits, and may be plantigrade. The proximal surface of the astragalus is generally flattened instead of being pulley-like as in Ungulata vera.

Suborder (1). Toxodontia.

This suborder includes some very aberrant extinct South American ungulates, which have characters recalling the Proboscidea, both groups of Ungulata vera, and the Rodentia. The limbs are subplantigrade or digitigrade, and the digits are three, rarely five, in number, the third being most developed. The carpus resembles that of the Ungulata vera, in that the bones interlock and the magnum articulates with the scaphoid. In the tarsus, however, the bones do not interlock. The astragalus has a pulley-like proximal surface (except in *Astrapotherium*, in which it is flat), and articulates only with the navicular, not meeting the cuboid. The calcaneum has a large facet for articulation with the fibula, as in Artiodactyla. There is no alisphenoid canal, and the orbit is confluent with the temporal fossa. Some of the forms (e.g. *Nesodon*) referred to this group have the typical mammalian series of forty-four teeth, but in others the canines are undeveloped. In *Toxodon* all the cheek teeth have persistent pulps, while in *Nesodon* and *Astrapotherium* they are rooted. A clavicle is sometimes present (*Typotherium*), and the femur sometimes has a third trochanter (*Typotherium* and *Astrapotherium*), sometimes is without one (*Toxodon*).

The remains of these curious Ungulates have been found in beds of late Tertiary age in South America.

Suborder (2). Condylarthra.

This group includes some comparatively small extinct ungulates, which are best known from the Lower Eocene of Wyoming, though their remains have also been found in deposits of similar age in France and Switzerland. Their characters are little specialised, and they show relationship on the one hand to the Ungulata vera and on the other to the Hyracoidea. They also have characters allying them to the Carnivora. They generally have the typical mammalian series of forty-four teeth, the molars being brachydont and generally bunodont. The premolars are more simple than the molars. The limbs are plantigrade, and have five digits with rather pointed ungual phalanges. The os magnum, as in living Subungulates, articulates with the lunar, not reaching the scaphoid. The astragalus has an elongated neck, a pulley-like proximal and a convex distal articular surface, and does not articulate with the cuboid. The humerus has an ent-epicondylar foramen, and the femur has a third trochanter. The best known genus is *Phenacodus*; it is perhaps the most primitive ungulate whose skeleton is thoroughly well known, and is of special interest from the fact that it is regarded as the lowest stage in the evolutionary series of the horse. Its remains are found in the Lower Eocene of Wyoming.

Suborder (3). Hyracoidea.

This group of animals is very isolated, having no very close allies, either living or extinct. The digits are provided with flat nails, except the second digit of the pes, which is clawed. Canine teeth are absent, and the dental formula is usually given as i ½, c $^0/_0$, pm $^4/_4$, m $^3/_3$. The upper incisors are long and curved, and have persistent pulps as in Rodents; their terminations are, however, pointed, not chisel-shaped, as in Rodents. The lower incisors have pectinated edges. The grinding teeth have a pattern much like that in *Rhinoceros*. In the skull (fig. 83) the postorbital processes of the frontal and jugal almost or quite meet. The jugal forms part of the glenoid cavity for articulation with the mandible, and also extends forwards so as to meet the lachrymal. There is an alisphenoid canal. There are as many as twenty-one or twenty-two thoracic vertebrae, and the number of thoraco-lumbar vertebrae reaches twenty-eight or thirty. There are no clavicles, and the scapula has no acromion; the coracoid process is, however, well developed. The ulna is complete. In the manus the second, third and fourth digits are approximately equal in size, the fifth is smaller, and the first is vestigial. The femur has a slight ridge representing the third trochanter. The fibula is complete, but is generally fused with the tibia proximally. There is a complicated articulation between the tibia and astragalus, which has a pulley-like proximal surface. In the pes the three middle digits are well developed, but there is no trace of a hallux, and the fifth digit is represented only by a vestigial metatarsal.

The only representatives of the suborder are some small animals belonging to the genus *Procavia* (*Hyrax*), which is found in Africa and Syria; some of the species are by many authors placed in a distinct genus *Dendrohyrax*.

Suborder (4). Amblypoda.

This suborder includes a number of primitive extinct Ungulates, many of which are of great size. Their most distinguishing characteristics are afforded by the extremities. In the carpus the bones interlock a little more than is the case in most Subungulata, and the corner of the os magnum reaches the scaphoid, while the lunar articulates partially with both magnum and unciform, instead of only with the magnum. In the tarsus the cuboid articulates with both the calcaneum and the astragalus, which is remarkably flat. The manus and pes are short, nearly or quite plantigrade, and have the full number of digits. The cranial cavity is singularly small. Canine teeth are present in both jaws, and the grinding teeth have short crowns, marked by V-shaped ridges. The pelvis is large, the ilia are placed vertically, and the ischia do not take part in the ventral symphysis.

The best known animals belonging to this suborder are the Uintatheriidae (Dinocerata), found in the Upper Eocene of Wyoming. They are as large as elephants, and are characterised by the long narrow skull drawn out into three pairs of rounded protuberances, by the strong occipital crest, and by the very large upper canines.

Suborder (5). Proboscidea.

This suborder includes the largest of land mammals, the Elephants, and certain of their extinct allies. The limbs are strong, and are vertically placed; the proximal segment is the longest, and the manus and pes are pentedactylate and subplantigrade. The digits are all enclosed in a common integument, and each is provided with a broad hoof. The vertebral centra are much flattened and compressed, especially in the cervical region. The number of thoracic vertebrae is very great, reaching twenty. The skull (figs. 96 and 97) is extremely large, this being due to the great development of air cells, which takes place in nearly all the bones of the adult skull. In the young skull there are hardly any air cells, and the growth of the cranial cavity does not by any means keep pace with the growth of the skull in general. The supra-occipital is very large, and forms a considerable part of the roof of the skull. The nasals and jugals are short, and the premaxillae very large. The rami of the mandible meet in a long symphysis, and the ascending portion is very high. Canine teeth are absent, and the incisors have the form of ever-growing tusks composed mainly of dentine; in living forms they are present in the upper jaw only. The grinding teeth are large, and in living forms have a very complex structure and mode of succession. In some of the extinct forms, such as *Mastodon* and especially *Dinotherium*, the teeth are much more simple. In every case the teeth have the same general structure, consisting of a series of ridges of dentine, coated with enamel. In the more specialised forms the valleys between the ridges are filled up with cement. The acromion of the scapula has a recurved process, similar to that often found in rodents. Clavicles are absent. The radius and ulna are not ankylosed, but are incapable of any rotatory movement. All the bones of the extremities are very short and thick; the scaphoid articulates regularly with the trapezoid and the lunar with the magnum. The ilia are vertically placed, and are very much expanded; the ischia and pubes are small, and form a short symphysis. The femur has no third trochanter, and the tibia and fibula are distinct. The fibula articulates with the calcaneum, and the astragalus is very flat.

179

Here brief reference may be made to the Tillodontia, a group of extinct mammals found in the Eocene beds of both Europe and North America. They seem to connect together the Ungulata, Rodentia, and Carnivora.

The skull resembles that of bears, but the grinding teeth are of Ungulate type, while the second incisors resemble those of rodents, and have persistent pulps. The femur has a third trochanter, and the feet resemble those of bears in being plantigrade and having pointed ungual phalanges, differing, however, in having the scaphoid and lunar distinct.

Order 5. Rodentia.

The Rodents form a very large and well-defined group of mammals easily distinguishable by their peculiar dentition. Canines are absent, and the incisors are very large and curved, growing from persistent pulps. They are rectangular in section and are much more thickly coated with enamel on their anterior face than elsewhere; consequently, as they wear down they acquire and retain a chisel-shaped (scalpriform) edge. There is never more than one pair of incisors in the mandible, and except in the Hares and Rabbits, there is similarly only a single pair in the upper jaw. These animals are, too, the only rodents which have well developed deciduous incisors. There is always a long diastema separating the incisors from the grinding teeth. The grinding teeth, which are arranged in a continuous series, vary in number from two to six in the upper jaw, and from two to five in the lower jaw. The number of premolars is always below the normal, often they are altogether wanting, but generally they are $^1/_1$. Sometimes the grinding teeth form roots, sometimes they grow persistently.

The premaxillae are always large, and the orbits always communicate freely with the temporal fossae. The condyle of the mandible is elongated from before backwards, and owing to the absence of a postglenoid process to the squamosal, a backward and forward motion of the jaw can take place. The zygomatic arch is complete, but the jugal is short and only forms the middle of it. The palate is small, being sometimes, as in the hares, narrowed from before backwards, sometimes as in the mole-rats (Bathyerginae) narrowed transversely.

The thoraco-lumbar vertebrae are usually nineteen in number. Clavicles are generally present, and the acromion of the scapula is commonly very long. The feet are as a rule plantigrade, and provided with five clawed digits.

There are two main groups of Rodentia; the Duplicidentata, or Hares and Rabbits, which have two pairs of upper incisors, whose enamel extends round to the posterior surface; and the Simplicidentata, in which there is only a single pair of upper incisors, whose enamel is confined to the anterior surface. This group includes all the Rodents except the Hares and Rabbits.

Order 6. Carnivora.

The living Carnivora form a natural and well-marked group, but as is the case with so many other groups of animals, when their extinct allies are included, it becomes impossible to readily define them.

The manus and pes never have less than four well-developed digits, and these are nearly always provided with more or less pointed nails, generally with definite claws. The hallux and pollex are never opposable. The dentition is diphyodont and markedly heterodont. The teeth are always rooted, except in the case of the canines of the Walrus. The incisors are generally $^3/_3$, and are comparatively small, while the canines are large, pointed, and slightly recurved. The cheek teeth are variable, and are generally more or less compressed and pointed; sometimes their crowns are flattened and tuberculated, but they are never divided into lobes by deep infoldings of enamel. The squamosal is drawn out into a postglenoid process, and the mandible has a large coronoid process. The condyle of the mandible is transversely elongated, and the glenoid fossa is very deep; in consequence of this arrangement the mandible can perform an up and down movement only, any rotatory or back and fore movement being impossible. The jugal is large, and the zygomatic arch generally strong, while the orbit and temporal fossa are in most cases completely confluent. The scapula has a large spine. The clavicle is never complete and is often absent, this forming an important distinction between the skeleton of a Carnivore and of any Insectivore except *Potamogale*. The humerus often has an ent-epicondylar foramen, and the radius and ulna, tibia and fibula are always separate. The manus is often capable of the movements of pronation and supination, and the scaphoid, lunar and centrale are in living forms always united together.

The order Carnivora includes three suborders.

Suborder (1). Creodonta.

This suborder contains a number of extinct Carnivora, which present very generalised characters.

The cranial cavity is very small; and the fourth upper premolar and first lower molar are not differentiated as carnassial teeth, as they are in modern Carnivora. The Creodonta also differ from modern Carnivora in the fact that the scaphoid and lunar are usually separate, and that the femur has a third trochanter. The feet are plantigrade.

They resemble the Condylarthra, another very generalised group, in having an ent-epicondylar foramen.

They occurred throughout the Tertiary period in both Europe and North America, and have also been found in India. One of the best known genera is *Hyaenodon*.

Suborder (2). Carnivora vera or Fissipedia.

The skeleton is mainly adapted for a terrestrial mode of life, and the hind limbs have the normal mammalian position. In almost every case the number of incisors is $^3/_3$. Each jaw always has one specially modified *carnassial* or sectorial tooth which bites like a scissors blade against a corresponding tooth in the other jaw. In front of it the teeth are always more or less pointed, while behind it they are more or less broadened and tuberculated. In the manus the first digit, and in the pes the first and fifth digits are never longer than the rest, and the digits of both limbs are almost invariably clawed. Some forms are plantigrade, some digitigrade, some subplantigrade. The group includes all the ordinary terrestrial Carnivora, and is divided into three sections:—

Æluroidea, including the cats, civets, hyaenas, and allied forms.

Cynoidea, including the dog tribe.

Arctoidea, including the bears, raccoons, weasels, and allied forms.

Suborder (3). Pinnipedia.

In this suborder the limbs are greatly modified and adapted for a more or less purely aquatic life, the proximal and middle segments of the limbs are shortened, while the distal segment, especially in the leg, is much elongated and expanded. There are always five well-developed digits to each limb, and in the pes the first and fifth digits are generally larger than the others. The digits generally bear straight nails instead of claws, but even nails are sometimes absent. There is no carnassial tooth, and the teeth in other ways differ considerably from those of Carnivora vera. The incisors are always fewer than $3/3$; while the cheek teeth generally consist of four premolars and one molar, all of very uniform character, being compressed with conical crowns, and never more than two roots.

The suborder includes three families—Otariidae (Eared Seals), Trichechidae (Walrus), and Phocidae (Seals).

Order 7. Insectivora.

This order contains a large number of small generally terrestrial mammals. The limbs are plantigrade or subplantigrade, and are generally pentedactylate. All the digits are armed with claws, and the pollex and hallux are not opposable. The teeth are diphyodont, heterodont, and rooted. The cheek teeth have tuberculated crowns, and there are never less than two pairs of incisors in the mandible; often the incisors, canines, and premolars are not clearly differentiated from one another, and special carnassial teeth are never found. The cranial cavity is small, and the facial part of the skull is generally much developed; often the zygomatic arch is incomplete. Clavicles are well developed (except in *Potamogale*), and the humerus generally has an ent-epicondylar foramen. The femur frequently has a ridge representing the third trochanter. There are two suborders:

Suborder (1). Dermoptera.

This suborder includes only a very aberrant arboreal genus *Galeopithecus*, remarkable for its greatly elongated limb bones, and peculiar dentition. The incisors of the lower jaw are deeply pectinated or divided by several vertical fissures, the canines and outer upper incisors have two roots. Ossified inter centra occur in the thoraco-lumbar region of the vertebral column.

Suborder (2). Insectivora vera.

This suborder includes all the ordinary Insectivora, such as moles, shrews and hedgehogs. The upper and lower incisors are conical, not pectinated.

Order 8. Chiroptera.

This order is perhaps the best marked and most easily defined of all the orders of mammals. The anterior limbs form true wings and the whole skeleton is modified in relation to flight.

The anterior limbs are vastly larger than the posterior; for all the bones except the carpals are much elongated, and this applies specially to the phalanges of all the digits except the pollex.

The pollex is clawed and so is sometimes the second digit; the other digits of the manus are without nails or claws. The teeth are divisible into the four usual types and the series never exceeds $i\ 2/3\ c\ 1/1\ pm\ 3/3\ m\ 3/3 \times 2$, total 38. The milk teeth are quite unlike the permanent teeth. The orbit is not divided by bone from the temporal fossa. The vertebral column is short, and in old animals the trunk vertebrae have a tendency to become partially fused together. The cervical vertebrae are remarkably wide, and the development of spinous processes is everywhere slight. The presternum has a prominent keel for the attachment of the pectoral muscles. The clavicles are very long and strong, and the scapula has a long spine and coracoid process. The ulna is vestigial, consisting only of a proximal end ankylosed to the radius. All the carpals of the proximal row—the scaphoid, lunar and cuneiform—are united, forming a single bone. The pelvis is very weak and narrow, and only in the Rhinolophidae do the pubes meet in a symphysis. The anterior caudal vertebrae are frequently united to the ischia. The fibula is generally vestigial, and the knee joint is directed backwards instead of forwards. The pes has five slender clawed digits, and the calcaneum is often drawn out into a spur which helps to support the membrane connecting the hind limbs with the tail.

There are two suborders of Chiroptera:

1. The Megachiroptera or Flying foxes, which almost always have smooth crowns to the molar teeth, and the second digit of the manus clawed.

2. The Microchiroptera including all the ordinary bats which have cusped molar teeth, and the second digit of the manus clawless.

Order 9. Primates.

The dentition is diphyodont and heterodont, the incisors generally number $^2/_2$, and the molars, except in the Hapalidae (Marmosets), are $^3/_3$. The cheek teeth are adapted for grinding, and the molars are more complex than the premolars. A process from the jugal meets the postorbital process of the frontal completing the postorbital bar.

The clavicle is well developed, and the radius and ulna are never united. The scaphoid and lunar of the carpus, and commonly also the centrale, remain distinct from one another. As a rule both manus and pes have five digits, but the pollex may be vestigial. The pollex is opposable to the other digits, and so is the hallux except in Man; the digits are almost always provided with flat nails. The humerus has no ent-epicondylar foramen and the femur has no third trochanter.

The order Primates is divisible into two suborders:

Suborder (1). Lemuroidea.

The skull has the orbit communicating freely with the temporal fossa beneath the postorbital bar (except in *Tarsius*). The lachrymal foramen is external to the margin of the orbit. Both pollex and hallux are well developed. In the pes the second digit is terminated by a long pointed claw, and so is also the third in *Tarsius*. The lumbar region of the vertebral column is long, sometimes including as many as nine vertebrae. Besides the Lemurs the group includes the aberrant *Tarsius* and *Chiromys*.

Suborder (2). Anthropoidea.

The skull has the orbit almost completely shut off from the temporal fossa, and the lachrymal foramen is situated within the orbit. The pollex is sometimes vestigial or absent. The second digit of the pes has a flattened nail except in the Hapalidae, in which all the digits of the pes except the hallux are clawed.

The Anthropoidea are divided into five families:

1. Hapalidae or Marmosets.
2. Cebidae or American Monkeys.
3. Cercopithecidae or Old World Monkeys.
4. Simiidae or Anthropoid Apes.
5. Hominidae or Men.

CHAPTER XXI.
THE SKELETON OF THE DOG (Canis familiaris).

I. EXOSKELETON.

The exoskeleton of the dog includes three sets of structures: 1. hairs, 2. claws, 3. teeth. **Hairs** and **claws** are epidermal exoskeletal structures, while **teeth** are partly of dermal, and partly of epidermal origin.

1. **Hairs** are delicate epidermal structures which grow imbedded in little pits or follicles in the dermis. Specially large hairs forming the **vibrissae** or **whiskers** grow attached to the upper lip.

2. **Claws** are horny epidermal sheaths, one of which fits on to the pointed distal phalanx of each digit. They are sharply curved structures, and being in the dog non-retractile, their points are commonly much blunted by friction with the ground. The claws of the pollex, and of the hallux when it is present, however do not meet the ground, and therefore remain comparatively sharp.

3. **Teeth**. Although as regards their mode of origin, teeth are purely exoskeletal or tegumentary structures, they become so intimately connected with the skull that they appear to belong to the endoskeleton.

Each tooth, as has been already described, consists of three distinct tissues, dentine and cement of dermal origin, and enamel of epidermal origin.

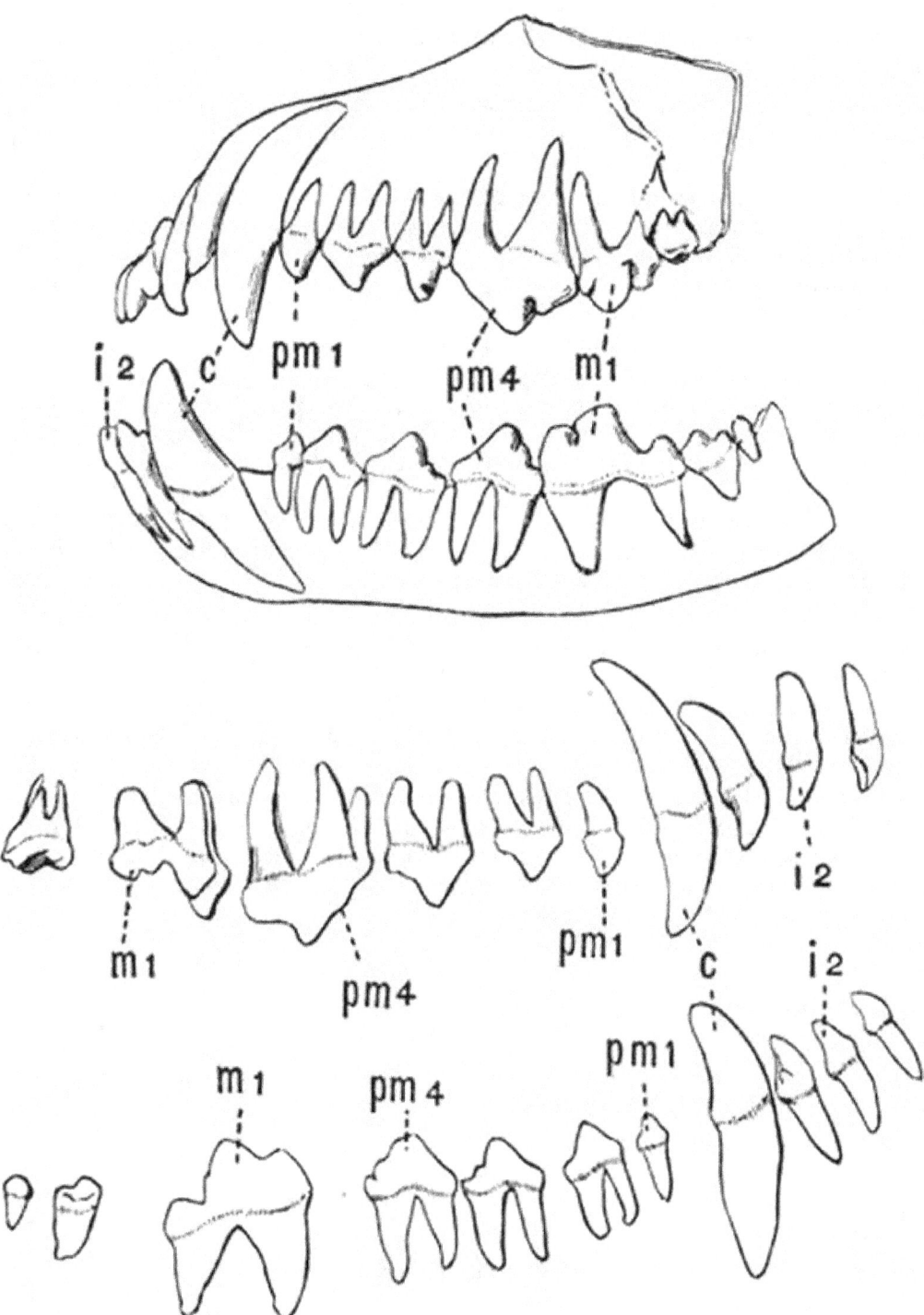

Fig. 68. Dentition of a Dog (*Canis familiaris*) × ½. (Camb. Mus.)

i 2. second incisor.

c. canine.

pm 1, *pm* 4. first and fourth premolars.

m 1. first molar.

The teeth of the dog (fig. 68) form a regular series arranged along the margins of both upper and lower jaws, and imbedded in pits or **alveoli** of the maxillae, premaxillae, and mandibles. They are all fixed in the bone by tapering roots, and none of them grow from persistent pulps.

They are divisible into four distinct groups, the **incisors**, **canines**, **premolars** and **molars**. There are three incisors, one canine and four premolars on each side of each jaw. But while there are three molars on each side of the lower jaw, the last is wanting in the upper jaw. The dentition of the dog may then be represented by the formula

$i\ ^3/_3\ c\ ^1/_1\ pm\ ^4/_4\ m\ ^2/_3 \times 2 = 42$.

In each jaw there is one large specially modified tooth called the **carnassial**, the teeth in front of this are more or less pointed and compressed, while those behind it are more or less flattened and tuberculated.

Teeth of the upper jaw.

The first and second **incisors** are small teeth with long conical roots and somewhat chisel-shaped crowns. Surrounding the base of the crown there is a rather prominent ridge, terminated laterally by a pair of small cusps. This ridge, the **cingulum**, serves to protect the edge of the gums from injury by the hard parts of food. The third incisor is a good deal like the others but larger, and has the cingulum well developed though not terminated by lateral cusps. All the incisors are borne by the premaxillae, the remaining teeth by the maxillae.

The **canine** is a large pointed tooth, slightly recurved and with a long tapering root.

The **premolars** are four in number, and in all the cingulum is fairly well seen. The first is a very small tooth with a single tapering root, the second and third are larger and have two roots, while the fourth, the **carnassial**, is much the largest and has three roots. Each of the second, third and fourth premolars has a stout blade, the middle portion of which is drawn out into a prominent cone; the posterior part of the fourth premolar forms a compressed ridge, and at the antero-internal edge of the tooth there is a small inner tubercle.

The two **molar** teeth are of very unequal size. The first, which has two anterior roots and one posterior, is wider than it is long, its outer portion being produced into two prominent cusps, while its inner portion is depressed. The second molar is a small tooth resembling the first in its general appearance, but with much smaller outer cusps.

Teeth of the lower jaw.

The three **incisors** of the lower jaw have much the same character as the first two of the upper jaw; while the **canine** is identical in character with that of the upper jaw.

The four **premolars** gradually increase in size from the first to the last, but none are very large. The first premolar is a single-rooted tooth resembling that of the upper jaw; the second, third and fourth are two-rooted, like the second and third of the upper jaw, which they closely resemble in other respects.

The first **molar** forms the **carnassial** (fig. 84, V), and with the exception of the canine, is much the largest tooth of the lower jaw; it is a two-rooted tooth, with a long compressed bilobed blade, and a posterior tuberculated talon or heel. The second molar is much smaller, though likewise two-rooted, while the third molar is very small and has only a single root. All the teeth except the molars are preceded in the young animal by temporary **milk teeth**. These milk teeth, though smaller, are very similar to the permanent teeth by which they are ultimately replaced.

II. ENDOSKELETON.

1. The Axial skeleton.

This includes the vertebral column, the skull, and the ribs and sternum.

A. The Vertebral column.

This consists of a series of about forty vertebrae arranged in succession so that their centra form a continuous rod, and their neural arches a continuous tube, surrounding a cavity, the **neural canal**.

The vertebrae may be readily divided into five groups:—

1. The **cervical** or neck vertebrae.

2. The **thoracic** or chest vertebrae which bear ribs.

3. The **lumbar** vertebrae which are large and ribless.

4. The **sacral** vertebrae which are fused with one another and united with the pelvis.

5. The **caudal** or tail vertebrae which are small.

Except in the sacral region the vertebrae are movably articulated to one another, while their centra are separated from one another by cartilaginous **intervertebral discs**.

General characters of a vertebra.

Take as a type the **fourth lumbar vertebra**. It may be compared to a short tube whose inner surface is smooth and regular, and whose outer surface is thickened and drawn out in a variety of ways. The basal part of the vertebra is the **centrum** or body which forms the thickened floor of the neural canal. Its two ends are slightly convex and are formed by the **epiphyses**, two thin plates of bone which are at first altogether distinct from the main part of the centrum, but fuse with it as the animal grows older; its sides are drawn out into a pair of strong **transverse processes**, which project forwards, outwards, and slightly downwards. The **neural arch** forms the sides and roof of the neural canal, and at each end just above the centrum bears a pair of **intervertebral notches** for the passage of the spinal nerves, the posterior notches being considerably deeper than the anterior. The neural arch is drawn out into a series of processes. Arising from the centre of the dorsal surface is a prominent median **neural spine** or **spinous process**, which projects upwards and slightly forwards; its anterior edge is vertical, while its posterior edge slopes gradually. At the two ends of the neural arch arise the two pairs of **zygapophyses** or articulating surfaces, which interlock with those of the adjacent vertebrae. The anterior or **prezygapophyses** look inwards, and are large and concave; they are borne upon a pair of large blunt outgrowths of the neural arch, the **metapophyses**. The posterior or **postzygapophyses** are slightly convex and look outwards and downwards; they are borne upon backwardly projecting outgrowths of the neural arch. Lastly there are a pair of minute projections arising from the posterior end of the neural arch, below the postzygapophyses. These are the **anapophyses**. In young individuals the development of all the processes of the various vertebrae is less marked, and the epiphyses are obviously distinct.

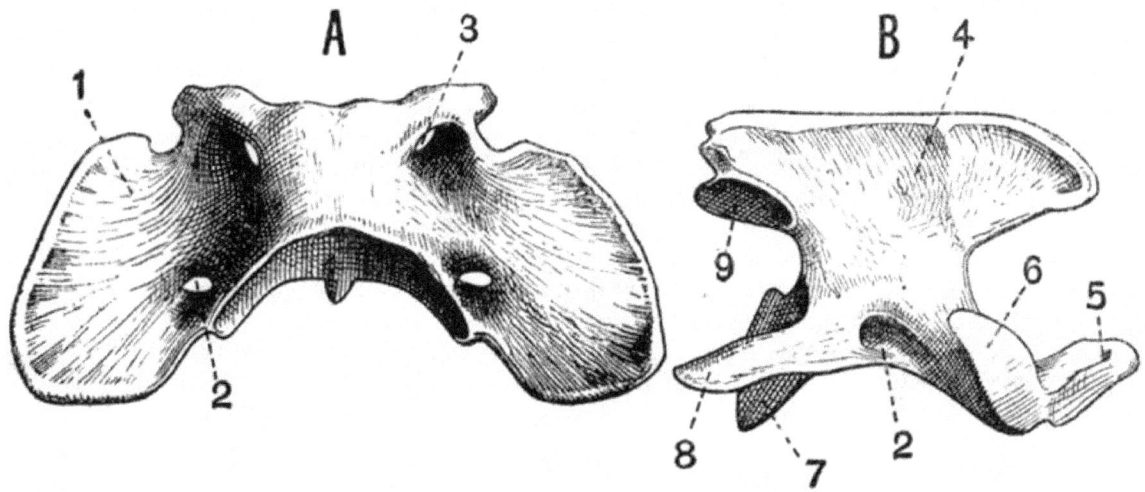

Fig. 69. A, atlas and B, axis vertebra of a Dog (*Canis familiaris*) (after von Zittel).

1. transverse process of atlas.
2. vertebrarterial canal.
3. foramen for exit of spinal
nerve.
4. neural spine.
5. odontoid process.

6. anterior articulating surface
of centrum.
7. centrum.
8. transverse process of axis.
9. postzygapophysis.

The Cervical vertebrae.

These are seven in number, as in almost all mammals. They are characterised by the fact that they have small ribs fused with them, forming transverse processes perforated by canals through which the vertebral arteries run.

The first, or **atlas** vertebra (fig. 69, A), differs much from all the others; it is drawn out into a pair of wide wing-like transverse processes (fig. 69, A, 1), and forms a ring surrounding a large cavity. This cavity is during life divided into two parts by a transverse ligament; the upper cavity is the true neural canal, while the lower lodges the **odontoid process** of the second vertebra, which is the detached centrum of the atlas. The neural arch is broad and regular; it has no spinous process, and is perforated in front by a pair of foramina for the passage of the first spinal nerves. The mid-ventral portion of the atlas is rather thick, and bears a minute backwardly-projecting hypapophysis. The bases of the broad transverse processes are perforated by the **vertebrarterial canals** (fig. 69, A, 2). The atlas bears at each end a pair of large articulating surfaces; those at the anterior end articulate with the condyles of the skull, and are very deeply concave; those at the posterior end for articulation with the axis, are nearly as large, but are flattened. The atlas ossifies from three centres, one forming the mid-ventral portion, the others the two halves of the remainder.

The second, or **axis** vertebra (fig. 69, B), also differs much from the other cervicals. The long and broad centrum has a very flat dorsal surface, and is produced in front into the conical **odontoid process** (fig. 69, B, 5), and bears a pair of very large convex outwardly-directed surfaces for articulation with the atlas. At its posterior end it is drawn out into a pair of small backwardly-directed spines, the transverse processes; these are perforated at their bases by the vertebrarterial canals. The neural arch is deeply notched in front and behind for the passage of the spinal nerves, and is drawn out above into a very long compressed neural spine (fig. 69, B, 4), which projects a long way forwards, and behind becomes bifid and thickened,

185

bearing a pair of flat downwardly directed postzygapophyses. In the young animal the odontoid process is readily seen to ossify from a centre anterior to that forming the anterior epiphysis of the axis.

The remaining five cervical vertebrae, the third to the seventh inclusive, have rather flattened wide centra, obliquely truncated at either end. The neural spine progressively increases in size as the vertebrae are followed back. The transverse processes vary considerably; those of the third are divided into a thicker backwardly-, and a more slender forwardly-projecting portion; those of the fourth and fifth mainly extend downwards, and that of the sixth is divided into a horizontal portion and a downwardly-projecting **inferior lamella**. All the cervical vertebrae except the seventh have the bases of the transverse processes perforated by the vertebrarterial canals. The prezygapophyses in each case look upwards and slightly inwards, while the postzygapophyses look downwards and slightly outwards.

The Thoracic vertebrae.

The **thoracic vertebrae** are twelve or thirteen in number, and all bear movably articulated ribs. As a group they are characterised by their comparative shortness, and in the case of the first eight or nine by the great length of the backwardly-sloping neural spine. The posterior thoracic vertebrae approach in character the succeeding lumbar vertebrae.

As type of the anterior thoracic vertebrae, take any one between the second and sixth inclusive. The centrum is short, and has its terminations vertically truncated. At the top of the centrum, at both anterior and posterior ends on each side, is a demi-facet (fig. 70, A, 4), which, together with that on the adjacent vertebra, forms an articulating surface for the capitulum of the rib. The neural arch is small and deeply notched behind for the passage of the spinal nerve. It is drawn out above into a very long neural spine (fig. 70, A, 1), whose base extends back over the succeeding vertebra and bears the downwardly-directed postzygapophyses (fig. 70, A, 6). The summit of the neural arch is deeply notched in front, and on each side of the notch are the prezygapophyses, which look almost vertically upwards. The transverse processes are short and blunt, and are flattened below (fig. 70, A, 3) for the articulation of the tubercula of the ribs.

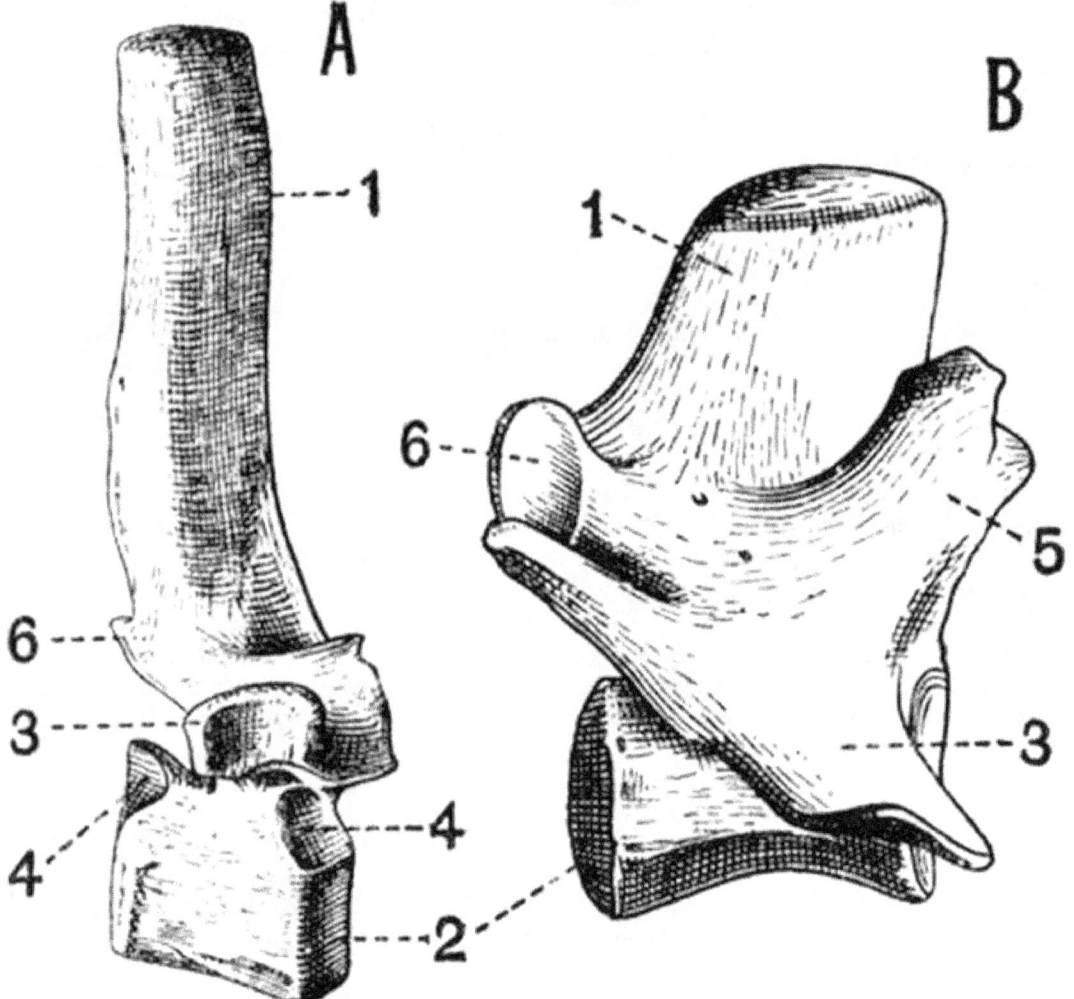

Fig. 70. A, second thoracic, and B, second lumbar vertebra of a Dog (*Canis familiaris*) SEEN FROM THE RIGHT SIDE (after von Zittel).

186

1. neural spine.

2. centrum.

3. transverse process bearing in A the facet for articulation with the tuberculum of the rib.

4. facet for articulation with the capitulum of the rib.

5. metapophysis.

6. postzygapophysis.

The posterior three or four thoracic vertebrae differ much from the others. The centra are longer, the neural spines short and not directed backwards, the articular facets for the heads of the ribs are confined to the anterior end of the centrum of each vertebra, not overlapping on to the preceding vertebra. The transverse processes are small and irregular, and metapophyses and anapophyses are developed. The prezygapophyses also look more inwards, and the postzygapophyses more outwards than in the more typical thoracic vertebrae.

The Lumbar vertebrae.

The **lumbar vertebrae** are seven in number, and their general characteristics have been already described. As a group they are characterised by their large size, and the great development of the transverse processes, metapophyses and neural spines.

The Sacral vertebrae.

Three vertebrae are commonly found fused together, forming the **sacrum**; the divisions between the three being indicated by the foramina for the exit of the spinal nerves.

Of these three vertebrae, the first is much the largest, and is firmly united to the ilium on each side by a structure formed by the transverse processes and expanded ribs. In the adult this structure forms one continuous mass, but in the young animal a ventral portion formed by the rib is clearly distinguishable from a dorsal portion formed by the transverse process. All three have low neural spines. The anterior sacral vertebra bears a large pair of prezygapophyses, while the posterior one bears a small pair of postzygapophyses.

The Caudal vertebrae.

The **caudal vertebrae** are about nineteen in number. The earlier ones have well-developed neural arches, transverse processes, and zygapophyses, but as the vertebrae are followed back they gradually lose all their processes, and the neural arch as well, becoming at about the thirteenth from the end reduced to simple cylindrical centra.

B. The Skull.

The **skull** consists of the following three parts: (*a*) the cranium, with which are included the skeletal supports of the various special sense organs, and the bones of the face and upper jaw; (*b*) the lower jaw or mandible, which is movably articulated to the cranium, and (*c*) the hyoid.

(*a*) The Cranium.

The cranium is a compact bony box, forming the anterior expanded portion of the axial skeleton. It has a longitudinal axis, the **craniofacial** axis around which the various parts are arranged, and this axis is a direct continuation of that of the vertebral column. Similarly the cavity of the cranium is a direct continuation of the spinal canal. The posterior part of the craniofacial axis, which has relations only with the cranium, is called the **basicranial axis**.

In the dog as in the other types previously described, the skull in its earliest stages is cartilaginous, containing no bone. In the adult, however, the cartilage is to a great extent replaced by bone, and in addition to this cartilage bone, membrane bone is largely developed, and intimately united with the cartilage bone to form one complete whole.

In the description of the dog's skeleton, as in those of the previous types, the names of the membrane bones are printed in italics, while those of the cartilage bones are printed in thick type.

Most of the numerous foramina perforating the skull walls will be described after the bones have been dealt with.

For purposes of description the cranium may be further subdivided into:—

1. The cranium proper or brain case.

2. The sense capsules.

3. The upper jaw.

1. The Cranium Proper or Brain Case.

Taking the membrane and cartilage bones together, they are seen to be more or less arranged in three segments, which however must not be regarded as homologous with the segments forming the vertebral column.

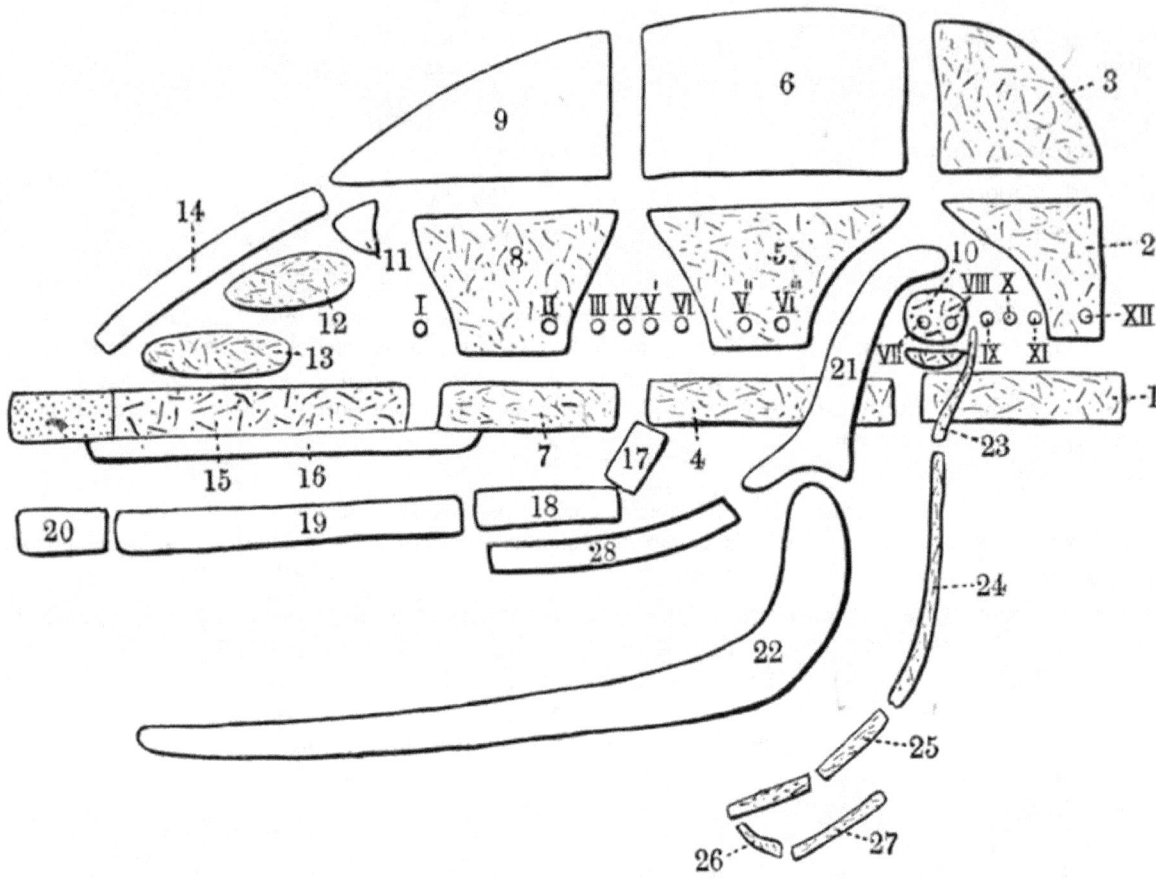

Fig. 71. Diagram of the relations of the principal bones in the
Mammalian Skull (modified after Flower).
Cartilage is dotted. Cartilage bones are marked by dots and dashes,
membrane bones are left white.

1. basi-occipital.

2. exoccipital.

3. supra-occipital.

4. basisphenoid.

5. alisphenoid.

6. parietal.

7. presphenoid.

8. orbitosphenoid.

9. frontal.

10. periotic, immediately below

which is the tympanic.

11. lachrymal.

12. ethmo-turbinal.

13. maxillo-turbinal.

14. nasal.

15. mesethmoid.

16. vomer.

17. pterygoid.

18. palatine.

19. maxillae.

20. premaxillae.

21. squamosal.

22. mandible.

23. tympano-hyal.

24. stylo-hyal.

25. epi-hyal.

26. basi-hyal. Between this and

the epi-hyal is the cerato-hyal.

27. thyro-hyal.

28. jugal.

Nerve exits are indicated by Roman numerals.

The **occipital segment** is the most posterior of the three, and consists of four cartilage bones, which in the adult are commonly completely fused together. They surround the great**foramen magnum** (fig. 75, 2) through which the brain and spinal cord communicate. Forming the lower margin of the foramen magnum is a large flat unpaired bone, the **basi-occipital** (fig. 75, 5). Above this on each side are the **exoccipitals**, whose sides are drawn out into a pair of downwardly-directed **paroccipital processes**, which are applied to the tympanic bullae. The inner side of each exoccipital is converted into the large rounded **occipital condyle** (fig. 72, 13) by which the skull articulates with the atlas vertebra. The dorsal boundary of the foramen magnum is formed by a large unpaired flat bone, the **supra-occipital** (figs. 72 and 75, 1), which is continuous with a small bone, the *interparietal*, prolonged forwards between the parietal bones of the next segment.

In old animals the interparietal forms the hind part of a prominent ridge running along the mid-dorsal surface of the skull and called the **sagittal crest**, while the junction line of the occipital and parietal segments forms a prominent **occipital crest**.

The plane in which the bones of the occipital segment lie is called the occipital plane; the angle that it makes with the basicranial axis varies much in different mammals.

The **parietal segment** consists of both cartilage and membrane bones. It is formed of five bones, which are in contact with those of the occipital segment on the dorsal and ventral surfaces, while laterally they are separated by the interposition of the auditory bones, and to some extent of the squamosal. The **basisphenoid** (fig. 75, 6), an unpaired bone forming the ventral member of this segment, is the direct continuation of the basi-occipital. It tapers anteriorly, but is rather deep vertically, its upper or dorsal surface bearing a depression, the **sella turcica**, which lodges the pituitary body of the brain. From the sides of the basisphenoid arise the **alisphenoids** (fig. 75, 11) a pair of bones of irregular shape generally described as wing-like; each gives off from its lower surface a **pterygoid plate**, which is united in front with the palatine, and below with the pterygoid. The alisphenoids are united above with a pair of large nearly square bones, the *parietals* (fig. 73, 2), which meet one another in the mid-dorsal line. The line of junction is frequently drawn out into a strong ridge, which forms the anterior part of the **sagittal crest**.

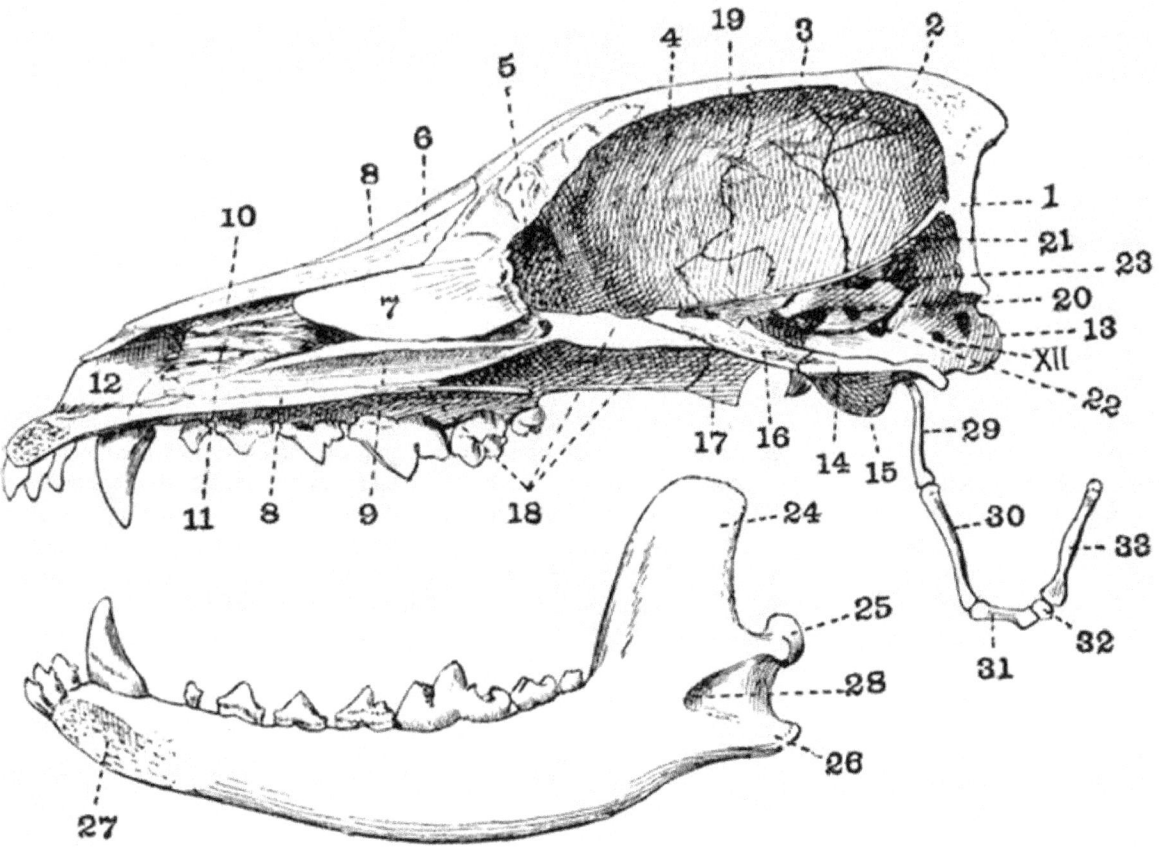

Fig. 72. Vertical longitudinal section taken a little to the left of the middle line through the skull of a Dog (*Canis familiaris*) × ³/₅. (Camb. Mus.)

1. supra-occipital.	18. palatine.
2. interparietal.	19. alisphenoid.
3. parietal.	20. internal auditory meatus.
4. frontal.	21. tentorium.
5. cribriform plate.	22. foramen lacerum posterius.

189

6. nasal.

7. mesethmoid.

8. maxillae.

9. vomer.

10. ethmo-turbinal.

11. maxillo-turbinal.

12. premaxillae.

13. occipital condyle.

14. basi-occipital.

15. tympanic bulla.

16. basisphenoid.

17. pterygoid.

23. floccular fossa.

24. coronoid process.

25. condyle.

26. angle.

27. mandibular symphysis.

28. inferior dental foramen.

29. stylo-hyal.

30. epi-hyal.

31. cerato-hyal.

32. basi-hyal.

33. thyro-hyal.

XII. condylar foramen.

The **frontal segment**, which surrounds the anterior part of the brain, is closely connected along almost its whole posterior border with the parietal segment.

Its base is formed by the **presphenoid** (fig. 75, 12), a very deep unpaired bone, narrow and compressed ventrally, and with an irregular dorsal surface. The presphenoid is continuous with a second pair of wing-like bones, the **orbitosphenoids**. Each orbitosphenoid meets the alisphenoid behind, but the relations of the parts in this region are somewhat obscured by a number of large foramina piercing the bones, and also by an irregular vacuity, the **foramen lacerum anterius** or **sphenoidal fissure**, which lies between the orbitosphenoid and alisphenoid, separating the lateral parts of the parietal and frontal segments, in the same way as the space occupied by the auditory bones separates the lateral parts of the occipital and parietal segments. The orbitosphenoids pass obliquely forwards and upwards, and are united above with a second pair of large membrane bones, the *frontals* (fig. 73, 3). The outer side of each frontal is drawn out into a rather prominent rounded **postorbital process** (fig. 73, 10), from which a ridge converges backwards to meet the sagittal crest. The anterior part of the frontal is produced to form the long nasal process, which is wedged in between the nasal and maxillae.

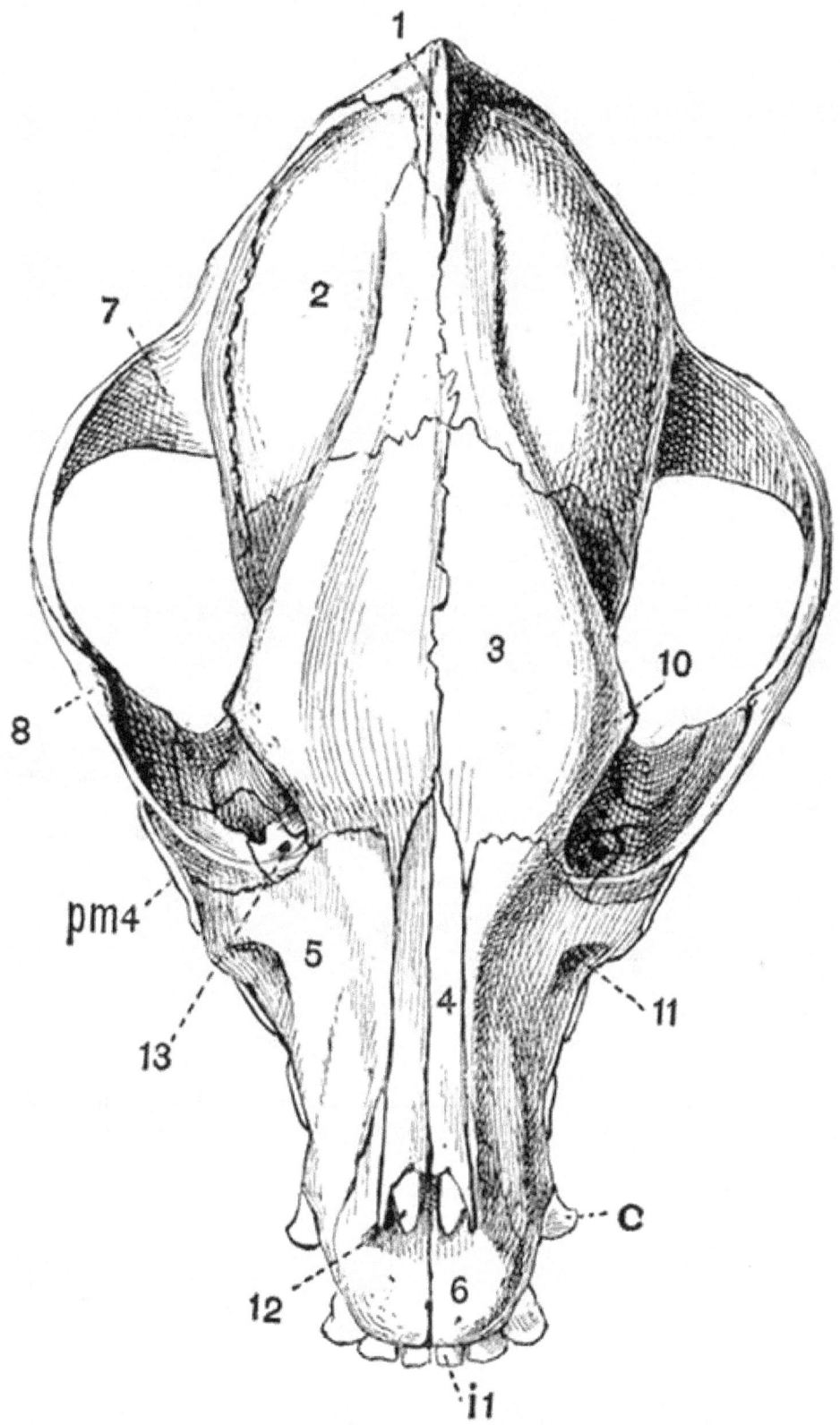

Fig. 73. Dorsal view of the cranium of a Dog
(*Canis familiaris*) × $^2/_3$.

1. supra-occipital.   10. postorbital process of frontal.

2. parietal.   11. infra-orbital foramen.

3. frontal.

4. nasal.

5. maxillae (facial portion).

6. premaxillae.

7. squamosal.

8. jugal.

12. anterior palatine foramen.

13. lachrymal foramen.

i 1. first incisor.

c. canine.

pm 4. fourth premolar.

The cranial cavity is continuous in front with the **nasal** or **olfactory cavities**, but the passage is partially closed by a screen of bone, the **cribriform plate** (fig. 72, 5), which is placed obliquely across the anterior end of the cranial cavity, and is perforated by a number of holes through which the olfactory nerves pass. The plane of the cribriform plate is called the **ethmoidal plane**, and as was the case also with the occipital plane, the angle that it makes with the basicranial axis varies much in different mammals, and is of importance. The **olfactory fossa** in which lie the olfactory lobes of the brain, is partially separated from the **cerebral fossa**, or cavity occupied by the cerebral hemispheres, by ridges on the orbitosphenoids and frontals. The presphenoid is connected in front with a vertical plate formed partly of bone, partly of unossified cartilage; this plate, the **mesethmoid** (fig. 72, 7), separates the two olfactory cavities which lodge the olfactory organs. Its anterior end always remains unossified, and forms the septal cartilage of the nose.

The brain case may then, to use the words of Sir W.H. Flower, be described as a tube dilated in the middle and composed of three bony rings or segments, with an aperture at each end, and a fissure or space at the sides between each of them.

2. The Sense capsules.

Each of the three special sense organs, of hearing, of sight, and of smell, is in the embryo provided with a cartilaginous or membranous protecting capsule; and two of these, the auditory and olfactory capsules, become afterwards more or less ossified, and intimately related to the cranium proper.

(1) Bones in relation to the Auditory capsules.

These bones lie on each side wedged into the vacuity between the lateral parts of the occipital and parietal segments; they are three in number, the **periotic**, the *tympanic* and the *squamosal*.

The **periotic** is the most important of them, as it replaces the cartilaginous auditory capsule of the embryo, and encloses the essential organ of hearing. It commences to ossify from three centres corresponding to the pro-otic, epi-otic and opisthotic of lower skulls, such as those of the Turtle and Crocodile.

These ossifications however very early combine to form a single bone, the **periotic**, which nevertheless consists of two portions, the **petrous** and the **mastoid**, differing considerably from one another.

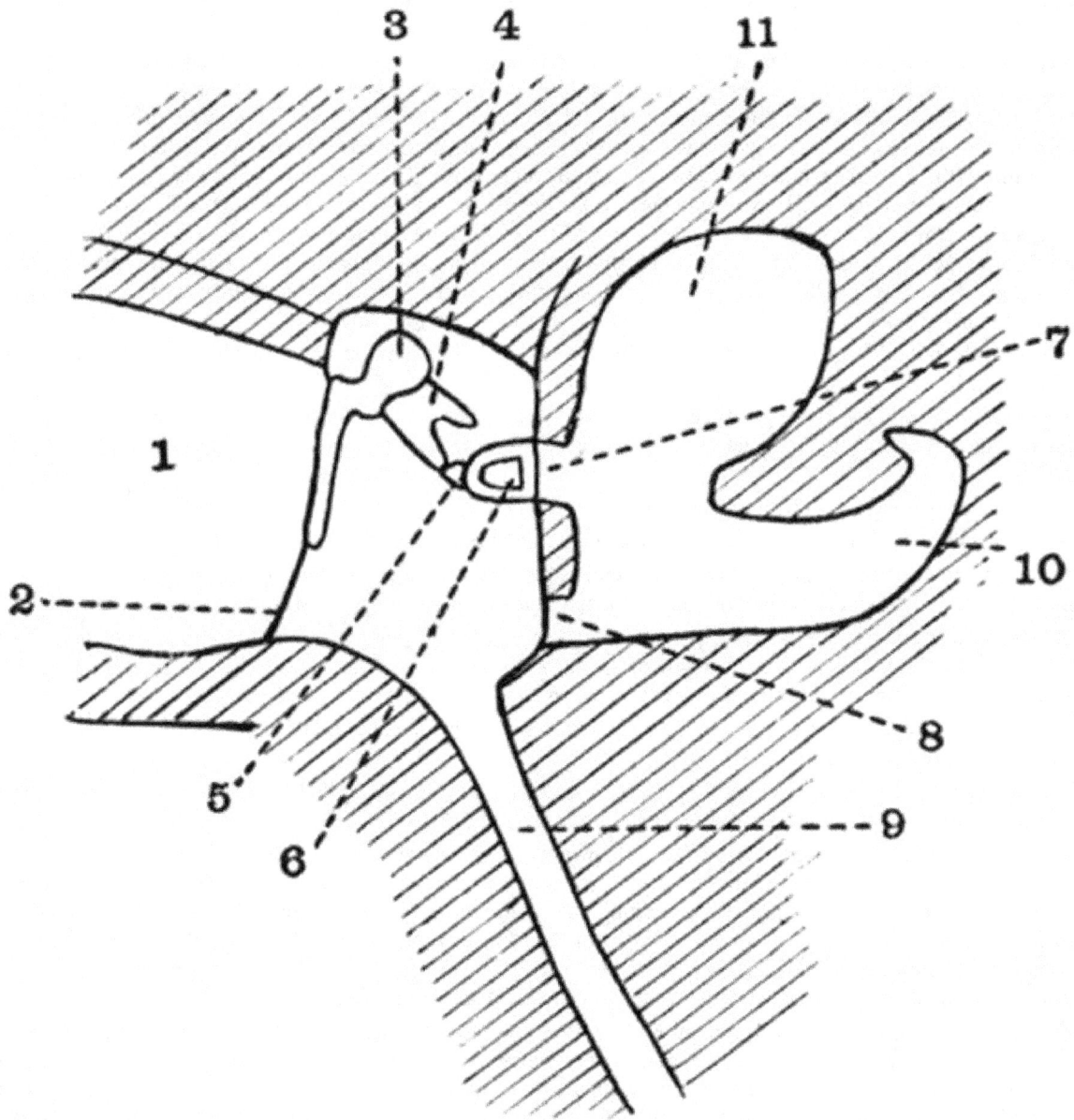

Fig. 74. Diagram of the mammalian tympanic cavity and associated parts (modified from Lloyd Morgan).

1. external auditory meatus.	7. fenestra ovalis.
2. tympanic membrane.	8. fenestra rotunda.
3. malleus.	9. Eustachian tube.
4. incus.	10. cavity occupied by the cochlea.
5. lenticular.	11. cavity occupied by the
6. stapes.	membranous labyrinth.

The **petrous portion** lies dorsally and anteriorly, and is much the more important of the two, as it encloses the essential part of the auditory organ. It forms an irregular mass of hard dense bone, projecting into the cranial cavity, and does not appear on the external surface at all. The **mastoid portion** lies ventrally and posteriorly, is smaller, and formed of less dense bone than is the petrous portion, from which it differs also in the fact that it appears on the surface of the skull, just external to the exoccipital. The petrous portion bears a ridge, which together with a ridge on the supra-occipital, and the **tentorium** (fig. 72, 21), a transverse fold of the dura mater, separates the large cerebral fossa from the **cerebellar fossa**, which is much smaller than the cerebral fossa and lies behind and partly beneath it. The plane of the tentorium is called the **tentorial plane**, and the angles that it makes with the basicranial axis and with the occipital and ethmoidal planes vary much in different mammals.

The periotic has its inner surface marked by important depressions, while both inner and outer surfaces are pierced by foramina. At about the middle of its inner surface are seen two deep pits, one lying immediately above the other. Of these the more ventral is a foramen, the **internal auditory meatus** (fig. 72, 20), through which the VIIth (facial) and VIIIth (auditory) nerves leave the cranial cavity, the facial nerve passing through the bone and afterwards leaving the skull by the **stylomastoid foramen** (fig. 75, VII), while the auditory passes to the inner ear. The more dorsal of the two pits is not a foramen but the **floccular fossa** (fig. 72, 23) which lodges the floccular lobe of the cerebellum. In some skulls another wide and shallow but fairly prominent depression is seen dorsal to and slightly behind the floccular fossa, this also lodges part of the cerebellum. Behind the internal auditory meatus, between the periotic and exoccipital is seen the internal opening of the **foramen lacerum posterius** (fig. 72, 22). The shape of this opening varies. The ventro-anterior border of the periotic is marked by a deep notch, the sides of which sometimes unite, converting it into a foramen.

On the outer side of the periotic, and clearly seen only after the removal of the tympanic, are two holes, the **fenestra ovalis** and the **fenestra rotunda**.

The *tympanic* (figs. 72, 15 and 75, 4) is a greatly expanded boat-shaped bone, which forms the auditory bulla and lies immediately ventral to the periotic; it is separated from the periotic by the **tympanic cavity** into which the fenestra rotunda and the fenestra ovalis open.

There are several other openings into the tympanic cavity.

(*a*) On the external surface is a large oval opening, the **external auditory meatus** bounded by a thickened rim.

(*b*) Into the outer and anterior part of the cavity the outer end of the **Eustachian tube** opens; while the inner end passes through a foramen (fig. 75, 22) just external to the foramen lacerum medium, on its way to open into the pharynx.

(*c*) The internal carotid artery also enters the tympanic cavity by a canal which commences in the foramen lacerum posterius, and passes forwards to open on the inner side of the bulla. The artery then passes forwards, and barely appearing on the ventral surface of the cranium, enters the brain cavity through the foramen lacerum medium (fig. 75, 9).

Immediately behind the tympanic, between it and the mastoid process of the periotic and the paroccipital process of the exoccipital is the **stylomastoid foramen** (fig. 75, VII).

Within the tympanic cavity are four small bones, the **auditory ossicles** (cp. fig. 74), called respectively the **malleus**, **incus**, **lenticular** and **stapes**; these together form a chain extending from the fenestra ovalis to the tympanic membrane.

The **malleus** has a somewhat rounded head (fig. 100, B, 1) which articulates with the incus, while the other end of the bone is drawn out into a long process, the **manubrium**, which lies in relation to the tympanic membrane. The head is also more or less connected by a thin plate of bone, the **lamella**, to another outgrowth, the **processus longus**. The incus (fig. 100, B, 3) is somewhat anvil-shaped, and is drawn out into a process which is connected with the **lenticular**, a nodule of bone interposed between the incus and the stapes, with which it early becomes united. The **stapes** (fig. 100, B, 2) is stirrup-shaped, consisting of a basal portion from which arise two **crura**, which meet and enclose a space, the **canal**.

The *squamosal* (fig. 73, 7) is a large bone occupying much of the side wall of the cranial cavity, and articulating above with the parietal, and behind with the supra-occipital, while in front it overlaps the frontal and alisphenoid. But though it occupies so large a space on the outer wall, it forms very little of the internal wall of the skull, but is really like a bony plate attached to the outer surface of the cranial wall. The squamosal is drawn out into a strong forwardly-directed **zygomatic process** which meets the jugal or malar. The ventral side of the zygomatic process is hollowed out, forming the **glenoid fossa** (fig. 75, 8), a smooth laterally elongated surface with which the lower jaw articulates, while the hinder edge of the glenoid fossa is drawn out into a rounded **postglenoid process** (fig. 75, 23). The articulation is such as to allow but little lateral play of the lower jaw.

(2) **Bones in relation to the Optic capsules.**

The only bone developed in relation to the optic capsule on each side is the *lachrymal*. This is a small membrane bone lying between the frontal and palatine behind, and the maxillae and jugal in front. It is perforated by a prominent **lachrymal foramen** (fig. 73, 13) which opens within the orbit.

(3) **Bones in relation to the Olfactory capsules.**

In connection with the **olfactory capsules**, five pairs of bones are developed, two pairs being membrane bones, and three pairs cartilage bones.

Of membrane bones, the *nasals* (fig. 73, 4) are a pair of long narrow bones, lying closely side by side, and forming the main part of the roof of the olfactory chamber. Their posterior ends overlap the frontals, and the outer margin of each is in contact with the nasal process of the frontal, and with the maxillae and premaxillae.

Lying immediately ventral to the nasals, and on each side of the perpendicular mesethmoid, are the **ethmoid** or **turbinal** bones, which have a curious character, being formed of a number of delicate plates intimately folded on one another. The posterior pair of these bones, the **ethmo-turbinals** (fig. 72, 10), are the larger, and form a mass of intricately folded lamellae

194

attached behind to the cribriform plate, and passing laterally into two thin plates of bone, which abut on the maxillae. The uppermost lamella of each ethmo-turbinal is larger than the others and more distinct. It is sometimes distinguished as the **naso-turbinal**, and forms an imperfect lower boundary to a canal, which is bounded above by the nasals. In front of and somewhat below the ethmo-turbinals, lie another pair of bones of similar character, the **maxillo-turbinals** (fig. 72, 11).

The last bone to be mentioned in connection with the olfactory capsules is a membrane bone, the *vomer* (fig. 72, 9). This is a slender vertically-placed bone, whose anterior part lies between the maxillo-turbinals, while behind it extends beyond the mesethmoid, so as to underlie the anterior part of the presphenoid. The anterior part of the vomer forms a kind of trough, while further back in the region of the ethmo-turbinals it sends out a pair of strong lateral plates, each of which, passing below the ethmo-turbinal, joins the side wall of the nasal cavity, and forms a partition dividing the nasal cavity into a lower **narial passage** and an upper **olfactory chamber**.

The Jaws.

In the embryo both upper and lower jaws are formed of cartilaginous bars, but in the adult not only has the cartilage entirely disappeared, but even cartilage bone is absent, the jaws being formed of membrane bone.

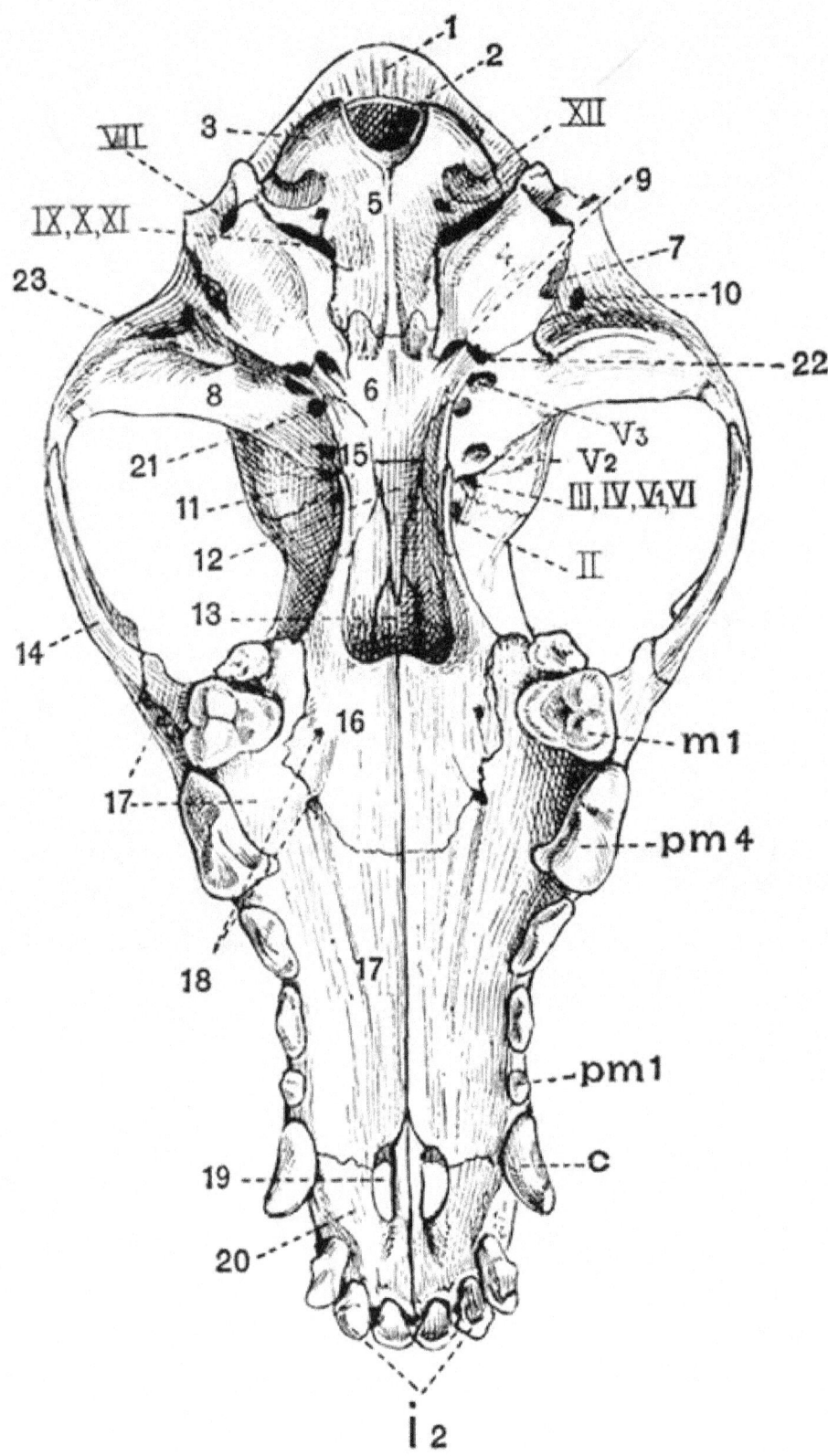

Fig. 75. Ventral view of the cranium of a Dog
(*Canis familiaris*) × ³/₅. (Camb. Mus.)

1. supra-occipital.

17. maxillae (palatal portion).

2. foramen magnum.

18. posterior palatine foramina.

3. occipital condyle.

4. tympanic bulla.

5. basi-occipital.

6. basisphenoid.

7. external auditory meatus.

8. glenoid fossa.

9. foramen lacerum medium and
anterior opening of carotid
canal.

10. postglenoid foramen.

11. alisphenoid.

12. presphenoid.

13. vomer.

14. jugal.

15. pterygoid.

16. palatal process of palatine.

19. anterior palatine foramen.

20. premaxillae.

21. alisphenoid canal.

22. Eustachian foramen.

23. postglenoid process of
squamosal.

II. optic foramen.

III, IV, V_{1}, VI. foramen lacerum
anterius.

V_{2}. foramen rotundum.

V_{3}. foramen ovale.

VII. stylomastoid foramen.

IX, X, XI. foramen lacerum
posterius.

XII. condylar foramen.

i 2. second incisor.

c. canine.

pm 1, *pm* 4. first and fourth
premolars.

m 1. first molar.

3. The Upper jaw.

The bones of the upper jaw are closely connected with those of the cranium proper and olfactory capsules. The most posterior of them is the *pterygoid* (fig. 75, 15), a thin vertically placed plate of bone, which articulates above with the basisphenoid, the presphenoid, and the strong pterygoid process of the alisphenoid. The ventral end of the pterygoid is drawn out into a small backwardly-projecting **hamular process**. In front the pterygoid articulates with the *palatine*, a much larger bone, consisting of (1) a vertical portion, which passes up to meet the orbitosphenoid and frontal, and sends inwards a plate which meets the presphenoid and vomer, forming much of the roof of the posterior part of the narial passage; and (2) a strong horizontal portion, the **palatal process** (fig. 75, 16), which passes inwards and meets its fellow in the middle line, forming the posterior part of the bridge of bone supporting the hard palate. The palatal process is continuous in front, with a large bone, the *maxillae*, which, like the palatine, consists of vertical and horizontal portions. The vertical, or **facial portion** (fig. 73, 5), is the largest, and constitutes the main part of the side of the face in front of the orbit, forming also the chief part of the outer wall of the nasal cavity. It is continuous in front with the premaxillae, above with the nasal and frontal, and behind with the lachrymal, jugal, and palatine. The horizontal, or **palatal portion** (fig. 75, 17), forms the anterior part of the bony plate supporting the hard palate, and meets its fellow in a long straight symphysis. The junction line between the palatal and facial portions is called the **alveolar border**, and along it are attached the canine, premolar, and molar teeth.

The anterior part of the upper jaw on each side is formed by a small bone, the *premaxillae*, which bears the incisor teeth. It, like the maxillae, has a palatal portion (fig. 75, 20), which meets its fellow in the middle line, and an ascending portion, which passes backwards as the **nasal process**, tapering regularly and lying between the nasal and the maxillae. The two premaxillae form the outer and lower borders of the anterior nares. The last bone to be mentioned in connection with the upper jaw and face is the *jugal* or *malar* (figs. 73, 8, and 75, 14), a strong bone which forms the anterior half of the zygomatic arch. It is firmly united in front to the maxillae, and behind meets the zygomatic process of the squamosal, being drawn out dorsally into a short **postorbital process** at the point of meeting. This process lies immediately below the postorbital process of the frontal, and if the two met, as they do in some mammals, they would partially shut off the orbit from a larger posterior cavity, the **temporal fossa**. In the living animal a ligament unites the two postorbital processes.

(*b*) The Lower jaw or Mandible.

This consists of two elongated symmetrical halves, the **rami**, which are united to one another at the median symphysis in front, while behind they diverge considerably, and each articulates with the glenoid surface of the corresponding squamosal. In young animals the rami are united at the symphysis by fibrous tissue, but in old animals they sometimes become fused together. The upper or alveolar border bears the teeth, and behind them is drawn out into a high laterally compressed coronoid process (fig. 72, 24), which is hollowed on its outer surface. Immediately behind the coronoid process is the transversely elongated **condyle** (fig. 72, 25), which fits into the glenoid cavity in such a way as to allow free up and down movement of the jaw, with but little rolling motion. The posterior end of the jaw below the condyle forms a short rounded process, the **angle** (fig. 72, 26). Two prominent foramina are to be seen in the lower jaw. These are firstly the **inferior dental foramen** (fig. 72, 28), which lies on the inner surface below the coronoid process; through it an artery and a branch of the fifth nerve enter to supply the teeth, and secondly the **mental foramen**, which lies on the outer side near the anterior end, and through which a branch of the same nerve emerges.

(*c*) The Hyoid.

The **Hyoid** of the dog consists of a transverse median piece, the **basi-hyal** (fig. 72, 32), from which arise two pairs of **cornua**. The **anterior cornu** is much the longer of the two, and consists principally of three short separate ossifications, placed end to end and called respectively the **cerato-hyal**, **epi-hyal**, and **stylo-hyal**. All of them are short rods of bone, contracted in the middle, and expanded at the ends, where they are tipped with cartilage. The cerato-hyal (fig. 72, 31) lies next to the basi hyal. The stylo-hyal is terminated by a much smaller bone, the **tympano-hyal**, which lies in a canal between the tympanic and periotic, and is ankylosed to the periotic just to the anterior and inner side of the stylomastoid foramen.

The **posterior cornu** of the hyoid is much smaller than the anterior; it consists of a short bone, the **thyro-hyal** (fig. 72, 33), which connects the basi-hyal with the thyroid cartilage of the larynx.

Foramina of the skull.

The foramina, or apertures perforating the walls of the skull, are very numerous, and may either be due to holes actually penetrating the bone, or may be small vacuities between the margins of two elsewhere contiguous bones.

They may be divided into two groups, the first including

I. The holes through which the **twelve cranial nerves** leave the cranial cavity.

a. The most anterior of these nerves, the olfactory, leaves the skull by a number of small holes piercing the **cribriform plate** (fig. 72, 5).

b. The second, or optic, passes out by a large hole, the **optic foramen** (fig. 75, II) piercing the orbitosphenoid. The optic foramen is the most anterior of the three prominent holes seen within and immediately behind the orbit.

c. The third, fourth, and sixth nerves, i.e. those supplying the eye muscles, and with them the first or ophthalmic branch of the large fifth or trigeminal nerve, pass out by a large hole, the **foramen lacerum anterius** (fig. 75, III, IV, V_1, VI), which, as has been already mentioned, lies between the orbitosphenoid and alisphenoid.

d. Immediately behind the foramen lacerum anterius, the alisphenoid is perforated by a prominent round hole, the **foramen rotundum** (fig. 75, V_2), through which the second branch of the trigeminal nerve passes out.

e. A quarter of an inch further back there is another prominent hole, the **foramen ovale** (fig. 75, V_3), through which the third branch of the trigeminal nerve leaves the cranium.

f. The seventh or facial nerve, as already mentioned, leaves the cranial cavity and enters the auditory capsule, through an opening in the periotic called the **internal auditory meatus**, while it finally leaves the skull by the **stylomastoid foramen** (fig. 75, VII), which lies between the tympanic bulla, the paroccipital process, and the mastoid portion of the periotic.

g. The eighth or auditory nerve on leaving the cranial cavity, passes with the facial straight into the auditory capsule through the **internal auditory meatus** (fig. 72, 20). It is then distributed to the organ of hearing.

h. The ninth, tenth and eleventh nerves leave the skull through the **foramen lacerum posterius** (fig. 75, IX, X, XI), a large space lying between the auditory bones and the exoccipital.

i. Finally, the twelfth nerve, the hypoglossal, passes out through the prominent **condylar foramen** (fig. 75, XII), which perforates the exoccipital just behind the foramen lacerum posterius.

II. Other Openings in the Skull.

a. The **anterior narial opening** lies at the anterior end of the skull, and is bounded by the premaxillae and nasals. In the natural condition it is divided into two by a vertical partition, formed by the **narial septum**, the anterior unossified part of the mesethmoid.

b. Penetrating the middle of the maxillae at the side of the face is the rather large **infra-orbital foramen** (fig. 73, 11), through which part of the second branch of the trigeminal nerve passes out from the orbit to the side of the face.

c. Several foramina are seen perforating the anterior part of the orbit. The most dorsal of these, perforating the lachrymal bone, is the **lachrymal foramen** (fig. 73, 13). Lying below and slightly external to this is a large foramen, through which part of the second branch of the trigeminal enters on its way to the infra-orbital foramen and so to the side of the face. Lastly, lying below these, and perforating the palatine, are two closely apposed foramina, the **internal orbital foramina**, through which part of the first or ophthalmic branch of the trigeminal nerve leaves the orbit, passing into the nasal cavity.

d. The anterior part of the palate between the premaxillae and the maxillae is perforated by a pair of long closely apposed apertures, the **anterior palatine foramina** (fig. 75, 19). They transmit part of the trigeminal nerve.

e. Towards the posterior part of the palate are two pairs of small **posterior palatine foramina** (fig. 75, 18). These perforate the palatine and transmit branches of the trigeminal nerve and certain blood-vessels.

f. The **posterior narial opening** is bounded chiefly by the palatines.

g. The **alisphenoid canal** (fig. 75, 21) is a short canal penetrating the base of the alisphenoid bone, and transmitting the external carotid artery. It lies between the foramen rotundum and the foramen ovale.

h. Between the auditory bulla and the foramen ovale are seen two openings. The more external of these is the opening of the **Eustachian canal** (fig. 75, 22), which communicates with the tympanic cavity. The more internal is the **foramen lacerum medium** (fig. 75, 9), through which the internal carotid enters the cranial cavity.

i. The **external auditory aperture** (fig. 75, 7) is a large opening with rough edges at the outer side of the tympanic bulla.

j. Between it and the glenoid surface of the squamosal is the **postglenoid** foramen (fig. 75, 10) through which a vein passes out.

k. Lastly, there is the great **foramen magnum** (fig. 75, 2), between the occipital condyles. Through it the brain and spinal cord communicate.

C. The Ribs and Sternum.

These, together with the thoracic vertebrae, form the skeletal framework of the thorax. Each rib is a curved rod, which at its dorsal end is movably articulated to the vertebra, and at its ventral end is either connected with the sternum, or ends freely. In the dog there are thirteen pairs of ribs, nine pairs of which are directly connected with the sternum, while the remaining four end freely and are known as **floating ribs**. Each rib is obviously divided into two parts, a dorsal or **vertebral part**, and a ventral or **sternal part**. The vertebral portion, which forms about two-thirds of the whole rib, is a flattened, regularly curved rod, completely ossified. Its dorsal end is rounded, forming the **head or capitulum**, which articulates with a concave surface furnished partly by the corresponding vertebra and partly by the vertebra next in front. The last three or four however articulate with one vertebra only. A short way behind the capitulum on the dorsal side of the rib is a rounded outgrowth, the **tubercle** or **tuberculum**, by means of which the rib articulates with the transverse process. The portion of the rib between the head and the tubercle is known as the **neck**. The **sternal portion** of the rib (fig. 76) is a short bar of calcified or imperfectly ossified cartilage, about one-third of the length of the corresponding bony portion. The anterior sternal ribs are somewhat more cartilaginous than the posterior ones. The vertebral portions increase in length from the first which is very stout, and has the capitulum and tuberculum very distinct, to about the eighth or ninth; afterwards they gradually diminish in size. The first nine to eleven have the capitula and tubercula separate, afterwards they gradually merge together.

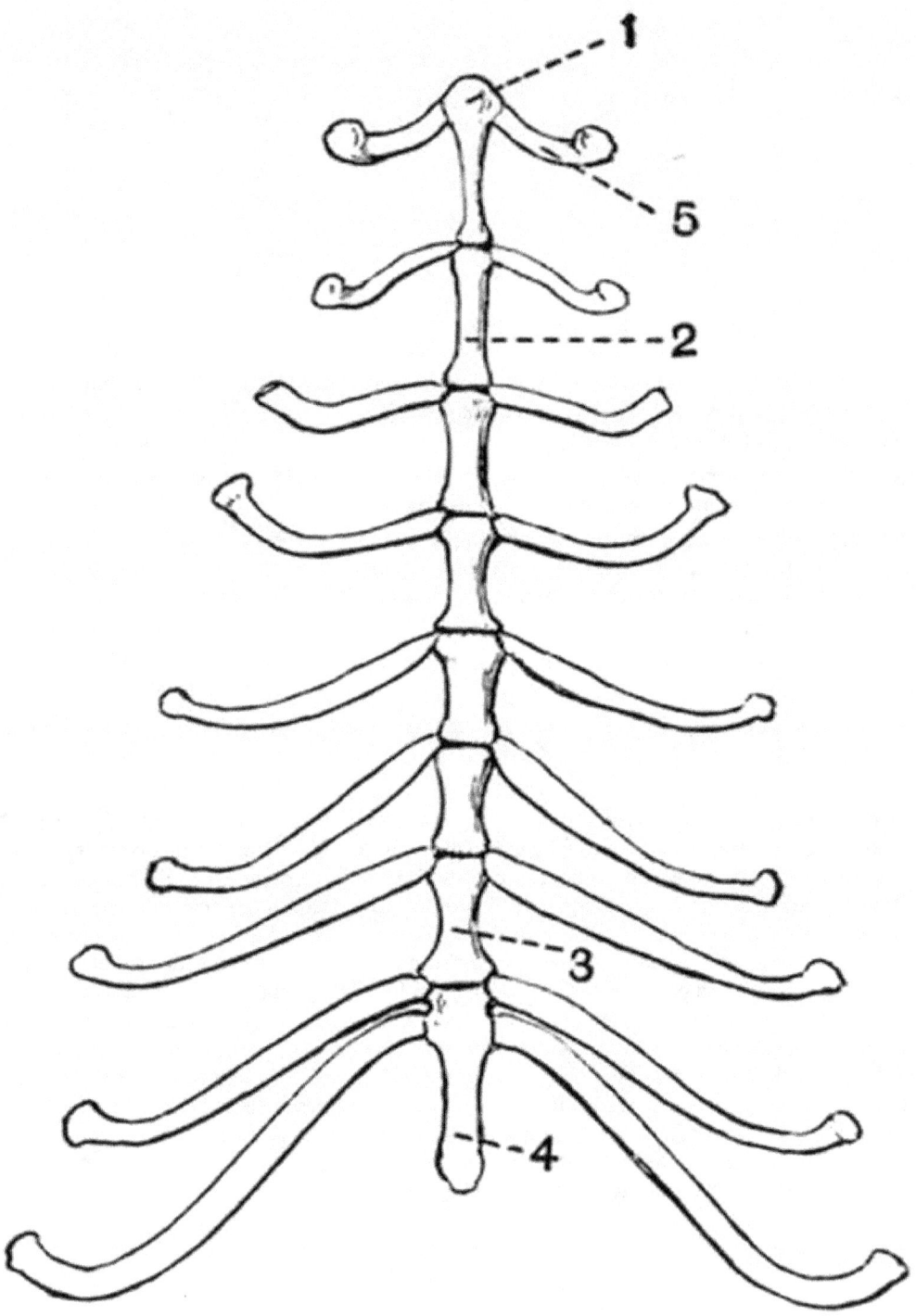

Fig. 76. Sternum and sternal ribs of a Dog (*Canis familiaris*) × ½.

1. presternum.

2. first sternebra of mesosternum.

3. last sternebra of mesosternum.

The Sternum.

4. xiphisternum. The flattened cartilaginous plate terminating the xiphisternum is not shown.

5. first sternal rib.

This is an elongated cylindrical structure lying in the mid-ventral wall of the thorax, and is divided into eight segments or **sternebrae**. The anterior segment, the **presternum** (fig. 76, 1) or **manubrium sterni** is expanded in front; the next six segments, which, together form the **mesosternum** are elongated, somewhat contracted in the middle and expanded at the ends. The last segment or **xiphisternum** (fig. 76, 4) is long and narrow, and terminates in a flattened expanded plate of cartilage. The first pair of sternal ribs articulate with the sides of the presternum, and the remaining pairs between the successive sternebrae. Between the last sternebra and the xiphisternum two pairs articulate. Development shows that the sternum is formed by the union in the middle line of two lateral portions; this can be well seen in the presternum and xiphisternum of the puppy, but no traces of this median division remain in the adult dog.

2. The Appendicular Skeleton.

The appendicular skeleton consists of the bones of the anterior and posterior limbs, and of their respective supports, the pectoral and pelvic girdles.

The Pectoral Girdle.

The **pectoral girdle** lies external to the ribs, and has no bony attachment to the axial skeleton. In almost all Mammalia it is, as compared with that in Sauropsids, very incomplete; and in the dog it is even more reduced than in the majority of Mammalia. The dorsal portion or **scapula** is well developed, but the ventral portion is almost entirely absent.

The **scapula** is somewhat triangular in shape, the apex being directed downwards and forwards, and being expanded to form the shallow **glenoid cavity** with which the head of the humerus articulates. The inner surface of the scapula is nearly flat, while the outer is drawn out into a very prominent ridge, the **spine**, which, arising gradually near the dorsal end, runs downwards, dividing the surface into two nearly equal parts, the **prescapular** and **postscapular fossae**, and ends in a short blunt process, the **acromion**. The anterior border of the scapula is somewhat curved, and is called the **coracoid border**; it is terminated ventrally by a slight blunt swelling, the **coracoid process**, which ossifies from a different centre from the rest of the scapula, and is probably the sole representative of the **coracoid**. The dorsal or **suprascapular border** of the scapula is rounded, while the posterior or **glenoid border** is nearly straight. The clavicle or collar bone, which in a large proportion of mammals is well seen, in the dog is very imperfectly developed; it is short and broad, and is suspended in the muscles, not reaching either the scapula or sternum.

The Anterior Limb.

The anterior limb of the dog is divisible into the usual three portions, the **brachium** or **upper arm**, the **antibrachium** or **fore-arm**, and the **manus** or **wrist** and **hand**.

The **brachium** or **upper arm** includes only a single bone, the **humerus**.

The **humerus** is a stout elongated bone, articulating by its large proximal **head** (fig. 77, 1) with the glenoid cavity of the scapula, and at its distal end by the **trochlea** with the bones of the fore-arm. The head passes on its inner side into an area roughened for the attachment of muscles and called the **lesser tuberosity** (fig. 77, 2); while in front it is divided by the shallow **bicipital groove** from a large roughened area, the **greater tuberosity** (fig. 77, 3), which is continued as a slight roughened ridge, extending about one-third of the way down the outer side of the shaft. This ridge, which in many animals is much more strongly developed than it is in the dog, is called the **deltoid ridge**. The **trochlea** (fig. 77, 5) at the distal end of the bone is a pulley-like surface, elevated at the sides and grooved in the middle. It articulates with the radius and ulna of the fore-arm. On each side of it are slight roughened projections, the **internal** and **external condyles** (fig. 77, 7). In the cat and many other animals there is a foramen, the **ent-epicondylar foramen** above the internal condyle, but in the dog this is not developed. Passing up the shaft from the external condyle is a slight ridge, the **supinator** or **ectocondylar ridge**; this is better developed in many mammals. Immediately above the trochlea in front and behind are the deep **supra-trochlear fossae**, which communicate with one another through the **supra-trochlear foramen** (fig. 77, 8). The posterior of these, the **olecranon fossa**, is much the deeper, and receives the olecranon process of the ulna when the arm is extended. The head and tuberosities of the humerus ossify from one centre, the shaft from a second, and the trochlea and condyles from a third.

The **fore-arm** or **antibrachium** contains two bones, the **radius** and **ulna**; they are immovably articulated with one another, but not fused. The pre-axial bone, the **radius** (fig. 77, B), which lies more or less in front of the ulna, is external to the ulna at its proximal end, and at its distal end is internal to that bone. It articulates with the external portion of the trochlea, while the ulna articulates with the internal portion. It is a straight bone with its distal end slightly larger than its proximal end. The proximal end articulates with the trochlea, the distal end with the bones of the carpus.

The postaxial bone, the **ulna** (fig. 77, C), has the proximal end much enlarged, forming the **olecranon** (fig. 77, 11), and tapers gradually to the distal end. Near its proximal end the ulna is marked by a deep **sigmoid notch**, which bears on its inner side a concave surface (fig. 77, 12) for articulation with the trochlea. The pointed proximal end of the sigmoid notch is called the **coronoid process**. Somewhat in front of and below the sigmoid notch is a smaller hollow (fig. 77, 13), with which the radius articulates.

201

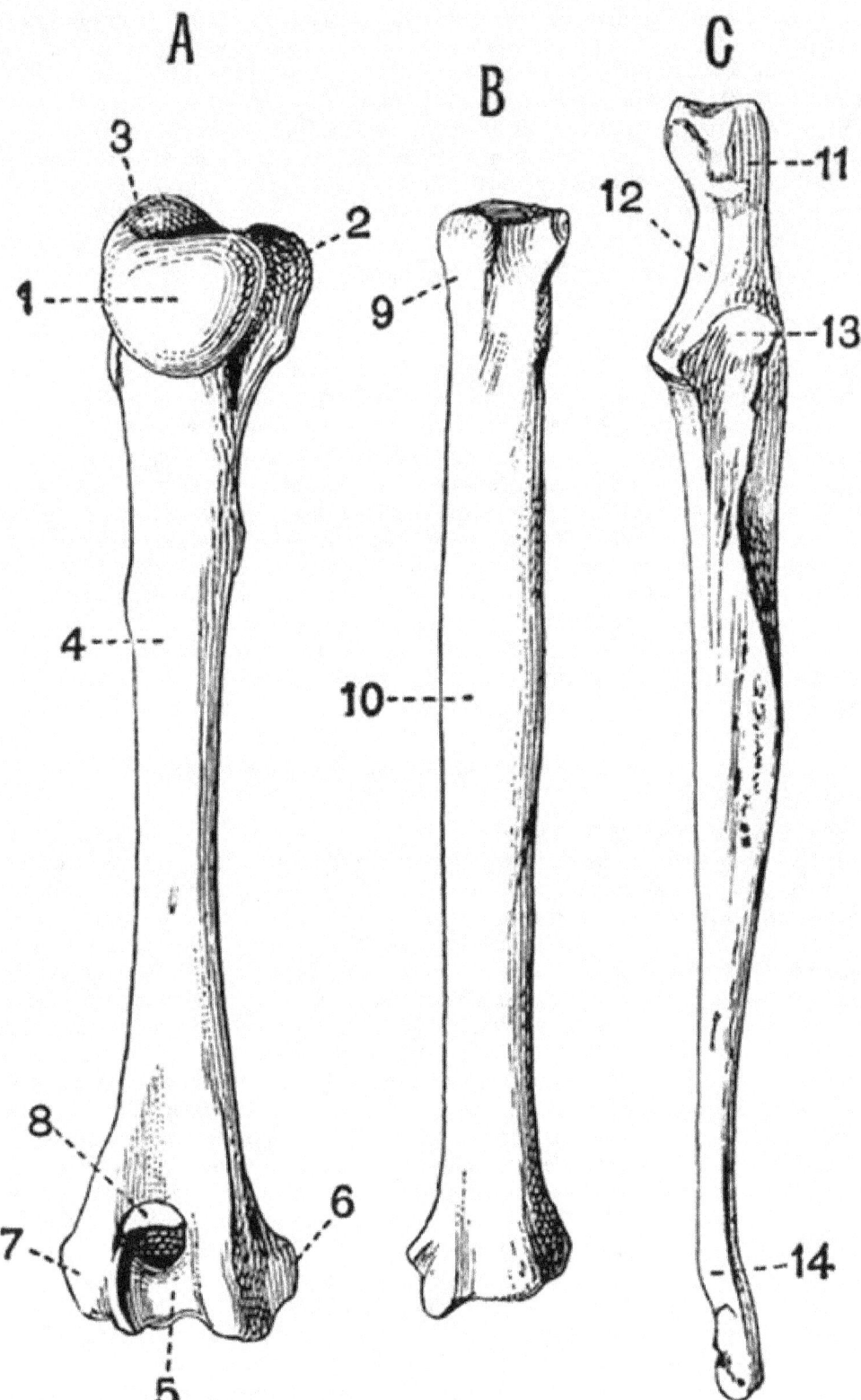

Fig. 77. Bones of the left upper arm and fore-arm of a Dog
(*Canis familiaris*) × ½.
A, humerus (seen from the posterior side); B, radius, C, ulna, both seen
from the anterior side.

1. head.

2. lesser tuberosity.

3. greater tuberosity.

4. shaft of the humerus.

5. trochlea.

6. internal condyle.

7. external condyle.

8. supra-trochlear foramen.

9. proximal end of the radius.

10. shaft of the radius.

11. olecranon.

12. surface for articulation with the trochlea.

13. surface for articulation with the radius.

14. distal end of the ulna.

In the young animal the ends of both radius and ulna are seen to ossify from centres different from those forming the shafts. The epiphyses forming both ends of the radius, and the distal end of the ulna are large, while that at the proximal end of the ulna is small, and forms only the end of the olecranon.

The **Manus** is divided into

a. The **carpus** or **wrist**, formed of a group of small bones.

b. The **hand**, which includes firstly some elongated bones, the **metacarpals**, forming what corresponds to the palm of the hand, and secondly the phalanges, which form the **fingers**.

The **Carpus** or **wrist**. The carpus of the dog consists of seven small bones, arranged in a proximal row of three, and a distal row of four. It differs much from the simpler type met with in the newt. The largest bone of the proximal row is the **scapho-lunar** (fig. 80, 1), formed by the fused **scaphoid** (radiale), **lunar** (intermedium), and **centrale**; it has a large convex proximal surface for articulation with the radius, and articulates distally with the trapezium, trapezoid, and magnum, and internally with the cuneiform. The **cuneiform** (ulnare) (fig. 80, 2) has a posterior rounded surface articulating with the ulna; it articulates in front with the unciform, and internally with the **pisiform** (fig. 80, 7), which is a comparatively large sesamoid bone on the ulnar side of the carpus. Frequently also there is a small sesamoid bone on the radial side of the carpus. The **trapezium** (carpale 1), **trapezoid** (carpale 2), and **magnum** (carpale 3) (fig. 80, 5) are all small bones, and support respectively the first, second, and third metacarpals. The **unciform** (carpalia 4 and 5) (fig. 80, 6) is larger, and supports the fourth and fifth metacarpals.

The hand has five **digits**, each consisting of an elongated **metacarpal**, followed by **phalanges**, the last of which, the **ungual phalanx**, is pointed and curved, and bears the claw. Each of the metacarpals is seen in the young animal to have its distal end formed by a prominent epiphysis, and each of the phalanges, except those bearing the claws, has a similar epiphysis at its proximal end.

The **pollex** (fig. 80, A, I) is far shorter than the other digits, and normally does not touch the ground in walking. It has only two phalanges, while each of the other digits has three. A pair of small sesamoid bones are developed on the ventral or flexor side of the metacarpo-phalangeal articulations of all the digits except the pollex. Frequently similar sesamoid bones occur also on the dorsal side of the phalangeal articulations.

The Pelvic Girdle.

The **pelvic girdle** consists of two halves, which lie nearly parallel to the vertebral column.

Each half is firmly united to its fellow in a ventral symphysis behind, and is in front expanded and united to the sacrum. Each half or **innominate bone** is seen in the young animal to consist of four distinct parts, the **ilium** or dorsal element, the **pubis** or anterior ventral element, the **ischium** or posterior ventral element, and a small fourth part, the **acetabular** or **cotyloid** bone, wedged in between the three others. These parts, though all distinct in the young animal, are in the adult so completely fused that their respective boundaries cannot be distinguished. At about the middle of the outer surface of the innominate bone is a very deep cavity, the **acetabulum** (fig. 78, A, 1) with which the head of the femur articulates; all the bones except the pubis take part in its formation.

The **ilium** is a rather long bone, expanded in front and contracted behind; it forms about half the acetabulum. On its inner or **sacral surface** (fig. 78, 4) is a large roughened patch for articulation with the sacrum; its outer or **gluteal surface** is concave. The posterior part of the bone is flattened below, forming the narrow **iliac surface** (fig. 78, A, 5).

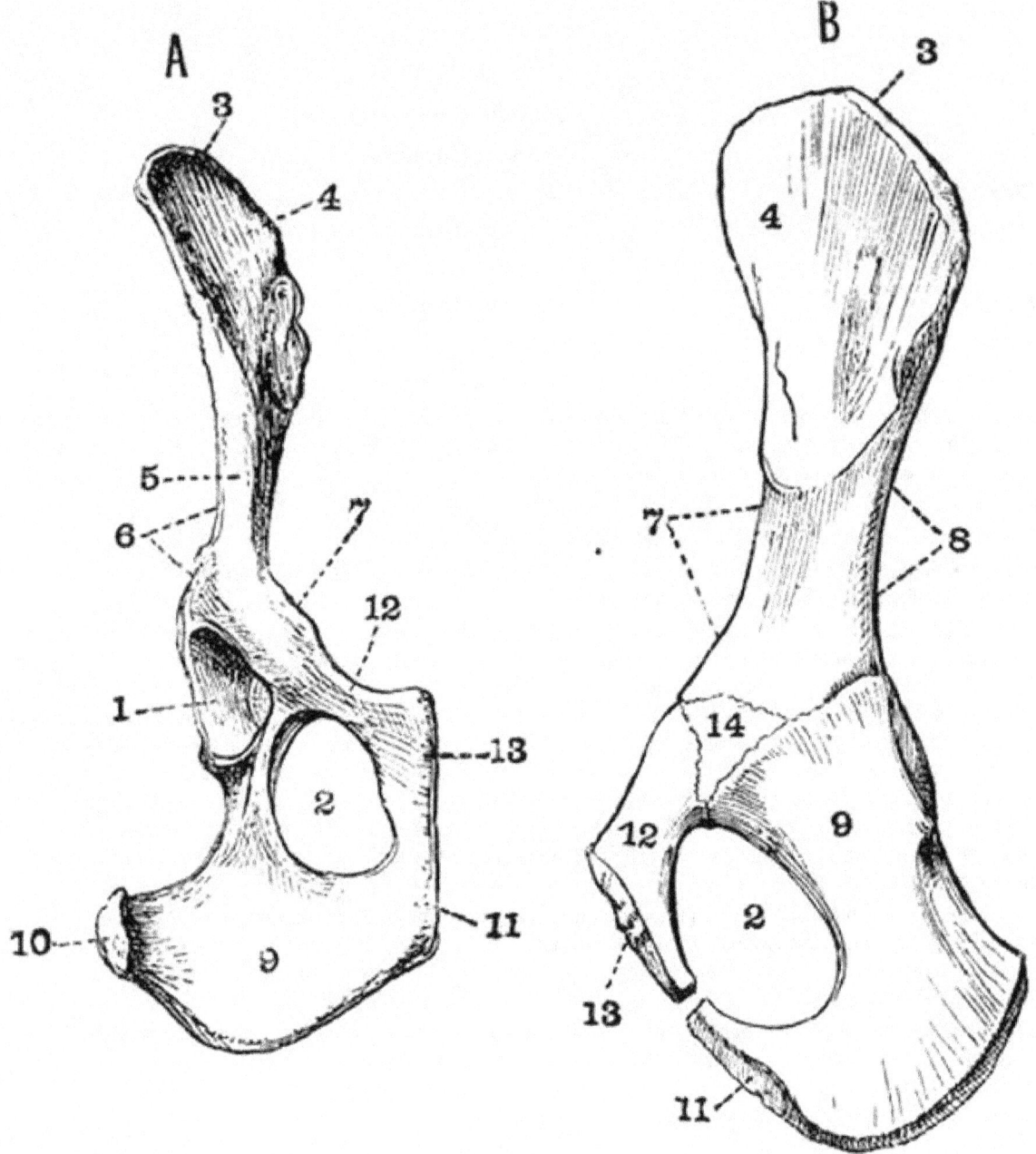

Fig. 78. Right innominate bone, A, of a full-grown Terrier, B, of
a Collie Puppy. × 1.

A is seen from the ventral side, B from the inner or sacral side.

1. acetabulum.

2. thyroid foramen.

3. supra-iliac border of ilium.

4. sacral surface.

5. iliac surface.

6. acetabular border.

7. pubic border.

8. ischial border.

9. ischium.

10. tuberosity of ischium.

11. ischial symphysis.

12. pubis.

13. pubic symphysis.

14. cotyloid or acetabular bone.

The **ischium** (fig. 78, 9) is a wide flattened bone forming the posterior part of the innominate bone. It meets the pubis ventrally, but is separated from it for the greater part of its length by the large **obturator** or **thyroid foramen** (fig. 78, 2). At its posterior end externally it bears a rather prominent roughened **ischial tuberosity** (fig. 78, A, 10). The ischium meets its fellow in a ventral symphysis, and forms about one-third of the acetabulum.

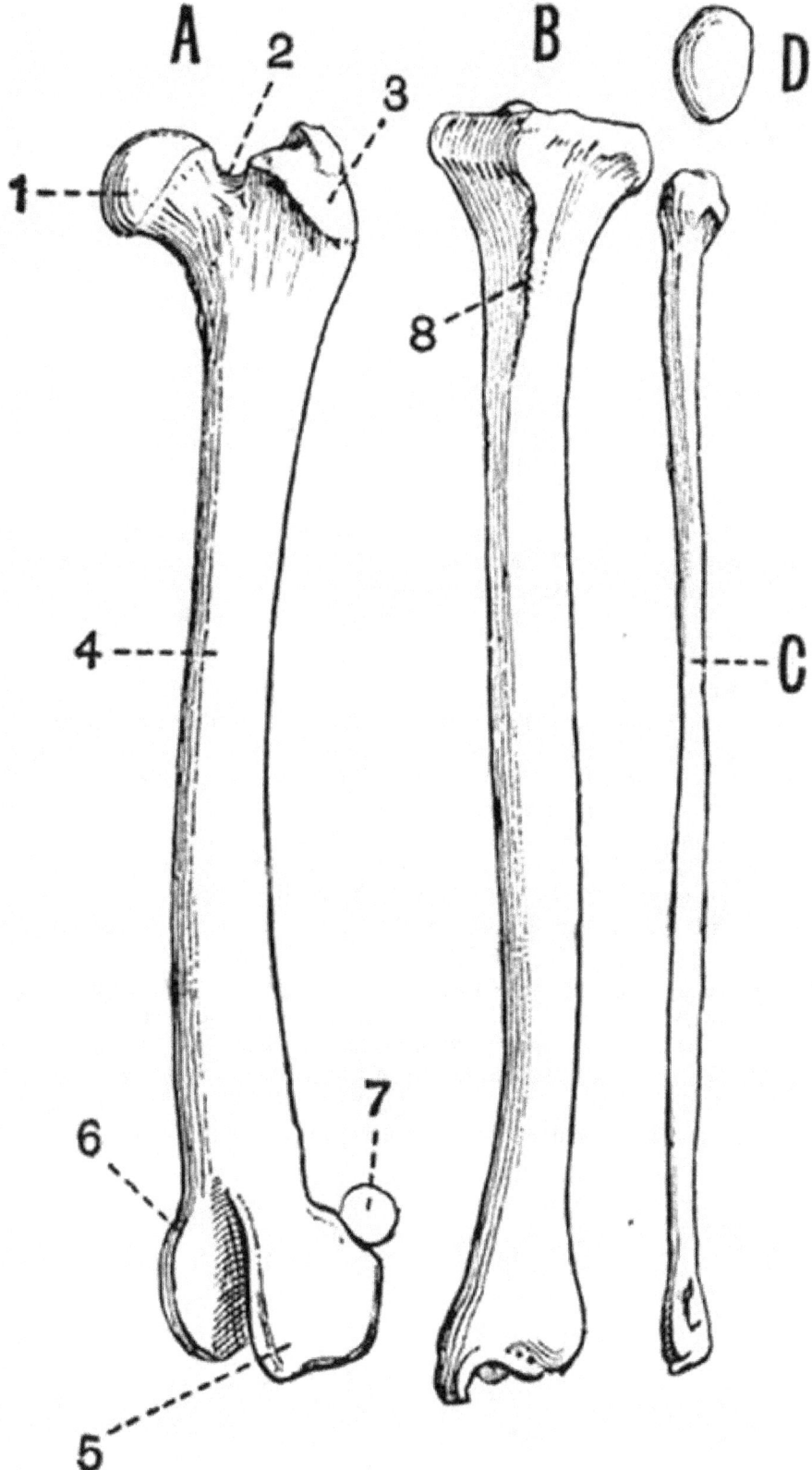

Fig. 79. Front view of the left leg bones of a Dog
(*Canis familiaris*) × ½.
A femur, B tibia, C fibula, D patella.

205

1. head of femur.	5. external condyle.
2. neck.	6. internal condyle.
3. great trochanter.	7. fabella.
4. shaft.	8. cnemial crest.

The **pubis** (fig. 78, 12) is smaller than either the ischium or ilium; it does not take part in the formation of the acetabulum, and like the ischium, meets its fellow in a ventral symphysis. The **acetabular bone** (fig, 78, B, 14) is small and triangular, and is wedged in between the other three. It forms about one-sixth of the acetabulum.

The Posterior Limb.

The **posterior limb**, like the anterior, is divisible into three parts; these are the **thigh**, the **crus** or **shin**, and the **pes**.

The **thigh** contains only a single bone, the **femur**.

The **femur** is a long straight bone with a nearly smooth shaft and expanded ends. The proximal end bears on its inner side the large rounded **head** (fig. 79, A, 1) which articulates with the acetabulum. External to the head and divided from it by a deep pit is a large rough outgrowth, the **great trochanter** (fig. 79, 3). The deep pit is the **trochanteric** or **digital fossa**. On the inner side below the head is a smaller roughened surface, the **lesser trochanter**. The lower or distal end of the bone bears two prominent rounded surfaces, the **condyles**, which articulate with the tibia. They are separated from one another by the deep **intercondylar notch**, which is continued above and in front as a shallow groove, lodging a large sesamoid bone, the **patella** or **knee-cap**. At the back of the knee-joint are a pair of smaller sesamoids, the **fabellae** (fig. 79, 7).

In the young animal there are three epiphyses to the shaft of the femur, one forming the head, one the great trochanter, and one the distal end.

The **crus** or **shin** contains two bones, the **tibia** and **fibula**. The **tibia** is a fairly thick straight bone, expanded at both ends, especially at the head or proximal end. The proximal end is triangular in cross section, and bears two facets for articulation with the condyles of the femur. The anterior surface of the proximal end of the tibia is marked by the strong **cnemial crest** (fig. 79, 8), which runs some way down the shaft. The distal end of the tibia articulates with the astragalus by an irregular, somewhat square surface.

The shaft of the tibia ossifies from one centre, the distal end from a second, and the proximal end from two more.

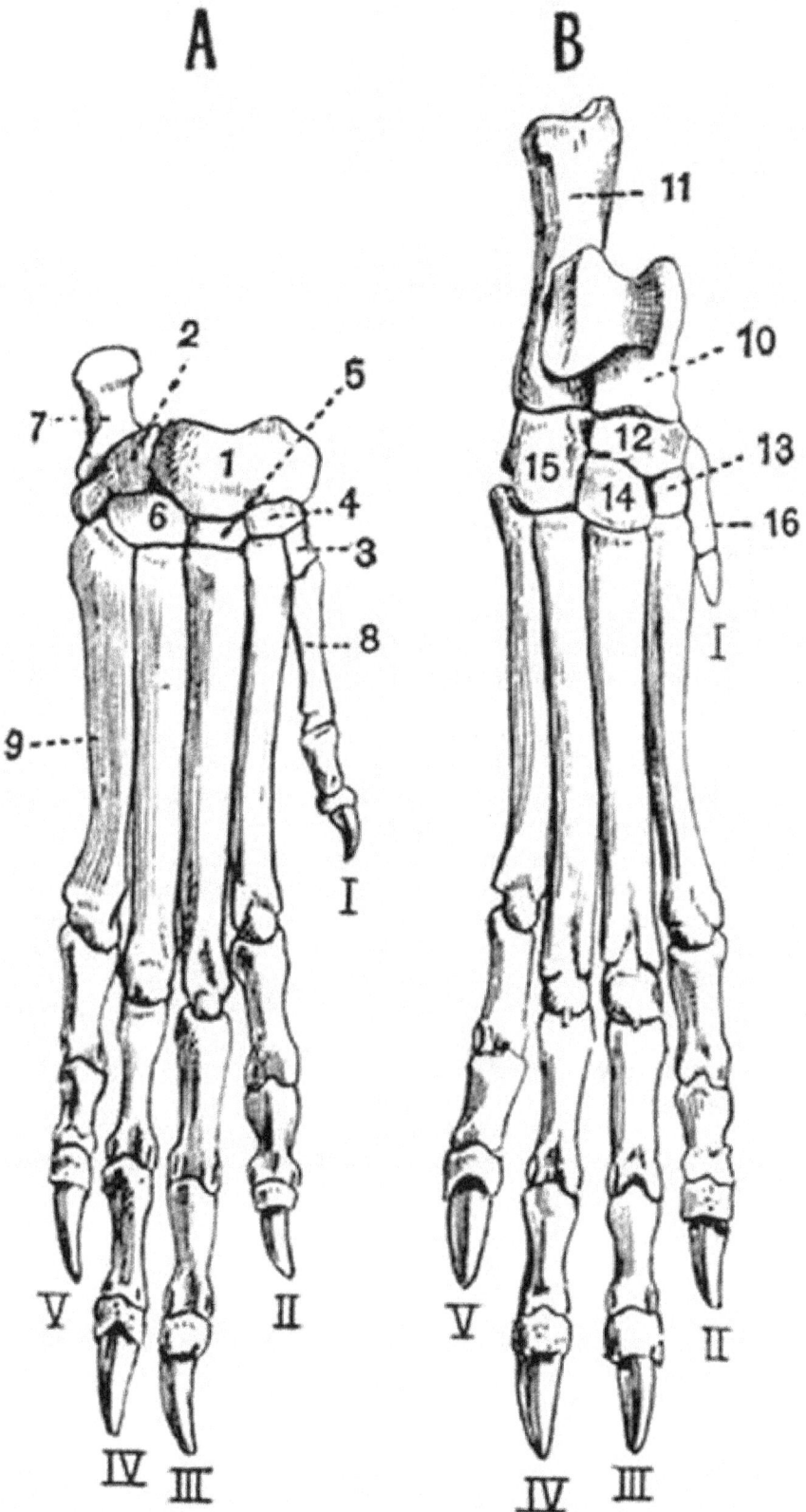

Fig. 80. A, right manus, B, right pes of a Dog
(*Canis familiaris*) × ½ (after von Zittel).

1. bone representing the fused
scaphoid, lunar and

9. fifth metacarpal.

10. astragalus.

centrale.

2. cuneiform.

3. trapezium.

4. trapezoid.

5. magnum.

6. unciform.

7. pisiform.

8. first metacarpal.

11. calcaneum.

12. navicular.

13. middle cuneiform.

14. external cuneiform.

15. cuboid.

16. first metatarsal.

The digits are numbered with

Roman numerals.

The **fibula** (fig. 79, C) is a distinct but very slender bone, somewhat expanded at both ends. It lies external to the tibia and articulates by its proximal end with the head of the tibia, and by its distal end with the calcaneum. Its shaft and proximal end ossify from one centre, and its distal end from a second.

The **Pes**.

The structure of the **pes** corresponds closely with that of the manus. It is divided into:—

a. The **tarsus** or **ankle** formed of a group of small bones.

b. The **foot**, which includes, firstly, some elongated bones, the **metatarsals**, forming what corresponds to the sole of the foot, and secondly the **phalanges**, which form the toes.

The **Tarsus**. The tarsus of the dog consists of seven bones arranged in two rows, of two and four respectively, with a **centrale** between them. The two bones of the proximal row are the **astragalus** and **calcaneum**.

The **astragalus** (fig. 80, 10) corresponds to the fused **tibiale** and **intermedium** of the typical tarsus. Its proximal end is much wider than its distal end, and forms a large rounded **condyle** articulating with the tibia, while its posterior end meets the navicular. It lies to the dorsal side of the foot.

The **calcaneum** (fibulare) (fig. 80, 11), the thickest bone in the pes, lies somewhat behind, and to the outer side of the astragalus. It articulates with the astragalus and fibula, and is drawn out behind into a long rounded process, which forms the heel, and is in the young animal terminated by an epiphysis. Between the proximal and distal rows of tarsals is the **navicular** (centrale) (fig. 80, 12), a somewhat flattened and square bone articulating with the astragalus.

The distal row of tarsals consists of four bones. The **internal cuneiform** (tarsale 1) is a smooth flattened bone lying to the inner side of the foot; it articulates with the first metatarsal and with the navicular. The **middle cuneiform** (tarsale 2) (fig. 80, 13) is a still smaller bone, lying external to the internal cuneiform. It articulates with the second metatarsal and with the navicular. The **external cuneiform** (tarsale 3) (fig. 80, 14) is a larger, somewhat square bone lying external to the middle cuneiform. It articulates with the third metatarsal and with the navicular. The **cuboid** (tarsalia 4 and 5) (fig. 80, 15) is a considerably larger bone lying to the outer side of the foot. It articulates with the fourth and fifth metatarsals and with the calcaneum.

The pes has sometimes five digits, sometimes four, the hallux being absent. Even when present the **hallux** (fig. 80, **B, I**) is commonly much reduced, and may be quite vestigial, and represented only by a small nodular metatarsal.

Each of the other digits consists of a long metatarsal, which in the young animal has a prominent epiphysis at its distal end, and of three phalanges. The proximal and middle phalanges have epiphyses at their proximal ends, while the distal phalanx is without epiphyses and is claw-shaped.

CHAPTER XXII.
GENERAL ACCOUNT OF THE SKELETON IN MAMMALIA.

THE EXOSKELETON AND VERTEBRAL COLUMN.

Epidermal Exoskeleton.

Hair, which forms the characteristic Mammalian exoskeleton, varies much in different animals, and in different parts of the same animal. A large proportion of mammals have the surface fairly uniformly covered with hair of one kind only. In some forms however there are two kinds of hair, a longer and stiffer kind alone appearing on the surface, and a shorter and softer kind forming the under fur. In most mammals hairs of a special character occur in certain regions, such as above the eyes, on the margins of the eyelids, and on the lips and cheeks, here forming the vibrissae or whiskers.

Sometimes as in *Hippopotamus*, *Orycteropus* and the Sirenia, the hair, though scattered over the whole surface, is extremely scanty, while in the Cetacea it is limited to a few bristles in the neighbourhood of the mouth, or may even be absent altogether in the adult. In most mammals the hairs are shed and renewed at intervals, sometimes twice a year, before and after the winter. The vibrissae or large hairs which occur in many animals upon the upper lip, and the mane and tail of Equidae are probably persistent.

In the hedgehogs, porcupines and *Echidna* certain of the hairs are modified and greatly enlarged, forming stiff spines. Similar spines occur in the young of *Centetes*, and in *Acanthomys* among the Muridae.

Several other forms of epidermal exoskeleton are met with in mammals, including:—

(*a*) **Scales**. These overlie the bony scutes of armadillos and occur covering the tail in several groups of mammals, such as beavers and rats. In the Manidae the body is covered by flat scales which overlap.

(*b*) The **horns** of Bovine Ruminants. These, which must on no account be confused with antlers, are hollow cases of hardened epidermis fitting on to bony outgrowths of the frontals. In almost every case they are unbranched structures growing continuously throughout life, and are very rarely shed entire. In the Prongbuck *Antilocapra* however they are bifurcated and are periodically shed. Horns are nearly always limited to a single pair, but the four-horned antelope *Tetraceros* has two pairs, the anterior pair being the smaller.

(*c*) The **horns of Rhinoceroses**. These are conical structures composed of a solid mass of hardened epidermal cells growing from a cluster of long dermal papillae. From each papilla there grows a fibre which resembles a thick hair, and cementing the whole together are cells which grow from the interspaces between the papillae. These fibres differ from true hairs in not being developed in pits in the dermis. Rhinoceros horns may be either one or two in number, and are borne on the fronto-nasal region of the skull. They vary much in length, the longest recorded having the enormous length of fifty-seven inches.

(*d*) *Nails*, *hoofs* and **claws**. In almost all mammals except the Cetacea, these are found terminating the digits of both limbs. **Nails** are more or less flattened structures, **claws** are pointed and somewhat curved. In most mammals the nails tend to surround the ends of the digits much more than they do in man. Sometimes the nail of one digit differs from that of all the others; thus the second digit of the pes in the Hyracoidea and Lemuroidea is terminated by a long claw, the other digits having flat nails. In the Felidae the claws are retractile, the ungual phalanx with claw attached folding back when the animal is at rest into a sheath, above, or by the side of the middle phalanx. In the Sloths and Bats enormously developed claws occur, forming hooks by which the animals suspend themselves. In *Notoryctes* the third and fourth digits of the manus bear claws of great size; similar claws occur in *Chrysochloris*, being correlated in each case with fossorial habits. The nail at its maximum development entirely surrounds the terminal phalanx of the digit to which it is attached, and is then called a **hoof**. Hoofs are specially characteristic of the Ungulata.

(*e*) **Spurs** and **beaks** are structures which are hardly represented among mammals, while so characteristic of birds. They are however both found in the Monotremata. In both *Echidna* and *Ornithorhynchus* the male has a peculiar hollow horny spur borne on a sesamoid bone articulated to the tibia. The jaws in *Ornithorhynchus* are cased in horny beaks similar to those of birds, and are provided with horny pads which act as teeth.

(*f*) **Horny plates** of a ridged or roughened character occur upon the anterior portion of the palate, and of the mandibular symphysis in all three genera of recent Sirenia; also upon the toothless anterior portion of the palate in Ruminants.

(*g*) The **baleen of whales** also belongs to the epidermal exoskeleton. It consists of a number of flattened horny plates arranged in a double series along the palate. The plates are somewhat triangular in form and have their bases attached to the palate at right angles to its long axis, while their apices hang downwards into the mouth cavity. The outer edge of each plate is hard and smooth, while the inner edge and apex fray out into long fibres which look like hair. At the inner edge of each principal plate are subsidiary smaller plates. The plates are formed of a number of fibres each developed round a dermal papilla in the same way as are the fibres forming the horns of *Rhinoceros*. Baleen and Rhinoceros horn likewise agree in that the fibres are bound together by less hardened epithelial cells, which readily wear away and allow the harder fibres to fray out. The greatest development of baleen occurs in the Northern Right whale, *Balaena mysticetus*, in which the plates number three hundred and eighty or more on each side, and reach a length of ten or twelve feet near the middle of the series.

Dermal Exoskeleton.

Mammals show two principal kinds of exoskeletal structures which are entirely or partially dermal in origin, viz. the bony scutes of armadillos, and teeth.

The **bony scutes of armadillos** are quadrate or polygonal in shape and are in general aggregated together, forming several shields protecting various regions of the body. The head is generally protected by a *cephalic* shield, the anterior part of the body by a *scapular*, and the posterior by a *pelvic* shield. The tail is also generally encased in bony rings, and scutes are irregularly scattered over the surface of the limbs. The mid-body region is protected by a varying number of bands of scutes united by soft skin, so as to allow of movement. Corresponding to each dermal scute is an epidermal plate. In *Chlamydophorus* the scutes are mainly confined to the posterior region where they form a strong vertically-placed shield which coalesces with the pelvis. The anterior part of the body is mainly covered by horny epidermal plates with very little ossification beneath. In the gigantic extinct Glyptodonts the body is covered with a solid carapace formed by the union of an immense number of plates, and there are no movable rings. The top of the head is defended by a similar plate, the tail is generally encased in an unjointed bony tube, and there is commonly a ventral plastron.

In *Phocaena phocaenoides* the occurrence of vestigial dermal ossicles has been described, and in *Zeuglodon* the back was probably protected by dermal plates.

Teeth.

Teeth are well developed in the vast majority of mammalia, and are of the greatest morphological and systematic importance, many extinct forms being known only by their teeth. Mammalian teeth differ from those of lower animals in various well-marked respects. (1) They are attached only to the maxillae, premaxillae and mandible, never to the palatines, pterygoids or other bones. (2) They frequently have more than one root. (3) They are always, except in some Odontoceti, placed in distinct sockets. (4) They are hardly ever ankylosed to the bone. (5) They are in most cases markedly heterodont. (6) They are commonly developed in two sets, the milk dentition and permanent dentition.

It sometimes happens that teeth after being formed are reabsorbed without ever cutting the gum. This is the case, for instance, with the upper incisors of Ruminants.

The form of mammalian teeth varies much, some are simple conical structures comparable to those of most reptiles, and these may either have persistent pulps, as in the case of the upper canines of the Walrus and the tusks of Elephants, or may be rooted as in most canine teeth. Some teeth have chisel-shaped edges, and this may be their original form, as in the human incisors, or may, as in those of Rodents, be brought about by the more rapid wearing away of the posterior edge, the anterior edge being hardened by a layer of enamel. Then, again, the crown may, as in the majority of grinding teeth, be more or less flattened. The various terms used in describing some of the forms of the surface of grinding teeth are defined on page 345.

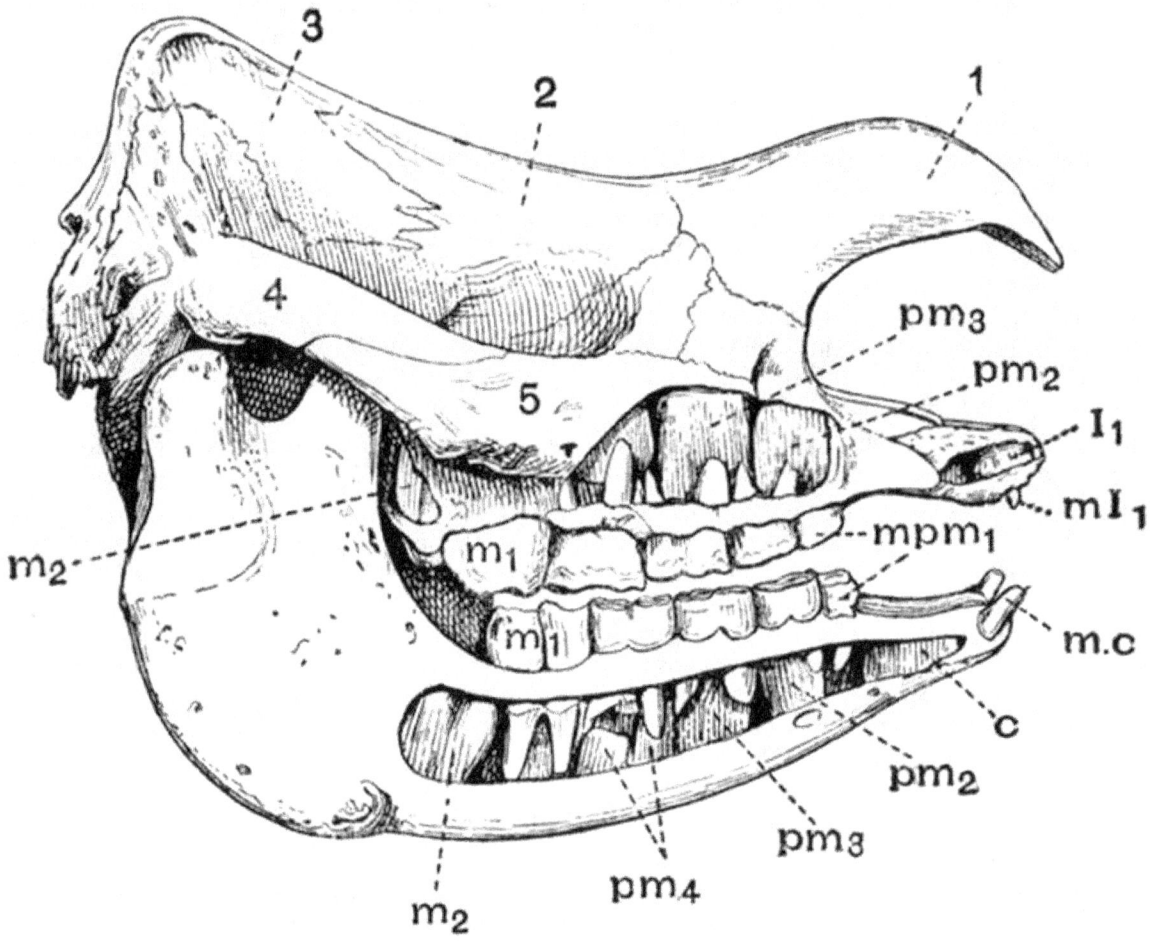

Fig. 81. Skull of a young Indian Rhinoceros (*R. unicornis*),
SHOWING THE CHANGE OF THE DENTITION × 1/7. (Brit. Mus.)

1. nasal.   *mc*. milk canine.

2. frontal.   *mpm_{1}*. milk premolar.

3. parietal.	*I_{1}*. first incisor.
4. zygomatic process of	*c.* canine.
squamosal.	*pm_{2}, pm_{3}, pm_{4}*. 2nd, 3rd and
5. jugal.	4th premolars.
ml_{1}. milk incisor.	m_1, m_2. first and second molars.

The teeth of the Aard Varks are compound, and differ completely from those of all other mammals (see p. 425).

As a rule, the higher the general organisation of an animal the better are its milk teeth developed, and the more do they form a reproduction on a small scale of the permanent set. This fact is well seen in the Primates, Carnivora and Ungulata. The method of notation by which the dentition of any mammal can be briefly expressed as a formula has been already described. The regular mammalian arrangement of teeth for each side is expressed by the formula

$i\ ^3/_3\ c\ ^1/_1\ pm\ ^4/_4\ m\ ^3/_3 \times 2$; total, 44.

Monotremata. In *Echidna* teeth are quite absent. In the young *Ornithorhynchus* functional molar teeth of a multi-tubercular type resembling those of some Mesozoic mammalia are present, but in the adult they disappear, their office being discharged by horny plates.

Marsupialia have a heterodont dentition, which has generally been regarded as almost monophyodont, the only tooth which has an obvious deciduous predecessor being the last premolar. The researches of Röse and Kükenthal tend to show that the teeth of Marsupials are developed in the same way as in other mammals, and are diphyodont. In the case of the premolars, teeth which are homologous with the permanent teeth of other mammals begin to develop as lateral outgrowths from the milk teeth, but afterwards become absorbed, so that the teeth which actually persist belong to the milk series. The last premolar, however, does as a rule develop and replace its milk predecessor; sometimes, however, as in *Didelphys*, it takes its place among the milk molars without replacing one of them.

The types of dentition characteristic of the different groups of placental mammals may mostly be paralleled among the Marsupials. Thus among the polyprotodont forms the Didelphyidae or opossums, and some of the Dasyuridae, such as *Sarcophilus* and *Thylacinus*, have a typical carnivorous dentition with small incisors, large canines, and molars with pointed compressed crowns. The dental formula of *Thylacinus*, is $i\ ^4/_3\ c\ ^1/_1\ pm\ ^3/_3\ m\ ^4/_4$, total 46.

In *Myrmecobius* five or six molar teeth occur on each side, and the total number of teeth reaches fifty-two or fifty-six. The teeth bear rows of tubercles, and resemble those of the Multituberculate mesozoic Mammals, more than do those of any other living form. Calcified teeth have recently been described in *Myrmecobius* earlier than the functional or milk set. This would relegate the milk teeth of mammals in general to a second series, and the permanent teeth to a third. In *Notoryctes* the dental formula is given as $i\ ^3/_2\ c\ ^1/_1\ pm\ ^2/_3\ m\ ^4/_4$, total 40. The canines are small, and the anterior molars have strongly developed cusps, and much resemble those of *Chrysochloris* (Insectivora).

Among the diprotodont types the Phascolomyidae, or Wombats, have a dentition recalling that of the Rodents. All the teeth grow from persistent pulps, and the incisors have enamel only on the anterior surface as in Rodents. The dental formula is $i\ ^1/_1\ c\ ^0/_0\ pm\ ^1/_1\ m\ ^4/_4$, total 24. There are indications of a vestigial second pair of incisors.

The Macropodidae, or Kangaroos, have a herbivorous dentition with the formula $i\ ^3/_1\ c\ ^{(0-1)}/_0\ pm\ ^2/_2\ m\ ^4/_4$. The incisors are sharp and cutting, and are separated by a long diastema or gap from the molars, which have their crowns marked by ridges or cusps. There are indications of several vestigial incisors.

Coenolestes, a remarkable form recently described from America, belongs to the diprotodont section, and is the only living member of the section known outside the Australian region. An exceptional dentition is seen in the case of the extinct *Thylacoleo*, in which the functional teeth are reduced to two pairs; one pair of large cutting incisors and one of compressed sharp-edged premolars.

Edentata. Some Edentata, viz. the ant-eaters (Myrmecophagidae) are, as far as is known, absolutely toothless at all stages of their existence; being the only mammals except *Echidna* in which no tooth germs have been discovered; others, viz. the Manidae, though showing foetal tooth germs, are quite toothless in post-foetal life; others, viz. some of the armadillos, have the largest number of teeth met with in land mammals. The teeth are homodont except in the Aard Varks, and grow from persistent pulps. In the sloths (Bradypodidae) and the Megatheriidae, there are five pairs of teeth in the upper and four in the lower jaw. The teeth of sloths consist of a central axis of vasodentine, surrounded firstly by a thin coating of hard dentine, and secondly by a thick coating of cement.

In no living Edentate have the teeth any enamel; it has, however, been described as occurring in certain early Megatheroid forms from S. America, and an enamel organ has also been discovered in an embryo *Dasypus*. In the Armadillos (Dasypodidae) the number of teeth varies from $^8/_8$ or $^7/_7$ in *Tatusia*, to upwards of $^{25}/_{25}$ in *Priodon*, which therefore may have upwards of a hundred teeth, the largest number met with in any land mammal. In *Tatusia* all the teeth except the last are preceded by two-rooted milk teeth. The Aard Varks are diphyodont, and milk teeth are also known in a species of *Dasypus*,

211

but with these exceptions Edentates are, as far as is known, monophyodont. In *Glyptodon* the teeth are almost divided into three lobes by two deep grooves on each side.

The Aard Varks (Orycteropodidae) are quite exceptional as regards their teeth, which are cylindrical in shape, and are made up of a number of elongated denticles fused together. Each denticle contains a pulp cavity from which a number of minute tubes radiate outwards. These teeth are diphyodont and somewhat heterodont, eight to ten pairs occur in the upper jaw and eight in the lower, but they are not all in place at one time. The last three teeth in each jaw are not preceded by milk teeth.

Sirenia. The teeth of Sirenia show several very distinct types, the least modified being that of the extinct Halitheriidae, which have large incisors in the upper jaw, and five or six pairs of tuberculated grinding teeth in each jaw, the anterior ones being preceded by milk teeth.

In both the living genera the dentition is monophyodont. In *Manatus* the dentition is $i\ ^2/_2\ pm$ and $m\ ^{11}/_{11}$. The incisors are vestigial, and disappear before maturity. The grinding teeth have square enamelled crowns marked by transverse tuberculated ridges. They are not all present in the jaw at the same time. In *Halicore* the upper jaw bears a pair of straight tusklike incisors; in the male these have persistent pulps and project out of the mouth; in the female they soon cease to grow and are never cut. They are separated by a long diastema from the grinding teeth which have tuberculated crowns and are $^5/_5$ or $^6/_6$ in number, but are not all in place at once. Several other pairs of slender teeth occur in the young animal, but are absorbed or fall out before maturity. In *Rhytina* teeth are altogether absent.

Cetacea.

Archaeoceti. Zeuglodon has the following dentition, $i\ ^3/_3\ c\ ^1/_1\ pm$ and $m\ ^5/_5$, total 36. The incisors and canines are simple and conical; the cheek teeth are compressed and have serrated cutting edges like those in some seals.

In the *Mystacoceti*, or whalebone whales, calcified tooth germs probably belonging to the milk dentition are present in the embryo, but they are never functional, and are altogether absent in the adult. The anterior of these germs are simple, the posterior ones are originally complex, but subsequently split up into simple teeth like those of the anterior part of the jaw. Hence according to Kükenthal, who described these structures, the Cetacean dentition was originally heterodont.

In the living *Odontoceti* the dentition is homodont and monophyodont. In some cases traces occur of a replacing dentition which never comes to maturity, and renders it probable that the functional teeth of the Odontoceti are really homologous with the milk teeth of other mammals. Some of the dolphins afford the apparently simplest type of mammalian dentition known. The teeth are all simple, conical, slightly recurved structures, with simple tapering roots and without enamel. The dentition is typically *piscivorous*, being adapted for seizing active slippery animals such as fish. The prey is then swallowed entire without mastication. Sometimes the teeth are excessively numerous, reaching two hundred or more (fifty to sixty on each side of each jaw) in *Pontoporia*. This multiplication of teeth is regarded by Kükenthal as due to the division into three parts of numbers of trilobed teeth similar to those of some seals.

In the Sperm whale, *Physeter*, the lower jaw bears a series of twenty to twenty-five stout conical recurved teeth, while in the upper jaw the teeth are vestigial and remain imbedded in the gum. An extinct form, *Physodon*, from the Pliocene of Europe and Patagonia is allied to the Sperm whale, but has teeth in both jaws. In the Killer *Orca*, the teeth number about $^{12}/_{12}$, and are very large and strong. In some forms the teeth are very much reduced in number; thus in *Mesoplodon* the dentition consists simply of a pair of conical teeth borne in the mandible. In the Narwhal *Monodon* the dentition is practically reduced to a single pair of teeth, which lie horizontally in the maxillae, and in the female normally remain permanently in the alveoli. In the male the right tooth remains rudimentary, while the left is developed into an enormous cylindrical tusk marked by a spiral groove. Occasionally both teeth develop into tusks, and there is reason for thinking that two-tusked individuals are generally or always female. In the extinct *Squalodon* the dentition is decidedly heterodont, and the molars have two roots. The dental formula is

$i\ ^3/_3\ c\ ^1/_1\ pm\ ^4/_4\ m\ ^7/_7$, total 60.

It is probable that the homodont condition of modern Odontoceti is not primitive, but due to retrogressive evolution.

Ungulata.

Just as in the Cetacea a piscivorous dentition is most typically developed, so the Ungulata are, as a group, the most characteristic representatives of a *herbivorous* dentition in its various forms.

Ungulata vera.

Artiodactyla. As regards the living forms, the Artiodactyla can be readily divided into two groups, namely those with bunodont and those with selenodont teeth. It has, however, been shown that selenodont teeth always pass through an embryonic bunodont stage. The bunodont type is best seen in Pigs and Hippopotami and such extinct forms as *Hyotherium*. In *Hippopotamus* the dental formula is $i\ ^{(2-3)}/_{(1-3)}\ c\ ^1/_1\ pm\ ^4/_4\ m\ ^3/_3$.

The incisors and canines of *Hippopotamus* are very large and grow continuously. The genus *Sus*, which affords a good instance of an *omnivorous* type of dentition, has the regular unmodified Mammalian dental formula $i\ ^3/_3\ c\ ^1/_1\ pm\ ^4/_4\ m\ ^3/_3$, total 44. The canines, specially in the male, are large and have persistent pulps, and the upper canines do not have the usual downward direction but pass outwards and upwards. In the Wart Hog, *Phacochaerus*, they are enormously large, but a still more extraordinary development of teeth is found in *Babirussa*. In the male *Babirussa* the canines, which are without enamel, are long, curved and grow continuously. Those of the upper jaw never enter the mouth, but pierce the skin of the face and curve backwards over the forehead. The dental formula of *Babirussa* is $i\,^2/_3\,c\,^1/_1\,pm\,^2/_2\,m\,^3/_3$, total 34.

The Wart Hog has a very anomalous dentition, for as age advances all the teeth except the canines and last molars show signs of disappearing; both pairs of persisting teeth are however very large.

Various extinct Ungulata such as *Anoplotherium* have teeth which are intermediate in character between the bunodont and selenodont types. *Anoplotherium* has the regular mammalian series of forty-four teeth. The crowns of all the teeth are equal in height, and there is no diastema—an arrangement found in no living mammal but man.

We come now to the selenodont Artiodactyla.

The Tylopoda—camels (Camelidae) and Llamas (Aucheniidae) when young have the full number of incisors, but in the adult the two upper middle ones are lost. The molars are typically selenodont and hypsodont. In the Camel the dental formula is $i\,^1/_3\,c\,^1/_1\,pm\,^3/_2\,m\,^3/_3$, total 34. The upper incisors, canines and first premolars of the Camel are very small teeth, and the first premolar is separated by a long diastema from the others.

The Tragulina or Chevrotains have no upper incisors, while the canines are largely developed, especially in the male.

The Ruminantia or Pecora are very uniform as regards their dentition. The upper incisors are always absent, for though their germs are developed they are reabsorbed without ever becoming visible, and as a rule the upper canines are absent too, while the lower canines are incisiform. The grinding teeth are typically selenodont, and in the lower jaw form a continuous series separated by a wide diastema from the canines. The dental formula is usually

$i\,^0/_3\,c\,0\text{-}^1/_1\,pm\,^3/_3\,m\,^3/_3$.

The canines are largely developed in the male Muskdeer (*Moschus*) and in *Hydropotes*.

Perissodactyla. The premolars and molars have a very similar structure and form a continuous series of large square teeth with complex crowns. The crowns are always constructed on some modification of the bilophodont plan, as is easily seen in the case of the forms with brachydont teeth, but in animals like the Horse, in which the teeth are very hypsodont, this arrangement is hard to trace. All four premolars in the upper jaw are preceded by milk teeth, while in Artiodactyla the first has no milk predecessors.

In the Tapiridae the grinding teeth are brachydont and the lower ones are typically bilophodont. The last two upper molars have the transverse ridges united by an outer longitudinal ridge. The dentition is $i\,^3/_3\,c\,^1/_1\,pm\,^4/_3\,m\,^3/_3$, total 42.

In some of the extinct Perissodactyles such as *Lophiodon*, the dentition is brachydont and bilophodont, the grinding teeth in general resembling the posterior upper molars of the Tapir. The same type of brachydont tooth is seen in *Palaeotherium* but the transverse ridges are crescentic instead of straight, and are separated from one another by shallow valleys without cement. Some of the Palaeotheridae have the regular series of forty-four teeth.

A complete series of forms is known showing how from the simple brachydont teeth of the Palaeotheridae, were derived the complicated hypsodont teeth of the Equidae. The increase in depth of the tooth was accompanied by increase in the depth and complexity of the enamel infoldings, and of the cement filling them.

Both upper and lower grinding teeth of the Equidae are much complicated by enamel infoldings, but their derivation from the bilophodont type can still be recognised. The diastema in front of the premolars is longer in the living Equidae than in their extinct allies. In the adult horse the dental formula is $i\,^3/_3\,c\,^1/_1\,pm\,^3/_3\,m\,^3/_3$, total 40, with often a vestigial first upper premolar (fig. 82, *pm* 1). The last molar is not more complex than the others, and in the female the canine is quite vestigial. The incisors are large and adapted for cutting and have the enamel curiously folded in forming a deep pit. The milk dentition is $di\,^3/_3\,dc\,^0/_0\,dpm\,^3/_3$, total 24. The last milk premolar is not more complex than the premolar that succeeds it. The horse affords an excellent instance of a typically *herbivorous* type of dentition, the cutting incisors, reduced canines and series of large square flat-crowned grinding teeth being most characteristic.

In *Rhinoceros* the grinding teeth are much like those of *Lophiodon*, having an outer longitudinal ridge from which two crescentic transverse ridges diverge. The upper premolars are as complex as the molars, and there are no canines; in some species incisors also are absent. The dental formula is

$i\,^{(0-2)}/_{(0-1)}\,c\,^0/_{(0-1)}\,pm\,^4/_4\,m\,^3/_3$.

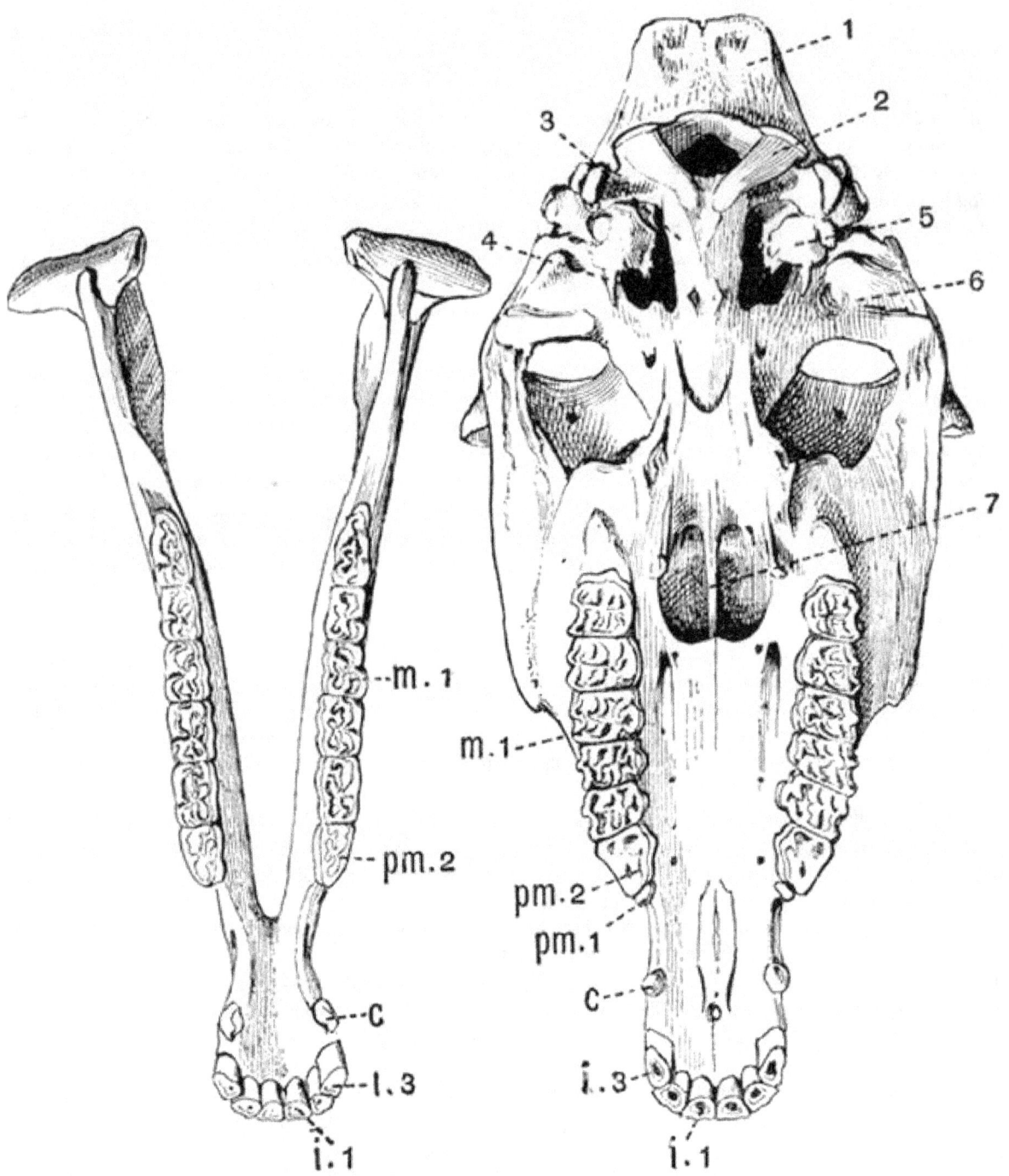

Fig. 82. Palatal aspect of the cranium and mandible of a Donkey (*Equus asinus*) × $\frac{1}{5}$. (Camb. Mus.)

1. supra-occipital.

2. occipital condyle.

3. basi-occipital.

4. vacuity representing the confluent foramen lacerum posterius and foramen lacerum medium.

5. auditory bulla.

6. glenoid surface.

7. vomer.

i 1, *i* 3. first and third incisors.

c. canine.

pm 1, *pm* 2. first and second premolars.

m 1. first molar.

Among the Titanotheriidae *Palaeosyops* has very brachydont teeth whose crowns have been described as *buno-selenodont*, the inner pair of columns being bunodont, the outer, selenodont. Similar grinding teeth occur in *Chalicotherium*. Some of the Titanotheriidae have the regular mammalian series of forty-four teeth.

Subungulata.

Toxodontia. Nesodon has the regular dental formula; its grinding teeth are rooted and the upper ones resemble those of Rhinoceros. The second upper and third lower incisors form ever-growing tusks. There is a marked difference between the deciduous and permanent dentition. *Astrapotherium* likewise has large rooted cheek teeth of a rhinocerotic type, and each jaw bears a pair of permanently growing tusks, those of the lower jaw being the canines. The dental formula is

$i\ ^1/_3\ c\ ^0/_1\ pm\ ^2/_1\ m\ ^3/_3$, total 28.

In *Toxodon* the upper incisors and molars are large and curved and all the teeth have persistent pulps. In *Typotherium* there are no tusks, but the upper incisors are chisel-like, recalling those of Rodents.

The *Condylarthra* have brachydont, generally bunodont teeth, with the premolars simpler than the molars. They generally have the regular dental formula.

Hyracoidea. The dental formula of *Procavia* is usually given as $i\ ^1/_2\ c\ ^0/_0\ pm\ ^4/_4\ m\ ^3/_3$, total 34; in young individuals however there occur a second pair of upper incisors which early fall out. The upper incisors resemble those of Rodents in being long and curved and growing from persistent pulps. They are however triangular in transverse section, not rectangular, having two antero-lateral faces covered with enamel and a posterior face without enamel. Their terminations are pointed, not chisel-shaped as in Rodents. The lower incisors (fig. 83, *i* 1) are pectinate or partially divided by vertical fissures, and the grinding teeth are of the rhinocerotic type.

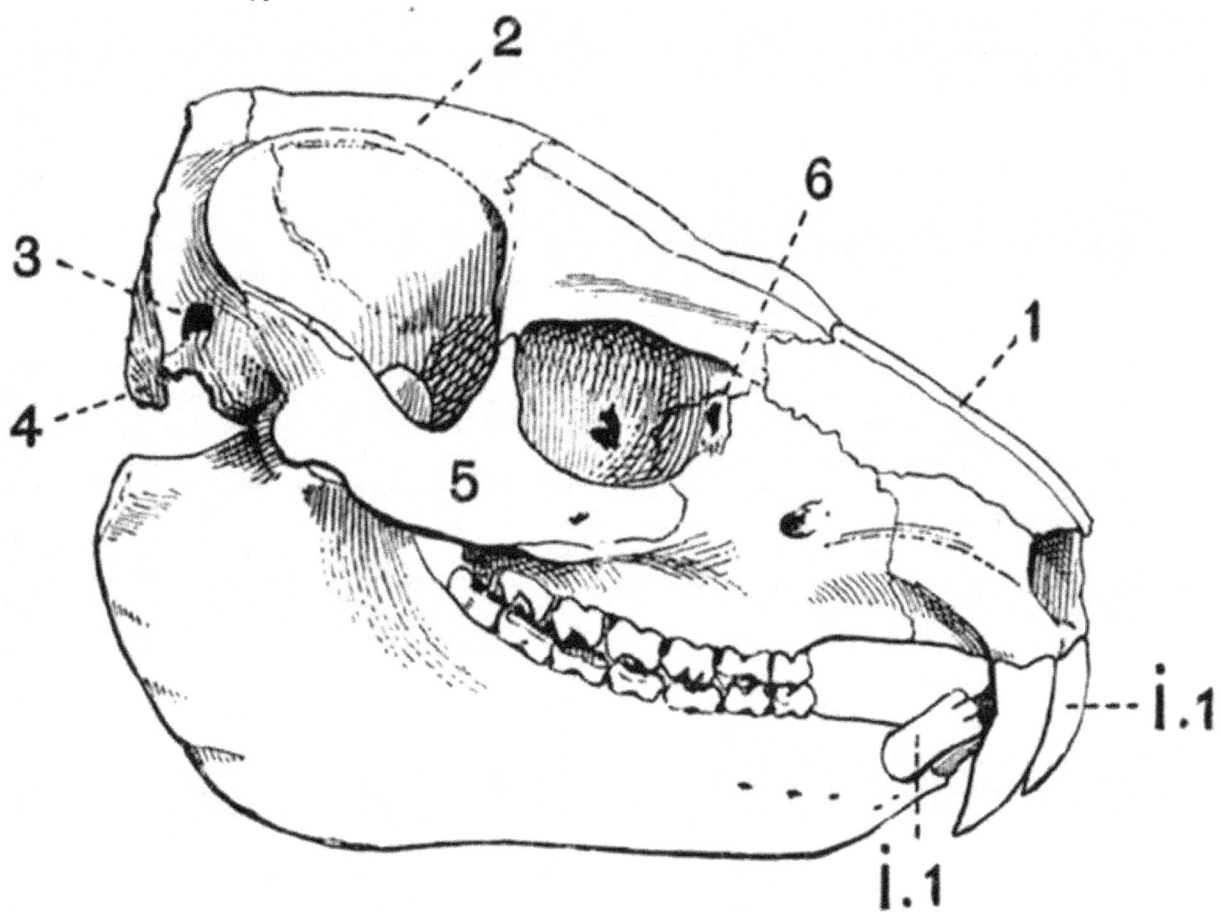

Fig. 83. Skull of *Procavia (Dendrohyrax) dorsalis* × $^2/_3$. (Camb. Mus.)

1. nasal.

2. parietal.

3. external auditory meatus.

5. jugal.

6. lachrymal foramen.

i 1. first incisor.

215

4. paroccipital process of the

exoccipital.

Amblypoda. Two of the best known forms belonging to this extinct group differ much as regards dentition. For while *Coryphodon* has the regular dental formula, and the canines of both jaws of moderate size, in *Uintatherium* the dentition is very specialised, there are no upper incisors, and the upper canines form a pair of enormous tusks. The grinding teeth form a continuous series marked by **V**-shaped ridges and the dental formula is $i\ ^0/_3\ c\ ^1/_1\ pm\ ^3/_3\ m\ ^3/_3$ total 34.

Proboscidea. The incisors are composed entirely of dentine and have the form of conical tusks projecting greatly from the mouth. In living forms they are confined to the upper jaw, in some species of the extinct *Mastodon* however they occur in the lower jaw also. In *Dinotherium* they are probably absent from the upper jaw, but form a pair of downwardly and backwardly-directed tusks growing from the elongated symphysis of the mandible.

The grinding teeth in the various Proboscidea show a very remarkable series of modifications. In *Dinotherium* they are bilophodont or else are marked by three straight transverse ridges. The dental formula is $i\ ^{0?}/_1\ c\ ^0/_0\ pm\ ^2/_2\ m\ ^3/_3$, and the teeth have the normal method of succession. In *Mastodon* as in *Dinotherium* the grinding teeth are marked by transverse ridges, but the ridges are subdivided into conical or mammillary cusps, and similar cusps often occur between the ridges. These cusps are covered with very thick enamel and the spaces between them are not filled up with cement. There are six of these grinding teeth for each side of each jaw but only three are in place at once. The first three are milk teeth as they may be succeeded vertically by others.

In the true Elephants the number and depth of the enamel folds is much increased, and the spaces between the folds are filled up with cement. A very complete series of extinct forms is known with teeth intermediate in character between those of *Mastodon* and those of the Mammoth and living elephants. The dental formula of *Elephas* is

$di\ ^1/_0\ i\ ^1/_0\ c\ ^0/_0\ dm\ ^{3—4}/_{3—4}\ m\ ^3/_3.$

Sir W.H. Flower describes the mode of succession of teeth in Elephants as follows: "As regards the mode of succession that of modern Elephants is as before mentioned very peculiar. During the complete lifetime of the animal there are but six molar teeth on each side of each jaw with occasionally a rudimentary one in front, completing the typical number of seven. The last three represent the true molars of ordinary mammals, those in front appear to be milk molars which are never replaced by permanent successors, but the whole series gradually moves forwards in the jaw, and the teeth become worn away and their remnants cast out in front while development of others proceeds behind. The individual teeth are so large and the processes of growth and destruction by wear take place so slowly, that not more than one or portions of two teeth are ever in place and in use on each side of each jaw at one time, and the whole series of changes coincides with the usual duration of the animal's life. On the other hand the *Dinotherium*, the opposite extreme of the Proboscidean series, has the whole of the molar teeth in place and use at one time, and the milk molars are vertically displaced by premolars in the ordinary fashion. Among Mastodons transitional forms occur in the mode of succession as well as in structure, many species showing a vertical displacement of one or more of the milk molars, and the same has been observed in one extinct species of Elephant (*E. planifrons*) as regards the posterior of these teeth."

In the Tillodontia the grinding teeth are of Ungulate type, while the second incisors are large and grow from persistent pulps, so as to resemble those of Rodents.

Rodentia have a most characteristic and very constant dentition, the common dental formula being

$i\ ^1/_1\ c\ ^0/_0\ pm\ ^{(0—1)}/_{(0—1)}\ m\ ^3/_3$, total 18 or 20.

The incisors always have chisel-like edges and persistent pulps, and are separated by a wide diastema from the premolars. Canines are always absent, and there are generally three grinding teeth not preceded by milk teeth; their surface may be grooved, or may be bunodont. Teeth are most numerous in the Duplicidentata (Hares and Rabbits), in which the formula is $i\ ^2/_1\ c\ ^0/_0\ pm\ ^3/_2\ m\ ^3/_3$, total 28, and fewest in *Hydromys* and certain other forms, in which the formula is $i\ ^1/_1\ c\ ^0/_0\ pm\ ^0/_0\ m\ ^2/_2$, total 12. The hares and rabbits are the only rodents which have well developed deciduous incisors, though a vestigial milk incisor has been described in the Mouse (*Mus musculus*). The last upper molar of *Hydrochaerus* is very complicated, its structure approaching that of the teeth of Elephants.

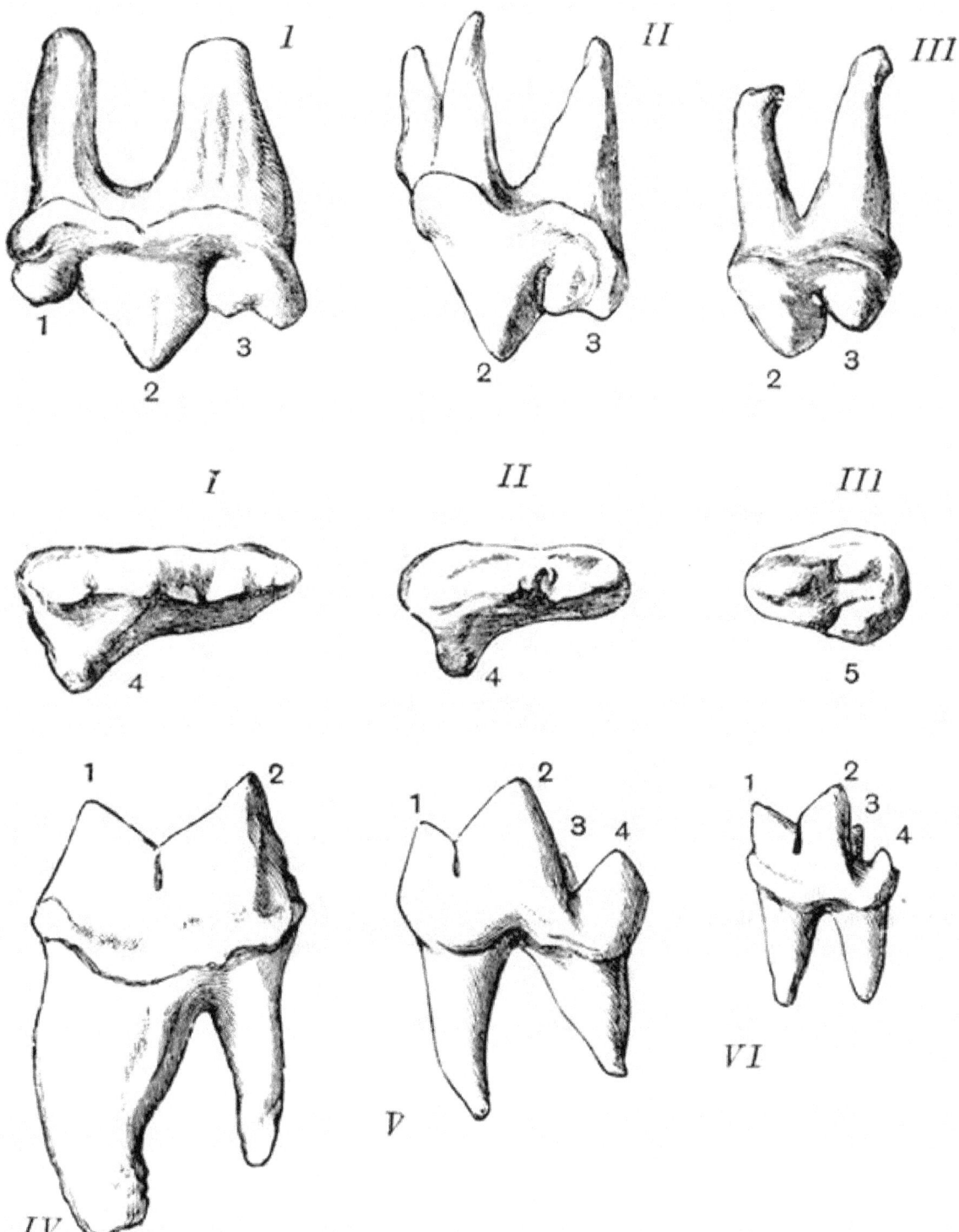

Fig. 84. Carnassial or Sectorial Teeth of Carnivora (from Flower).

Upper sectorial teeth of I. *Felis*, II. *Canis*, III. *Ursus*. 1. anterior, 2. middle, 3. posterior cusp of blade, 4. inner lobe supported on distinct root, 5. inner lobe posterior in position and without distinct root, characteristic of the Ursidae.

Lower sectorial teeth. 1. *Felis*, 2. *Canis*, 3. *Herpestes*. 1. anterior, 2. posterior lobe of blade, 3. inner tubercle, 4. heel.]

Carnivora have the teeth rooted and markedly diphyodont and heterodont. The canines are greatly developed, and the incisors are small.

217

In *Carnivora vera* the incisors are almost always $^3/_3$. The fourth upper premolar and first lower molar are differentiated as carnassial teeth (see p. 436), and retain fundamentally the same characters throughout the suborder. The upper carnassial (fig. 84, I. II. III.) consists of a more or less compressed, commonly trilobed blade borne on two roots, with an inner tubercle borne on a third root. The lower carnassial has only two roots; its crown consists of a bilobed blade with generally an inner cusp, and a heel or talon (fig. 84, 4) behind the blade.

The most thoroughly carnivorous type of dentition is seen in the Æluroidea, and especially in the cat tribe (Felidae). In the genus *Felis* the dental formula is *i* $^3/_3$ *c* $^1/_1$ *pm* $^3/_2$ *m* $^1/_1$, total 30. The incisors are very small, so as not to interfere with the action of the large canines, the lower carnassial is reduced to simply the bilobed blade (fig. 84, IV), and the cheek teeth are greatly subordinated to the carnassial. The extinct *Machaerodus* has the upper canines comparable in size to those of the Walrus.

The Civets and Hyaenas have a dentition allying them closely to the cats. The hyaena-like *Proteles* has, however, the grinding teeth greatly reduced.

In the Cynoidea the general dentition is *i* $^3/_3$ *c* $^1/_1$ *pm* $^4/_4$ *m* $^2/_3$, total 42. This differs from the regular mammalian dentition only in the absence of the last upper molar. The upper carnassial tooth (fig. 84, II.) consists of a larger middle and smaller posterior lobe with hardly any trace of an anterior lobe. The lower carnassial (fig. 84, V.) is typical, consisting of a bilobed blade with inner cusp and posterior talon.

The dentition of the Cynoidea is most closely linked with that of the Arctoidea by means of fossil forms.

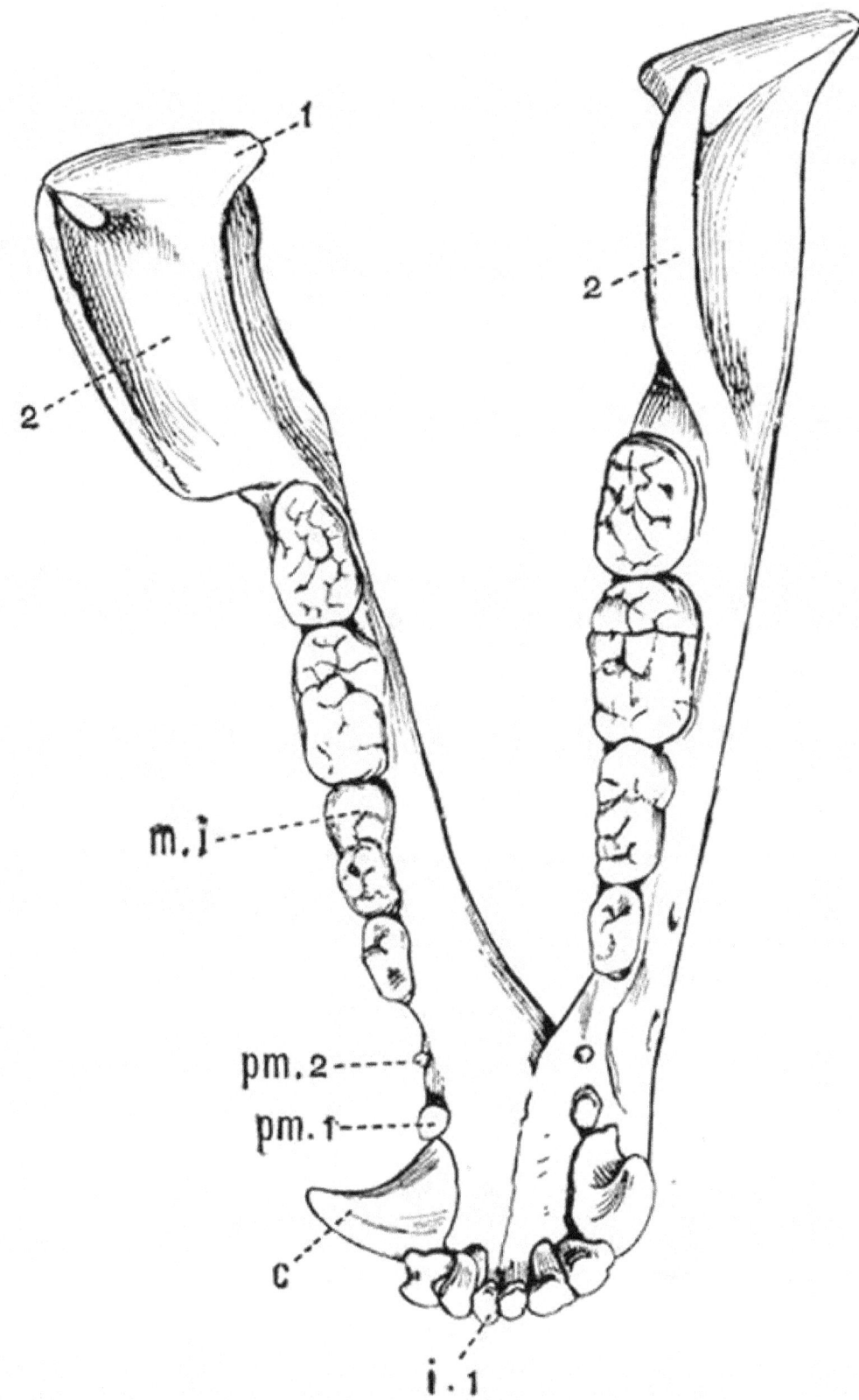

Fig. 85. Mandible of Isabelline Bear (*Ursus isabellinus*) × ½.
(Camb. Mus.)

1. condyle.   *m* 1. first molar. The dotted

219

2. coronoid process.

i 1. first incisor.

c. canine.

pm 1, *pm* 2. first and second

premolars.

line is pointing to the posterior

half of the tooth.

This specimen has only

three premolars, there

should be four.

In the Arctoidea the dentition is not so typically carnivorous as in the Æluroidea and Cynoidea. In the bears, Ursidae, the molars have broad flat tuberculated crowns (fig. 85). The dental formula in *Ursus* is $i\ ^3/_3\ c\ ^1/_1\ pm\ ^4/_4\ m\ ^2/_3$, total 42. The upper carnassial (fig. 84, III.) differs from that of the Æluroidea and Cynoidea in having no inner lobe supported on a third root. In the large group of Mustelidae there are generally two molars in the lower and one in the upper jaw. The grinding teeth commonly have large, flattened, more or less tuberculated crowns, and the upper molar may be as large or much smaller than the carnassial.

In the *Creodonta* there are no specially differentiated carnassial teeth.

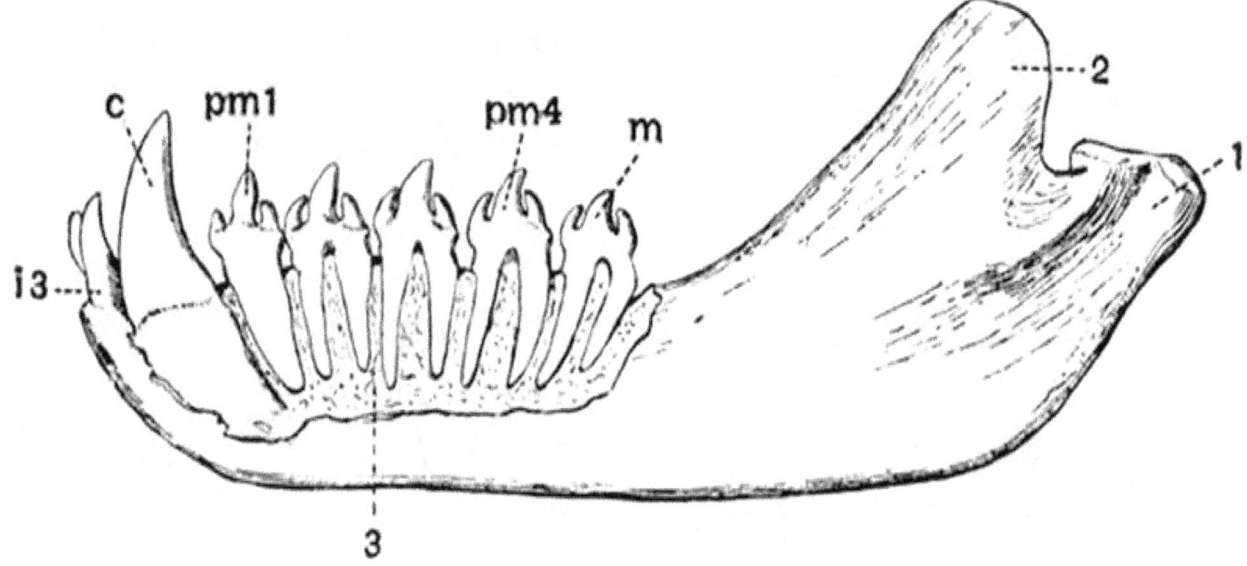

Fig. 86. Left mandibular ramus of the Sea Leopard (*Ogmorhinus leptonyx*)
WITH THE ROOTS OF THE TEETH EXPOSED × ¹/₃. (Camb. Mus.)

1. condyle.

2. coronoid process.

i 3. third incisor.

c. canine.

pm 1, *pm* 4. first and fourth

premolars.

m. molar.

In the *Pinnipedia* the dentition differs considerably from that of the Carnivora vera. The milk dentition is always vestigial, and the teeth are frequently absorbed before birth. There are four premolars and one molar, forming an uniform series of cheek teeth, all of which except in the Walrus have compressed and pointed, never flattened, crowns. There is no special carnassial tooth, and the incisors are always fewer than ³/₃. In *Otaria* the dentition is

$i\ ^3/_2\ c\ ^1/_1\ pm\ ^4/_4\ m\ 1$ or $^2/_1$, total 34 or 36.

In the Walrus the upper canines form immense tusks. The other teeth are all small and one-rooted, and the molars have flat crowns. In the true seals the dentition is strikingly piscivorous, the cheek teeth often having accessory cusps (fig. 86).

The Insectivora are diphyodont and heterodont, having well-developed rooted teeth. The canines are usually weak, the incisors pointed, and those of the two jaws often meet like a pair of forceps. The crowns of the molars are characteristically studded with short cusps. Some genera, such as *Gymnura* and the mole, *Talpa*, have the regular mammalian dentition. In the hedgehog, *Erinaceus*, the dentition is

$i\,^3/_2\ c\,^1/_1\ pm\,^3/_2\ m\,^3/_3$, total 36.

In the genus *Sorex* (Shrews) the teeth differ in the following two marked respects from those of most other Monodelphia, (1) they are monophyodont, (2) the lower incisors sometimes become fused to the jaws. Most Insectivora have square molar teeth, but in *Potamogale*, *Chrysochloris*, *Solenodon* and the Centetidae the molar teeth are triangular in section. Four molars occur in *Centetes*.

In the aberrant genus *Galeopithecus* the dentition is $i\,^2/_3\ c\,^1/_1\ pm\,^2/_2\ m\,^3/_3$, total 34. The upper incisors are placed at some distance from the anterior end of the jaw, and the outer upper incisors and canines of both jaws have two roots,—a very unusual character. The lower incisors are deeply grooved or pectinated in the same way as are the lower incisors of *Procavia*. The upper incisors and canines of both jaws bear many cusps, and are very similar in appearance to the cheek teeth of some Seals.

The dentition of the Chiroptera is diphyodont and heterodont, and the dental formula never exceeds

$i\,^2/_3\ c\,^1/_1$ pm $^3/_3\ m\,^3/_3$, total 38.

The milk teeth are very slender and have sharp recurved cusps; they are quite unlike the permanent teeth. The permanent teeth are of two types. In the Insectivorous forms the molar teeth are cusped, and resemble those of Insectivora. In the blood-sucking Vampire bat *Desmodus*, the teeth are peculiarly modified; the canines and the single pair of upper incisors are much enlarged and exceedingly sharp, while all the other teeth are much reduced in size.

In the Frugivorous bats the molar teeth have nearly always smooth crowns. The dental formula in the chief genus *Pteropus* is $i\,^2/_2\ c\,^1/_1\ pm\,^3/_3\ m\,^2/_3$, total 34.

The Primates have a diphyodont and heterodont dentition, generally of an omnivorous type, with cheek teeth adapted for grinding. The incisors are generally $^2/_2$, and the molars, except in the Hapalidae, are $^3/_3$. In the Lemurs the upper canines are large, and the lower incisors slender and directed almost horizontally forwards. The Aye Aye, *Chiromys*, has the following singular dentition: $i\,^1/_1\ c\,^0/_0\ pm\,^1/_0\ m\,^3/_3$, total 18. The incisors much resemble those of rodents having persistent pulps, and enamel only on the anterior face.

In Man and in the Anthropoid and Old World Apes the dental formula is always $i\,^2/_2\ c\,^1/_1\ pm\,^2/_2\ m\,^3/_3$, total 32.

In the Cebidae there is an extra premolar in each jaw bringing the number up to 36. In the Hapalidae, as in the Cebidae, there is a third premolar, but the molars are reduced to $^2/_2$. Man is the only Primate that has the teeth arranged in a continuous series. In all the others there is a gap or diastema of larger or smaller size between the incisors and canines. In all except man also the canines are enlarged, especially in the males.

The Exoskeletal structures of mammals may be summarised in the following table:

I. *Epidermal exoskeletal structures.*

1. Hairs (*a*) ordinary hair,

   (*b*) vibrissae and bristles,

   (*c*) spines of hedgehog, porcupine, *Echidna*,

   Centetes, *Acanthomys*.

2. Scales { of Manidae,

   { on tails of rats, beavers, &c.

3. Horns of Rhinoceros.

4. Horns of Bovine Ruminants.

5. Nails, claws, hoofs.

6. Spurs of male *Ornithorhynchus* and *Echidna*.

7 Horny beak and teeth of *Ornithorhynchus*.

8 Horny pads on jaws of Sirenians and
. Ruminants.

9 Baleen of whales.
.

1 Enamel of teeth.
0.

II. *Dermal exoskeletal structures.*

1 Dentine and cement of
. teeth.

2 Bony scutes of Armadillos.
.

ENDOSKELETON.
Vertebral Column.
Cervical Vertebrae.

The cervical vertebrae of all mammals have certain characters in common. However long the neck may be, the number of cervical vertebrae, with very few exceptions, is seven. Movable ribs are generally absent, and if present are small and do not reach the sternum. The transverse processes are generally wide but not long, and are perforated near the base by the vertebrarterial canals, through which the vertebral arteries pass; they generally bear downwardly-directed inferior lamellae which are sometimes as in the seventh human cervical seen to ossify from centres distinct from those forming the rest of the transverse process, and are really of the nature of ribs. The atlas and axis always differ much from the other vertebrae.

We may pass now to the special characters of the cervical vertebrae in the different groups. In Monotremes and Marsupials the number of cervical vertebrae is always seven. With the exception of the atlas of *Echidna* the cervical vertebrae of Monotremes are without zygapophyses. In Monotremes the transverse processes ossify from centres distinct from that forming the body, and remain suturally connected with the rest of the vertebra until the adult condition is reached. The method of the ossification of the atlas in Marsupials varies considerably, thus in some forms such as the Wombats (*Phascolomys*) there is an unossified gap in the middle of the inferior arch of the atlas, which may remain permanently open; in *Thylacinus* this gap is filled up by a distinct heart-shaped piece of bone, while in *Didelphys* and *Perameles* the atlas is ossified below in the same way as in other mammals. In *Notoryctes* the second to sixth cervical vertebrae are ankylosed together.

The cervical vertebrae of the Edentata have some remarkable peculiarities. In the three-fingered Sloth, *Bradypus*, there are nine cervical vertebrae, all except the last of which have their transverse processes perforated by the vertebrarterial canals. In a two-fingered sloth, *Choloepus hoffmanni*, there are only six cervical vertebrae. In the Megatheriidae, Anteaters (Myrmecophagidae), Pangolins (Manidae), and Aard Varks (Orycteropodidae), the cervical vertebrae are normal, but in the Armadillos (Dasypodidae), and still more in the Glyptodonts, several of them are commonly fused together. The fusion affects not only the centra, but also the neural arches, so that the neural canals form a continuous tube.

In the Glyptodonts there is a complex joint at the base of the neck to allow the partial retraction of the head within the carapace. This arrangement recalls that in Tortoises.

As a rule the Sirenia possess seven short cervical vertebrae, not fused together and not presenting any marked peculiarities. In *Manatus* however there are only six cervical vertebrae and they are very variable.

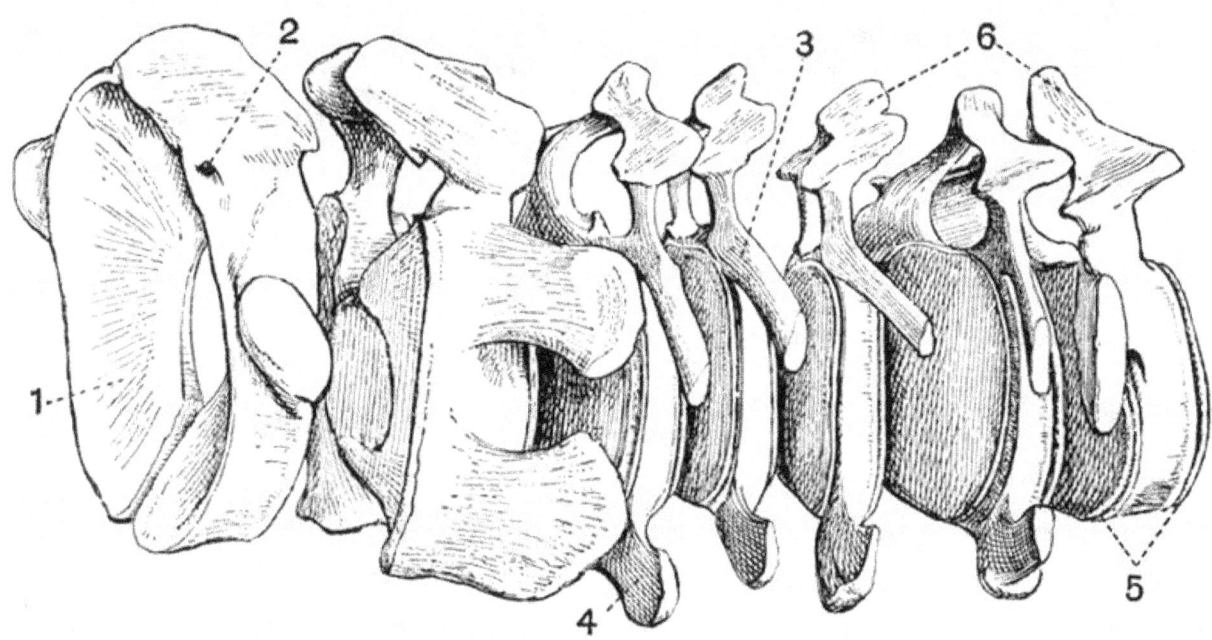

Fig. 87. Cervical vertebrae of a young Fin Whale
(*Balaenoptera musculus*) × ¹/₁₀. (Camb. Mus.)

1. surface on the atlas for articulation

with the occipital condyle

of the skull.

2. foramen for exit of the first

spinal nerve.

3. upper transverse process.

4. lower transverse process.

In the fresh specimen these two

transverse processes are united by

cartilage, in adult individuals the

whole transverse process is ossified.

5. epiphyses of centrum.

6. neural spine.

In the Cetacea there are invariably seven cervical vertebrae, but they are always very short and are frequently even before birth fused together by their centra into one continuous mass (see fig. 67). Sometimes the last one or two are free. In the Rorquals (*Balaenoptera*) however, the cervical vertebrae are quite separate and distinct (fig. 87), and in the fluviatile Odontoceti, *Platanista, Inia,* and *Pontoporia*, and also in *Beluga* and *Monodon*, though very short they are free. In *Physeter* the first vertebra is free while the others are fused. An odontoid process is not commonly present even in Cetaceans with free cervical vertebrae, but a very short one occurs in the Rorquals. The cervical vertebrae of Rorquals give off on each side two transverse processes (fig. 87, 3 and 4) which enclose between them a wide space. These processes are not completely ossified till the animal is adult.

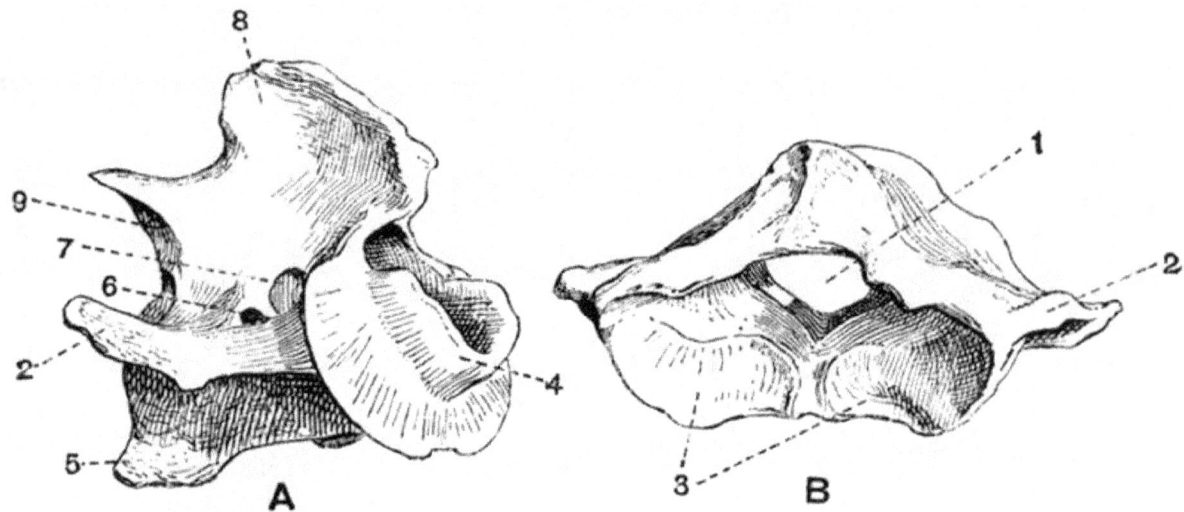

Fig. 88. Atlas (B) and axis (A) vertebrae of an Ox
(*Bos taurus*) × ¼. (Camb. Mus.)

1. neural canal.

2. transverse process.

3. surfaces for articulation with

the occipital condyles of

the skull.

4. spout-like odontoid process.

5. hypapophysis.

6. anterior opening of the

vertebrarterial canal.

7. foramen for the exit of the

second spinal nerve.

8. neural spine.

9. postzygapophysis.

In all Ungulata the number of cervical vertebrae is seven. Among the Artiodactyla two forms of the odontoid process of the axis occurs; in the Suina and Tragulina it is conical, in the Ruminantia and Tylopoda it is spout-like (fig. 88, 4). The atlas in the Suina and to a less extent in the Ruminantia has long flattened transverse processes, and the remaining cervical vertebrae are opisthocoelous. Those of the Giraffe and Llama (fig. 103) are noticeable for their great length. In the Tylopoda the posterior half of the vertebrarterial canal is confluent with the neural canal.

The Perissodactyla have remarkably opisthocoelous cervical vertebrae. Those of *Macrauchenia* have the posterior half of the vertebrarterial canal confluent with the neural canal as in Tylopoda. In the Proboscidea they are short flattened discs slightly opisthocoelous; the axis and seventh vertebra and to a less extent the sixth have high neural spines.

In the Rodentia the atlas generally has broad wing-like transverse processes, and the axis a large and long neural spine, while the odontoid process is much developed. In the Jerboas (*Dipus*) all the cervical vertebrae except the atlas are fused together, a condition recalling that in armadillos.

In the Carnivora the wings of the atlas are well developed (fig. 69, A, 1), and it is deeply cupped for articulation with the condyles of the skull. The axis has a long odontoid process and a high compressed neural spine (fig. 69, B, 4). The third to sixth cervical vertebrae have large transverse processes with prominent perforated inferior lamellae, whose ventral margins in the third and fourth vertebrae diverge as they pass backwards, while in the fifth they are parallel and in the sixth convergent. The transverse processes of the seventh vertebra have no inferior lamellae and are not perforated. Metapophyses are often developed.

In the Insectivora the cervical vertebrae vary considerably. The neural spines except in the case of the axis are generally very small and in the Shrews and Moles the neural arches are exceedingly slender.

In the Chiroptera all the cervical vertebrae are broad and short with slender neural arches.

Primates. In Man the cervical vertebrae have short blunt transverse processes and small often bifid neural spines. The neural and vertebrarterial canals are large. The atlas forms a ring surrounding a large cavity, and has a very slender inferior arch and small transverse processes. Traces of a pro-atlas have been described in *Macacus* and *Cynocephalus*. The axis has a prominent spine and odontoid process and short transverse processes. In most Primates the cervical vertebrae are very similar to those of man, but the inferior lamellae of the transverse processes are better developed. In the Anthropoid Apes the neural spines are as a rule much elongated.

Thoraco-lumbar, or Trunk vertebrae.

In the Monotremata there are nineteen thoraco-lumbar vertebrae, sixteen (*Echidna*) or seventeen (*Ornithorhynchus*) of which bear ribs. The transverse processes are very short and do not articulate with the ribs, which are united to the centra only.

In the Marsupialia there are always nineteen thoraco-lumbar vertebrae, thirteen of which generally bear ribs. The lumbar vertebrae frequently have large metapophyses and anapophyses, these being specially well seen in the Kangaroos and Koala (*Phascolarctus*).

The Edentata are very variable as regards their trunk vertebrae. The two genera of Sloths differ much as regards the number, for while *Bradypus* has only nineteen, fifteen or sixteen of which bear ribs, *Choloepus* has twenty-seven, twenty-four of which are thoracic, and bear ribs. In *Bradypus* a small outgrowth from the transverse process articulates with the neural arch of the succeeding vertebra. In both genera the neural spines are all directed backwards.

In the Megatheriidae as in the sloths the neural spines are all directed backwards, and in the lumbar region additional articulating surfaces occur, better developed than are those in *Bradypus*.

In the ant-eaters (Myrmecophagidae) there are seventeen or eighteen thoraco-lumbar vertebrae, all of which except two or three bear ribs. The posterior thoracic and anterior lumbar vertebrae articulate in a very complex fashion, second, third, and fourth pairs of zygapophyses being progressively developed in addition to the ordinary ones, as the vertebrae are followed back.

In the Armadillos the lumbar vertebrae have long metapophyses which project upwards and forwards and help to support the carapace. In *Glyptodon* almost all the thoraco-lumbar vertebrae are completely ankylosed together.

In the Manidae there are no additional zygapophyses but the normal ones of the lumbar and posterior thoracic regions are very much developed, the postzygapophyses being semi-cylindrical and fitting into the deep prezygapophyses of the succeeding vertebra.

In the Sirenia the number of lumbar vertebrae is very small; in the dugong there are nineteen thoracic and four lumbar, and in the manatee seventeen thoracic and two lumbar.

In the Cetacea the number of thoracic vertebrae varies from nine in *Hyperoödon* to fifteen or sixteen in *Balaenoptera*, and the number of lumbar vertebrae from three in *Inia* to twenty-four or more in *Delphinus*. The lumbar vertebrae are often very loosely articulated together and the zygapophyses sometimes as in the Dolphins are placed high up on the neural spines. The centra are large, short in the anterior region but becoming longer behind. The epiphyses are prominent, and so are the neural spines and to a less extent the metapophyses. The transverse processes are well developed, anteriorly they arise high up on the neural arch, but when the vertebral column is followed back they come gradually to be placed lower down, till in the lumbar region they project from the middle of the centra. This can be well traced in the Porpoise (*Phocaena*). In the Physeteridae the transverse processes of the anterior thoracic vertebrae are similar to those of most Cetacea, but when followed back, instead of shifting their position on the vertebrae, they gradually disappear, and other processes gradually arise from the point where the capitulum of the rib articulates.

Ungulata. In the Ungulata vera the thoraco-lumbar vertebrae are slightly opisthocoelous. The anterior thoracic vertebrae commonly have exceedingly high backwardly-projecting neural spines (fig. 89, 1); but those of the lumbar and posterior thoracic vertebrae often point somewhat forwards so that the spines all converge somewhat to a point called the *centre of motion* (cp. fig. 101). In the Artiodactyla there are always nineteen thoraco-lumbar vertebrae, and in the Perissodactyla twenty-three.

Procavia sometimes has thirty thoraco-lumbar vertebrae, a greater number than occurs in any other terrestrial mammal; twenty-two of these are thoracic and eight lumbar. In *Phenacodus* the convergence of the neural spines to a centre of motion is well seen.

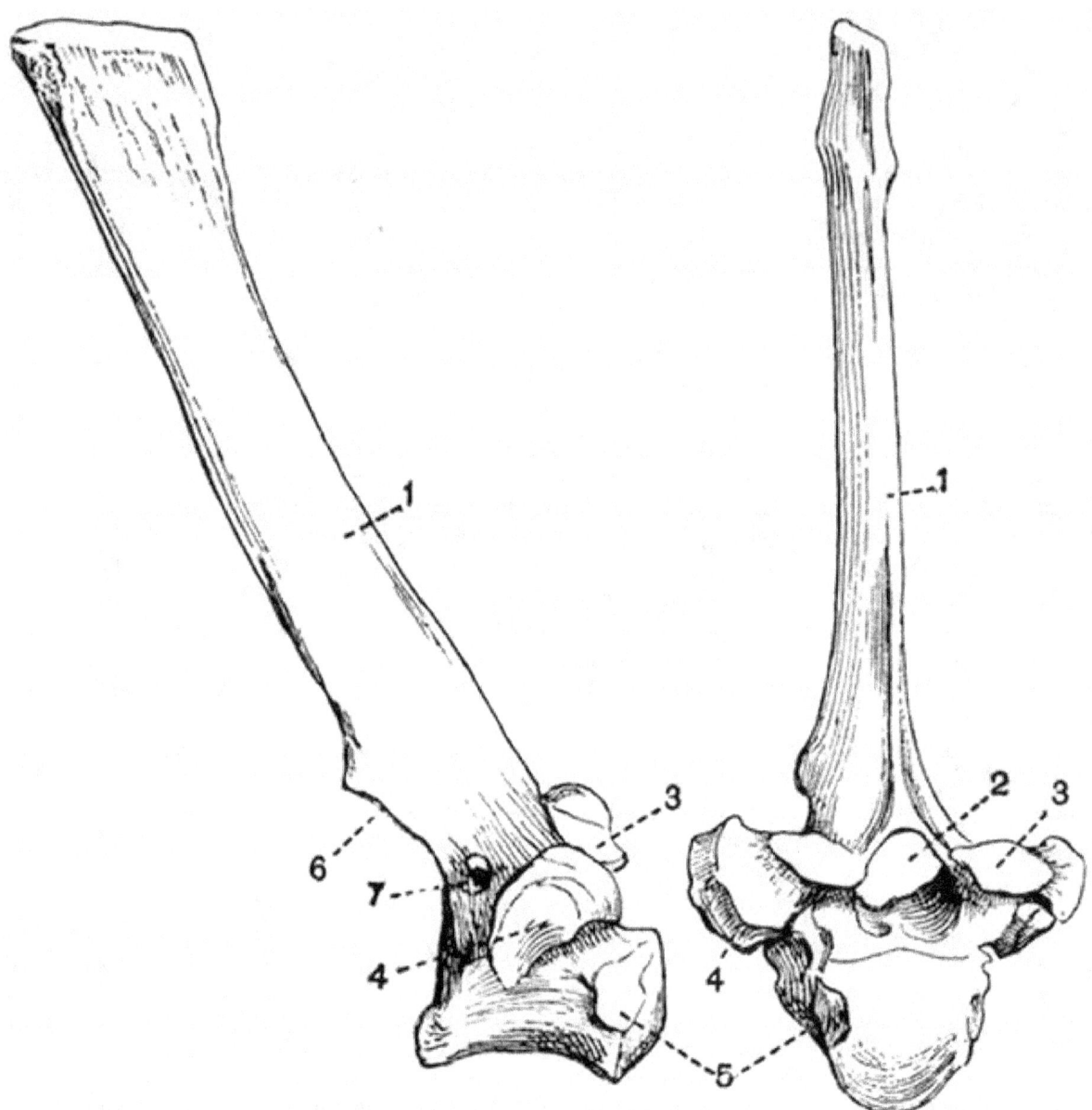

Fig. 89. First and second thoracic vertebrae of an Ox
(*Bos taurus*) × ¹/₃. (Camb. Mus.)

1. neural spine.

2. neural canal.

3. prezygapophysis.

4. facet for articulation with the

tuberculum of the rib.

5. facet for articulation with the

capitulum of the rib.

6. postzygapophysis.

7. foramen for exit of spinal

nerve.

In the Proboscidea there are twenty-three thoraco-lumbar vertebrae, of which nineteen or twenty bear ribs.

In the Rodentia there are generally nineteen thoraco-lumbar vertebrae but occasionally the number rises as high as twenty-five. In the Hares (Leporidae) the number is nineteen, twelve or thirteen of which are thoracic. The anterior thoracic vertebrae have short centra and high backwardly-directed neural spines, the lumbar vertebrae have large forwardly- and downwardly-directed transverse processes with expanded ends. Metapophyses, anapophyses and hypapophyses are all present. In the Agouti (*Dasyprocta*) the convergence of the neural spines to a centre of motion is very strongly marked.

In the Carnivora the trunk vertebrae are nearly always twenty or twenty-one in number; in the genera *Felis* and *Canis* thirteen of these are thoracic and seven lumbar. The anterior thoracic vertebrae have long backwardly-projecting neural spines, while the posterior thoracic and lumbar vertebrae have shorter and thicker neural spines which project slightly forwards. In the Pinnipedia there is no change in the direction of the neural spines, and anapophyses are but little developed.

In the Insectivora the number of trunk vertebrae varies much from nineteen—thirteen thoracic and six lumbar—in *Tupaia*, to twenty-four—nineteen thoracic and five lumbar—in *Centetes*. The development of the various processes varies in accordance with the habits of the animals, being great in the active forms, slight in the slowly moving or burrowing forms. In *Talpa* and *Galeopithecus* the intervertebral discs of the thoraco-lumbar region instead of being cartilaginous have ossified forming inter centra, a condition met with in very few mammals.

In the Chiroptera there are seventeen or eighteen thoraco-lumbar vertebrae, eleven to fourteen of which may bear ribs. The development of processes is slight.

Among Primates the number of trunk vertebrae is generally nineteen, of which twelve to fourteen bear ribs; in man and the Gorilla and Chimpanzee the number is, however, seventeen, and in the Orang (*Simia*) sixteen. In some of the Lemuroidea there are as many as twenty-three or twenty-four. In most cases the neural spines converge more or less to a centre of motion, and this is especially marked in some of the Lemurs; it does not occur in man and the anthropoid apes.

Sacral and caudal vertebrae.

At the posterior end of the trunk in all mammals a certain number of vertebrae are found fused together forming the sacrum. But of these only two or three answer to the definition of true sacral vertebrae in being united to the ilia by small ribs. The others which belong to the caudal series may be called pseudosacral vertebrae. In different individuals of the same species it sometimes happens that different vertebrae are attached to the pelvis and form the sacrum. Sometimes even different vertebrae are attached to the pelvis at successive periods in the life history of the individual. This is owing to a shifting of the pelvis and has been especially well seen in man. In young human embryos the pelvis is at a certain stage attached to vertebra 30, but as development goes on it becomes progressively attached to the twenty-ninth, twenty-eighth, twenty-seventh, twenty-sixth and twenty-fifth vertebrae. As the attachment to these anterior vertebrae is gained, the attachment to the posterior ones becomes lost, so that in the adult the pelvis is generally attached to vertebrae 25 and 26. But there are no absolutely pre-determined sacral vertebrae, as sometimes the pelvis does not reach vertebra 25, remaining attached to vertebrae 26 and 27; sometimes it becomes attached even to vertebra 24. This shifting of the pelvis is seen in *Choloepus* in a more marked degree even than in man.

Of the Monotremata, *Ornithorhynchus* has two sacral vertebrae ankylosed together, while *Echidna* has three or four.

In Marsupialia as a rule only one vertebra is directly united to the ilia, but one or two more are commonly fused to the first. In the Wombats there may be as many as four or five vertebrae fused together in the sacral region. In *Notoryctes* there is extensive fusion in the sacral region, six vertebrae, owing mainly to the great development of their metapophyses, being united with one another, and with the ilia, and the greater part of the ischia.

In most Edentata there is an extensive fusion of vertebrae in the sacral region. This is especially marked in the Armadillos and Megatheriidae, and to a less extent in the Sloths and Aard Varks.

In the Sirenia the vestigial pelvis is attached by ligament to the transverse processes of a single vertebra, which hence may be regarded as sacral.

In Cetacea there is no sacrum, the vestigial pelvis not being connected with the vertebral column.

In most Ungulata the sacrum consists of one large vertebra united to the ilia, and having a varying number of smaller vertebrae fused with it behind.

The same arrangement obtains in most Rodentia, but in the Beavers (Castoridae) all the fused vertebrae are of much the same size, the posterior ones having long transverse processes which nearly meet the ilia.

In Carnivora there may be two sacral vertebrae as in the Hyaena, three as in the Dog, four or five as in Bears and Seals.

In Insectivora from three to five are united, while in many Chiroptera all the sacral and caudal vertebrae have coalesced. Among Primates, in Man and Anthropoid Apes there are usually five fused vertebrae forming the sacrum, but of these only two or three are connected to the ilia by ribs. In most of the other Anthropoidea there are two or three fused vertebrae, and in the Lemuroidea two to five.

Free Caudal Vertebrae. The free caudal vertebrae vary greatly in number and character. When the tail is well developed, the anterior vertebrae are comparatively short and broad, with well-developed neural arches and zygapophyses; but as the tail is followed back, the centra gradually lengthen and become cylindrical, and at the same time the neural arches and all the processes gradually become reduced and disappear, so that the last few vertebrae consist of simple rod-like centra. Chevron bones are frequently well-developed.

Of the Monotremes *Echidna* has twelve caudal vertebrae, two of which bear irregular chevron bones. In *Ornithorhynchus* there are twenty or twenty-one caudal vertebrae with well-developed hypapophyses, but no chevron bones.

In Marsupials there is great diversity as regards the tail. In the Wombat and Koala the tail is small and without chevron bones. In most other Marsupials it is very long, having sometimes as many as thirty-five vertebrae in the prehensile-tailed opossums. In the Kangaroos the tail is very large and stout. Chevron bones are almost always present, and in *Notoryctes* are large and expanded.

Most Edentates have large tails with well-developed chevron bones. The length of the tail varies greatly from the rudimentary condition in Sloths to that in the Pangolins, one of which has forty-six to forty-nine caudal vertebrae—the largest number in any known mammal. Chevron bones are much developed, sometimes they are Y-shaped, sometimes as in *Priodon*, they have strong diverging processes. The caudal vertebrae of Glyptodonts, though enclosed in a continuous bony sheath, have not become ankylosed together.

The Sirenia have numerous caudal vertebrae with wide transverse processes. In the Cetacea also the tail is much developed, and the anterior vertebrae have large chevron bones and prominent straight transverse processes; the posterior caudal vertebrae, which in life are enclosed in the horizontally expanded tail fin, are without transverse processes.

In Ungulata the tail is simple, formed of short cylindrical vertebrae, which in living forms are never provided with chevron bones. The number of caudal vertebrae varies from four, sometimes met with in *Procavia*, to thirty-one in the Elephant. The tail is exceedingly long in *Anoplotherium* and in *Phenacodus*, in which there are thirty caudal vertebrae.

In Rodentia the tail is variable. In the Hares, Guinea pig (*Cavia*) and *Capybara* it is very small, in *Pedetes* and the Beaver it is very long and has well-developed chevron bones.

Most of the Carnivora except the Bears and Seals have very long tails, the greatest number of vertebrae, thirty-six, being met with in *Paradoxurus*. Bears have only eight to ten caudal vertebrae. Chevron bones are not often much developed.

In Insectivora the tail is very variable as regards length, the number of vertebrae varying from eight in *Centetes* to forty-three in *Microgale*.

In Chiroptera the tail is sometimes quite rudimentary, and as in *Pteropus*, composed of a few coalesced vertebrae, sometimes it is formed of a large number of slender vertebrae.

In Primates also the tail is very variable. In Man all the four caudal vertebrae are rudimentary and are fused together, forming the *coccyx*. In the Anthropoid apes, too, there are only four or five caudal vertebrae. In many monkeys of both the eastern and western hemispheres the tail is very long, having thirty-three vertebrae in *Ateles*, in which genus it is also prehensile. Chevron bones are present in all Primates with well-developed tails. In the Lemuroidea the number of caudal vertebrae varies from seven to twenty-nine.

CHAPTER XXIII.
GENERAL ACCOUNT OF THE SKELETON IN MAMMALIA
(*continued*).

THE SKULL AND APPENDICULAR SKELETON.

The Skull.

Monotremata. In both genera the cranium is thin-walled, has a fairly large cavity, and is very smooth and rounded externally. The sutures between many of the bones early become obliterated in a manner comparable to that in birds, and the facial portion of the skull is much prolonged.

In *Echidna* the face is drawn out into a gradually tapering rostrum, formed mainly by the premaxillae, maxillae and nasals. The zygomatic arch is very weak, and the palate extends very far back. The tympanic forms a slender ring. The mandible is extremely slight, with no ascending portion, and but slight traces of the coronoid process and angle. The hyoid has a wide basi-hyal and stout thyro-hyals, while the anterior cornua are slender, and include ossified epi-hyals and cerato-hyals.

In *Ornithorhynchus* the zygomatic arch is much stouter than in *Echidna*. The face is produced into a wide beak, mainly supported by the premaxillae, between whose diverging anterior ends there is a dumb-bell-shaped bone. The maxillae are flattened below, and each bears a large horny tooth, which meets a corresponding structure borne on a surface near the middle of the mandible. The mandible is considerably stouter than in *Echidna*, but the angle and coronoid process are but little developed. The infra-orbital foramen and the inferior dental and mental foramina of the mandible are all very large.

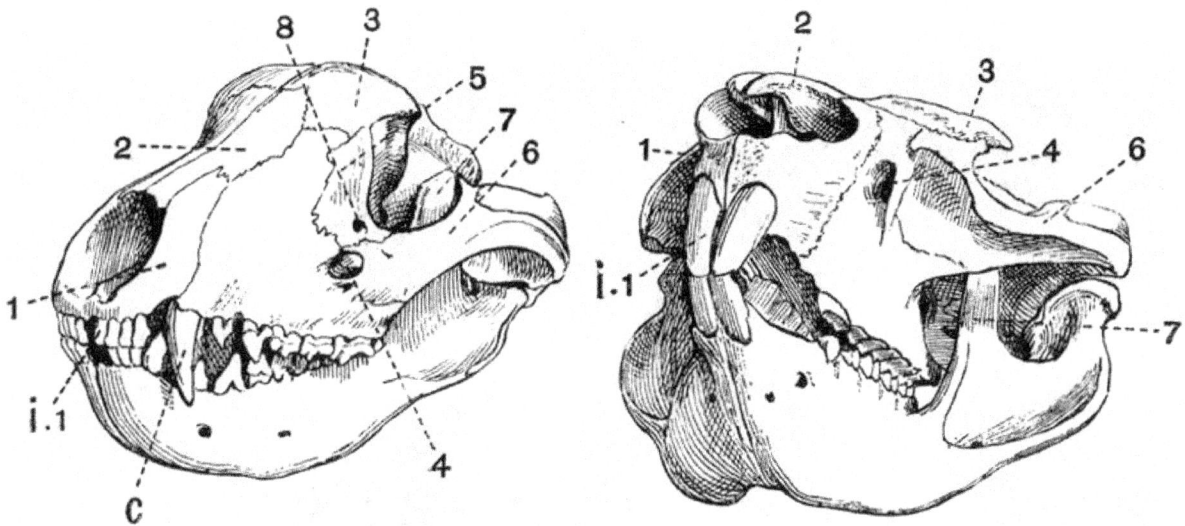

Fig. 90. Half front view of the skulls of a Tasmanian Wolf
(*Thylacinus cynocephalus*) (to the left) × $^3/_8$; and of a
hairy-nosed Wombat
(*Phascolomys latifrons*) (to the right) × $^3/_8$. (Camb. Mus.)

1. premaxillae.

2. nasal.

3. frontal.

4. infra-orbital foramen.

5. lachrymal.

6. jugal.

7. coronoid process of the

mandible.

8. lachrymal foramen.

i. 1. first upper incisor.

C. canine.

Marsupialia. The skulls of the various types of the Marsupials frequently bear a strong superficial resemblance to those of some of the different groups of placental mammals. Thus the skull of the Dasyuridae resembles that of the Carnivora, the resemblance being most marked between the skulls of *Thylacinus* and the dog. The skull of *Notoryctes* is strongly suggestive of that of an Insectivore, and that of other Marsupials such as the wombat, recalls equally the characteristic features of a Rodent's skull. But, however much they may differ from one another, the skulls of all Marsupials agree in the following respects. (1) The brain cavity, and especially the cerebral fossa, has a very small comparative size. (2) The nasals are always large, and the mesethmoid is extensively ossified, and terminated by a prominent vertical edge. (3) Processes from the jugal and frontal in living forms never meet and enclose the orbit, but the zygomatic arch is always complete. (4) The jugal always extends back to form part of the glenoid fossa. (5) The lachrymal canal opens either external to or upon the margin of the orbit, and the nasal processes of the premaxillae never quite reach the frontals. (6) The posterior part of the palate is commonly pierced by large oval vacuities. (7) The tympanic is small and never fused to the bones of the cranium. (8) The carotid canal perforates the basisphenoid and not the tympanic bulla. (9) The optic foramen and sphenoidal fissure are confluent. (10) In every case except *Tarsipes* the angle of the mandible is more or less inflected.

The skull of the extinct *Thylacoleo* differs from that of all other Marsupials in the fact that the postorbital bar is complete. The hyoid is constructed on much the same plan in all Marsupials. It consists of a small basi-hyal, a pair of broad cerato-hyals, and a pair of strong thyro-hyals. The epi-hyals and stylo-hyals are generally unossified.

Edentata. In Sloths (Bradypodidae) the sutures become early obliterated, the cranial portion of the skull is rather high, and the facial portion very short. The lachrymal is very small, and its canal opens outside the orbit. The zygomatic arch is incomplete, and the jugal (fig. 91, 5) is curiously forked, but in a manner differing in the two genera. The premaxillae are very small,—in *Bradypus* quite vestigial. The mandible is well developed, the angle being specially marked in *Bradypus*. In *Choloepus* the symphysial part is drawn out in a somewhat spout-like manner (fig. 91, 6). In both genera the thyro-hyals are ankylosed with the basi-hyal.

229

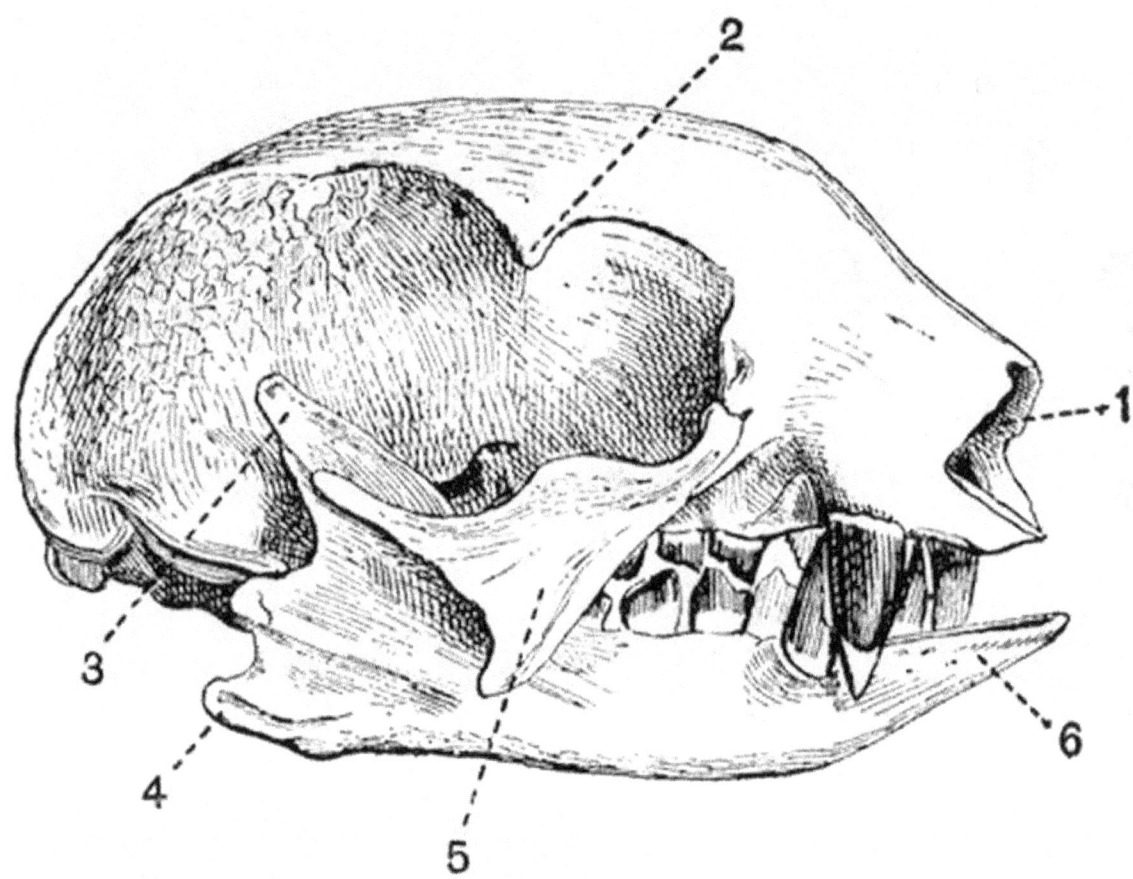

Fig. 91. Skull of a two-fingered Sloth
(*Choloepus didactylus*) × ½. (Camb. Mus.)

1. anterior nares.

2. postorbital process of the

frontal.

3. coronoid process.

4. angle of the mandible.

5. jugal.

6. spout-like prolongation of the

mandible.

In *Megatherium* the general appearance of the skull is distinctly sloth-like, but the facial portion is more elongated, partly owing to the development of a prenasal bone, and the zygomatic arch is complete. The mandible is very deep in the middle, and is drawn out into a long spout-like process in front.

Anteaters (Myrmecophagidae) have a much modified skull, and this is especially the case in the Great Anteater, *Myrmecophaga*. The skull is smooth and evenly-rounded, in these respects recalling that of *Echidna*, but it is longer and tapers much more gradually than in *Echidna*. The occipital condyles are remarkably large. The premaxillae are small, and the long rostrum is chiefly composed of the maxillae and nasals with the mesethmoid and vomer. The zygomatic arch is incomplete, and there is no trace of a separation between the orbit and the temporal fossa. The palate is much elongated, the pterygoids meeting in the middle line just like the palatines. The mandible is very long and slender, there being no definite coronoid process, and a short and slight symphysis. The hyoid arch is noticeable for the length of the anterior cornu.

In the Armadillos (Dasypodidae) the skull varies a good deal in shape, but the facial portion is always tapering and depressed. The zygomatic arch is complete. In *Dasypus* and *Chlamydophorus* the tympanic bulla is well ossified.

In the Glyptodontidae the skull is very short and deep; the zygomatic arch is complete, and has a long downwardly projecting maxillary process. The mandible is massive, and has a very high ascending portion.

In the Manidae the skull is smooth and rounded, the zygomatic arch is incomplete, and the orbit is inconspicuous. The palate is long and narrow, but the pterygoids do not take part in its formation. The mandible is slightly developed and has no angle or coronoid process.

In *Orycteropus* the zygomatic arch is complete, and there is a small postorbital process to the frontal. The mandible is well-developed, having a coronoid process and definite ascending portion, and the hyoid is well ossified.

230

Sirenia. The skull, and especially the brain case of all Sirenia, is remarkable for the general density of the component bones, which, though often very thick, are without air sinuses. It is noticeable also for the roughness of the bones, and the irregular manner in which they are united together.

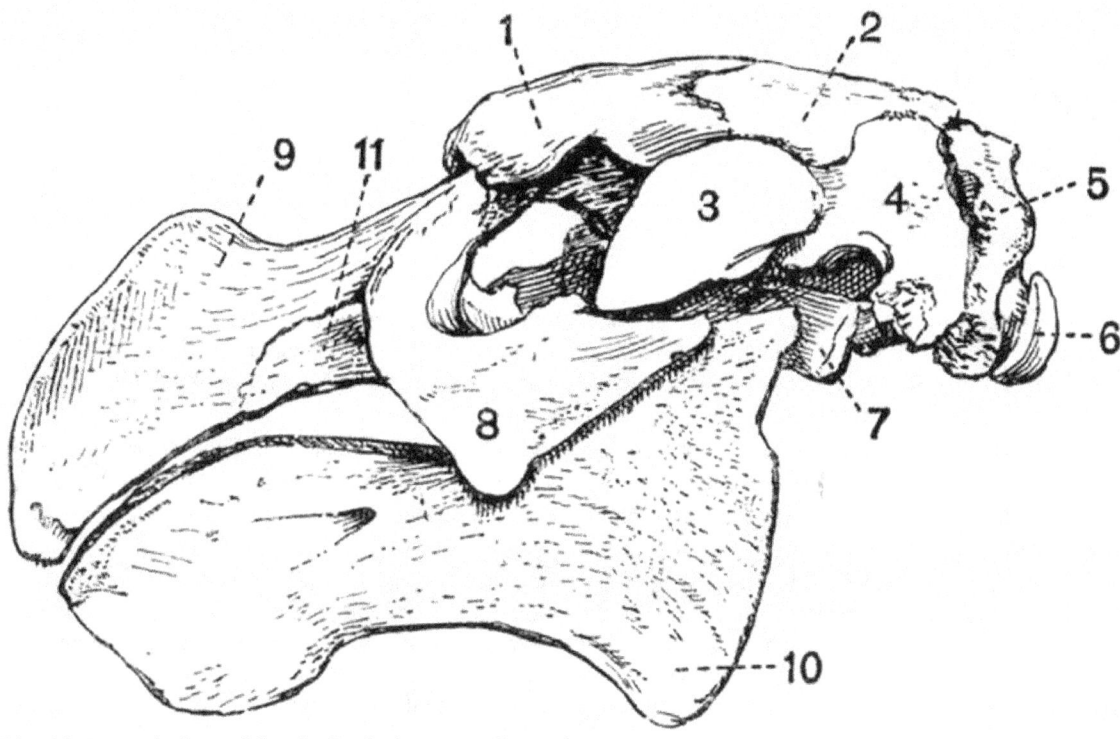

Fig. 92. Lateral view of the skull of *Rhytina stelleri* × ¹/₈. (Brit. Mus.)

1. frontal.

2. parietal.

3. zygomatic process of the

squamosal.

4. squamosal.

5. exoccipital.

6. occipital condyle.

7. pterygoid process of the

alisphenoid.

8. jugal.

9. premaxillae.

10. angle of the mandible.

11. maxillae.

The cranial cavity is decidedly small, the reduction being specially noticeable in the cerebral fossa, which is not much larger than the cerebellar fossa. The foramen magnum is large, and the dorsal surface of the cranium narrow. The zygomatic arch is very strongly developed, the squamosal (fig. 92, 4) being especially prominent, and being drawn out not only into the zygomatic process, but also into a large post-tympanic process which articulates with the exoccipital. At the side of the skull between the squamosal, supra-occipital and exoccipital, there is a wide vacuity in the cranial wall, partially filled up by the very large periotic, which is ankylosed to the tympanic, but is not united to any other bones of the skull. The foramen lacerum medium is confluent with the foramen lacerum anterius, and the two together form an enormous vacuity on the floor of the skull, bounded chiefly by the exoccipital, basi-occipital, alisphenoid and squamosal. The jugal (fig. 92, 8) is large and in *Manatus* sends up a strong process, which nearly or quite meets the postorbital process of the frontal, completing the orbit. In the other Sirenia the orbit is completely confluent with the very large temporal fossa. The lachrymal in *Manatus* is very small, but is larger in *Halicore*. The premaxillae (fig. 92, 9) are large, but smaller in *Manatus* than in the other genera, in all of which they are curiously bent down in front. Their upper margin forms the anterior border of a very large aperture lying high on the roof of the skull and extending back for a considerable distance. This aperture is formed by the union of the two anterior nares. The nasals are quite vestigial or absent, and the narial aperture is bounded above by the frontals; in its floor are seen the slender vomer and large mesethmoid. The palate is long and narrow, and formed mainly by the maxillae; behind it

231

there is a large irregular process formed by the union of the palatine, pterygoid, and pterygoid plate of the alisphenoid. The mandible is very massive and has a very high ascending portion, a rounded angle (fig. 92, 10), and a prominent coronoid process; the two rami are firmly ankylosed together. The hyoid consists principally of the broad flat basi-hyal; the anterior cornua are but slightly ossified, while the thyro-hyals are not ossified at all.

Cetacea. The skull in all Cetacea, especially in the Odontoceti, is a good deal modified from the ordinary mammalian type.

In the *Archaeoceti* this modification is less marked than in either of the other suborders. The nasals and premaxillae are a good deal larger than they are in living forms, and the anterior nares are placed further forward. The maxillae do not extend back over the frontals, and there is a well-marked sagittal crest.

In the *Mystacoceti* the skull is always quite bilaterally symmetrical, and is not so much modified from the ordinary mammalian type as in the Odontoceti. The parietals are not, as in the Odontoceti, separated by a wide interparietal, but meet; they are, however, hidden under the very large supra-occipital. The nasals are developed to a certain extent, and the nares, though placed very far back and near the top of the head, terminate forwardly-directed narial passages. Turbinal bones are also developed to some extent; this fact, and the occurrence of a definite though small olfactory fossa constituting important distinctions from the Odontoceti. The maxillae are large, but do not extend back to cover the frontals as in the Odontoceti. The zygomatic process of the squamosal is very large. The mandibular rami are not compressed, but are rounded and arched outwards, and never meet in a long symphysis.

Odontoceti. The skull departs widely from the ordinary mammalian type. The following description will apply to any of the following genera of the Delphinidae, *Phocaena, Globicephalus, Lagenorhynchus, Delphinus, Tursiops, Prodelphinus, Sotalia.*

The upper surface of the skull is more or less asymmetrical. The cerebral cavity is high, short and broad; and formed mainly by the cerebral fossa, the olfactory fossa being entirely absent. The supra-occipital (fig. 93, 3) is very large, and forms much of the posterior part of the roof of the skull. It has the interparietal (fig. 93, 7) fused with it, and completely separates the two parietals. The frontal (fig. 93, 10) is large and laterally expanded, forming the roof of the orbit, but is almost completely covered by an extension of the maxillae. The zygomatic arch is very slender, and is mainly formed by a rod-like process from the jugal (fig. 93, 15), the zygomatic process of the squamosal being short and stout.

The nasal passages are peculiarly modified, instead of passing horizontally forwards above the roof of the mouth, they pass upwards and even somewhat backwards towards the top of the skull (fig. 93, 23). They are bounded laterally by two processes from the premaxillae, the left of which is shorter than the right. The nasal cavities are narrow and without turbinals and the nasals (fig. 93, 19) are almost as much reduced as in Sirenia.

232

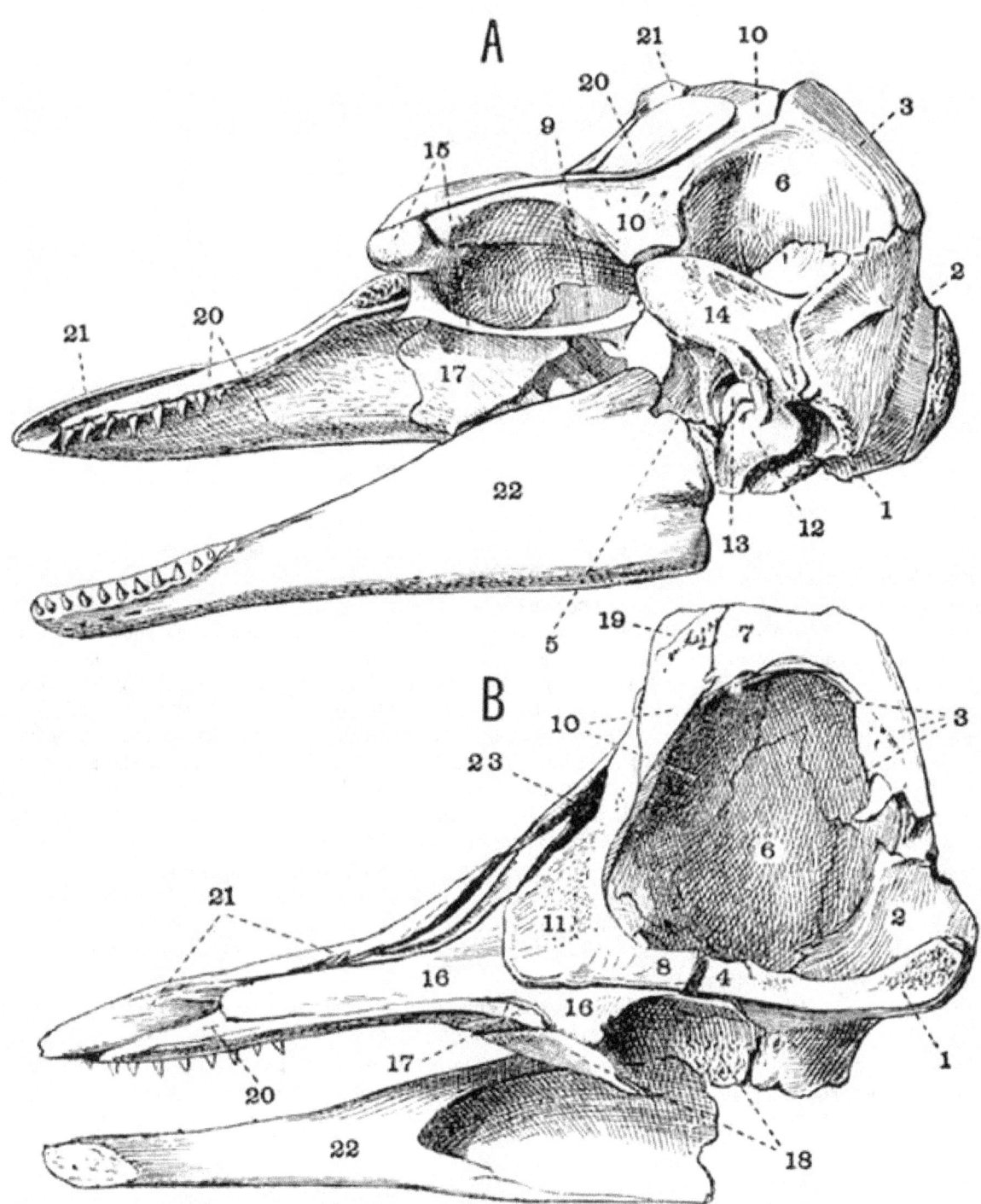

Fig. 93. A, Lateral view, and B, Longitudinal section of the skull
of a young Ca'ing Whale (*Globicephalus melas*) × ¹/₆. (Brit. Mus.)

1. basi-occipital.	13. periotic.
2. exoccipital.	14. squamosal.
3. supra-occipital.	15. jugal.
4. basisphenoid.	16. vomer.
5. alisphenoid.	17. palatine.

6. parietal.

7. interparietal.

8. presphenoid.

9. orbitosphenoid.

10. frontal.

11. mesethmoid.

12. tympanic.

18. pterygoid.

19. nasal.

20. maxillae.

21. premaxillae.

22. mandible.

23. anterior nares.

In front of the nasal openings the face is prolonged as a narrow beak or rostrum of varying length, formed by the maxillae and premaxillae surrounding the vomer and large mesethmoid (fig. 93, 11), which sends forwards a long partially cartilaginous process, and is fused behind with the presphenoid (fig. 93, 8). The basi-occipital (fig. 93, 1) too is fused with the basisphenoid. The foramen rotundum is confluent with the sphenoidal fissure, and the foramen ovale with the foramen lacerum medium and the foramen lacerum posterius. The palate is mainly formed by the maxillae; the premaxillae and palatines (fig. 93, 17), though both meet in symphyses, forming very little of it. The pterygoids vary in size in the different genera, sometimes as in *Lagenorhynchus* and *Delphinus* meeting in the middle line, sometimes as in *Phocaena* and *Globicephalus* (fig. 93, 18) being widely separated. The tympanic and periotic are not fused together, and the periotic has generally no bony union with the rest of the skull. The mandible is rather slightly developed, with the rami straight, compressed and tapering to the anterior end. The condyle is not raised at all above the edge of the ramus; the angle is rounded and the coronoid process is very small. *Platanista* has a curiously modified skull; the rostrum and mandible are exceedingly long and narrow, and arising from the maxillae are two great plates of bone which nearly meet above.

In the Physeteridae the skull is raised into a very prominent crest at the vertex behind the nares. In front of this in *Hyperoödon* a pair of ridges occur, formed by outgrowths from the maxillae. In the old male these ridges reach an enormous size and almost meet in the middle line. In *Physeter*, the Sperm whale, these ridges are not developed; the maxillae and premaxillae unite with the other bones of the crest enclosing an enormous half basin-shaped cavity, at the base of which are the very asymmetrical anterior narial apertures.

In all living Cetacea the hyoid has the same general shape, consisting firstly of a crescentic bone formed by the fusion of the thyro-hyals with the basi-hyal, and secondly of the anterior cornu formed principally by the strong stylo-hyal.

Ungulata. None of the distinctive characters separating the Ungulata from the other groups of mammals are drawn from the skull. But in the Ungulata vera as opposed to the Subungulata a distinguishing feature is found in the fact that the lachrymal and jugal form a considerable part of the side of the face, and that the jugal always forms the anterior part of the zygomatic arch, the maxillae taking no part in it.

Ungulata vera.

Artiodactyla. The skull in Artiodactyla differs from that in Perissodactyla in the fact that the posterior end of the nasal is not expanded and there is no alisphenoid canal.

The skulls in the different groups of Artiodactyla differ considerably from one another.

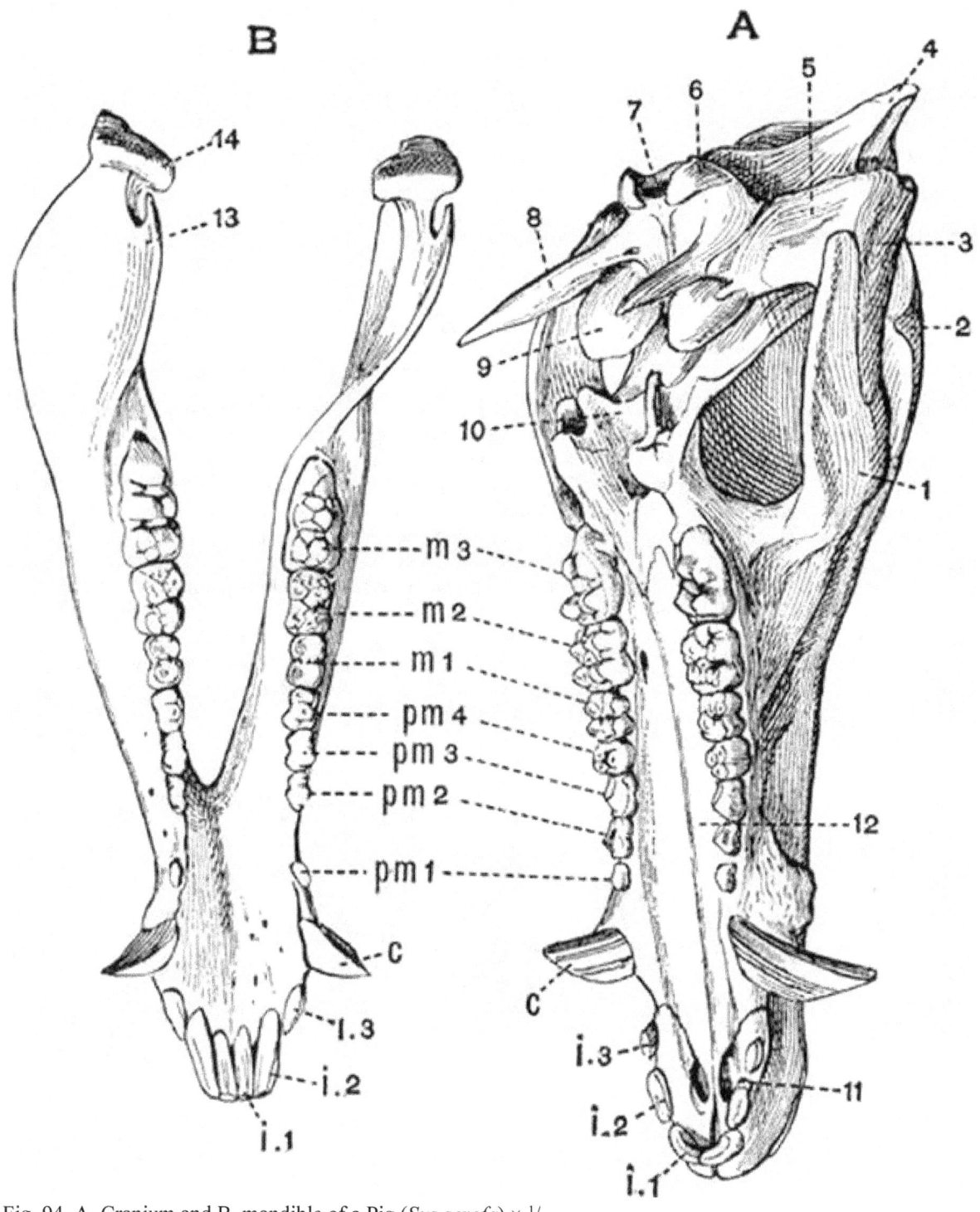

Fig. 94. A, Cranium and B, mandible of a Pig (*Sus scrofa*) × ¹/₅.
(Camb. Mus.)

1. jugal.

2. postorbital process of the

frontal.

3. zygomatic process of the

11. anterior palatine foramen.

12. palatal plate of maxillae.

13. coronoid process.

14. mandibular condyle.

235

squamosal.

4. supra-occipital.

5. glenoid cavity.

6. occipital condyle.

7. foramen magnum.

8. paroccipital process of the

exoccipital.

9. tympanic bulla.

10. pterygoid.

i 1, *i* 2, *i* 3. first, second, and third

incisors.

c. canine.

pm 1, *pm* 2, *pm* 3, *pm* 4. first,

second, third, and fourth

premolars.

m 1, *m* 2, *m* 3. first, second, and

third molars.

The skull of the Pig will be described as illustrative of the skull in the Suina. In the Pig as in most Artiodactyla the face is bent sharply down on the basicranial axis, the commencement of the vomer being situated below the mesethmoid instead of in front of it as in most skulls. The occipital region of the skull is small, and the line of junction of the supra-occipital and parietals is raised into a prominent occipital crest. The parietal completely fuses at an early stage with its fellow, and the exoccipital is drawn out into a long paroccipital process (fig. 94, A, 8). The frontal is large and broad and drawn out into a small postorbital process. The lachrymal too is large and takes a considerable part in forming the side of the face in front of the orbit, as does also the jugal, though to a less extent. The face is long and tapers much anteriorly. The nasals are long and narrow, as are the nasal processes of the premaxillae, which do not however reach the frontals. A prenasal ossicle is developed in front of the mesethmoid. The palate is long and narrow, the pterygoid (fig. 94, A, 10) is small, but the pterygoid process of the alisphenoid is prominent. The squamosal is small and has the tympanic fused with it; the tympanic is dilated below, forming a bulla (fig. 94, A, 9) filled with cancellous bone, and above forms the floor of a long upwardly-directed auditory meatus. The mandible has a high ascending portion and a small coronoid process (fig. 94, B, 13). The hyoid differs from that of most Ungulates, the stylo-hyal being very imperfectly ossified.

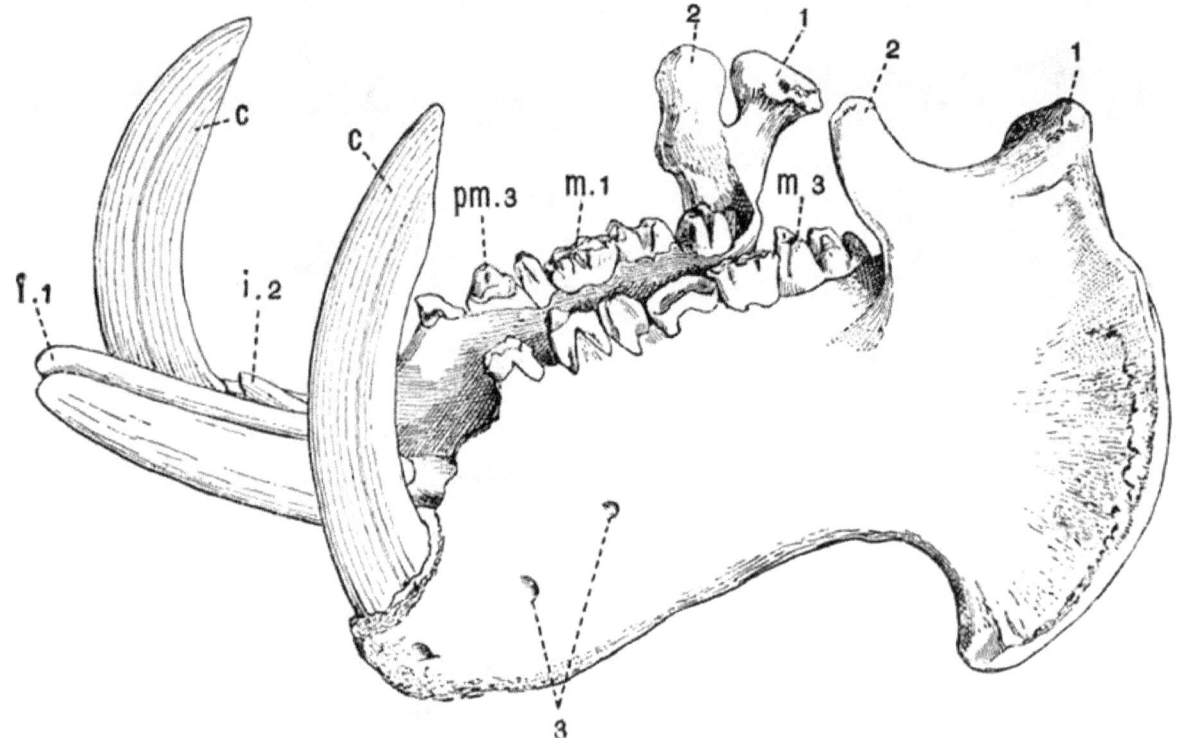

Fig. 95. Mandible of a Hippopotamus (*H. amphibius*) × ¹/₇.
(Camb. Mus.)
The second incisor of the left side is missing and the crowns of the grinding teeth are much worn.

1. condyle.

2. coronoid process.

3. mental foramina.

i 1, *i* 2. first and second incisors.

c. canine.

pm 3. third premolar.

m 1, *m* 3. first and third molar.

In *Hippopotamus* the skull though essentially like that of the pig is much modified in detail. The brain cavity is very small, while the jaws are immensely developed. The face contracts in front of the orbits and then expands again greatly, to lodge the enormous incisor and canine teeth. The postorbital bar is complete or nearly so, and the orbits project curiously outwards and slightly upwards; the lachrymal is thin and much dilated. The squamosal is drawn out into a postglenoid process, and the hamular process of the pterygoid is prominent. The tympanic bulla is filled with cancellous bone. The mandible is enormously large, the symphysis is long, the angle much expanded and drawn out into a process which projects outwards and forwards.

Among extinct forms related to the Suina, *Cyclopidius* is noticeable for having large vacuities in the lachrymo-nasal region, while *Cotylops* has the postorbital bar complete; both these forms are from the North American Miocene.

In the Tylopoda and Tragulina the skull resembles in most respects that of the Ruminants, shortly to be described; but it is allied to that of the Suina in having the tympanic bulla filled with cancellous bone. The tympanic bulla is better developed in the Tragulina than in most Ungulates.

Among Ruminants, the Bovidae, that large group including the Oxen, Sheep, and Antelopes, as a rule have the face bent on the basicranial axis much as in the Suina. The parietals are generally small and early coalesce, the frontals are large and are usually drawn out into horn cores, which are however absent in the skulls of some domestic varieties of sheep and oxen, and also in some of the earlier extinct forms of Bovidae. These horn cores are formed internally of cancellous bone, and on them the true epidermal horns are borne. In young animals there is a distinct interparietal, but this early fuses with the supra-occipital, and in the oxen also with the parietals. The occipital crest is generally well marked, but in the genus *Bos* becomes merged in a very prominent straight ridge running between the two horn cores; this ridge, which contains air cells communicating with those in the horn cores, is not nearly so well marked in *Bison*. There is often, as in *Gazella*, a vacuity on the side of the face between the nasal, frontal, lachrymal, and maxillae, but this is not found in oxen or sheep. The premaxillae are small, the nasals are long and pointed, and the turbinals are much developed. The Saiga antelope has a curiously specialised skull; the nasals are absent or have coalesced with the frontals and the anterior nares are enormously large. In all Ruminants the lachrymal is large and forms a considerable part of the side of the face; it often bears a considerable depression, the *suborbital* or *lachrymal fossa*, well seen in most of the smaller antelopes. The postorbital bar is complete, and the orbit is prominent and nearly circular. The palatines and pterygoids are moderately large, and the pterygoids have a backwardly-projecting hamular process. The squamosal is small, but has a postglenoid process. The tympanic is not fused to the periotic and has a small bulla not filled with cancellous bone. There is a large paroccipital process to the exoccipital and the mandible has a long slender coronoid process.

In the Cervidae and Giraffidae the face is not bent down on the basicranial axis as it is in the Bovidae. The frontals are drawn out, not into permanent horn cores as in the Bovidae, but into short outgrowths, the pedicels, upon which in the Cervidae long antlers are annually developed. These *antlers* are outgrowths of bone, and are covered during development by vascular integument, which dries up and peels off when growth is complete. Every year they are detached, by a process of absorption at the base, and shed. They may occur in both sexes, as in the Reindeer, but as a rule they are found only in the male. They are generally more or less branched, and are sometimes of enormous size and weight, as in the extinct *Cervus megaceros*. In young animals they are always simple, but become annually more and more complicated as the animal grows older.

In the Giraffe the frontals bear a small pair of bony cores, which are at first distinct, but subsequently become fused to the skull. In the allied *Sivatherium*, a very large form from the Indian Pliocene, the skull bears two pairs of bony outgrowths, a pair of short conical outgrowths above the orbits, and a pair of large expanded outgrowths on the occiput.

The opening of the lachrymal canal is commonly double and the lachrymal fossa is large in the Cervidae and the Giraffidae except *Sivatherium*. The vacuity between the frontal, lachrymal, maxillae, and nasal is specially large.

The hyoid of Ruminants is noticeable for the development of the anterior cornua, which include stout and short cerato-hyals and epi-hyals, long and strong stylo-hyals and large tympano-hyals which are more or less imbedded in the tympanics.

Perissodactyla. In the skull of Perissodactyles an alisphenoid canal is found and the nasals are expanded behind. Among the living animals belonging to this group the skull least modified from the ordinary type is that in *Rhinoceros*. In this form the skull is considerably elongated, the facial portion being very large. The occipital region is elevated, but the cranial cavity is small, the boundary line between the occipital and parietal regions being drawn out into a prominent crest, which is occupied by air cells. There is no postorbital process to the frontal, and the orbit is completely confluent with the temporal fossa. The nasals are fused together and are very strongly developed, extending far forwards, sometimes considerably beyond the premaxillae. In some extinct species, such as *Elasmotherium* and the Tichorhine Rhinoceros, *R. antiquitatis*, the mesethmoid is ossified as far forwards as the end of the nasals. The nasals are arched and bear one or two roughened surfaces to which the great nasal horns are attached. The premaxillae are very small and the pterygoids are slender. The palate is long, narrow, and deeply excavated behind. The postglenoid process of the squamosal is well developed, and generally longer than the paroccipital process of the exoccipital. The tympanic and periotic are both small and are fused together. The condyle of the mandible is very wide, the angle rounded, and the coronoid process moderately developed.

In the Titanotheriidae, a family of extinct Perissodactyla from the Miocene of North America, the occipital region is much elevated, as is also the fronto-nasal region, the nasals (perhaps only in the male) bearing a pair of blunt bony outgrowths. Between these two elevated regions the skull is much depressed. The cranial cavity is very small, the orbit confluent with the temporal fossa, and the zygomatic arch massive.

In *Tapirus* the orbit and temporal fossa are confluent. The nasals are small, wide behind and pointed in front, and are supported by the mesethmoid; the anterior nares are exceedingly large and their lateral boundaries are entirely formed by the maxillae. The postglenoid and post-tympanic processes of the squamosal are large. The periotic is not fused to the squamosal or to the small tympanic. The mandible is large and has the angle much developed and somewhat inflected.

Palaeotherium, which lived in early Tertiary times, has a skull much like that of the Tapir, especially as regards the nasal bones.

In the Horse and its allies (Equidae) the facial portion of the skull is very large as compared with the cranial portion, the nasals and nasal cavities being specially large. In the living species of the genus *Equus* there is no fossa between the maxillae and lachrymal, but it occurs in some extinct species. The lachrymal and jugal form a considerable part of the side of the face; and the orbit though small is complete and prominent. The postorbital bar is formed by a strong outgrowth from the frontal, which unites with a forward extension of the squamosal. The squamosal may extend forwards and form part of the wall of the orbit, a very unusual feature, as in most mammals the squamosal stops before the postorbital bar. The palate is narrow and excavated behind as in *Rhinoceros*; the palatines take very little part in its formation. The glenoid surface for the articulation of the mandible is very wide. The squamosal gives rise to small postglenoid and post-tympanic processes, and the exoccipital to a large paroccipital process. The tympanic and periotic are ankylosed together, but not to any other bones.

In the Subungulata, the lachrymal and jugal do not form any considerable part of the side of the face, and the maxillae commonly takes part in the formation of the zygomatic arch.

Toxodontia. The skull in the Toxodontia shows several Artiodactyloid features, while the manus and pes are of a more Perissodactyloid type. The Artiodactyloid features are (1) the absence of an alisphenoid canal, (2) the fact that the palate is not excavated behind, and that the palatines form a considerable part of it, and (3) the fusion of the tympanic to the squamosal and exoccipital, forming the floor of an upwardly directed auditory meatus. The frontal has a fairly well developed postorbital process, but the orbit is confluent with the temporal fossa. The premaxillae is well developed, as is the paroccipital process of the exoccipital, especially in *Typotherium*. The mandible has a rounded angle and a coronoid process of moderate size. In *Typotherium* the ascending portion is very massive.

Condylarthra. As far as is known the skull of these generalised Ungulates is depressed, and is frequently marked by a strong sagittal crest. The cranial cavity is small, the cerebral fossa in *Phenacodus* being exceptionally small. The orbit is completely confluent with the temporal fossa.

Hyracoidea. The skull of *Procavia* resembles that of Perissodactyles more than that of any other Ungulates, but differs strongly in the comparatively small size of its facial portion. The posterior portion of the cranium is rather high, the occipital plane being nearly vertical. There is a small interparietal. The nasals are wide behind, and the zygomatic arch is strongly developed, its most anterior part being formed by the maxillae. The jugal and parietal give rise to postorbital processes which sometimes meet, but as a rule the orbit is confluent with the temporal fossa; it is very uncommon for the parietal to give rise to a postorbital process, and even in *Procavia* the frontal often forms part of the process. The alisphenoid canal, and postglenoid and paroccipital processes are well developed. The tympanic bulla is large and the periotic and tympanic are fused together, but not as a rule to the squamosal. The ascending portion of the mandible is very high and broad, the angle rounded and the coronoid process moderate in size. The hyoid is singular, there is a large flat basi-hyal prolonged laterally into two broad flattened thyro-hyals. Articulating with its anterior end are two large triangular cerato-hyals, which are drawn out into two processes meeting in the middle line.

Amblypoda. In the Uintatheriidae (Dinocerata) the skull has a very remarkable character, being long and narrow and drawn out into three pairs of rounded protuberances, a small pair on the nasals, a larger pair on the maxillae in front of the orbits, and the largest pair on the parietals. The cranial cavity, and especially the cerebral fossa, is extraordinarily small. The orbit is not divided behind from the temporal fossa. The mandible has a prominent angle, and a long curved coronoid process; its symphysial portion bears a curious flattened outgrowth to protect the great upper canines.

In *Coryphodon* the skull is of a more normal character, being without the conspicuous protuberances. The cranial cavity though very small is not so small as in *Uintatherium*.

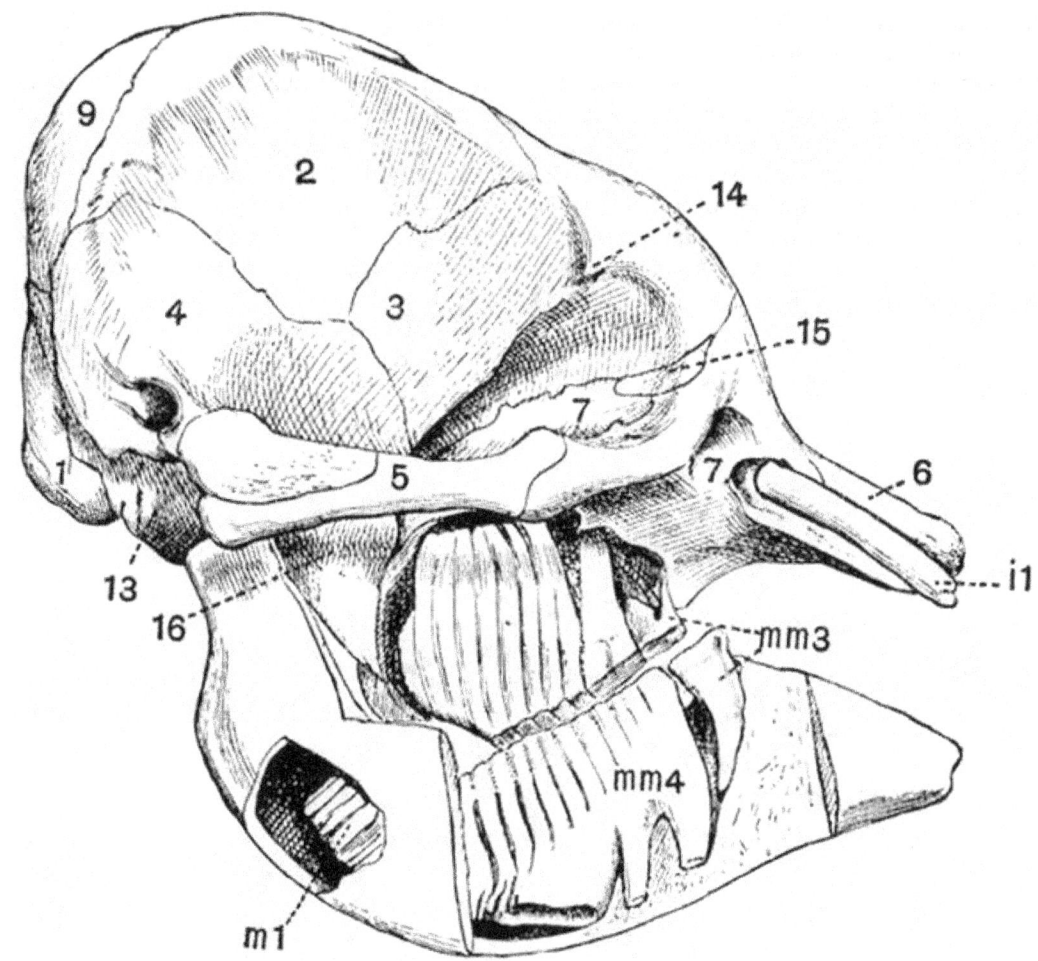

Fig. 96. Skull of a young Indian Elephant (*Elephas indicus*), SEEN from the right side, the roots of the teeth have been exposed. × ¹/₈. (Camb. Mus.).

1. exoccipital.

2. parietal.

3. frontal.

4. squamosal.

5. jugal.

6. premaxillae.

7. maxillae.

9. supra-occipital.

13. basi-occipital.

14. postorbital process of the

frontal.

15. lachrymal.

16. pterygoid process of the

alisphenoid.

i 1. incisor.

mm 3., *mm* 4. third and fourth

milk molars.

m 1. first molar.

Proboscidea. The character of the skull in the young elephant differs much from that in the old animal. In very young individuals the skull is of a normal character, and the cranial cavity is distinctly large in proportion to the bulk of the skull. But as the animal gets older, while its brain does not grow much, the size of its trunk and especially of its tusks increases greatly; and consequently the skull wall is required to be of very great superficial extent in order to afford space for the attachment of the muscles necessary for the support of these heavy weights. This increase in superficial extent is brought about without much increase in weight of bone by the development of an enormous number of air cells in nearly all the bones of the skull; sometimes, as in the case of the frontal, separating the inner wall of the bone from the outer, by as much as a foot. This development of air cells is accompanied by the obliteration of the sutures between the various bones. The most

239

noticeable point with regard to the cranial cavity is the comparatively large size of the olfactory fossa. The supra-occipital (figs. 96 and 97, 9) is large—exceedingly large in the adult skull; the parietals (figs. 96 and 97, 2) are also very large. The frontals send out small postorbital processes, but these do not meet processes from the small jugal, which forms only the middle part of the slender zygomatic arch, the anterior part being formed by the maxillae. The lachrymal (fig. 96, 15) is small and lies almost entirely inside the orbit. The anterior narial aperture (fig. 97, 8) is wide and directed upwards, opening high on the anterior surface of the skull. It is bounded above by the short thick nasals and below by the premaxillae. The narial passage is freely open, maxillo-turbinals not being developed. The palatine is well developed, the pterygoid is small and early fuses with the pterygoid process of the alisphenoid. The tympanic is united with the periotic but not with the squamosal, and forms a large auditory bulla. There are no paroccipital or postglenoid processes. The exoccipital is not perforated by the condylar foramen,—a very exceptional condition.

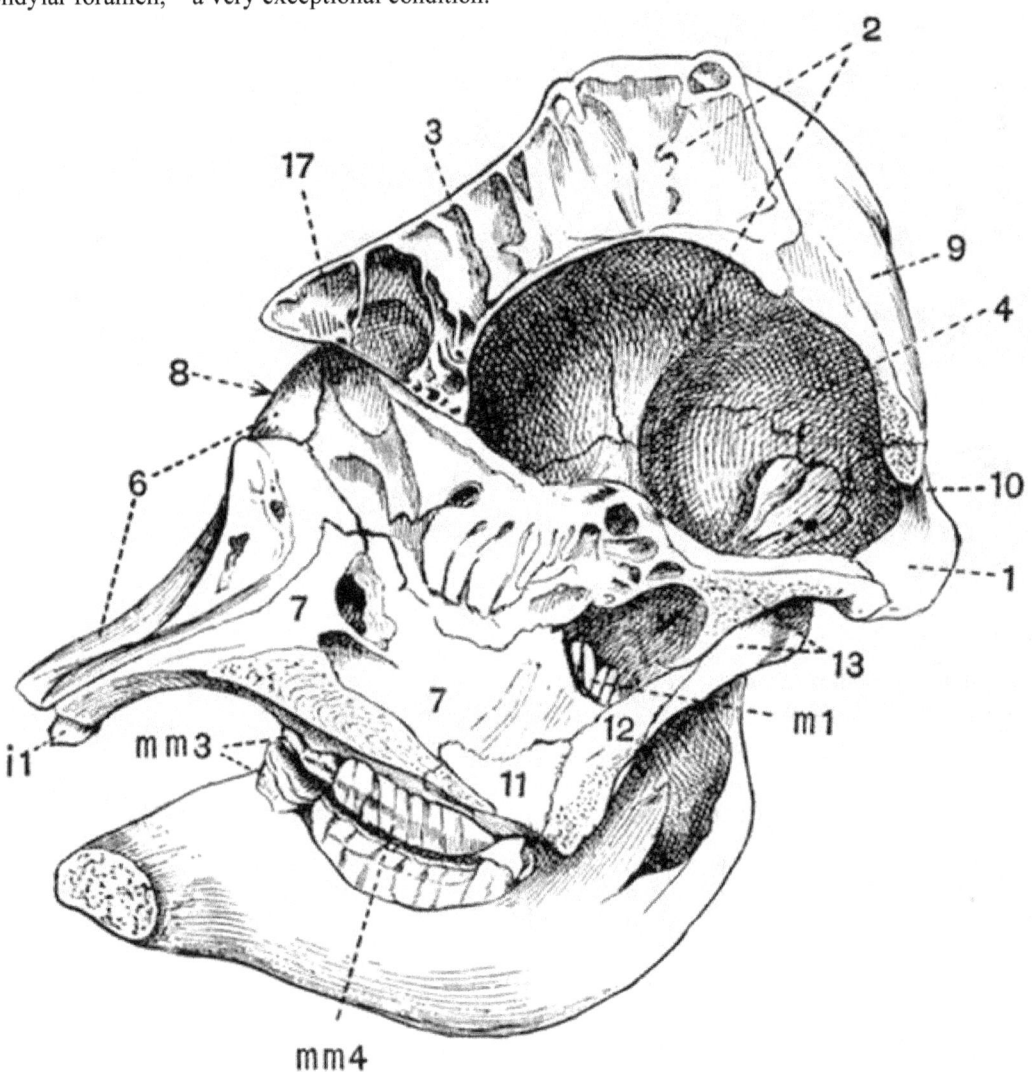

Fig. 97. Longitudinal section taken rather to the right of the middle line of the skull of a young Indian Elephant (*E. Indicus*) × $^1/_8$. (Camb. Mus.)

8. anterior nares.	12. pterygoid.
10. periotic.	17. nasal.
11. palatine.	Other numbers as in Fig. 96.

The mandible has a high ascending portion, is rounded off below and has no angle. The symphysial portion is long, narrow, and spout-like, and the coronoid process is small. The thyro-hyals are ankylosed with the basi-hyal, which is connected with the large forked stylo-hyals by ligament only.

Rodentia. The cranial cavity is depressed, elongated, and rather small, and the cerebral fossa lies entirely in front of the cerebellar fossa. The occipital plane is vertical or directed somewhat backwards, and the supra-occipital does not form much of the roof of the cranium. The paroccipital processes of the exoccipitals are generally of moderate size; in the Capybara (*Hydrochaerus*), however, they are very long, and are laterally compressed and directed forwards. The parietals are small, and often become completely fused together; there is sometimes a small interparietal. The frontals in most genera have no trace of a postorbital process; in Squirrels, Marmots and Hares, however, one occurs, but in no case does it meet a corresponding process from the zygomatic arch, so the orbit and temporal fossa are completely confluent. In Hares the postorbital process of the frontal is much flattened, and has an irregular margin. The temporal fossa is always small, and in *Lophiomys* is arched over by plates arising respectively from the parietal and jugal; a secondary roof is thus partially developed in a manner unique among mammals, but carried to a great extent in many Chelonia. The nasal bones and cavities are large, attaining their maximum development in the Porcupines (fig. 98, 1). The premaxillae is always very large, and sends back a long process which meets the frontal. The vomer is occasionally found persisting in two separate halves, a feature recalling the arrangement in Sauropsids. In many Rodents there is an enormous vacuity at the base of the maxillary portion of the zygomatic arch. It is sometimes as large as the orbit, and attains its maximum development in the Capybara and other Hystricomorpha; in the Marmots, Beavers, and Squirrels (Sciuromorpha), and in the Hares it is undeveloped. In *Lagostomus* the maxillae bears an upwardly directed plate of bone, shutting off from this vacuity a space which is the true infra-orbital foramen.

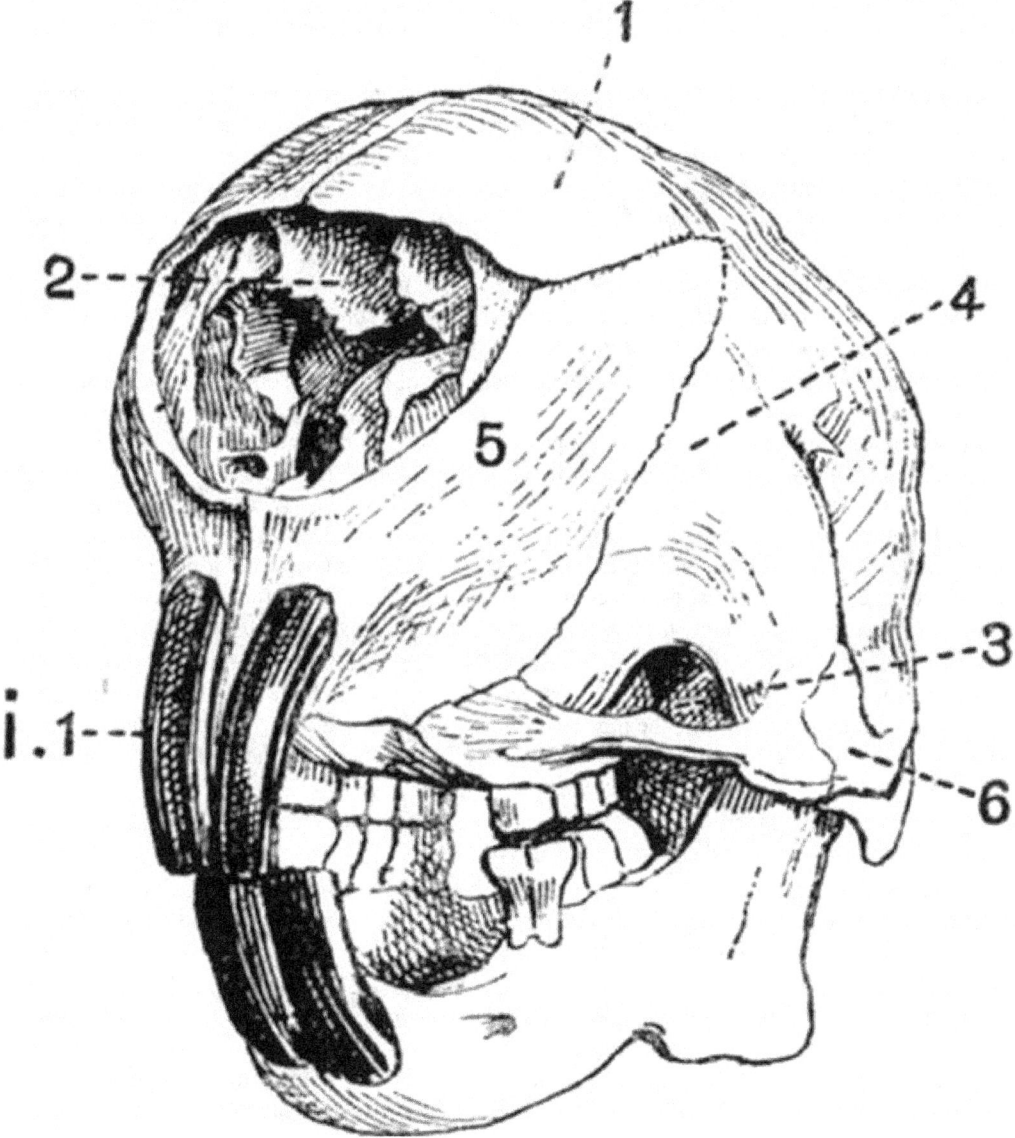

Fig. 98. Half front view of the skull of a Porcupine
(*Hystrix cristata*) × ½. (Camb. Mus.)

1. nasal.   5. premaxillae.

241

2. maxillo-turbinals.

3. infra-orbital vacuity.

4. maxillae.

6. jugal.

i 1. upper incisor.

The zygomatic arch is always complete, and in many cases the jugal extends back to form part at least of the glenoid surface for articulation with the mandible. In *Coelogenys* the jugal and maxillary portion of the zygomatic arch is greatly expanded and roughened, and the maxillary portion encloses a large cavity. The palate in Rodents is narrow, and the space between the incisor and molar teeth passes imperceptibly into the sides of the face. The anterior palatine foramina form long, rather narrow slits in this region. The bony palate between the grinding teeth is sometimes as in the Hares very short, sometimes as in the Capybara very long. The maxillae extends back beneath the orbit to unite with the squamosal. The pterygoid is always small, but sometimes has a well-marked hamular process which in *Hystrix, Lagostomus*, and some other genera unites with the tympanic bulla. The periotic is large, and fused with the tympanic, which forms a prominent bulla, and is generally drawn out into a tubular meatus. The bulla attains its maximum development in *Chinchilla* and *Dipus*.

The mandible is narrow and rounded in front, the two halves meeting in a long symphysis. The angle is generally drawn out into a long backwardly-projecting process, which is often pointed and directed upwards. In the Hares the angle is rounded. The coronoid process is never large.

There are a number of points in which the skull of the Duplicidentata (Hares and Rabbits) differs from that of other Rodents. (*a*) The sutures between the basi-occipital and basisphenoid, and between the basisphenoid and presphenoid remain open throughout life. (*b*) Much of the maxillae forming the side of the face in front of the orbit is fenestrated. (*c*) The optic foramina are united to form a single hole, much as in birds. (*d*) The coronoid process is slightly differentiated from the ascending portion of the mandible. The first two of these points have been thought to indicate degradation of the hares and rabbits as compared with higher mammals.

Carnivora. It is characteristic of the skull in Carnivora that the glenoid fossa is deep, and the postglenoid process (fig. 75, 23) well developed. The condyle of the mandible is much elongated transversely. The orbit and temporal fossa in the great majority of forms communicate freely, the postorbital bar being incomplete.

Carnivora vera. The axis of the facial portion of the skull is a direct continuation of that of the cranial portion. The cranial cavity though rather depressed is large, and generally long, though in Cats it is comparatively short and wide. The occipital plane is nearly vertical, and the exoccipitals are developed into fairly prominent paroccipital processes. The interparietal is commonly distinct, and the parietals unite in a long sagittal suture, which is often developed into a crest. The nasals (fig. 73, 4) are well developed, especially in Cats, and the nasal processes of the premaxillae do not nearly reach the frontals. A considerable part of the palate is formed by the palatine, and the maxillary portion is pierced by rather long anterior palatine foramina. The pterygoid has a hamular process. The zygomatic arch is strong, especially in Cats. Postorbital processes are developed on the frontal (fig. 73, 10) and jugal, but never form a complete postorbital bar. A carotid canal is well seen in the Ursidae, and to a less extent in the Felidae; in the Canidae there is an alisphenoid canal (fig. 75, 21).

The auditory bulla differs a good deal in the different groups. In the Bears (Ursidae) it is not much inflated, and is most prominent along its inner border; it is not closely connected with the paroccipital process. In the Cats it is very prominent, and its cavity is almost divided by a septum into two parts, the inner of which contains the auditory ossicles. The paroccipital process is closely applied to the bulla. In the Dogs the bulla is intermediate in character between that of the Cats and that of the Bears; it is partially divided by a septum, and is moderately expanded.

The mandible is well developed with a prominent angle (fig. 72, 26), and a large coronoid process. The hyoid consists of a broad basi-hyal, a long many-jointed anterior cornu and short thyro-hyals (fig. 72, 33).

The skull in the *Creodonta* is in most respects allied to that of the Canidae, but presents some ursine affinities. The tympanic bulla is fairly prominent, but has no well-developed septum. The cranial cavity is very small and narrow, the zygomatic arch standing away from it. The temporal fossa is of great size.

In the *Pinnipedia* the cranial cavity is large and rounded. The skull is much compressed in the interorbital region, and in correlation with this compression the ethmo-turbinals are little developed, while the maxillo-turbinals are large. The orbit is large, and the temporal fossa smaller than in the Carnivora vera. In the Walrus (*Trichechus*) the anterior part of the face is distorted by the development of the huge canines. The Otariidae have an alisphenoid canal. The tympanic bulla is small in *Otaria*, large in the Phocidae, and flattened in the Walrus. The hyoid is similar to that in Carnivora vera.

Insectivora. The skull varies much in the different members of the order Insectivora, but the following points of agreement are found. The cranial cavity is of small size, and is never much elevated. The facial part of the skull is generally considerably elongated, and the nasals and premaxillae are well developed. The zygomatic arch is usually slender or incomplete, and the coronoid process and angle of the mandible are commonly prominent.

In some Insectivora, such as *Galeopithecus*, *Tupaia*, and *Macroscelides*, the skull shows a higher type of structure than is met with in most members of the order. In these genera the cranial cavity is comparatively large, and the occipital plane is nearly vertical. The zygomatic arch is fairly strong, and the frontal and jugal give rise to postorbital processes which nearly or quite (*Tupaia*) meet. The tympanic bulla is well developed, and produced into a tubular auditory meatus, this being specially well marked in *Macroscelides*.

In the other Insectivora the cranial cavity is of smaller comparative size, and the orbit and temporal fossa are completely confluent, often without any trace of a postorbital bar. The occipital plane commonly slopes forwards. In the Hedgehogs (Erinaceidae) and Centetidae the tympanic is very slightly developed, forming a small ring. The zygomatic arch of Hedgehogs and *Gymnura* is very slender, the jugal being but little developed and the squamosal and maxillae meeting one another; in the Centetidae the jugal is absent and the arch is incomplete.

The Moles (Talpidae) have an elongated, depressed and rounded skull with a very slender zygomatic arch formed by the squamosal and maxillae. The nasals are fused together, and the mesethmoid is ossified very far forwards. In the Shrews (Soricidae) there is no zygomatic arch; the tympanic is ring-like, and the angle of the mandible is very prominent. The hyoid has a transversely extended basi-hyal, a long anterior cornu with three ossifications, and thyro-hyals which are sometimes fused to the basi-hyal.

Chiroptera. In the frugivorous Flying Foxes (Pteropidae) the skull is elongated, and the cranial cavity is large and arched, though considerably contracted in front. There are commonly strong sagittal and supra-orbital crests. The parietals take a great part in the formation of the walls of the cranial cavity, the supra-occipital and frontals being small. The frontal is drawn out into a long postorbital process, but the zygomatic arch, which is slender, and formed mainly by the squamosal and maxillae, gives rise to only a small postorbital process, so that the orbit and temporal fossa are confluent. There is no alisphenoid canal, and the tympanics are very slightly connected with the rest of the skull. The mandible has a large coronoid process, a rounded angle, and a transversely expanded condyle.

In Insectivorous Bats the skull is generally shorter and broader than in the Pteropidae. The cranial cavity is large and rounded, and has thin smooth walls. The zygomatic arch is slender, and postorbital processes are not generally well developed. The premaxillae is generally small, sometimes absent. The tympanics are ring-like and are not connected with the surrounding bones. The angle of the mandible is distinct. The hyoid in most respects resembles that of the Insectivora.

Primates. The characters of the skull differ greatly in the two suborders of Primates, the Anthropoidea and the Lemuroidea.

In the *Lemuroidea* the general relative proportions of the cranium and face are much as in most lower mammals, and the occipital plane forms nearly a right angle with the basicranial axis. The postorbital processes of the frontals are commonly continued as a pair of ridges crossing the roof of the cranium and meeting the occipital crest. Though the postorbital bar is complete, the orbit and temporal fossa communicate freely below it. The lachrymal canal opens outside the orbit, and the lachrymal forms a considerable part of the side of the face. The tympanic is developed into a large bulla. The hyoid apparatus much resembles that of the Dog.

In the *Anthropoidea* the skull differs greatly from that in the Lemuroidea. The cranial portion of the skull is very large as compared with the facial portion, though the comparative development varies, some monkeys, such as the baboons (Cynocephali) having the facial portion relatively large. The comparative size of the jaws does not vary inversely with the general development of the animal, some of the Cercopithecidae having comparatively larger jaws than some of the Cebidae. The great size of the cranial part of the skull is mainly due to the immense development of the cerebral fossa, which commonly completely overlaps the olfactory fossa in front, and the cerebellar fossa behind. This development also has the effect of making the ethmoidal and occipital planes lie, not at right angles to the basicranial axis, but almost in the same straight line with it. This is, however, not always the case, as the Howling Monkey (*Mycetes*) and also some of the very highest monkeys, the Gibbons (*Hylobates*), have the occipital plane nearly vertical to the basicranial axis. In adult Man the basi-occipital, exoccipitals and supra-occipital coalesce, forming the so-called occipital bone; while the basisphenoid, presphenoid, alisphenoids, orbitosphenoids and pterygoids form the sphenoid bone. The roof of the skull is partly formed by the large supra-occipital and frontals, but mainly by the parietals (fig. 99, 1), which in Man are of enormous extent.

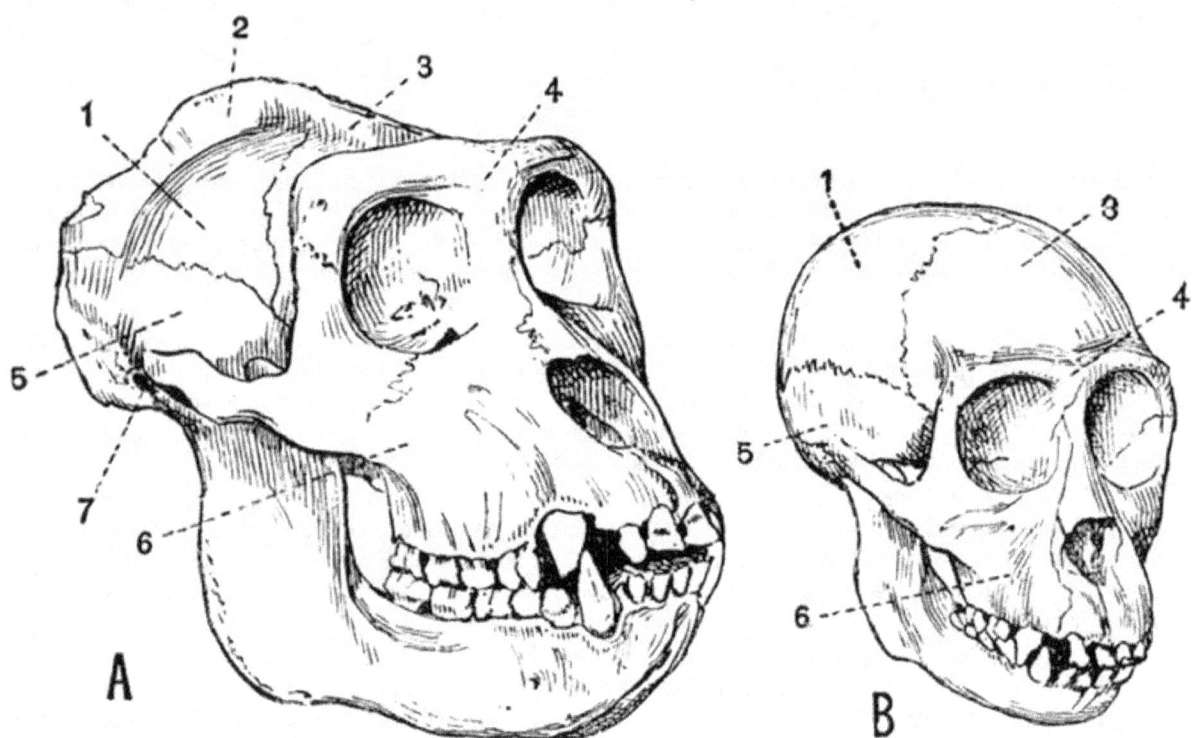

Fig. 99. Half front view of the skulls, *A* of an old, *B* of a
young Gorilla (*Gorilla savagei*) × ¼. (Camb. Mus.)

1. parietal.

5. squamosal.

2. sagittal crest.

6. maxillae.

3. frontal.

7. external auditory meatus.

4. supra-orbital ridge.

In Man and in most monkeys, at any rate when young (fig. 99, B), the roof of the skull is smooth and rounded, but in many forms, such as the Baboons, in the adult the supra-orbital and occipital ridges are much developed. In the Gorilla this is also the case with the sagittal crest (fig. 99, A, 2). The bones of the upper surface of the cranium interlock with wavy outlines. The nasals vary much in length, being much shorter in man than in most monkeys; they commonly become early fused together, as do also the frontals. The vomer is well developed, and the ethmo-turbinal always forms part of the boundary of the orbit. There are frequently, as in many Lemuroidea, a pair of more or less well-marked ridges, crossing the roof of the skull from the postorbital processes of the frontals to the occipital crest. The orbit is completely encircled by bone, and the alisphenoid assists the jugal and frontal in shutting it off from the temporal fossa, leaving however a communication between the two as the sphenomaxillary fissure. In most cases the frontals meet one another in the middle line between the mesethmoid and orbitosphenoid, but in Man, Simia, and some Cebidae this does not take place. In nearly all Cebidae the parietal and jugal meet one another, separating the frontal and alisphenoid on the skull wall; in Man and all Old World monkeys, on the other hand, the alisphenoid and frontal meet and separate the jugal and parietal. The premaxillae nearly always send back processes which meet the nasals. The palate is rather short and both the palatine and the premaxillae take a considerable part in its formation. The pterygoid plate of the alisphenoid is decidedly large, and there is no alisphenoid canal. There is never any great development either of the paroccipital process of the exoccipital, or of the postglenoid process of the squamosal. The periotic and tympanic are always fused together; in Cebidae they form a small bulla, but a bulla is not developed in any Old World forms. The periotic is large, especially the mastoid portion, which forms a distinct portion of the skull wall between the squamosal and exoccipital. In Man and still more in Old World monkeys, the external auditory meatus is drawn out into a definite tube, whose lower wall is formed by the tympanic; in the Cebidae the tympanic is ring-like. The perforation of the periotic by the carotid canal is always conspicuous.

The mandible is rather short and broad, and the angle formed by the meeting of the two rami is more obtuse than in most mammals. The coronoid process is fairly well developed, and the angle is more or less rounded. In most Primates the condyle is considerably widened, but this is not the case in Man. In *Mycetes* the mandible is very large, its ascending portions being specially developed. The hyoid of Primates is remarkable for the large expanded basi-hyal, which is generally concave above and convex below. The anterior cornu is never well ossified, but the thyro-hyal is always strong. In *Mycetes* the basi-hyal is enormously large, forming a somewhat globular thin-walled capsule.

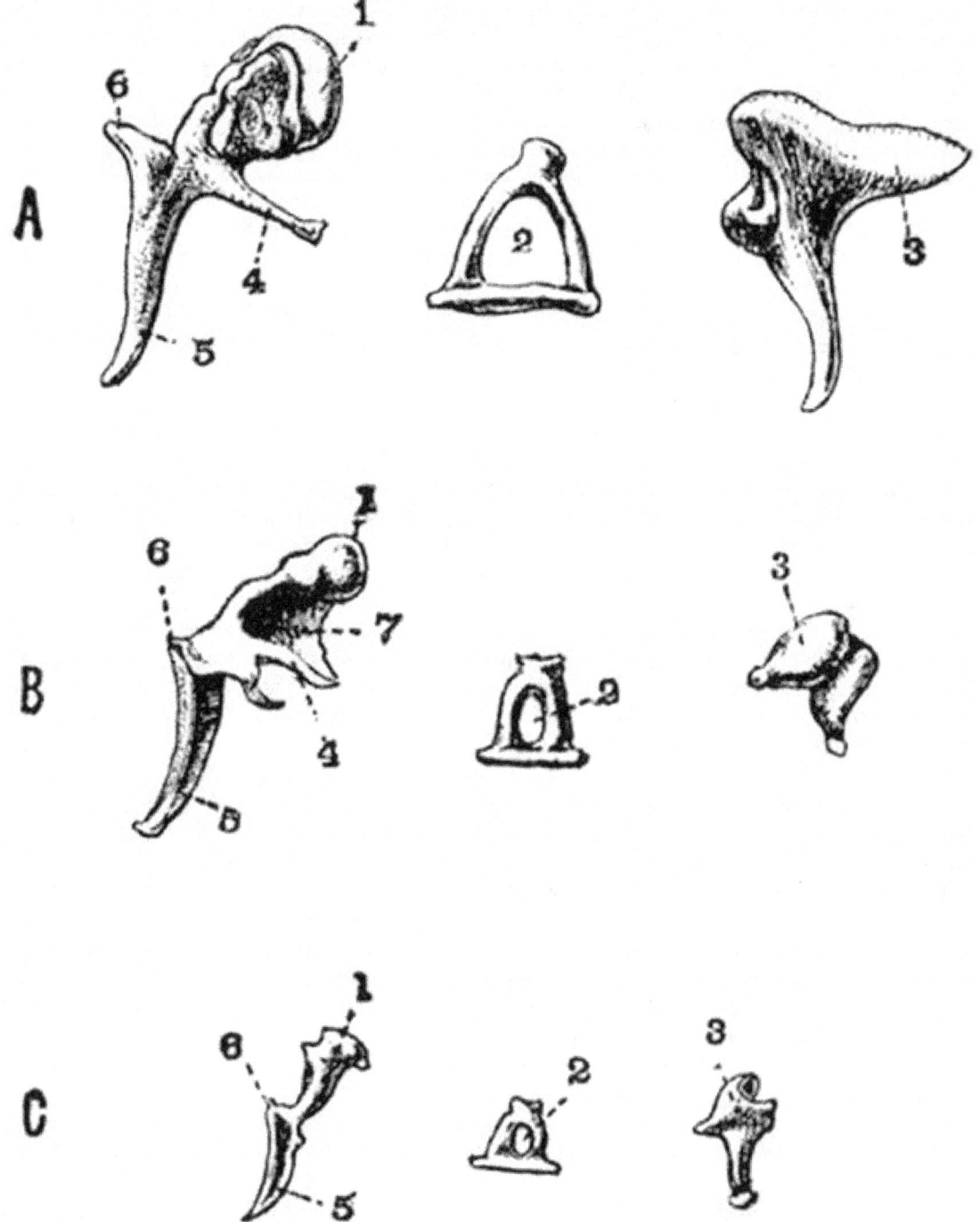

Fig. 100. Malleus, stapes and incus of
A. Man. *B.* Dog. *C.* Rabbit. (After Doran) x 1.

1. head of malleus.

2. canal of stapes.

3. incus.

5. manubrium of malleus.

6. processus brevis.

7. lamella.

4. processus longus (or gracilis).

Auditory ossicles.

There are in mammals four auditory ossicles forming a chain extending from the fenestra ovalis to the tympanic membrane. Three of these, the **malleus**, **incus** and **stapes**, are always distinct, while the fourth, the **lenticular**, is smaller than the others and is sometimes not distinct. The names are derived from human anatomy and indicate in the case of the first three a more or less fanciful resemblance respectively to a hammer, an anvil and a stirrup. The ossicles are homologous as a whole to the hyomandibular of fishes and to the columellar chain of Sauropsids and Amphibians. The malleus is homologous to the extra-columella of Crocodiles and the stapes to the columella. The **malleus** when typically developed consists of a rounded *head* (fig. 100, 1) which bears a surface articulating with the incus, and a short *neck* continued into a process, the *manubrium* (fig. 100, 5), which comes into relation with the tympanic membrane. From the junction of the neck and manubrium two processes are given off, a *processus longus* or *gracilis* (fig. 100, 4), which in the embryo is continuous with Meckel's cartilage, and a *processus brevis* (fig. 100, 6). The **incus** generally consists of a more or less anvil-shaped portion which articulates with the malleus, and of a process which is connected with the stapes by the small **lenticular**. The **stapes** is generally stirrup shaped, consisting of a basal portion from which arise two *crura* separated by a space the canal through which a branch of the pharyngeal artery runs The lenticular is frequently cartilaginous and sometimes is not developed at all.

The above is the arrangement of the auditory ossicles met with in the higher Mammalia, but in the lower Mammalia the characters approach more nearly to those met with in Sauropsids.

In Monotremes the ossicles, though distinctly mammalian in character, show a very low type of development. The incus is articulated, or often fused, with an outgrowth from the head of the malleus. The stapes is very much like a reptilian columella, having a single crus with no perforation.

In Marsupials the ossicles are of a low type, but not so low as the rest of the skeleton might have led one to expect, and all or almost all the points showing a low grade of development may be paralleled among the Monodelphia. The lowest Marsupials as regards the ossicles are the Peramelidae, whose ossicles are of a frail papery consistence. The Didelphyidae on the other hand have the most highly developed ossicles, the malleus much resembling that of many Insectivores, and the stapes having two definite crura separated by a canal.

In Edentates the character of the ossicles varies much. In Sloths the stapes approaches that of Sauropsids in its narrowness and the slight trace of a canal; this character is however still more marked in *Manis*, whose stapes is as Sauropsidan as that of Monotremes, and consists of a nearly circular basal plate bearing a column which does not show any sign of division into crura. The stapes of other Edentates, such as ant-eaters, aard varks, and most armadillos, is of a high type and has well-developed crura. *Priodon* has a lower type of stapes than *Dasypus* and *Tatusia*.

The ossicles of the Sirenia differ widely from those of all other mammals in their great density and clumsy form.

In Cetacea the ossicles are solid, though not so solid as in Sirenia, and their details vary much. The malleus is always firmly fused to the tympanic by means of the processus longus, and the manubrium is very little if at all developed. The incus has the stapedial end greatly developed, and the stapes has very thick crura with hardly any canal. The ossicles of the Mystacoceti are apparently less specialised than are those of the Odontoceti.

The auditory ossicles of the Ungulata do not present any characters common to all the members of the group.

Among Ruminants they are chiefly remarkable for the development of a broad lamellar expansion between the head and the processus longus of the malleus. In some cases the malleus of the foetus differs strikingly from that of the adult. Among Perissodactyla the Rhinoceros and Tapir have the malleus of a low type, recalling those of Marsupials; while in the Horse the head is well developed, and the malleus is of a higher type.

The ossicles of *Procavia*, which recall those of the Equidae, are chiefly remarkable for the small size of the body of the incus. In Elephants the ossicles are large and massive.

In the Rodentia (fig. 100, C) the malleus is generally characterised by a very broad manubrium. In many genera such as *Bathyergus*, and most of the Hystricomorpha such as *Hystrix*, *Chinchilla* and *Dasyprocta*, the malleus and incus are ankylosed together.

Carnivora. In Carnivora vera the most striking feature of the malleus is the occurrence of a broad lamellar expansion between the head and neck and the processus longus. This however does not occur in some Viverridae. In the Carnivora vera the incus and stapes are small as compared with the malleus, but in the Pinnipedia they are large. In the Pinnipedia the auditory ossicles have a very dense consistence, and except in the Otariidae are very large. The stapes frequently has no canal, or only a very small one.

In Insectivora the characters of the auditory ossicles are very diverse. Many forms such as shrews, moles, hedgehogs, and the Centetidae have a low type of malleus resembling that of Edentates. *Chrysochloris* has very extraordinary auditory ossicles. The head of the malleus is drawn out into a great club-shaped process, the incus is long and narrow, and differs much from the ordinary type.

246

In Chiroptera the ossicles and especially the malleus much resemble those of shrews. The stapes is always normal in character, never becoming at all columelliform.

Primates. In Man and the Anthropoid Apes the malleus has a rounded head, a short neck, and the manubrium, a processus longus and a processus brevis. The incus consists of an anvil-shaped portion from which arises a long tapering process. The stapes has diverging crura and consequently a wide canal. The crura in other monkeys do not diverge so much as in man and anthropoid apes. The New World monkeys have no neck to the malleus.

The Sternum.

In Monotremes and most Marsupials the sternum does not present any characters of special importance. The presternum is strongly keeled in *Notoryctes*.

The sternum in Edentates is very variable: in the Sloths it is very long, the mesosternum of *Choloepus* having twelve segments. In the ant-eaters and armadillos the presternum is broad and sometimes as in *Priodon* strongly keeled. In *Manis macrura* the xiphisternum is drawn out into a pair of cartilaginous processes about nine inches long.

In the Sirenia the sternum is simple and elongated, and of fairly equal width throughout, in the adult it shows no sign of segmentation. Its origin from the union of two lateral portions can be well seen in *Manatus*.

Two distinct types of sternum are met with in the Cetacea. In the Odontoceti the sternum consists of a broad presternum followed by three or four mesosternal segments, but with no xiphisternum. Indications of the original median fissure can be traced, and are very evident in *Hyperoödon*. In the Mystacoceti, on the other hand, the sternum consists simply of a broad flattened presternum which is sometimes more or less heart-shaped, sometimes cross-shaped. Only a single pair of ribs are united to it.

The sternum in Ungulata is generally long and narrow and formed of six or generally seven segments. The presternum is as a rule small and compressed, often much keeled, especially in the horse and tapir. The segments of the mesosternum gradually widen as followed back and the xiphisternum is often terminated by a cartilaginous plate.

In the Rodentia the sternum is long and narrow and generally has a large presternum, and a xiphisternum terminated by a broad cartilaginous plate.

In the Carnivora, too, the sternum (fig. 76) is long and narrow and formed of eight or nine pieces, all of nearly the same size. The xiphisternum generally ends in an expanded plate of cartilage.

In Insectivora the sternum is well developed but variable. The presternum is commonly large and is sometimes as in the Hedgehog (*Erinaceus*) bilobed in front, sometimes as in the Shrew (*Sorex*) trilobed. It is especially large in the Mole (*Talpa*) and is expanded laterally and keeled below.

In the Chiroptera the presternum is strongly keeled and so is sometimes the mesosternum.

Among Primates, in Man and the Anthropoid Apes the sternum is rather broad and flattened; the mesosternum consists of four segments which are commonly fused together and the xiphisternum is imperfectly ossified.

The Ribs.

Free ribs are borne as a rule only by the thoracic vertebrae; ribs may be found in other regions, especially the cervical and sacral, but these are almost always ankylosed to the vertebrae. As a general rule the first thoracic rib joins the presternum, while the succeeding ones are attached between the several segments of the mesosternum. Some of the posterior ribs frequently do not reach the sternum; they may then be attached by fibrous tissue to the ribs in front, or may end freely (*floating ribs*). There are generally thirteen pairs of ribs, and in no case do they have uncinate processes.

In Monotremes (fig. 102, B) each rib is divided not into two but into three parts, an intermediate portion being interposed between the vertebral and sternal parts. The sternal ribs are well ossified, and some are very broad and flat. The intermediate portions are unossified, those of the anterior ribs are short and narrow, but they become longer and wider further back.

In Marsupials there are almost always thirteen pairs of ribs, whose sternal portions are very imperfectly ossified. *Notoryctes* has fourteen pairs of ribs, eight of which are floating: the first rib is very stout, and is abruptly bent on itself to join the sternum. It has no distinct sternal portion. All the other ribs are slender.

Of the Edentates the Sloths have very numerous ribs; twenty-four pairs occur in *Choloepus*, and half of these reach the sternum. In the Armadillos there are only ten or twelve pairs of ribs, but the sternal portions are very strongly ossified. The first rib is remarkably broad and flat, and is not divisible into vertebral and sternal portions.

In the Sirenia there are a very large number of ribs noticeable for their great thickness and solidity, but not more than three are attached to the sternum.

Cetacea. In the Whalebone whales the ribs are remarkable for their very loose connection both with the vertebral column and with the sternum. The capitula are scarcely developed, and the attachment of the tubercula to the transverse processes is loose. The first rib is the only one connected with the sternum. In the Toothed whales the anterior ribs have capitula articulating with the centra, as well as tubercula articulating with the transverse processes; in the posterior ones, however,

247

only the tubercula remain. Seven pairs of well-ossified sternal ribs generally meet the sternum. In the Physeteridae most of the ribs are connected to the vertebrae by both capitula and tubercula.

In the Ungulata the ribs are generally broad and flattened, and this is especially the case in the genera *Bos* and *Bubalus* (fig. 101, 6). The anterior ribs are short and nearly straight, and sternal ribs are well developed. The Artiodactyla have twelve to fifteen pairs of ribs, the Perissodactyla eighteen or nineteen, and *Procavia* twenty to twenty-two. The Elephant has nineteen to twenty-one pairs, seven of which may be floating ribs.

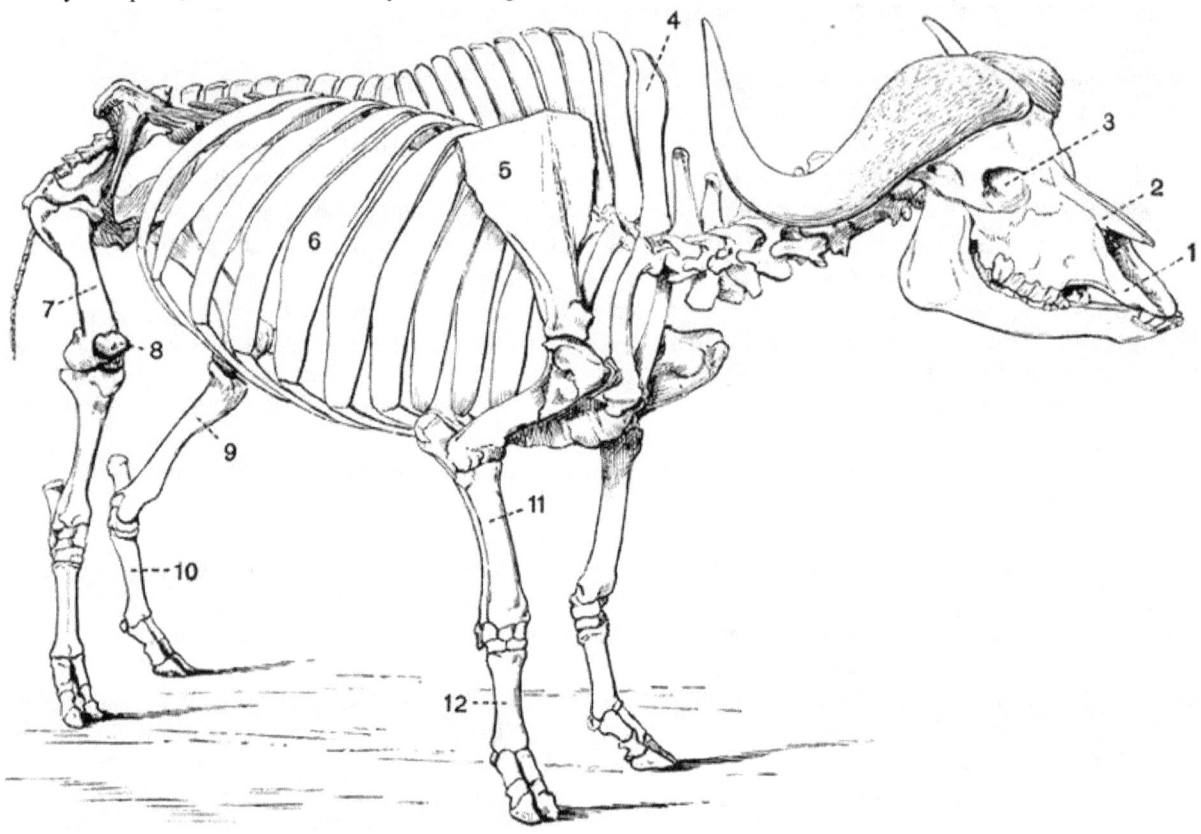

Fig. 101. Skeleton of a Cape Buffalo (*Bubalus caffer*).
The left scapula is omitted for the sake of clearness × $^1/_{17}$. (Brit. Mus.)

1. premaxillae.

2. nasal.

3. orbit.

4. neural spine of first thoracic

vertebra.

5. scapula.

6. rib.

7. femur.

8. patella.

9. tibia.

10. metatarsals.

11. radius.

12. metacarpals.

In the Rodentia there are generally thirteen pairs of ribs, which do not present any marked peculiarities.

The Carnivora have thirteen to fifteen pairs of ribs, whose vertebral portions are slender, nearly straight and subcylindrical, while their sternal portions are long and imperfectly ossified (fig. 76, 5). There is nothing that calls for special remark about the ribs, in either Insectivora or Chiroptera.

Primates. In Man and the Orang (*Simia*) there are generally twelve pairs of ribs; in the Gorilla and Chimpanzee (*Anthropopithecus*), and Gibbons (*Hylobates*), there are thirteen, in the Cebidae twelve to fifteen, and in the Lemuroidea twelve to seventeen pairs. The first vertebral rib is shorter than the others, and the sternal ribs generally remain cartilaginous throughout life, though in man the first may ossify.

Appendicular Skeleton.

The Pectoral Girdle.

By far the most primitive type of the pectoral or shoulder girdle is found in the Monotremata. The scapula (fig. 102, A, 1) is long and recurved, and has only two surfaces, one corresponding to the prescapular fossa, the other to the postscapular and subscapular fossae. The coracoid is a short bone attached above to the scapula and below to the presternum; it forms a large part of the glenoid cavity. In front of the coracoid there is a fairly large flattened epicoracoid (fig. 102, 6); there is also a large **T**-shaped interclavicle (fig. 102, 4), which is expanded behind and rests on the presternum. The clavicles rest on and are firmly united to the anterior border of the interclavicle. This shoulder girdle differs greatly from that of any other mammals, and recalls that of some Lacertilia.

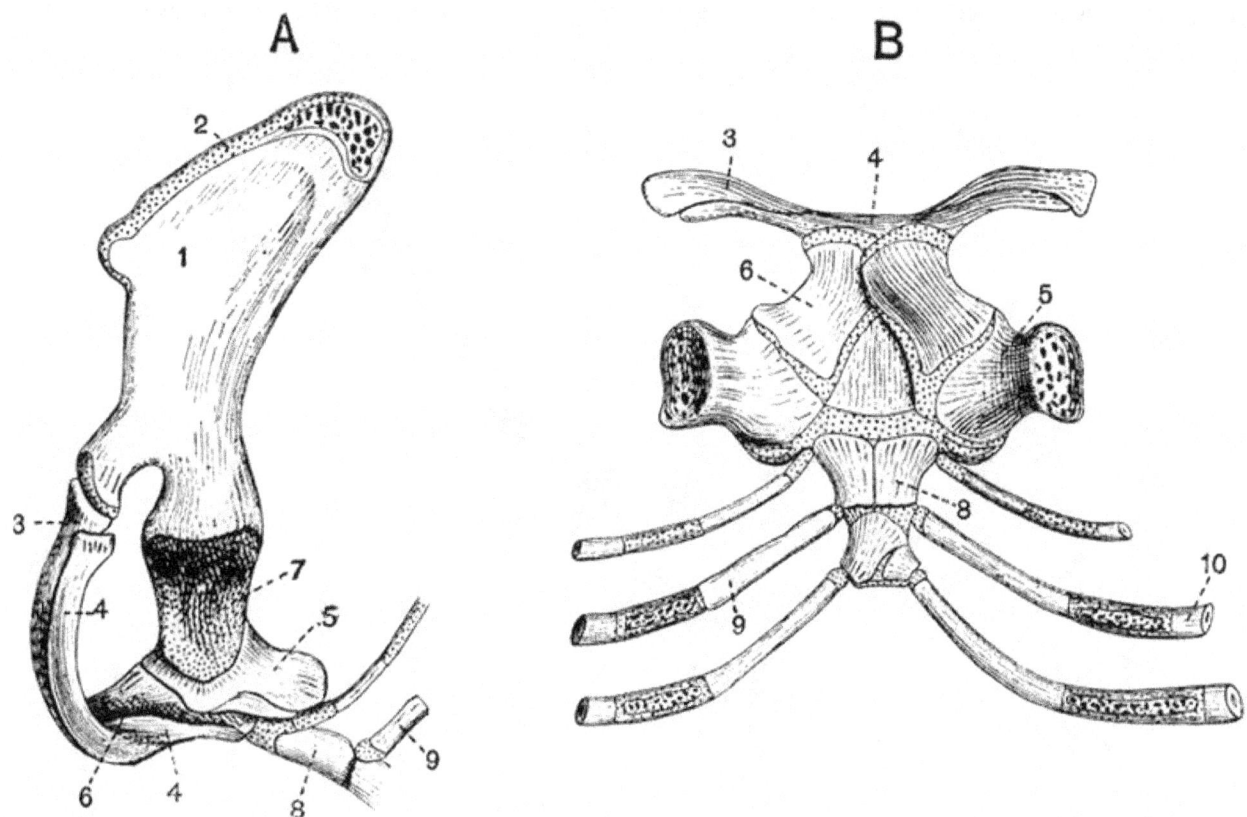

Fig. 102. *A*, Side view, *B*, Dorsal view of the shoulder girdle and part or the sternum of the Spiny Anteater (*Echidna aculeata*) × 1. (After Parker.)

1. scapula.	6. epicoracoid.
2. suprascapula.	7. glenoid cavity.
3. clavicle.	8. presternum.
4. interclavicle.	9. second sternal rib.
5. coracoid.	10. second vertebral rib.

In Marsupials, as in all mammals except the Monotremes, the shoulder girdle is much reduced; there are no epicoracoids and interclavicle, and the coracoid forms simply a small process on the scapula, ossifying from a centre separate from that giving rise to the rest of the bone. The scapula has a long acromion, and a clavicle is always present except in *Perameles*. Unossified remains of the precoracoids are found at either end of the clavicle. The scapula of *Notoryctes* has a very high overhanging spine, and there is a second strong ridge running along the proximal part of the glenoid border.

The shoulder girdle of the Edentata shows some very curious variations. In *Orycteropus* the scapula is of very normal form and the clavicle is well developed. In the Pangolins and Anteaters the scapula is very broad and rounded; there is no clavicle in the Pangolins, and generally only a vestigial one in Anteaters. In Armadillos, Sloths, and Megatheriidae, the acromion is very long and the clavicle is well developed. In the Sloths, *Megatherium*, and *Myrmecophaga*, a connection is formed between the coracoid, which is unusually large, and the coracoid border of the scapula, converting the coraco-scapula notch into a foramen. In *Bradypus* the clavicle is very small, and is attached to the coracoid, which sometimes forms a distinct bone.

In the Sirenia the scapula is somewhat narrow and curved backwards: the spine, acromion, and coracoid process are moderately developed, and there is no clavicle.

Cetacea. In nearly all the Odontoceti the scapula is broad and somewhat fan-shaped; the prescapular fossa is much reduced, and the acromion and coracoid process form flattened processes, extending forwards nearly parallel to one another. Some of the Mystacoceti, such as *Balaenoptera*, have a broad, fan-shaped scapula, with a long acromion and coracoid process, extending parallel to one another. Others, such as *Balaena*, have a higher and narrower scapula, with a smaller coracoid process.

In Ungulata the scapula is always high and rather narrow, and neither acromion nor coracoid process is ever much developed. In no adult Ungulate except *Typotherium* is there any trace of a clavicle, but a vestigial clavicle has been described in early embryos of sheep.

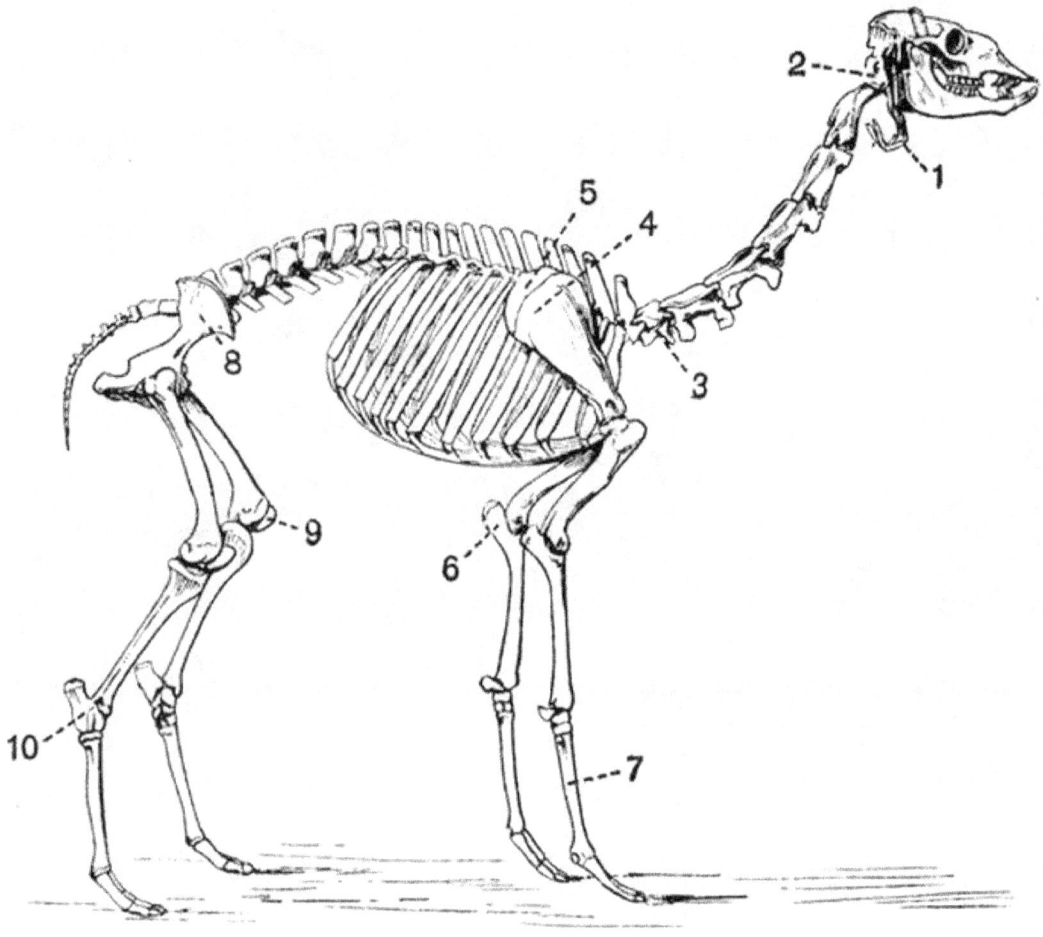

Fig. 103. Skeleton of a Llama (*Auchenia glama*) × ¹/₁₈.
(Brit. Mus.)

1. hyoid.

2. atlas vertebra.

3. seventh cervical vertebra.

4. scapula.

5. imperfectly ossified

suprascapula.

6. olecranon process of ulna.

7. metacarpals.

8. ilium.

9. patella.

10. calcaneum.

Ungulata vera. In the Ruminantia the suprascapular region (fig. 103, 5) is very imperfectly ossified, and when this is removed the upper border of the scapula is very straight (fig. 101, 5). The spine is prominent, and generally has a fairly well-marked acromion. In *Hippopotamus* the acromion is fairly prominent, but in the other Suina, though the spine is prominent, the acromion is not developed. The Perissodactyla have no acromion, but while the Equidae and *Hyracotherium* have the scapula long and slender, with the spine very slightly developed, the other living Perissodactyla have the spine prominent and strongly bent back at about the middle of its length.

Subungulata. *Typotherium* (Toxodontia) differs from all other known Ungulates in having well-developed clavicles; its scapula has a strong backwardly-projecting process, much like that in *Rhinoceros*.

Phenacodus (Condylarthra), has a curiously rounded scapula, with the coracoid and suprascapular borders passing imperceptibly into one another. The scapula resembles that of a carnivore more than does that of any existing Ungulate.

Procavia has a triangular scapula with a prominent spine and no acromion; there is a large unossified suprascapular region.

The scapula in the Proboscidea has a large rounded suprascapular border and a narrow, slightly concave glenoid border. The spine is large, and has a prominent process projecting backwards from about its middle. The spine lies towards the front end of the scapula, so that the postscapular fossa is much larger than the prescapular fossa.

In Rodentia the shoulder girdle is of a rather primitive type. The scapula is generally high and narrow, somewhat as in Ruminantia; it differs, however, from the Ruminant scapula in having a high acromion, which is often, as in the Hares and Rabbits, terminated by a long metacromion. The development of the clavicle varies, and sometimes it is altogether absent. It is frequently connected by cartilaginous bands or ligaments (fig. 104, 7 and 9), on the one hand with the scapula, and on the other with the sternum. These unossified bands are remains of the precoracoid. Epicoracoidal vestiges of the sternal ends of the coracoids (fig. 104, 11) are also often present.

In the Carnivora vera the scapula is large, and generally has rather rounded borders. The spine and acromion are well developed, and the prescapular and postscapular fossae are nearly equal in size. The coracoid is very small, and the clavicle is never completely developed, being often absent, as in the Bears and most of their allies. In the Seals (Phocidae) the scapula is elongated and curved backwards, and has a very concave glenoid border. In the Eared Seals (Otariidae) the scapula is proportionally much larger and wider, the prescapular fossa being specially large, and being traversed by a ridge, which converges to meet the spine.

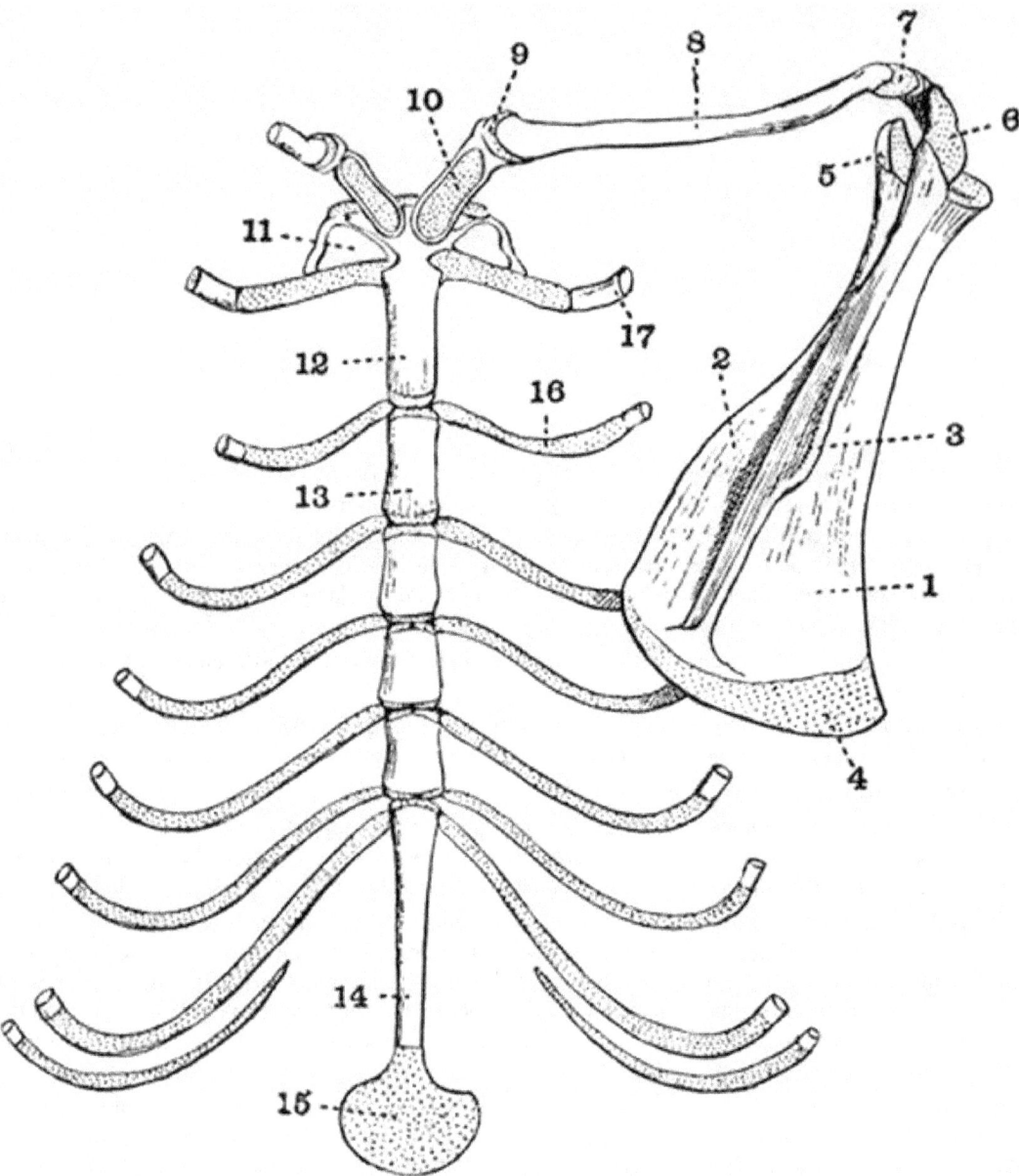

Fig. 104. Dorsal view of the sternum and right half of the shoulder-girdle of *Mus sylvaticus* × 4. (After Parker.)

1. postscapular fossa.

2. prescapular fossa.

3. spine.

4. suprascapular border unossified.

5. coracoid process.

6. acromion.

7. cartilaginous vestige of precoracoid at scapular end of

precoracoid at sternal end of clavicle.

10. omosternum.

11. epicoracoid.

12. presternum.

13. first segment of mesosternum.

14. xiphisternum.

15. cartilaginous termination of xiphisternum.

clavicle.

8. clavicle.

9. cartilaginous vestige of

16. 2nd sternal rib.

17. 1st vertebral rib.

In the Insectivora the shoulder girdle is well developed and, as in Rodents, remains are met with of various parts not generally seen in mammals. In the Shrews the scapula is long and narrow, and has a well-marked spine, whose end bifurcates, forming the acromion and metacromion. The clavicle is long and slender, and is connected with the sternum and acromion by vestiges of the precoracoid. Considerable remains of the sternal end of the coracoid are also found. In *Potamogale*, however, there are no clavicles. In the Mole the shoulder girdle is greatly developed, and of very remarkable form. The scapula is high and very narrow, with the spine and acromion very little developed. The other shoulder girdle element is an irregular bone, which articulates with the humerus and presternum, and is connected by ligaments with the scapula. This bone appears to represent both the coracoid and the clavicle, being formed partly of cartilage bone, partly of membrane bone.

In the Chiroptera the scapula is large and oval, and has a moderately high spine and a large acromion. The coracoid process is well developed and is often forked. The clavicles are also well developed, and vestiges of the precoracoid and of the sternal end of the coracoid are often found.

In Primates the clavicle and coracoid process are always well developed. In Man and the Gorilla the scapula has a long straight suprascapular border, a well-developed coracoid process and spine, and a large curved acromion. Vestiges of the precoracoid occur at each end of the clavicle. The shape of the scapula varies much in the lower Primates.

The Upper arm and Fore-arm.

In the Monotremata the humerus is short, very broad at each end and contracted in the middle. The radius and ulna are stout and of nearly equal size, while the ulna has a greatly expanded olecranon.

In the Marsupialia the humerus is generally a strong bone, broad at the distal end and having well marked deltoid and supinator ridges, which are specially large in *Notoryctes*. An ent-epicondylar or supracondylar foramen (fig. 105, 5) is almost always present except in *Notoryctes*. The radius and ulna are always distinct and well developed, and a certain amount of rotation can take place between them. The ulna of *Notoryctes* has an enormous hooked olecranon which causes the bone to be nearly twice as long as the radius.

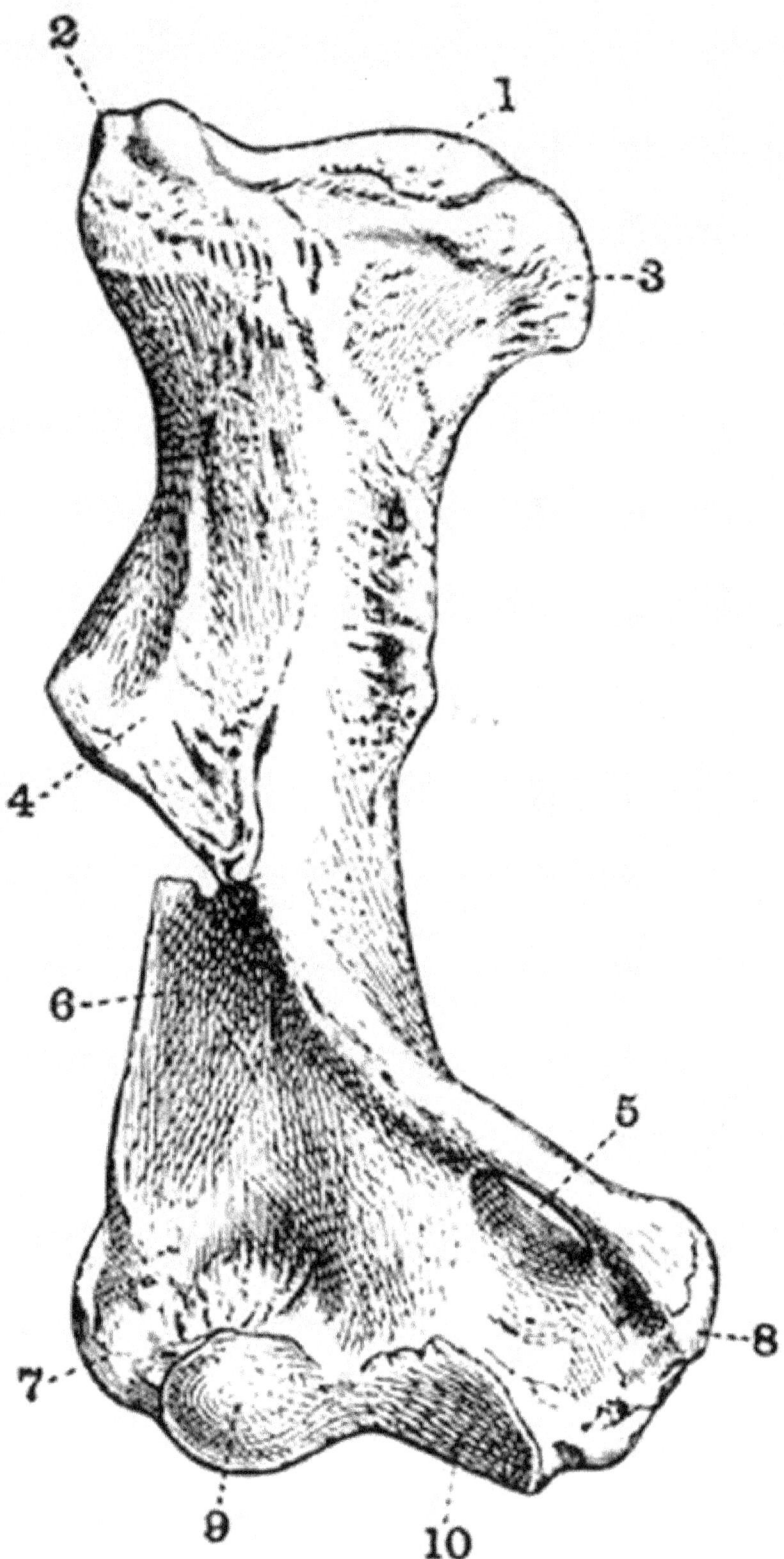

Fig. 105. Anterior surface of the right humerus of a Wombat
(*Phascolomys latifrons*). (After Owen.)

1. head.   6. supinator ridge.

2. greater tuberosity.

3. lesser tuberosity.

4. deltoid ridge.

5. ent-epicondylar

7. external condyle.

8. internal condyle.

9. articular surface for radius.

10. articular surface for ulna.

(supracondylar) foramen.

Edentata. The Sloths have long slender arm bones; the humerus is nearly smooth and has a very large ent-epicondylar foramen in *Choloepus*, but not in *Bradypus*. The radius and ulna can be rotated on one another to a considerable extent. The humerus in all other Edentates is very strong and has the points for the attachment of muscles much developed, especially in the Armadillos and Megatheriidae. An ent-epicondylar foramen is found in all living forms. The radius and ulna are well developed, but are not capable of much rotation.

In the Sirenia the humerus is well developed and of a normal character. It is expanded at each end and has a prominent internal condyle, a small olecranon fossa, and no ent-epicondylar foramen. In the Dugong and *Rhytina* there is a bicipital groove and the tuberosities are distinct, but in the Manatee there is no bicipital groove, and the tuberosities coalesce. The radius and ulna are about equally developed and ankylosed together at both ends.

In the Cetacea the arm bones are very short and thick. The humerus has a globular head, and a distal end terminated by two equal flattened surfaces to which the radius and ulna are united. There is no bicipital groove, and the tuberosities coalesce. The radius and ulna are flat expanded bones fixed parallel to one another, but the ulna has a definite olecranon. Scarcely any movement can take place between them and the humerus, and in old animals the three bones are often ankylosed together.

In the Ungulata vera the humerus is stout and rather short. The great tuberosity is always large and often overhangs the bicipital groove, it is especially large in *Titanotherium* (*Brontops*). There is never an ent-epicondylar foramen. The radius is always large at both ends, but the condition of the ulna is very variable. Sometimes, as in *Tapirus*, *Rhinoceros*, *Macrauchenia*, Suina and Tragulina, the ulna is well developed, and quite distinct from the radius; but in most forms, although complete, it is much reduced distally, and is fused to the radius. Sometimes, as in the Horse and Giraffe, it is reduced to the olecranon and to a very slender descending process which does not nearly reach the carpus. In the Tylopoda, though the ulna is complete and its distal end is often distinct, it has coalesced with the radius throughout its whole length; the olecranon is generally very large.

Subungulata. In the large Condylarthra the humerus has an ent-epicondylar foramen, and the radius and ulna are stout bones nearly equal in size.

In *Procavia* the humerus is rather long, and has a very prominent greater tuberosity, and a large supra-trochlear fossa, but no ent-epicondylar foramen.

In the Proboscidea the humerus is marked by a greatly developed supinator ridge, and is very long, longer than the radius and ulna. The ulna has a remarkable development, having its distal end larger than that of the radius, it has also a larger articular surface for the humerus than has the radius.

In Rodentia the humerus varies much in its development according to the animal's mode of life. In the Hares it is long and straight, with a small distal end, and a slight deltoid ridge. In the Beaver on the other hand the deltoid and supinator ridges are considerably developed. There is generally a large supra-trochlear fossa, but no ent-epicondylar foramen.

Carnivora. In the Carnivora vera the humerus has large tuberosities, a prominent deltoid ridge and a deep olecranon fossa. The shaft is generally curved, and an ent-epicondylar foramen is often found, though not in the Canidae, Hyaenidae, and Ursidae. The radius and ulna are never united. The radius (fig. 77, B) has a very similar development throughout its whole length, while the ulna has a large olecranon (fig. 77, C, 11) and a shaft tapering somewhat towards the distal end.

In the Pinnipedia the arm bones are very strongly developed. The humerus has a very prominent deltoid ridge, and the proximal end of the ulna and distal end of the radius are much expanded.

In the Insectivora the arm bones are well developed, and the radius and ulna, though sometimes united, are generally distinct; as a rule there is an ent-epicondylar foramen, but this is absent in the Hedgehog. The Mole has an extraordinary humerus, very short and curved, and much flattened and expanded at both ends. It articulates both with the scapula and coraco-clavicle. The ulna has a greatly developed olecranon.

In the Chiroptera both humerus and radius are exceedingly long and slender; the ulna is reduced to little more than the proximal end and is fused to the radius. There is no ent-epicondylar foramen.

All Primates have the power of pronation and supination of the fore-arm, by the rotation of the distal end of the radius round that of the ulna.

In Man and the Anthropoid Apes the humerus is long and straight, and has a globular head; neither of the tuberosities, nor the deltoid nor supinator ridges are much developed. The olecranon fossa is deep and there is no ent-epicondylar foramen. The radius is curved and has a narrow proximal, and expanded distal end, the ulna is straighter than the radius and has the distal end much smaller than the proximal; the olecranon is not much developed.

In the lower Primates, although the radius and ulna are always quite separate, the power of pronation and supination is not nearly so great as in the higher forms. In most of the Cebidae and Lemurs an ent-epicondylar foramen occurs.

The Manus.

The Manus is divisible into two parts, viz. the carpus or wrist, and the hand which is composed of the metacarpals and phalanges. The carpal bones are always modified from their primitive arrangement, sometimes more, sometimes less. One modification however is always found in mammals, viz. the union of carpalia, 4 and 5 to form the *unciform* bone. Two sesamoid bones are commonly developed, one on each side of the carpus, the *pisiform* or one on the ulnar side being much the larger and more constant: it has been suggested that these represent respectively vestiges of a prepollex and a post-minimus digit.

One or more of the five digits commonly present may be lost, and sometimes all are lost except the third. The terminal or ungual phalanges of the digits are commonly specially modified to support nails, claws, or hoofs. There are as a rule two small sesamoid bones developed on the ventral or flexor side of the metacarpo-phalangeal articulations, and sometimes similar bones occur on the dorsal or extensor side.

Monotremata. In *Echidna* the carpus is broad, the scaphoid and lunar are united and there is no centrale. The pisiform is large and several other sesamoid bones occur. Each of the five digits is terminated by a large ungual phalanx. In *Ornithorhynchus* the manus is more slender, but the general arrangement is the same as in *Echidna*.

Marsupialia. The carpus has no centrale and the lunar is generally small or absent. Five digits are almost always present. In *Choeropus* however the only two functional digits are the second and third, which have very long closely apposed metacarpals; the fourth digit is vestigial, but has the normal number of phalanges, while the first and fifth are absent. The manus in *Notoryctes* is extraordinarily modified, the scaphoid and all the distal carpalia are apparently fused, the first, second, and fifth digits are very small, the third and fourth, though having only one phalanx apiece, bear each an enormous claw. Lying on and obscuring the ventral surface of the manus is a large bone, probably a sesamoid.

Among the Edentata there is a great diversity in the structure of the manus, the centrale is however always wanting, and except in *Manis* the scaphoid and lunar are distinct. In the Sloths the manus is very long, narrow, and curved, and terminated by two or three long hooked claws, borne by the second and third, or the second, third and fourth digits. The fifth digit is absent, and the fourth is represented only by a small metacarpal. In the Anteaters the third digit is greatly developed and bears a long hooked claw. In *Myrmecophaga* all five digits are fairly well though irregularly developed, in *Cycloturus* the first, fourth, and fifth, are vestigial. In the Armadillos the manus is broad, and has strongly developed ungual phalanges. The digits, though almost always five in number, vary much in their relative arrangement. In *Dasypus* they are regular, but are remarkably irregular in Priodon. The pollex is absent in Glyptodonts and in *Megatherium.* In *Megatherium* the fifth digit is clawless while the second, third, and fourth bear enormous claws. In the Manidae the scaphoid and lunar are united; five digits are present, the third and fourth being very large, and all being terminated by deeply cleft ungual phalanges. In *Orycteropus* the pollex is absent, while the other digits are terminated by pointed ungual phalanges.

In Sirenia the general structure of the manus is quite of the ordinary mammalian type. In *Manatus* most of the bones of the carpus are distinct, but in *Halicore* many, especially those of the distal row, have coalesced. The digits are always five in number and have the normal number of flattened phalanges.

In the Cetacea, on the other hand, the manus is much modified by the fact that the number of phalanges may be greatly increased above the normal number of three, thirteen or fourteen sometimes occurring in each digit. These are believed to be duplicated epiphyses. In the Mystacoceti the manus remains largely cartilaginous, in the Odontoceti it is better ossified, and the phalanges commonly have epiphyses at both ends. In *Physeter* the carpal bones also have epiphyses. The carpus generally consists of six bones arranged in two rows of three each. Five digits are generally present, but sometimes as in *Balaenoptera musculus*, there are four, the third being suppressed. Their relative development varies much. The Sperm Whale which till recently was placed in the entrance hall of the Natural History Museum at South Kensington has one phalanx to the first digit, four to the second, five to the third, four to the fourth, and three to the fifth. Generally the manus is short and broad, but sometimes, as in *Globicephalus*, it is much elongated owing to the great development of the second and third digits.

Ungulata. The manus of the members of this great order is of very great classificatory and morphological importance. All the members agree in having the scaphoid and lunar distinct, and in almost every case the ends of the digits are either encased in hoofs or provided with broad flat nails. It is by means of characters derived from the manus and pes that the group is subdivided into the Ungulata vera and the Subungulata.

In the Ungulata vera the manus is never plantigrade, and there are not more than four digits, the pollex being almost always completely suppressed: in *Cotylops* among extinct Artiodactyla however a vestigial pollex is found. The centrale is absent, and the magnum articulates freely with the scaphoid, and is separated from the cuneiform by the unciform and lunar. All the bones of the carpus interlock strongly, and the axis of the third digit passes through the magnum and between the scaphoid and lunar.

There is a very strong distinction between the manus of the suborders Artiodactyla and Perissodactyla. In the Artiodactyla the axis of the manus passes between the third and fourth digits, which are almost equally developed and, except in the Hippopotami and some extinct forms such as *Anoplotherium*, have their ungual phalanges flattened on their contiguous surfaces.

In all *Artiodactyla* the third and fourth digits are large, but a gradual reduction in the second and fifth can be well traced. Thus in the Suina the second and fifth digits, though smaller than the third and fourth, are well developed and all four metacarpals are distinct. In the Tragulina too all four metacarpals are developed, and in *Dorcatherium* the third and fourth commonly remain distinct as in the Suina. In the other Artiodactyla however the third and fourth metacarpals are almost always united, though indications of their separate origin remain. In some Ruminantia, such as many Deer, the second and fifth digits are reduced to minute splint bones attached to the proximal end of the fused third and fourth metacarpals, and to small hoof-bearing phalanges, sometimes attached to splint-like distal vestiges of the metacarpals, sometimes altogether unconnected with any other skeletal structures. In some other Ruminants, such as the Sheep and Oxen, the only remnants of the second and fifth digits are nodules of bone supporting the hoofs, and in others, such as the Giraffe, *Anoplotherium commune*, some Antelopes and the Tylopoda, all traces of these digits have disappeared. The Camels differ from all living Ungulata vera in not having the distal phalanges completely encased in hoofs, and from all except the Hippopotami in placing a considerable amount of the manus on the ground in walking.

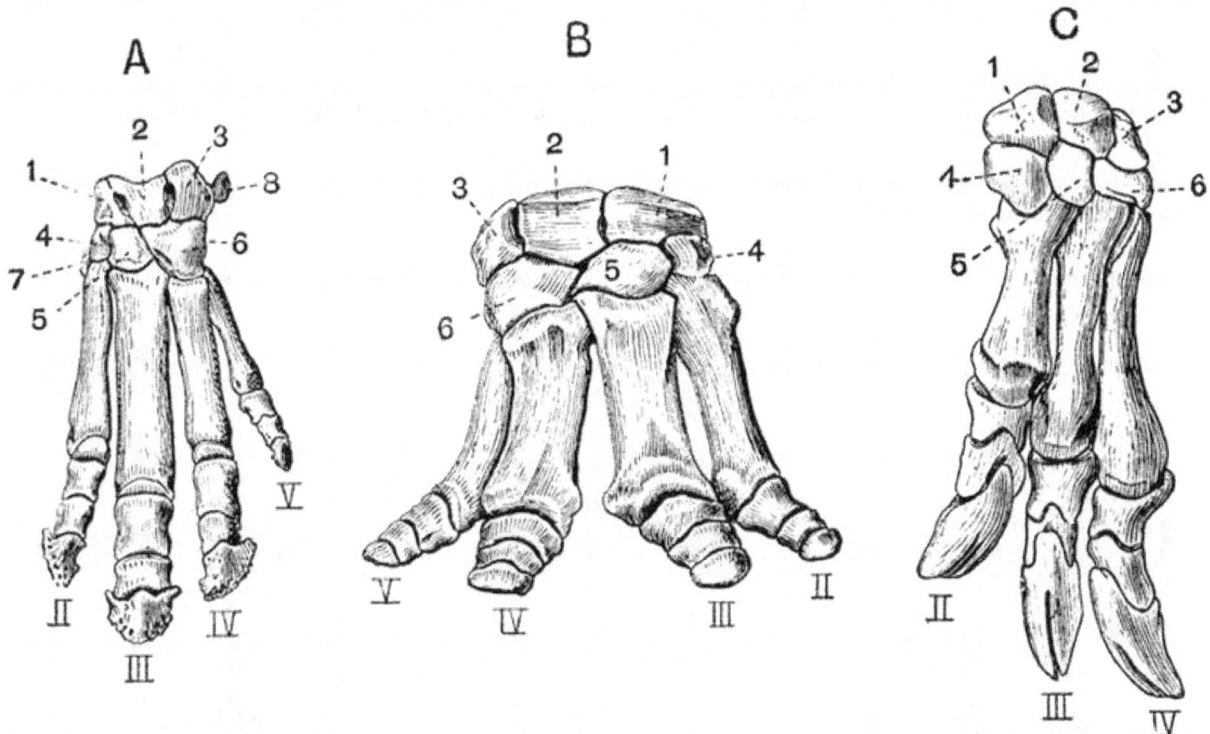

Fig. 106. Manus of Perissodactyles.

A. Left Manus of *Tapirus*. (After von Zittel.)

B. Right Manus of *Titanotherium*. (After Marsh.)

C. Left Manus of *Chalicotherium gigantium*. (After Gervais.)

1. scaphoid.	6. unciform.
2. lunar.	7. trapezium.
3. cuneiform.	II, III, IV, V. second, third,
4. trapezoid.	fourth and fifth digits.
5. magnum.	

While the manus of the Artiodactyla is symmetrical about a line drawn between the third and fourth digits, that of the *Perissodactyla* is symmetrical about a line drawn through the middle of the third digit, which is larger than the others and has its ungual phalanx evenly rounded and symmetrical in itself. The most reduced manus in the whole of the mammalia is found in the Horse and its allies, in which the third digit, terminated by a very wide ungual phalanx, is the only one functional. Small splint bones representing the second and fourth metacarpals are attached to the upper part of the third metacarpal. In *Hipparion* and other early horse-like animals the second and fourth digits, though very small and functionless, are complete and are terminated by small hoofs. In *Rhinoceros* the second and fourth digits are equally developed and nearly as large as the third, and reach the ground in walking, a vestige of the fifth is also present. In the Tapir (fig. 106, A) and *Hyracotherium* the fifth digit is fully developed but is scarcely functional. In *Titanotherium* (*Brontops*) (fig. 106, B) it is nearly as well developed as any of the others, and there is little or no difference between the relative development of the third and fourth digits.

The Chalicotheriidae, though distinctly Perissodactyles in various respects such as their cervical vertebrae and teeth, differ not only from all other Perissodactyles, but from almost all other Ungulates, in the very abnormal character of their manus. For while the carpus and metacarpus are like those of ordinary Perissodactyles, the phalanges resemble those of Edentates, each second phalanx having a strongly developed trochlea, and each distal one being curved, pointed and deeply cleft at its termination (fig. 106, C).

The Macraucheniidae, while agreeing with Perissodactyles in having only three digits, with the limb symmetrical about a line drawn through the middle of the third, have a carpus which approaches closely to the subungulate condition, the magnum articulating regularly with the lunar, and only to a slight extent with the scaphoid.

In the Subungulata the manus sometimes has five functional digits, and a considerable part of it rests on the ground in walking. The bones of the carpus retain their primitive relation to one another, the magnum articulating with the lunar, but not with the scaphoid. This character does not however hold in the Toxodontia, for in most of the animals belonging to this group the magnum does articulate with the scaphoid. The corner of the scaphoid just reaches the magnum also in Amblypoda.

As far as is known the *Toxodontia* generally have three, sometimes five digits to the manus, and the third is symmetrical in itself—a Perissodactyloid feature.

In *Phenacodus* (fig. 107, B) (*Condylarthra*) all five digits are well developed, the pollex being the smallest. The carpal bones retain their primitive arrangement, the magnum articulating with the lunar and not with the scaphoid. There is no separate centrale.

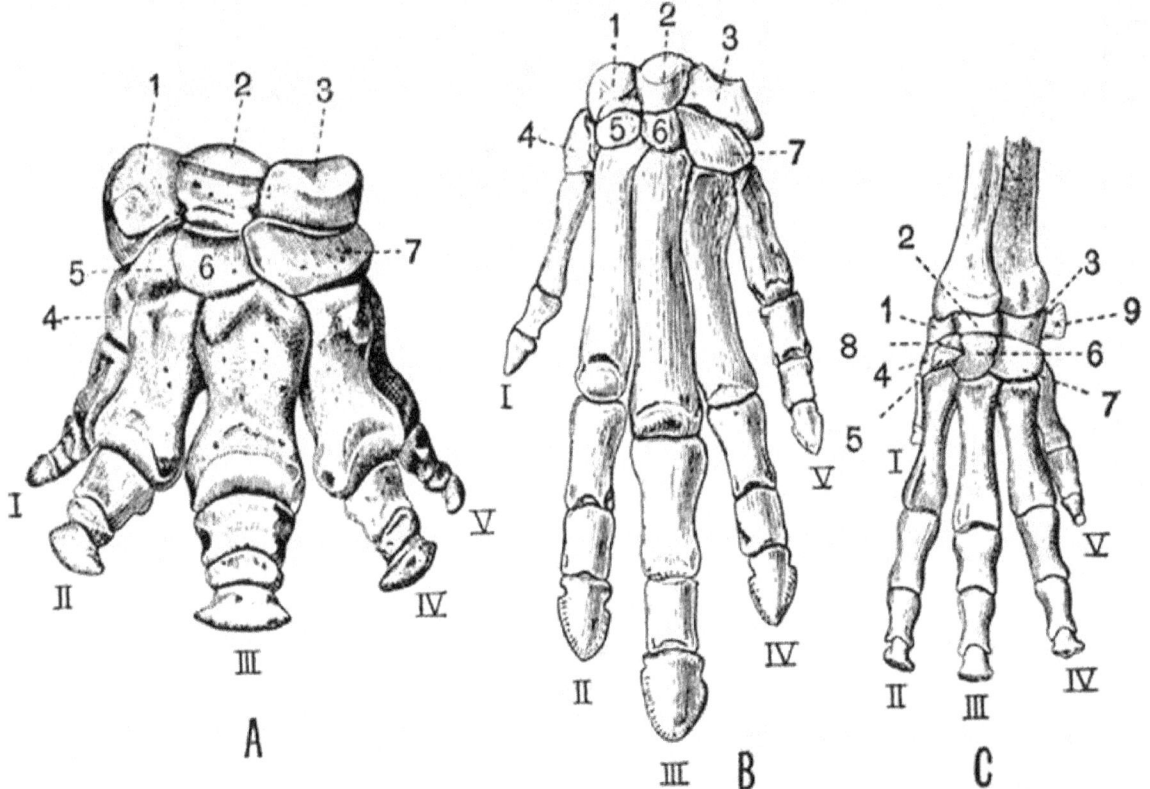

Fig. 107. Left manus of

A. Coryphodon hamatus. (After Marsh.) × $^1/_5$.

B. Phenacodus primaevus. (After Cope.) × $^1/_3$.

C. Procavia (Dendrohyrax) arboreus. (After von Zittel.) × $^6/_7$.

1. scaphoid.	7. unciform.
2. lunar.	8. centrale.
3. cuneiform.	9. pisiform.
4. trapezium.	I, II, III, IV, V. first, second,
5. trapezoid.	third, fourth and fifth
6. magnum.	digits respectively.

In the *Hyracoidea* (fig. 107, C) the manus is very similar to that in *Phenacodus*, but a centrale is present and the pollex is much reduced.

The manus of the *Amblypoda*, such as *Coryphodon* (fig. 107, A) and *Uintatherium*, is short and broad, with five well developed digits and large carpal bones. The carpals however interlock to a slight extent, and the corner of the magnum reaches the scaphoid.

In the *Proboscidea* the manus is very short and broad, with large somewhat cubical carpals which articulate by very flat surfaces and do not interlock at all. All five digits are present, and none of them are much reduced in size. The manus in Proboscidea and in *Coryphodon* is subplantigrade.

In the Tillodontia the manus is plantigrade and has pointed ungual phalanges, in this respect approaching the Carnivora. It differs however from that of all living Carnivora in having the scaphoid and lunar distinct.

In Rodentia the manus nearly always has five digits with the normal number of phalanges: the pollex may however be very small as in the Rabbit, or absent as sometimes in the Capybara. The scaphoid and lunar are generally united, and a centrale may be present or absent. In *Pedetes caffer* the radial sesamoid is double and the distal bone bears a nail-like horny covering. In *Bathyergus* the pisiform is double. It is upon these facts that the contention for the former existence of prehallux and post-minimus digits has partly been based.

In living Carnivora the scaphoid, lunar and centrale are always united, forming a single bone. All five digits are present, but as a rule in Carnivora vera the pollex is small, and in *Hyaena* is represented only by a small metacarpal. Sometimes, as in Cats and Dogs, the manus is digitigrade, sometimes, as in Bears, plantigrade. The ungual phalanges are large and pointed, and in forms like the Cats, whose claws are retractile, they can be folded back into a deep hollow on the ulnar side of the middle phalanx; a small radial sesamoid is often present.

In Pinnipedia the manus is large and flat and the digits are terminated by ungual phalanges which are blunt (sea lions and walrus), or slightly curved and pointed (seals). The pollex is nearly or quite as long as the second digit, and as a rule the digits then successively diminish in size.

The Creodonta differ from living Carnivora in the fact that the scaphoid and lunar are usually separate.

In Insectivora the scaphoid and lunar are sometimes united, sometimes separate, and a separate centrale is usually present. There are generally five digits, but sometimes the pollex is absent. In the Mole the manus is greatly developed and considerably modified. It is very wide, its breadth being increased by the great development of the radial sesamoid which is very large and sickle-shaped. The ungual phalanges are also large and are cleft at their extremities.

In the Chiroptera the manus is greatly modified for the purpose of flight. The pollex is short and is armed with a rather large curved claw, the other digits are enormously elongated, the elongation in the case of the Insectivorous bats being mainly due to the metacarpals, and in the Frugivorous bats to the phalanges. In the Frugivorous bats the second digit is clawed as well as the pollex, in other bats this claw is always absent, and so is often the ungual phalanx, the middle phalanx then tapering gradually to its termination.

In Primates as a rule the manus is moderately short and wide. The carpus has the scaphoid and lunar distinct, and generally also the centrale; sometimes however, as in Man, the Gorilla, Chimpanzee, and some Lemurs, the centrale has apparently fused with the scaphoid. There are almost always five well-developed digits, but in the genera *Colobus* and *Ateles* the pollex is vestigial.

The magnum in man is the largest bone of the carpus. The pisiform also is well developed, but there is no radial sesamoid. In Man, the Gorilla, Chimpanzee, and Orang, the carpus articulates only with the radius, in most Primates it articulates also with the ulna. The third digit of the Aye-Aye (*Chiromys*) is remarkable for its extreme slenderness.

The Pelvic Girdle.

259

The pelvic girdle in all mammals except the Sirenia and Cetacea consists of two halves, usually united with one another at the symphysis in the mid-ventral line, and connected near their upper ends, with the sacral vertebrae. Each half forms one of the *innominate* bones, and includes at least three separate elements, a dorsal bone, the ilium, and two ventral bones, the ischium and pubis. Very often a fourth pelvic element, the acetabular or cotyloid bone, occurs.

In the Monotremata the pelvis is short and broad, and the pubes and ischia meet in a long symphysis. The acetabulum is perforated in *Echidna* as in birds, but not in *Ornithorhynchus*. A pair of elongated slender bones project forwards from the edge of the pubes near the symphysis; these are sesamoid bones formed by ossifications in the tendons of the external oblique abdominal muscles, and are generally called *marsupial bones*.

In the Marsupialia the ilia are generally very simple, straight, and narrow, while the pubes and ischia are well developed and meet in a long symphysis. Marsupial bones are nearly always prominent, but are not developed in *Thylacinus* or *Notoryctes*. The ischium often has a well-marked tuberosity and in Kangaroos the pubis bears a prominent pectineal process on its anterior border close to the acetabulum. The pelvis in *Notoryctes* differs much from that in all other Marsupials, the ilium and ischium being ankylosed with six vertebrae in a manner comparable to that of many Edentates.

In the Edentata the pelvis is generally well developed, but the symphysis is very short. In the Sloths the pelvis is rather weak and slender, the obturator foramina are very large and the ischia do not meet in a symphysis. In the Megatheriidae the pelvis is exceedingly wide and massive, and is firmly ankylosed with a number of vertebrae. In the Armadillos, Glyptodonts, Anteaters, and Pangolins it is much developed and firmly united to the vertebral column by both the ilia and the ischia. In *Orycteropus* however the ischium does not become united to the vertebral column, and the pubis generally has a strongly developed pectineal process.

In the Sirenia the pelvis is quite vestigial. In the Dugong it consists on each side of two slender bones, one of which represents the ilium and the other the ischium and pubis; the two bones are placed end to end and are commonly fused together. The ilium is attached by ligament to the transverse process of one of the vertebrae. In the Manatee each half of the pelvis is represented by a triangular bone connected by ligaments with its fellow and with the vertebral column. In neither Manatee nor Dugong is there any trace of an acetabulum but one can be made out in *Halitherium*.

In the Cetacea the pelvis is even more vestigial than in the Sirenia, consisting simply of a pair of small straight bones which probably represent the ischia, and lie parallel to and below the vertebral column at the point where the development of chevron bones commences.

In Ungulata vera the pelvis is generally rather long and narrow. The ilium is flattened and expanded in front (fig. 103, 8), but becomes much narrower and more cylindrical before reaching the acetabulum. Both pubis and ischium contribute to the symphysis which is often very long. The ischia are large and have prominent tuberosities, especially in Artiodactyles. In most Ruminantia there is a deep depression, the supra-acetabular fossa above the acetabulum, but this is not found in the Suina or Tylopoda.

Subungulata. In *Procavia* the pelvis is long and narrow, and bears resemblance to that in Artiodactyles.

The Proboscidea have a very large pelvis set nearly at right angles to the vertebral column; the ilium is very wide, having expanded iliac and gluteal surfaces, and a narrow sacral surface. The pubes and ischia are rather small, but both meet their fellows in the symphysis. *Uintatherium* (suborder Amblypoda) also has a large and vertically placed pelvis (fig. 108) with a much expanded ilium. The pelvis however differs from that of the Proboscidea in the fact that the ischia do not meet in a ventral symphysis.

In many Rodentia the ilia have their gluteal, iliac, and sacral surfaces of nearly equal extent; in the Hares, however, the gluteal and iliac surfaces are confluent. The pubes and ischia are always well developed and sometimes, as in the Hares, the acetabular bone also. In these animals the pubis does not take part in the formation of the acetabulum, and the ischium bears on its outer side a well-marked ischial tuberosity.

In the Carnivora the pelvis is long and narrow. The iliac surfaces (fig. 78, A, 5) are very small and the sacral large; the crest or supra-iliac border is formed by the union of the sacral and gluteal surfaces. The symphysis is long and includes part of both pubis and ischium. The ischial tuberosity (fig. 78, A, 10) is often well marked, and sometimes as in *Viverra* the acetabular bone is distinct. In the Pinnipedia the pelvic symphysis is little developed, or sometimes not developed at all, and the obturator foramina are remarkably large.

In some Insectivora such as *Galeopithecus*, there is a long pelvic symphysis, in others such as *Erinaceus* and *Centetes*, it is very short, in others again such as *Talpa* and *Sorex*, there is no pelvic symphysis. The acetabular bone is exceptionally large in *Talpa* and *Sorex*.

In Chiroptera the pelvis is small and narrow, and in the great majority of cases the two halves do not meet in a ventral symphysis. The pubis has a strongly developed pectineal process, which occasionally unites with a process from the ilium enclosing a large pre-acetabular foramen.

Primates. In Man and the Anthropoid Apes the pelvis is very large and wide, and the ilium has much expanded iliac and gluteal surfaces. The symphysis is rather short and formed by the pubis alone. The acetabulum is deep and the obturator foramen large, and there is frequently a well-marked ischial tuberosity. In the lower Anthropoidea the ilium is long and narrow and has a small iliac surface. The ischial tuberosities are large in the old world monkeys.

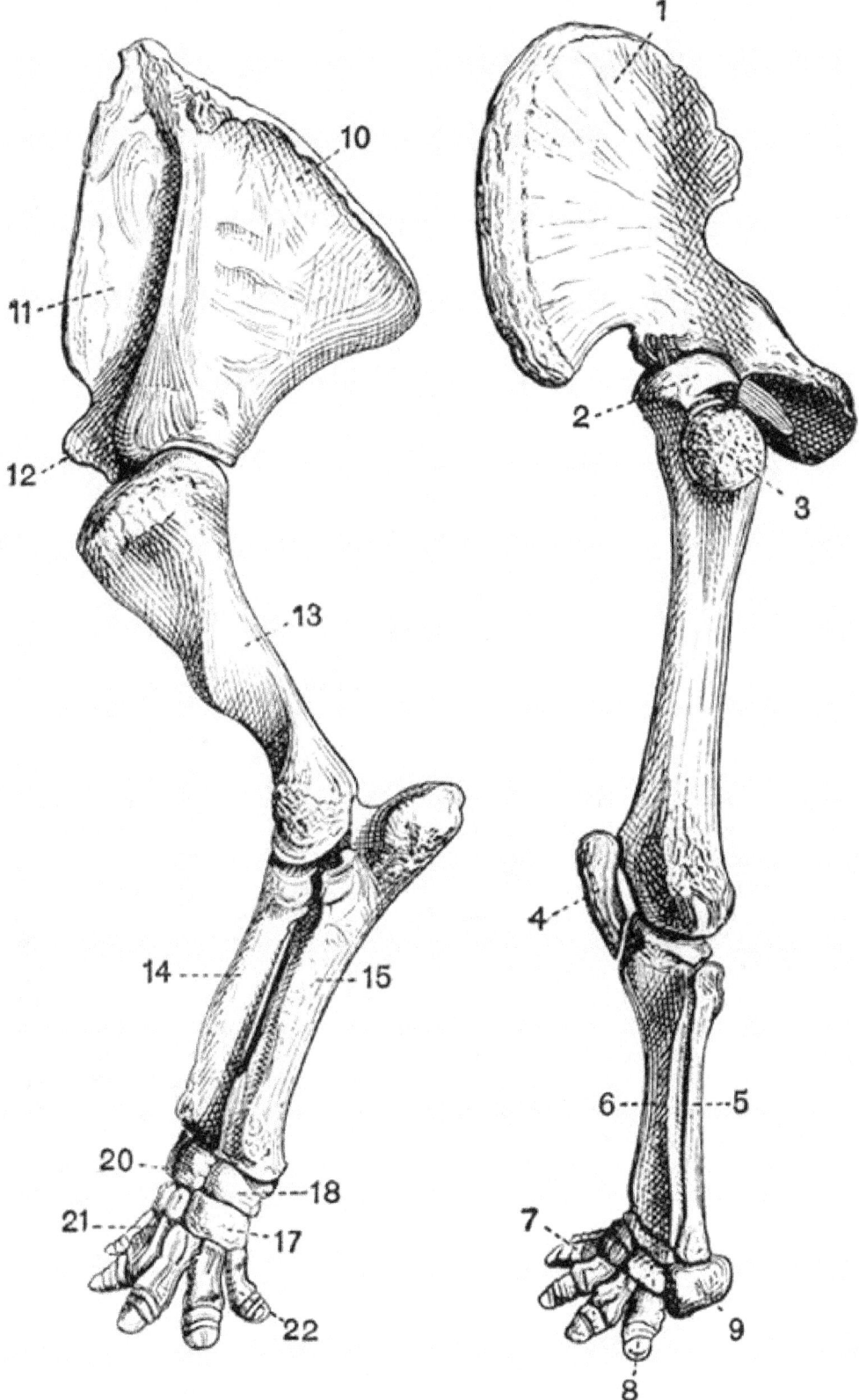

Fig. 108. Left anterior and posterior limb and limb girdle of *Uintatherium mirabile*. The anterior limb is to the left, the posterior to the right × ¹/₁₀. (From casts, Brit. Mus.)

261

1. ilium.

2. head of femur.

3. great trochanter.

4. patella.

5. fibula.

6. tibia.

7. second digit of pes.

8. ungual phalanx of fifth

digit of pes.

9. calcaneum.

10. postscapular fossa.

11. prescapular fossa.

12. coracoid process.

13. humerus.

14. radius.

15. ulna.

17. unciform.

18. cuneiform.

20. lunar.

21. first metacarpal.

22. fifth metacarpal.

The Thigh and Shin.

In the Monotremata the femur is short, rather narrow in the middle, and expanded at each end. The great and lesser trochanters are large and about equally developed, but there is no third trochanter. The fibula is very large and is expanded at its proximal end, forming a flattened plate much resembling an olecranon. The patella is well developed.

In the Marsupialia there is no third trochanter to the femur, the fibula is well developed but not the patella as a general rule. *Notoryctes* has a femur with a prominent ridge extending some little way down the shaft from the great trochanter; the tibia has a remarkably developed crest, and the fibula has its proximal end much expanded and perforated; there is an irregularly shaped patella closely connected with the proximal end of the tibia.

Edentata. In the Sloths the leg bones are all long and slender. The femur has no third trochanter, and the fibula is complete and nearly equal in size to the tibia. In the Megatheriidae the leg bones are extraordinarily massive, the circumference of the shaft of the femur in *Megatherium* equalling or exceeding the length of the bone. There is no third trochanter in *Megatherium*. In most of the remaining Edentata the leg bones are strongly developed. The femur in the Armadillos and Aard Varks has a strong third trochanter, and the tibia and fibula are both large and are commonly ankylosed together at either end. The limb bones are very massive also in the Glyptodonts.

Sirenia. In no living Sirenian is there any trace of a hind limb, but in *Halitherium* a vestigial femur is found, which articulates with the pelvis by a definite acetabulum.

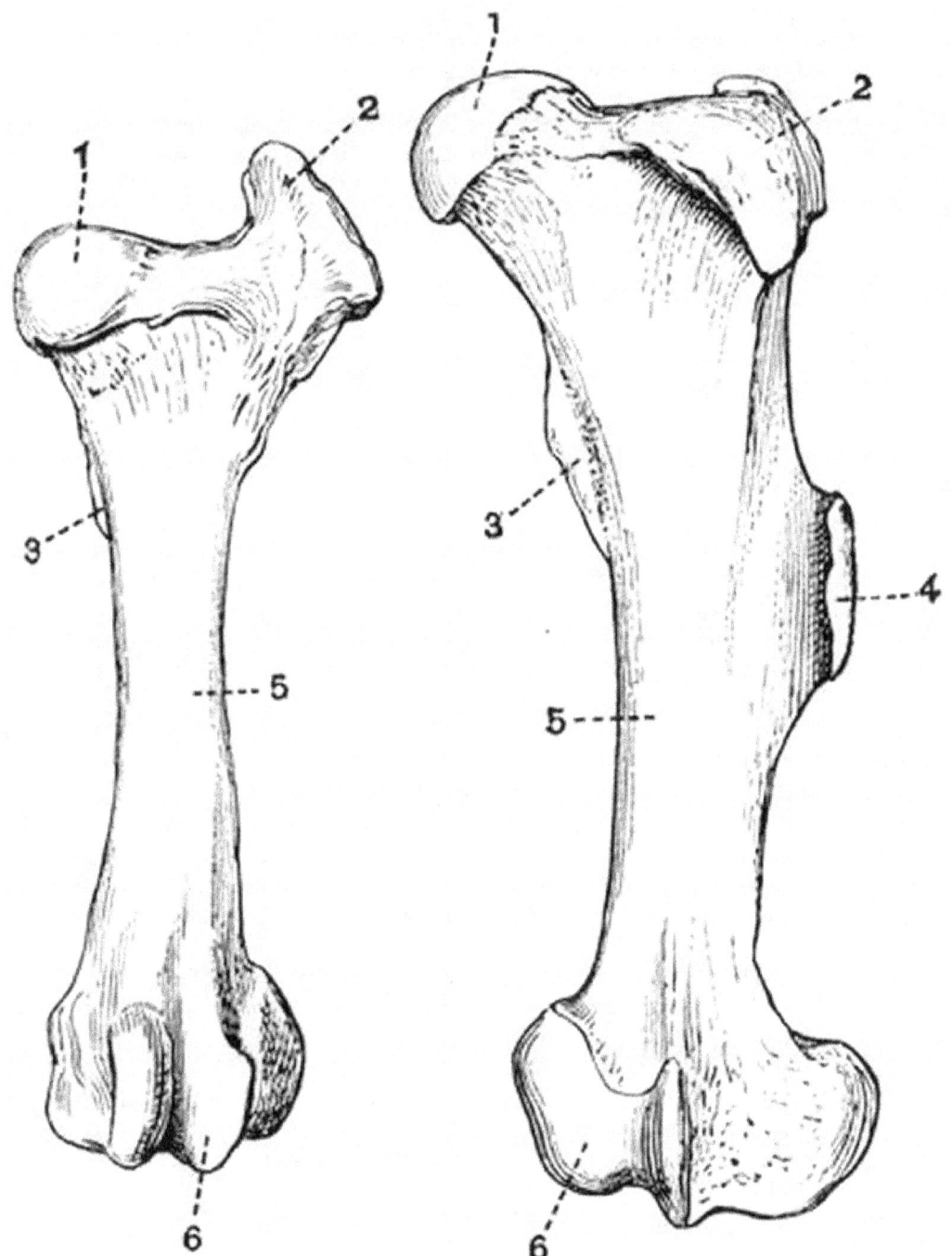

Fig. 109. Left femur of an Ox (*Bos taurus*) (to the left) and of
a Sumatran Rhinoceros (*R. sumatrensis*) (to the right). × $^1/_6$.
(Camb. Mus.)

1. head.   4. third trochanter.

2. great trochanter.   5. shaft.

3. lesser trochanter.   6. condyles.

In the Mystacoceti among the Cetacea small nodules of bone or cartilage occur connected with the vestigial pelvis, and may represent the femur and tibia. No trace of the skeleton of the hind limb is known in the Odontoceti.

In the Ungulata vera the femur is noticeable for the size of the great trochanter (fig. 109, 2); there is no definitely constricted neck separating the head from the rest of the bone, and the lesser trochanter (fig. 109, 3) is not very prominent.

All Perissodactyles except the Chalicotheriidae show a strongly marked third trochanter, but this is absent in all known Artiodactyles. The development of the fibula in general corresponds to that of the ulna. In *Rhinoceros, Macrauchenia, Tapirus* and the Suina it is distinct and fairly well developed; in the Tragulina on the other hand it is vestigial, being reduced to the proximal end only. In the Ruminantia and Tylopoda also, it is much reduced forming merely a small bone attached to the distal end of the tibia, sometimes, as in the Red deer a slender vestige of the proximal end also is preserved quite detached from the distal portion; in the Horse this proximal portion is all that there is found of the fibula. The progressive diminution of the fibula can be well seen in the series of forms that are regarded as the ancestors of the Horse. The patella of the Ungulata vera is well ossified, but fabellae are not usually found.

Subungulata. Of the Toxodontia, *Toxodon* has no third trochanter while *Typotherium* and *Astrapotherium* have one. In the Condylarthra the femur has well-marked lesser and third trochanters, and the fibula and patella are well developed. In the Hyracoidea there is a slight ridge on the femur in the place of the third trochanter, the fibula is complete, but is generally fused to the tibia at its proximal end.

Of the Amblypoda, *Coryphodon* has a third trochanter, but *Uintatherium* has none; in this respect, in the vertical position and general appearance (fig. 108) of the limb, and in the articulation of the fibula with the calcaneum, the leg of *Uintatherium* closely approaches that of the Proboscidea.

In the Proboscidea the femur is very long and straight, the development of trochanters is slight, and the fibula though slender is complete and articulates with the calcaneum.

A third trochanter is found in the Tillodontia.

In Rodentia the femur is variable, the great trochanter is generally large and so sometimes is the third as in the Hares. In most Rodents as in the Beaver the fibula is distinct, sometimes as in the Hares it is united distally with the tibia. The patella is well developed, and so too are the fabellae as a general rule.

Carnivora. In the Carnivora vera the femur (fig. 79, A) is generally rather straight and slender, and has a very distinct head. The fibula (fig. 79, C) is always distinct and there is generally a considerable interval between it and the tibia. Fabellae (fig. 79, 7) are commonly present.

In the Pinnipedia the femur is short, broad and flattened, having a prominent great trochanter. The fibula is nearly as large as the tibia, and the two bones are generally ankylosed together at their proximal ends.

The Creodonta differ from all living Carnivores in having a femur with a third trochanter.

In the Insectivora a third trochanter is sometimes developed. The fibula is sometimes distinct, sometimes fused distally with the tibia, thus differing from that of a Carnivore.

In Chiroptera the femur is straight, slender and rather short, with a small but well-developed head. The fibula may be well developed or quite vestigial or absent. Owing to the connection of the hind limb with the wing membrane the knee joint is directed backwards.

In Primates the femur is rather long and slender, having a nearly spherical head and large great trochanter. The tibia and fibula are always distinct and well developed. Fabellae are not found in the highest forms but are generally present in the others.

The Pes.

The skeleton of the pes is in most respects a counterpart of that of the manus. Just as in the manus if one digit is absent it is the pollex, so in the pes it is the hallux. But while in the manus the third digit is always well developed, however much the limb may be modified, in the pes any of the digits may be lost. In all mammals the tibiale and intermedium fuse to form the *astragalus*, and the fourth and fifth tarsalia to form the *cuboid*. Sesamoid bones are considerably developed. In almost every case the phalanges and first metatarsal have epiphyses only on their proximal ends, while the remaining four metatarsals have epiphyses only on their distal ends.

In the Monotremata all the usual tarsal bones are distinct, and the five digits have the normal number of phalanges. Several sesamoid bones are developed, the most important one, found only in the male, being articulated to the tibia and bearing the curious horny spur. The ungual phalanges of the pes like those of the manus, are deeply cleft at their extremities. In the Echidnidae the pes is turned outwards and backwards in walking.

In the Marsupialia the pes is subject to great modifications, but in every case the seven usual tarsal bones are distinct. In the Didelphyidae the foot is broad, all five digits are well developed, and the hallux is opposable to the others. In the Dasyuridae the foot is narrow, and the hallux may be very small, or as in *Thylacinus* completely absent. In *Notoryctes* the pes is much less abnormal than the manus, and all five digits have the usual number of phalanges. The fifth metatarsal has a curious projecting process, and there is a large sesamoid above the hallux. In the Wombats (Phascolomyidae) the foot is short and broad, the digits are all distinct, and the hallux is divaricated from the others.

In the remaining marsupials the second and third metacarpals and digits are very slender, and are enclosed within a common integument. This condition is known as *syndactylism*, and its effect is to produce the appearance of one toe with two claws. In the Kangaroos (Macropodidae) the pes is very long and narrow, owing to the elongation of the metacarpals. The fourth digit is greatly developed, the fifth moderately so, while the hallux is absent, and the second and third digits are very small. The Peramelidae have the foot constructed on the same plan as in the Kangaroos, and in one genus *Choeropus* the same type of foot is carried to a greater extreme than even in the Kangaroos. Thus the fourth digit is enormously developed, the second and third are small, and the fifth smaller still, while the hallux is absent. In the Phalangers and Koalas though the second and third toes are very slender, the hallux is well developed and opposable.

Edentata. In the Sloths the pes much resembles the manus, being long and narrow, but in both genera the second, third and fourth digits are well developed. Most of the other Edentates have a but little modified pes with the normal number of tarsal bones and the complete series of digits. In *Cycloturus* however the hallux is vestigial and it is absent in Glyptodonts. *Megatherium* has a greatly modified pes, the hallux is absent, and the second digit vestigial, while the third is very large, having an enormous ungual phalanx. The calcaneum too is abnormally large.

No trace of the pes occurs in either Sirenia or Cetacea.

In the Ungulata the pes like the manus is subject to much variation and is of great morphological importance.

In the Ungulata vera the pes is never plantigrade and never has more than four digits, the hallux being absent. The cuboid always articulates with the astragalus, and the tarsal bones strongly interlock. As was the case also with the manus, the pes is formed on two well-marked types characteristic respectively of the Artiodactyla and Perissodactyla.

Artiodactyla. Just as in the manus, the third and fourth digits are well and subequally developed; their ungual phalanges have the contiguous sides flat, and the axis of the limb passes between them, and between the cuboid and navicular. The astragalus has both the proximal and distal surfaces pulley-like, and articulates with the navicular and cuboid by two facets of nearly equal size. The calcaneum articulates with the lower end of the fibula if that bone is fully developed.

In the Suina four toes are developed, and though in the Peccaries the third and fourth metatarsals are united, they are all distinct in most members of the group, as are all the tarsal bones. In the Hippopotami the four digits are of approximately equal size, and the middle ones do not have the contiguous faces of their ungual phalanges flattened.

In the Tragulina the cuboid, navicular, and two outer cuneiforms are united forming a single bone; all four metatarsals are complete and the two middle ones are united. In the Tylopoda and *Anoplotherium commune* only the third and fourth digits are developed, their metatarsals are free distally, but are elsewhere united. In the Ruminantia the cuboid and navicular are always united and so are the second and third cuneiforms, while in *Cervulus* all four bones are united together. The third and fourth metatarsals in Ruminants are always united in the same way as are the third and fourth metacarpals, while the second and fifth are always wanting. In Deer the second and fifth digits are usually each represented by three small phalanges, but in the Giraffe and most Bovidae the bones of these digits are wanting.

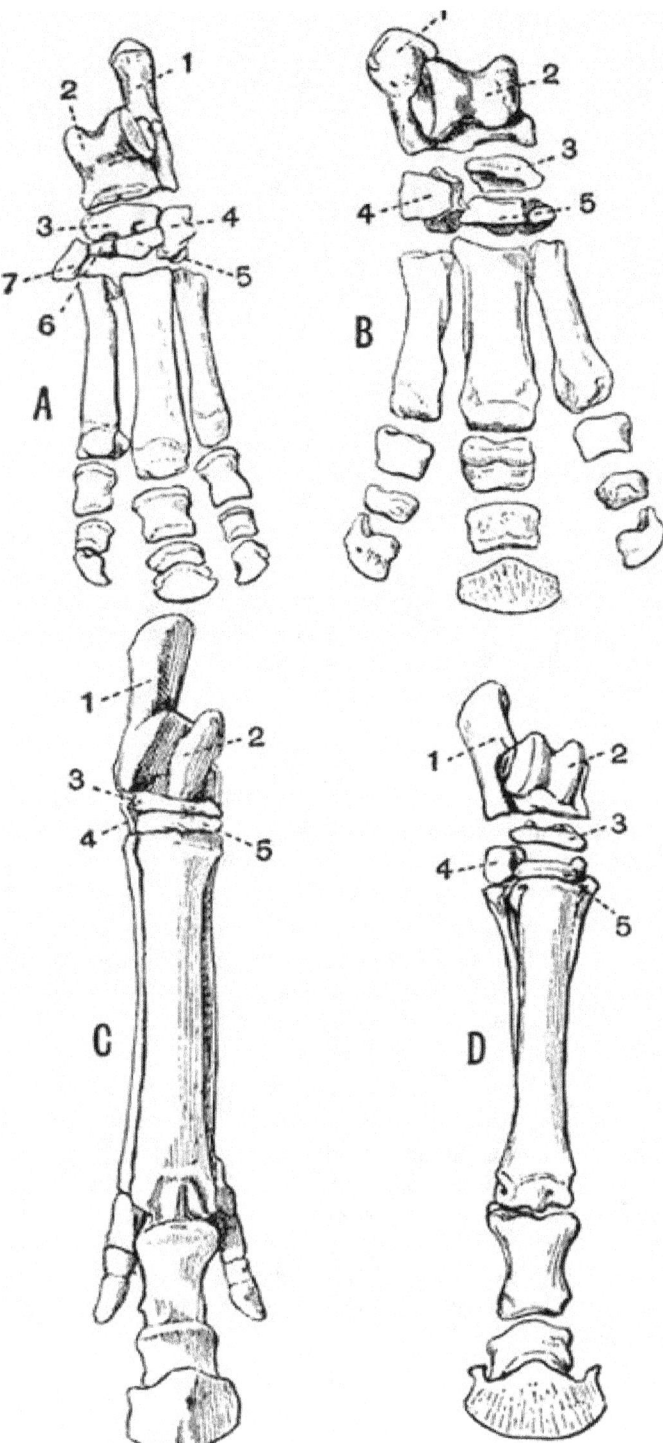

Fig. 110. *A.* Left pes of a Tapir (*Tapirus americanus*). × ¹/₆.

B. Right pes of a Rhinoceros (*R. sumatrensis*). × ¹/₈.

C. (Cast of) right pes of *Hipparion gracile.* × ¹/₇.

D. Right pes of a Horse (*Equus caballus*). × ¹/₁₀. (All Camb. Mus.)

1. calcaneum.   5. external cuneiform.

2. astragalus.   6. middle cuneiform.

3. navicular.   7. internal cuneiform.

4. cuboid.

In the *Perissodactyla* the pes like the manus is symmetrical about a line drawn through the third digit; this line when continued passes through the external cuneiform, navicular and astragalus. The astragalus has its distal portion abruptly truncated, and the facet by which it articulates with the cuboid is much smaller than that by which it articulates with the navicular. The calcaneum does not articulate with the fibula. The tarsus in *Macrauchenia* like the carpus differs from that of other Perissodactyles and resembles that of Subungulates in having the bones arranged in lines with little or no interlocking. The calcaneum resembles that of Artiodactyles in having a small facet for articulation with the fibula. *Tapirus* (fig. 110, A), *Rhinoceros* (fig. 110, B) and *Titanotherium* have a short and broad foot with the usual tarsal bones and three well-developed digits,—a number never exceeded by any Perissodactyle. From this tridactylate limb a series of stages is exhibited by various extinct forms leading gradually to the condition met with in the Horse (fig. 110, D) in which the third toe is greatly developed, while the second and fourth are reduced to slender metatarsals attached to the proximal half of the third metatarsal.

In *Chalicotherium* and *Agriochoerus* the pes has the same abnormal characters as the manus, the digits being clawed and the ungual phalanges in *Chalicotherium* deeply cleft.

In the Subungulata the pes is sometimes plantigrade and pentedactylate, the cuboid sometimes does not articulate with the astragalus, and the tarsal bones sometimes do not interlock.

In *Typotherium* (*Toxodontia*) the hallux is absent and the other four digits are well developed; in *Toxodon* and *Nesodon* the pes is tridactylate. The tarsal bones have the regular Subungulate arrangement, the cuboid not articulating with the astragalus. The calcaneum articulates with the fibula as in Artiodactyles. The astragalus in most forms, but not in *Astrapotherium*, resembles that of the Ungulata vera in having a grooved proximal surface.

In *Phenacodus* (*Condylarthra*) the tarsus is very little modified, five digits are present, the first and fifth being small and not reaching the ground.

In *Procavia* only the three middle digits are present with a vestige of the fifth metacarpal.

In the *Amblypoda* the pes (fig. 108) is very short and broad, all five digits are functional, and at any rate in *Coryphodon* plantigrade, the hallux being the smallest. The astragalus is very flat, and the tarsals interlock to a slight extent, the cuboid articulating with both calcaneum and astragalus.

The pes in the *Proboscidea* much resembles that in the Amblypoda, but differs in that the astragalus does not articulate with the cuboid, the tarsals not interlocking at all.

In the Rodentia the structure of the foot is very variable. In Beavers the foot is very large, all five digits being well developed; the fifth metatarsal articulates with the outer side of the fourth metatarsal, and not with the cuboid, and there is a large sesamoid bone on the tibial side of the tarsus. In the Rats, Porcupines and Squirrels, there are five digits, in the Hares only four, and in the Capybara and some of its allies only three. In the Jerboa (*Dipus*) a curious condition of the pes is met with, as it consists of three very long metatarsals fused together and bearing three short toes, each formed of three phalanges. *Lophiomys* differs from all other Rodents in having the hallux opposable.

Carnivora. In the Carnivora vera the pes is regular and shows little deviation from the normal condition. All the usual tarsal bones are present, but sometimes as in the Dogs, Cats, and Hyaenas, the hallux is vestigial. Sometimes as in the Bears the pes is plantigrade, sometimes as in the Cats and Dogs it is digitigrade. In this respect and in the character of the ungual phalanges, the pes closely corresponds with the manus. In the Sea Otter (*Latax*) the foot is large and flattened and approaches in character that of the Pinnipedia.

In the Pinnipedia the pes differs much from that in the Carnivora vera. In the Seals in which the foot cannot be used for walking, and is habitually directed backwards, the first and fifth digits are much longer and stouter than any of the others. In the Sea Lions which can use the pes for walking, the digits are all of nearly the same length, and in the Walrus the fifth is somewhat the longest.

In the Insectivora the pes is almost always normal, and provided with five digits.

In the Chiroptera the pes is pentedactylate, and the digits are terminated by long curved ungual phalanges. In some genera the toes have only two phalanges. The calcaneum is sometimes produced into a long slender process which helps to support the membrane between the leg and the tail.

Among the Primates Man has the simplest form of pes. In Man all five digits are well developed, the hallux being considerably the largest. Sesamoid bones occur only under the metatarso-phalangeal joint of the hallux.

In the other Primates the internal cuneiform has a saddle-shaped articulating surface for the hallux, which is obliquely directed to the side of the foot and opposable to the other digits. Two sesamoid bones are usually developed below each metatarso-phalangeal joint, and one below the cuboid. The second digit in Lemurs, and all except the hallux in *Chiromys* have pointed ungual phalanges; in all other cases the ungual phalanges are flat. In some of the Lemuroidea, especially *Tarsius*, the tarsus is curiously modified by the elongation of the calcaneum and navicular.

267

www.ingramcontent.com/pod-product-compliance
Lightning Source LLC
Chambersburg PA
CBHW080802180526
45168CB00006B/2296